BEACHES OF THE SOUTH AUSTRALIAN COAST AND KANGAROO ISLAND

A guide to their nature, characteristics, surf and safety

ANDREW D SHORT

Coastal Studies Unit
School of Geosciences
University of Sydney
Sydney NSW 2006

SYDNEY UNIVERSITY PRESS

COPYRIGHT © AUSTRALIAN BEACH SAFETY AND MANAGEMENT PROGRAM

Coastal Studies Unit   and   **Surf Life Saving Australia Ltd**
School of Geosciences F09   1 Notts Ave
University of Sydney   Locked bay 2
Sydney NSW 2006   Bondi Beach NSW 2026

Short, Andrew D
    Beaches of the South Australian Coast and Kangaroo Island   0-9586504-2-X
    A guide to their nature, characteristics, surf and safety   Published 2004

Other books in this series by A D Short:
- *Beaches of the New South Wales Coast, 1993*   0-646-15055-3
- *Beaches of the Victorian Coast and Port Phillip Bay, 1996*   0-9586504-0-3
- *Beaches of the Queensland Coast: Cooktown to Coolangatta, 2000*   0-9586504-1-1
- *Beaches of the Western Australian Coast: Eucla to Roebuck Bay, 2005*   0-9586504-3-8

Forthcoming books:
    Beaches of the Tasmania Coast and Islands (publication. 2006)   1-920898-12-3
    Beaches of the Northern Australian Coast: The Kimberley, Northern Territory and Cape York
        1-920898-16-6
    Beaches of the New South Wales Coast (2nd edition)   1-920898-15-8

Published by:
Sydney University Press
University of Sydney
www.sup.usyd.edu.au

Printed by:
University of Sydney Printing Service
University of Sydney

Copies of all books in this series may be purchased online from:

http://www.sup.usyd.edu.au/marine

South Australia beach database:
Inquiries about the South Australian beach database should be directed to Surf Life Saving Australia.
at info@slsa.asn.au

Cover photograph	South Port Surf Life Saving Club is located on the sand spit at the mouth of the Onkaparinga River. Note the bar and trough along the ebach produced by occasional higher south swell (A D Short)
Cover	Jacqui Owen
Book Layout	Werner Hennecke

TABLE OF CONTENTS

Australian Beach Safety and Management Program

To Julia, who I met at Kingston SE and also loves the South Australian coast.

FOREWORD

South Australia is home to a stunning array of beaches that are keenly visited by thousands of people each year. These beautiful coastal waters are home to some of the most unique and biologically diverse marine species in the world and they attract a vast range of recreational pursuits such as surfing, diving, sailing and fishing.

Volunteer lifesavers and lifeguards have made a commendable contribution to ensuring these coastal activities are pursued safely – and the publication of the *Australian Beach Safety and Management Program for Beaches of the South Australian Coast and Kangaroo Island* is another important contribution. Its pages detail the main characteristics of 1,788 beaches, including beach names, locations, physical characteristics, access and facilities and comments on their physical hazards and suitability for swimming, surfing and fishing.

Never before has a book provided the public with such a complete understanding of the health risks linked with swimming or fishing on South Australia's coastline. Its publication follows more than 20 years of research and gives lifesavers, local councils, recreational enthusiasts and coastal planners a greater knowledge of the coast.

As well as beach dangers, the book also provides a valuable insight into South Australia's marine waters, considered among the most biologically diverse in the world. Many of the state's marine flora and fauna occur nowhere else and some, such as the state's marine emblem the Leafy Seadragon, are under threat.

As a resident of South Australia and the Federal Environment Minister, I hope this book will encourage safer coastal visits and also help to foster a collective care for its long-term environmental future.

On behalf of the Federal Government I congratulate the University of Sydney and Surf Life Saving Australia for producing such a comprehensive guide and commend it to all those who use the coastline and wish to see its natural beauty preserved for generations to come.

Robert Hill

SENATOR ROBERT HILL
Leader of the Government in the Senate
Minister for the Environment and Heritage

PREFACE

This book is the fourth in a series of books produced by the Australian Beach Safety and Management Program (ABSMP). One of the aims of the program is to develop a better understanding of the location, type, characteristics, nature, hazards and public risk along all of Australia's 9000 plus beaches. To this end the program is developing a database on the beaches of each state and the Northern Territory, with the databases for New South Wales, Victoria, Queensland, South Australia and the Northern Territory now complete. It is also publishing books summarising the main characteristics of each beach on a state by state basis. This book is the fourth in the series with *Beaches of the New South Wales Coast* published in 1993, *Beaches of the Victorian Coast and Port Phillip Bay* in 1996, *Beaches of the Queensland Coast: Cooktown to Coolangatta* in 2000, while the *Beaches of Northern Australia: The Kimberley, Northern Territory and Cape York*, the fifth in the series, is well underway.

The ABSMP was initiated in New South Wales in 1986 as the New South Wales Beach Safety Program in collaboration with Andrew May of Surf Life Saving New South Wales (SLSNSW), with Chris Hogan as Research Officer. In 1990 with support from the Australian Research Council and in cooperation with Surf Life Saving Australia Ltd, the project was expanded to encompass all states and the Northern Territory.

The Coastal Studies Unit commenced scientific investigations on the South Australian coast in 1978 with an aim to conduct field experiments in high energy surf zones. Goolwa Beach was selected for the highly successful field experiment conducted in 1980. At the same time Rob Thomas of the Coastal Management Branch hearing of our work was instrumental in Andy Short and Patrick Hesp undertaking a study of the coastal geomorphology of the entire South East District published in 1980. This work led to two further studies for the Branch on the Eyre Peninsula and Kangaroo Island, both in collaboration with Doug Fotheringham of the Branch. The reports were published in 1986. In total the three reports covered 2594 km or 70 % of the South Australian coast and provided the author with an excellent overview of the South Australian coast and beaches.

Studies of the South Australian coast specifically for the present project commenced in March 1990 with an inspection of Yorke Peninsula beaches with Chris Hogan, followed by fieldwork in the South East, Fleurieu and Kangaroo Island in January 1995, and Eyre Peninsula in 1995. Aerial flights of the coast to photograph the beaches were undertaken along the South East coast in 1978, the Eyre Peninsula in 1987, the Adelaide, Fleurieu and lower Yorke Peninsula coast using the SLSA helicopter in March 1990, the South East and Kangaroo Island coast in 1995 (in sea fog) and 1996 and the western Eyre Peninsula in 1997.

In compiling a book of this magnitude there will be errors and omissions, particularly with regard to the names of beaches, many of which have no official name, and many local factors. If you notice any errors or wish to comment on any aspects of the book please communicate them to the author at the Coastal Studies Unit, University of Sydney, Sydney NSW 2006, phone (02) 9351 3625, fax (02) 9351 3644, email: A.Short@csu.usyd.edu.au or via Surf Life Saving South Australia (08) 8356 5544 or Surf Life Saving Australia (02) 9597 5588. In this way we can update the beach database and ensure that future editions are more up to date and correct.

Andrew D Short
Narrabeen Beach, January 2001

ACKNOWLEDGEMENTS

This book is the result of numerous field trips to South Australia by land and air, and while written in Sydney many people from both states have contributed to the final publication.

Four major field trips specifically to investigate the beach systems were undertaken. The first was in March 1990 along the Yorke Peninsula and Adelaide coast accompanied by Chris Hogan. The second was along the western Eyre Peninsula in April 1991 accompanied by my family. The third in January 1995 got to more inaccessible sections of the western Eyre Peninsula coast accompanied by Rob Tucker and Doug Fotheringham from the Coastal Management Branch; and finally a trip along the South East and Kangaroo Island coast in January 1996 accompanied again by my family. Most of the Eyre Peninsula was covered in previous field work in the mid-1980's with Doug Fotheringham and Ralph Buckley.

The Coastal Management Branch lent their excellent colour aerial photograph series during the writing of the book. These photographs provided an invaluable record of beach conditions along the entire coast. I particularly acknowledge the assistance of Doug Fotheringham. I am also indebted to my former colleague, Professor J L Davies, who generously provided a complete set of black and white aerial photographs of the entire coast, as he did for the three previous books.

The author's first flight along the South East Coast was undertaken in 1978 with Peter Cowell as we searched for a high energy beach to conduct field experiments. In 1987 I was invited to join Doug Fotheringham in a flight of parts of the Spencer Gulf and entire Eyre Peninsula funded by the Coastal Management Branch.

Flights specifically for this project commenced in March 1990 with an SLSA helicopter flight of the lower Yorke Peninsula, Adelaide and Fleurieu coasts. This was followed in January 1995 using an Australian Aerial Patrol (AAP) Cessna piloted by Dean Franklin and Arthur Badger with a flight along the South East, Fleurieu Peninsula and Kangaroo Island coast, the latter unfortunately partly hidden by summer sea fog. The foggy parts were reflown in January 1996. Finally in October 1997, AAP's Harry Mitchell and Dean Franklin flew the author along the western Eyre Peninsula coast while he photographed every beach. This last flight was the first leg of a flight of the entire Western Australian and Northern Territory coasts, part of the larger ABSMP program.

On the ground the project has received support from both SLSSA and the Coastal Management Branch. At SLSSA Elaine Farmer and Shane Daw have always strongly supported the program including financial assistance for some of the field investigations. Within the Coastal Management Branch Rob Tucker and Doug Fotheringham have both participated in joint field work as well as providing access to the invaluable aerial photograph collection.

At Surf Life Saving Australia (SLSA) where the project is based it received full support from Scott Derwin (CEO of SLSA, 1990-1995), during which time the project was supervised by Darren Peters and Stephen Leahy. The project also received an Australian Research Grant from 1990-1992 which enabled it to expand nationally. Since 1996 CEO Greg Nance has supervised the project at SLSA and has provided tremendous support leading to an expansion of the SLSA-University of Sydney collaboration with additional project funding from a joint Australian Research Council Collaborative Research Grant (1996-1998) and a Strategic Partnership with Industry- Research and Training (SPIRT) Grant (1999-2002).

Also at SLSA the Project Research Officer Katherine McLeod has been an essential component of ABSMP since 1996. She has worked full-time on the project ensuring all the data collected is recorded and managed in the ABSMP database. Katherine also drafted most of the figures in the book and edited the entire text. At the University of Sydney, Cathi Greve drafted additional figures and Cathi and Werner Hennecke did an excellent and speedy job at laying out the entire text and 286 figures to produce the following book. All photographs are by the author.

Finally, as the entire South Australian beach database was compiled, and the book was written, at my home office, I thank my wife and children for putting up with its intrusion into our home life.

ABSTRACT

This book is about the entire South Australian Coast, including Kangaroo Island and a few of the major islands. It begins with three chapters that provide a background to the physical nature and evolution of the South Australian coast and its beach systems. Chapter 1 covers the geological evolution of the coast and the role of climate, wave, tides and wind in shaping the present coast and beaches. Chapter 2 presents in more detail the twelve types of beach systems that occur along the South Australian coast, while chapter 3 discusses the types of beach hazards along the coast and the role of Surf Life Saving South Australia in mitigating these hazards. Finally the chapters 4 and 5 present a description of every one of the 1454 mainland beaches, as well as 218 beaches on Kangaroo Island and 83 beaches on five major islands, in all 1788 beaches. The description of each beach covers its name, location, physical characteristics, access and facilities, with specific comments on its surf zone character and physical hazards, as well as its suitability for swimming, surfing and fishing. Based on the physical characteristics each beach is rated in terms of the level of beach hazards from the least hazardous rated 1 (safest) to the most hazardous 10 (least safe). The book contains 286 figures which include 238 photographs, which illustrate all beach types, as well as beach maps and photographs of all beaches patrolled by surf lifesavers and many other popular beaches.

Keywords: beaches, surf zone, rip currents, beach hazards, beach safety, South Australia

---

Australian Beach Safety and Management Program (ABSMP)

Awards

NSW Department of Sport, Recreation and Racing
Water Safety Award – Research 1989
Water Safety Award – Research 1991

Surf Life Saving Australia
Innovation Award 1993

International Life Saving
Commemorative Medal 1994

New Zealand Coastal Survey
In 1997 Surf Life Saving New Zealand adopted and modified the ABSMP in order to compile a similar database on New Zealand beaches.

Great Britain Hazard Assessment
In 2002 the Royal National Lifeboat Institute adopted and modified the ABSMP in order to compile a similar database on the beaches of Great Britain.

Hawaiian Ocean Safety
In 2003 the Hawaiian Lifeguard Association adopted ABSMP as the basis for their Ocean Safety survey and hazard assessment of all Hawaiian beaches.

Handbook of Drowning 2005
This handbook was a product of the World Congress on Drowning held in Amsterdam in 2002. The handbook endorses the ABSMP approach to assessing beach hazards as the international standard.

---

1   THE SOUTH AUSTRALIAN COAST

The South Australian coast consists of 3,273 km of mainland coast, with another 458 km of coast on Kangaroo Island and an additional 473 km on 106 islands, providing a total coastline of 4,204 km. This long coast also spans a wide range of coastal environments, ranging from some of the most exposed, highest energy beaches in the world along the South East, parts of Kangaroo Island and the Eyre Peninsula (Fig. 1.1a), to the protected, more tidally influenced sand flats of the gulfs and protected bays (Fig.1.1b).

For coastal management purposes the South Australian coast is divided into seven protection districts. Table 1.1 lists the districts together with the extent of sand and rock coast and the number and length of sandy beaches. The beaches range in length from Australia's longest continuous beach, the 210 km long Coorong (Cape Jaffa to Murray Mouth), to many small pockets of sand just tens of metres in length. The average length of all the beaches is 1.3 km.

This book is concerned with the entire South Australian coast. In particular with the 1,755 beach systems which make up 62% of the mainland coast and 34% of the Kangaroo Island shore and 52% of the five other islands discussed in detail. In this book the term 'South Australia' will be taken to include just the mainland coast, unless otherwise stated. All islands, including Kangaroo Island, will be referred to separately.

This book is a product of the *Australian Beach Safety and Management Program*, a cooperative project of the University of Sydney's Coastal Studies Unit, Surf Life Saving Australia and Surf Life Saving South Australia. It is part of the most comprehensive study ever undertaken of beaches on any part of the world's coast. In South Australia it has investigated every beach on the mainland coast and Kangaroo Island. The project has already

published similar books on the beaches of the Queensland, New South Wales and Victorian coasts.

Figure 1.1   *a) A high energy beach and dune system extending to the west of Point Bell. b) Whyalla beach is a typical low energy gulf beach fronted by sand flats averaging 1 km in width.*

This book begins by examining the nature of the South Australian coast, including its geological evolution, climate and ocean processes (Chapter 1). Chapter 2 details the types of beaches that occur around the coast,

Table 1.1   Coastline characteristics of the seven South Australian coastal districts and islands

SA coastal districts	Number beaches	Beaches	Average beach length (km)	Sand coast (km)	Rock/calcarenite coast (km)	Total coast (km)
1. South East	148	1-148	2.6	382	28	410
2. Fleurieu Peninsula	66	149-214	1.1	71	68	139
3. Metropolitan	19	215-233	2.9	55	11	66
4. Yorke Peninsula	315	234-548	1.5	463	172	635
5. Spencer Gulf	57	549-605	2.3	131	142	273
6. Eyre Peninsula	849	606-1454	1.1	918	808	1726
Mainland subtotal	*1454*		*1.4*	*2020*	*1253*	*3273*
7. Kangaroo Island	218	1-218	0.7	156	302	458
Subtotal	*1672*		*1.3*	*2176*	*1555*	*3731*
Islands (in book)*	83		0.9	75	69	144
Other 101 islands	-		-	-	-	329
Total	**1755**		**1.3**	**2251**	**1624**	**4204**

* Thistle, East, Flinders, Eyre, St Peter

while Chapter 3 looks at beach usage and physical hazards. The bulk of the book is devoted to a description of every beach located on the mainland coast (Chapter 4) and Kangaroo Island (Chapter 5), including five larger islands. Information is provided on each beach's name, location, access, facilities, physical characteristics and surf conditions. Specific comments are made regarding each beach's suitability for swimming, surfing and fishing, together with a beach hazard rating from 1 (least hazardous) to 10 (most hazardous).

1.1   Beach systems

A *beach* is a wave deposited accumulation of sediment, usually sand, but occasionally cobbles and boulders. They extend from the upper limit of wave swash, approximately 3 m above sea level, out across the surf zone or sand flats and seaward to the depth to which average waves can move sediment shoreward. On the exposed, high wave energy sections of the South Australian coast, this means they usually extend to depths between 30 and 50 m and as much as several kilometres offshore. In the gulfs, in lee of Kangaroo Island and in several large bays protected by headlands and reefs, wave height and energy decrease substantially and the beaches range from the more exposed, with wind waves averaging less than 1 m, to essentially wave-less, sandy tidal flats. The low energy beaches and sand flats usually terminate at low tide. As waves decrease, tides and tide range become increasingly important, particularly in the upper gulfs where tides can reach heights of over 3 m.

To most of us however, the beach is that part of the dry sand we sit on, or cross to reach the shoreline and the adjacent surf zone. It is an area that has a wide variety of uses and users (Table 1.2). This book will focus on the dry or subaerial beach plus the surf zone or area of wave breaking, typical of the beaches illustrated in Figure 1.1.

Most South Australians live within an hour of a beach and even those who live inland often travel to the coast for their holidays. Many are frequent beachgoers and have their favourite beach. The beaches most of us have been to, or go to, are close by our home or holiday area. They are usually at the end of a sealed road, with a car park and other facilities. Often they are patrolled by lifesavers or professional lifeguards. These popular, developed beaches, however, represent only a minority of the state's beaches. In South Australia there are 19 surf lifesaving clubs and four beaches patrolled by professional lifeguards (Table 1.3) and most of these are located on the Adelaide region. These however represent just over 1% of all South Australian beaches. The vast majority are unpatrolled and many are extremely hazardous. Furthermore only 230 (14%) of the beaches have sealed road access, while 513 (31%) lie at the end of a gravel road. Offroad (4WD) vehicles are required to reach 694 beaches (42%) and 172 (10%) have no vehicle access and are only accessible on foot or by boat. Finally 63 beaches (4%) are completely inaccessible by vehicle or on foot.

Table 1.2   Types of South Australian beach users and their activities

Type	User	Location
Passive	sightseer, tourist	road, car park, lookout
Passive-active	sunbakers, picnickers beach sports	dry beach
Active	beachcombers, joggers	dry beach, swash zone
Active	fishers, swimmers	swash, inner surf zone
Active	surfers, water sports	breakers & surf zone
Active	skis, kayaks, windsurfers	breakers & beyond
Active	IRB, fishing boats	beyond breakers

Table 1.3   South Australian beaches - lifesaving facilities and access

	Mainland coast	Kangaroo Island	Total
Surf Lifesaving Clubs	19	0	19
Lifeguards (SLSSA)	4	0	4
Sealed road access	210	20	230
Gravel road access	463	50	513
Dirt road access (4WD)	605	89	694
Foot access only	124	48	172
No access	52	11	63
Total beaches	**1454**	**218**	**1672**

1.2   Evolution and Geology of the Coast

Beaches are a part of the coastal environment. They always occur at the shoreline, with part extending landward as the dry beach, often merging into sand dunes or bluffs and part extending seaward as sand flats or the surf, with the nearshore zone beyond. For a beach to form however, a number of parameters and processes are required. These include contributions from all the world's major spheres; the lithosphere or geology, which supplies sediment and boundaries; the atmosphere, which contributes the climate; the hydrosphere or ocean, source of waves and tides; and the biosphere, the source of the marine and dune biota, and in South Australia the majority of the beach and dune sediments (Fig. 1.2). The remainder of this chapter outlines the nature of the South Australian coastal environment and the contribution of each of these spheres to the evolution and nature of South Australia's beaches.

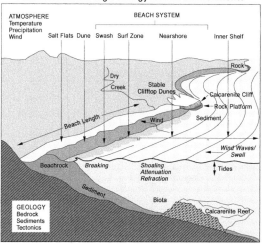

Figure 1.2   *The coastal environment is the most dynamic part of the earth's surface. It contains elements of all four spheres that make up the earth, namely the atmosphere, hydrosphere or ocean, the lithosphere or geology and the biosphere. As the four spheres interact at the coast, they produce a wide spectrum of coastal environments, ranging from rocky coast to sand flats to calcarenite reefs to sandy beaches.*

1.2.1 South Australian coastal geology

The beaches and geology of a region are inseparable, as without the underlying bedrock geology, there would be no beaches. The geology of a coast and often its hinterland, is essential for two reasons. Firstly, beaches are of a finite size; they have a certain width, height and depth below. This means they must rest on some other surface and this surface is usually the bedrock geology. On the South Australian coast, where beaches average 1.3 km in length, many are bounded by prominent headlands, bluffs and rock reefs, while others occupy drowned coastal valleys and

depressions. The bedrock or geology therefore provides what is called the coastal boundary within which, or on which, the beach rests. The overall shape and orientation of all major headlands is dependent on the ancient pre-Quaternary rocks (> 2 million years old). In addition in South Australia, more recent Pleistocene dune and beach calcarenite (dunerock/beachrock) blankets most exposed bedrock headlands and outcrops along the beach and in the nearshore as rocks, reefs and islets.

The South Australian and Kangaroo Island coast consists of 1,555 km (42%) of rocky coast composed of both bedrock and calcarenite. Calcarenite rocks (usually overlying bedrock) dominate along the high energy coastal sections such as the South East, south and western Kangaroo Island, lower Yorke Peninsula and the Eyre Peninsula, west of Cape Catastrophe. Bedrock is more prevalent along the lower energy Fleurieu Peninsula, the gulf shores and northern Kangaroo Island. Furthermore the calcarenite forms substantial reefs along and off the coast which act to modify the breaker wave regime and thereby the beach length and type. Finally, there are also 106 bedrock-based islands along the coast which have an approximate total coastline of 475 km (Appendix 1.1). Most of these islands are also wholly or partially capped by calcarenite.

The structure of the South Australian coast is therefore a function of the ancient bedrock geology, the blanketing of much of the open coast by Pleistocene calcarenite and the deposition of more recent unconsolidated sediments that make up the beaches and dunes.

Geology also plays an often major role in ultimately supplying terrigenous (i.e. terra or land-based) sediment to the coast. This material is reworked by waves, tides and winds into beaches, tidal flats and dunes. This sediment originates in the coastal hinterland and river catchments and is supplied to the coast by rivers and creeks, as well as through cliff and bluff erosion. These sediments are predominantly quartz sand grains, while locally they may be coarser cobbles and boulders and enriched by other minerals. This is partially the case in South Australia, where terrigenous sediments dominate 30% of the beaches. However the fact that 70% of the beaches are dominated by non-terrigenous sediments points to the other major source of sediments along the South Australian and much of the southern and Western Australian coasts – that is, shelf and seagrass biota producing carbonate detritus.

The reason for both the low proportion of terrigenous sediments and the dominance of calcareous sediments is twofold. Firstly, the lack of terrigenous sediments can be attributed to the long period of aridity across southern Australia dating back well into the Tertiary, that is, for some tens of millions of years. This is manifest in the present arid to semi-arid climate and in the almost complete absence of streams and major rivers. The only proper river in South Australia is the Murray (which rises in the eastern highlands), while there are some small streams along the more humid sections of the Fleurieu Peninsula and Kangaroo Island.

The lack of terrigenous sand is however matched by an abundance of calcareous sand in the gulfs and along the more exposed sections of the open coast. The origins of this sand are the biotic communities that inhabit the rich seagrass meadows in the gulfs, and on the open coast, the rich temperate carbonate province that inhabits the inner continental shelf. The latter is coupled with a strong wave environment that can both erode these materials, break and abrade them into sand size particles and transport them shoreward to build the massive beach systems, backed in many places by wind built calcareous dune systems. The South Australian mainland coast contains approximately 1736 km^2 of calcareous dunes, which, with an average height of 15 m, represent approximately 2.6 km^3 of calcareous sediments - a massive transfer of shelf organic detritus onto the coast. It equates to 800 m^3 of sand for every metre of the entire coastline. On high energy sections backed by massive dune systems this amount can reach as high as 100 000m^3/m of beach. The nature and sources of this material are discussed in section 1.7.

Furthermore, this blanketing of the coastal fringe by calcareous sediments has been occurring each time sea level has risen during the Pleistocene (see this section); at least 20 times. The net result has been a gradual layering of these calcareous sediments at the coast. Where the coast is stable-stationary they have built up layer after layer, each layer separated by a few tens of thousands of years, to form massive layered cliffs up to 150 m high (Fig. 1.3). Where the coast has been slowly rising, as in the South East, each successive sea level rise has built a new beach and dune system, a coastal sand barrier, in front of the previous, usually separated by an interbarrier depression (Fig. 1.4). Today these are known in the South East as the ranges (the barriers) and avenues (the interbarrier depressions). They not only parallel the modern coast for up to 200 km, but they also extend up to 300 km inland, all the way to Mildura, where the oldest systems are more than 2 million years old.

1.2.2   Geological evolution

Much of South Australia is part of the great Australian Craton, an ancient expanse of crystalline rocks that covers much of the central and western parts of the continent. The eastern boundary of the craton, known as the Tasman Line, lies along the eastern side of the Adelaide Geosyncline, highlighted by the Flinders-western Mount Lofty Ranges and the north coast of Kangaroo Island. That part of the craton that occupies much of the central South Australian region is called the Gawler Craton (Fig. 1.5). This summary of the coastal geology is largely derived from *The Geology of South Australia* by Drexel and

Preiss (1995). This is an excellent, well illustrated publication available from the Geological Survey of South Australia.

Figure 1.3   *A 100 m high calcarenite cliff on the lower Eyre Peninsula. Each successive layer of dune calcarenite has been deposited during successive high sea levels, with calcrete soils and lithification of the sand occurring during the intervening periods of lower sea level. This sequence represents several hundred thousand years of sedimentation.*

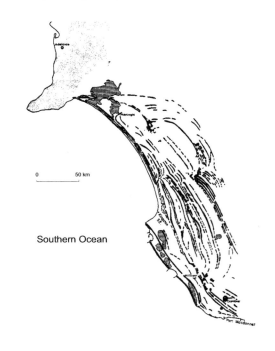

Figure 1.4   *The South East region of South Australia, highlighting the Quaternary barrier systems that extend more than 100 km inland. The inner barriers date to about 700 000 years old and have been stranded inland by the slowly rising land surface.*

1.2.2.1   *Gawler Craton (2300 to 600 Ma)*

The Gawler Craton is an ancient crystalline basement dating back in parts more than 2300 Ma (Ma = million years). Most of the craton consists of granites, with some pockets of younger volcanics and sediments (ca 1500 Ma). The craton dominates the coastal basement from the western coast of Yorke Peninsula across the Eyre Peninsula as far as the Eucla Basin, west of Cape Adieu.

The craton is concealed by platform basins; in the west, the Eucla Basin and in the gulfs, the Pirie-Torrens Basin and St Vincent Basin. Along the eastern boundary the basins have been deformed into within-basin fold belts, the most prominent being the Adelaide Fold Belt. To the east of the fold belt and the Tasman Line lies the newer Tasman Fold Belt.

In South Australia the deformation of the craton began in the lower Proterozoic (2000 Ma), with the Middleback sedimentation comprising in places 10 000 m of metamorphosed sediments of the Cleve Metamorphics, capped by the iron deposits of the Hutchinson Group evident at Iron Baron and Iron Knob. All these rocks were folded into a series of mountain chains during the Kimban Orogeny in the early Carpentarian (1800 Ma).

Subsequently the Charleston Phase (1600 Ma) intruded a massive zone of granite across the northeastern Eyre Peninsula, the largest in the Moonabie Range. During the Wartakan Phase of folding and acid volcanism (1500 Ma) there was an extrusion of a great volume of basalt in the Gawler Ranges. Subsequent to this, the vast bulk of the craton has been stable apart from minor tectonism at the end of the Adeladian phase (600 Ma).

1.2.2.2 *The Adelaide Geosyncline (1400 to 560 Ma)*

The Adelaide Fold Belt is part of the Adelaide Geosyncline. The geosyncline extends from the Mount Lofty ranges northward to the Flinders, Peak and Denison Ranges and west to the Torrens Hinge Line, which runs north paralleling the east shores of Yorke Peninsula. The geosyncline contains sedimentation dating back to 1400 Ma, terminating about 560 Ma. It represents an area of gentle platform downwarping bordered by rising basement areas, with sediments from the rising region deposited in the trough. In places, as much as 2 500 m of sediment was slowly deposited over an 800 Ma period.

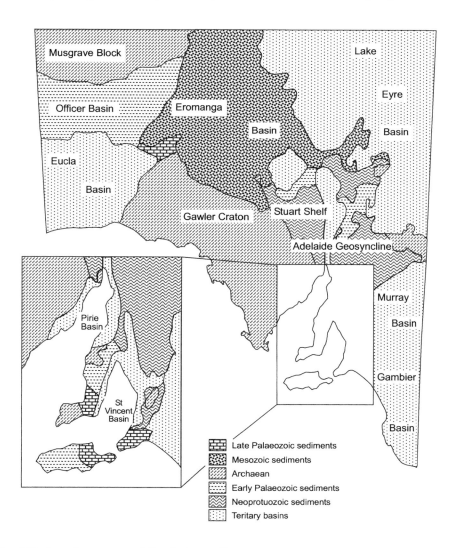

Figure 1.5   *Simplified geology map of South Australia.*

1.2.2.3   The Adelaide Fold Belt (600-250 Ma)

Sedimentation continued in the geosyncline until the Late Cambrian (500 Ma). During the Late Neoproterozoic to Cambrian (570-490 Ma) the Kanmantoo Marginal Basin filled with deepwater turbiditic sediments. These were deposited over earlier continental crust and newly created crust in South Australia and western Victoria. At the same time the shallow water Stansbury Basin occupied Yorke Peninsula and Gulf St Vincent.

The Kanmantoo Group sediments cover most of Kangaroo Island and the eastern Mount Lofty Ranges. To the east they also form the floor of much of the Murray Basin. The Adelaide sediments occupy the northern coast of Kangaroo Island and the western Mount Lofty ranges.

In the Late Cambrian (~490-470 Ma), westward contraction caused overthrusting of the deformed basin rocks over the edge of the Gawler Craton during the Delamerian Orogeny, giving rise to the Delamerian Highlands, including the Fleurieu Peninsula and early Kanmantoo Fold Belt. At the same time, the Neoproterozoic continental rift (the Adelaide Rift) produced the folding and uplift of the Adelaide Fold Belt. The folded sediments of the orogeny are clearly evident today in the prominent basin and ridge topography of the Flinders and Mount Lofty Ranges.

During the Middle Cambrian to Ordovician (500-430 Ma) the Kanmantoo Fold Belt experienced post-folding collapse and lithospheric expansion which resulted in subsidence, enabling deposition of transitional shallow-water to continental sediments and the accumulation of igneous rocks. Subsequent orogenic contraction during the Ordovician Benambran Orogeny converted the Adelaide-Kanmantoo region into a neocraton.

While the Kanmantoo Fold Belt was rising, there was a wide foreland shelfal basin with turbiditic and minor pelagic sedimentation adjacent to the rise. Subsequent to this episode the South Australian region remained continental, with no preserved sedimentation during the Silurian to Middle Devonian (434-378 Ma) and into the Middle Devonian to Early Carboniferous (384-330 Ma), when the Kanmantoo Fold Belt achieved its final form.

During the Early Permian (298-270 Ma) an ice sheet covered the southeast. Glacial striate and grooved rocks formed under the moving ice are preserved on the rock platforms at Hallett Cove. Following the retreat of the ice sheet in the Late Permian (270-250 Ma), sedimentation occurred in the Murray, St Vincent and Eucla basins, outcropping as the Cape Jervis beds in the lower Fleurieu, northeast Kangaroo Island and Yorke Peninsula. The beds are composed of glacial tills, sands and clays, and are prone to gully erosion.

1.2.2.4   Gondwanaland breakup (205-65 Ma)

The Australian continent started to break away from the rest of Gondwanaland during the Jurassic-Cretaceous (205-140 Ma). The rifting commenced in the northwest shelf region, then proceeded southward and anticlockwise during the Cretaceous (140-65 Ma), causing rifting between India and Australia, and later eastward between Australia and Antarctica (150 Ma) (Fig. 1.6). By the Late Cretaceous, seafloor spreading had commenced in the east between the Lord Howe Rise and the mainland (75-55 Ma).

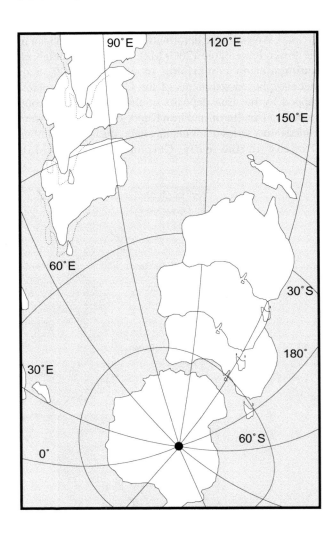

Figure 1.6   *During the past 120 million years the Australian plate, containing Australia, New Guinea and half of New Zealand, has been moving northward at a rate of several centimetres per year. This figure shows the movement of Australia and India as they have both moved northward away from Antarctica, the core of the once supercontinent - Gondwanaland.*

During the Late Cretaceous, the rifting across the southern Australian coast produced buckling of the plate and a series of basins (Eucla, Otway and Gippsland). They began filling with marine sediments, none of which are exposed today. In the South East Otway Basin, marine sediments 3000 m thick underlie the Tertiary Gambier units.

1.2.2.5 Cainozoic (65 Ma to present)

The beginning of the Cainozoic was marked in South Australia with the formation of the Murray and St Vincent Basins, continued sedimentation in the Gambier Embayment of the Otway Basin and uplift of the Eucla Basin. The basins were shallow, resulting in limited penetration of the Tertiary seas.

In the **Gambier Embayment** marine sedimentation, up to 1400 m thick, commenced in the Eocene, with the Gambier Limestone deposited during the Oligocene to Early Miocene (35-20 Ma). The rocks form the edge of the low coastal plain between the SA-Victoria border and Port MacDonnell and are superficially covered by the Pleistocene Bridgewater dune and barrier formations.

The large **Murray Basin** formed in the continental interior the result of basin subsidence caused by foreland loading, brought on by east-west oriented contraction affecting the Australian Plate. It reaches the coast between Kingston SE and Victor Harbor, while extending well inland to encompass much of the low lying sections of the Murray River catchment in South Australia, Victoria and New South Wales. During the Eocene, Oligocene and early Miocene (55-20 Ma), marine deposits occurred during eustatic sea level rises, alternating with continental facies during low sea levels. These sediments are only exposed in the cliffs of the Murray River. During the

Quaternary it has been capped by a 300 km wide series of calcareous barrier ridges extending as far inland as Mildura. The ridges have been reworked by strong winds during the dry glacial maxima periods, leading to the formation of glacial period desert dunes.

St Vincent Gulf experienced marine sedimentation beginning in the Mid Eocene and continuing up the early Miocene (40-20 Ma). These sediments are exposed along the east and west coast of St Vincent Gulf.

Along the **western Spencer Gulf** between Cowell and Port Lincoln, Cainozoic outwash sediments dominate the shoreline and immediate hinterland. These sediments have been derived from erosion of the adjacent cratonic uplands during the Tertiary and Quaternary. The sediments are generally weakly consolidated and are composed of sand, clay and gravel. Where they are exposed to wave erosion they form low, unconsolidated cliffs fronted by rock strandflats.

The **Eucla Basin** sedimentation occurred in two phases: the lower Wilson Bluff limestones deposited in shallow seas during the late Eocene (35 Ma) and the capping Nullarbor Limestone deposited during the Early Miocene (20 Ma). Subsequent uplift of the Eucla Basin formed the Nullarbor Plain and dramatically exposed the limestone to wave erosion along the limestone Nullarbor Cliffs.

1.2.2.6 Quaternary (2 Ma to present)

Quaternary marine deposits are the most recent, geologically brief (2 Ma to present) and thinnest (usually <20 m), yet are the most prevalent on the South

Table 1.4   South Australian coastal geology (major source: Drexel and Preiss, 1995)

Geological structure and provinces (1-8) (east to west)	Age (million years)	Coastal Location	Coastal rock types/morphology
Kanmantoo Fold Belt			
1. Otway Basin	50-20	Glenelg R.-Cape Banks	limestone/low coastal plain
Otway-Murray Basin	170-40	Cape Banks-Middleton	calcareous sands & dunerock/flat
2. Kanmantoo Trough	500-330	E Fleurieu Peninsula	greywacke, siltstones/hilly
Kanmantoo Trough	500-330	Kangaroo Island	greywacke, siltstones/hilly
Adelaide Fold Belt			
3. Adelaide Geosyncline	1400-560	W Fleurieu Peninsula	sandstone, quartzite/hilly
4. St Vincent Basin		NE St Vincent Gulf	calcareous sands/flat
Adelaide Geosyncline	1400-560	E Yorke Peninsula	sandstone, quartzite/hilly
Australian Craton			
5. Gawler Craton	2300-600	S & W Yorke Peninsula	granites and metasedimentary/gently undulating
6. Pirie Trough		N Spencer Gulf	calcareous sands/flat
7. Gawler Craton	2300-600	Eyre Peninsula	granites and metasedimentary/gently undulating
8. Eucla Basin	35-20	Nullarbor cliffs	limestone/flat plain

Australian coast. The reason for this is that all have been deposited at and close to present shorelines and for the most part form a relatively thin calcareous layer over older rocks. Furthermore, they have not only been deposited by waves across the inner continental shelf and shoreline, but have also been blown inland by the prevailing westerly winds. Finally, in the South East continued uplift has stranded these deposits up to 300 km inland, forming the world's largest Quaternary coastal barrier sequence, overlying part of the Gambier Embayment and straddling the Otway and Murray basins.

Beginning in the south, Quaternary marine and dune calcareous deposits formed the massive barrier ridges of the South East, dominating the modern shoreline from Cape Banks to Middleton, including the entire Coorong barrier system. They blanket the entire southern and western shore and cliffs of Kangaroo Island. They dominate the gulfs, beginning at Seaview in Adelaide, forming beach-dune ridges and north of Adelaide they form the extensive sand and salt flat deposits of northern Gulf St Vincent. They blanket the exposed foot of Yorke Peninsula and again form the sand and salt flats of upper Spencer Gulf. On the Eyre Peninsula they form a near continuous deposit of beach and dune systems beginning at the southern tip, Cape Catastrophe and extending all the way to Head of Bight; and finally at the Merdayerrah Sandpatch, they recommence and extend into Western Australia, all the way to Twilight Cove. In addition, calcareous deposits cap all the exposed islands, including Thistle and Flinders Islands.

The evolution of these massive calcareous deposits has been detailed by Short and Hesp (1984), Short and Fotheringham (1986) and Short, et al. (1986). On the open coast, the deposits are the result of wave transport of inner shelf calcareous detritus to the shoreline to be deposited as modern beaches. The strong onshore winds have then blown the sand inland to form, at times, massive dune systems. This has not only occurred during the present sea level stillstand, but during each of the 20 or so previous high stands of the sea over the past 2 Ma (Fig. 1.7). As a consequence, in places the deposits have been layered one atop the other forming cliffs up to 150 m high. In the gulf the sediments are sourced from the seagrass meadows and slowly reworked onto the intertidal sand flats by episodic wave activity.

During periods of low sea level, all the calcareous deposits are subject to pedogenesis (soil forming processes) which acts to partially cement the sand grains together, forming dune calcarenite (i.e. calcareous dunerock) and beach calcarenite in these deposits. As a consequence, when sea level returns some tens of thousands of years later to a high stand, the former sandy beach and dunes have been chemically transformed into more resilient dune and beach rock, as typifies much of the exposed South Australian coast.

Finally, much of the interior of the northern Eyre and Yorke peninsulas and parts of the eastern Eyre Peninsula shoreline south of Whyalla are covered by Pleistocene longitudinal dunes, last active during the late glacial maximum 18 thousand years ago.

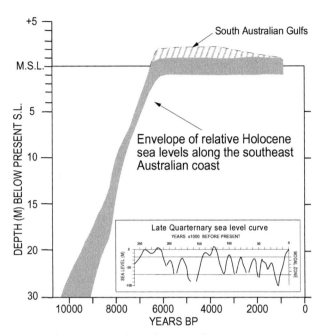

Figure 1.7   *Two plots showing the recent rise in sea level, between 10 000 and 6000 years ago, obtained from evidence along the southeast Australian coast; together with (insert) the regular oscillations in global sea level that have taken place over the past 250 000 years. Along parts of the upper gulfs, sea level stood up to 1-2 m higher about 4000 years ago. The longer-term oscillations in sea level are a result of fluctuating earth surface temperature and its impact on the growth and decay of continental ice sheets, or ice ages.*

1.3   Formation of the South Australian coast

The South Australian coastline is a product of four factors. First was the separation from Gondwanaland beginning across the south about 100 Ma. This defined the location of the coast roughly along the edge of the present continental shelf. When initially formed, the Southern Ocean was a narrow gap between Australia and Antarctica. Now 100 Ma years later, at a northern drift rate of about 5 cm yr^{-1}, it has opened up to form the 5,500 km wide ocean between the two continents.

Second, folding associated with the Adelaide Geosyncline, produced the higher Flinders-Mount Lofty-Kangaroo Island ranges, as well as the downwarping and formation of the two troughs that form St Vincent and Spencer Gulfs.

The third factor has been the buckling of the southern edge of the Australian continent, due in large part to the continental rifting. This produced a series of basins – Otway, Murray and Eucla. The subsequent infilling and uplift of these basins added new structure to the modern coastline.

The final factor has been the eustatic (glacially driven) changes in sea level, on the order of 120 m, over the past 2 Ma (Fig. 1.7). These have resulted in major sea level transgressions (rising sea level) that have drowned the continental shelf and low-lying coastal areas, including the two gulfs and all bays. These are followed by sea level regressions (falling sea level) which expose the continental shelf to a depth of up to 120 m, thereby linking all the offshore islands to the coast. The gulfs are drained dry and the shoreline moves out toward the edge of the continental shelf, in places hundreds of kilometres south of the present shoreline.

1.4   Coastal geological provinces

The South Australian coast consists of eight geological provinces (Table 1.4) composed of rocks ranging in age from 2300 to 2 Ma. This section briefly describes each province beginning in the east in the Otway Basin, then extending west to the Eucla Basin.

1.4.1   Kanmantoo Fold Belt: Otway & Murray Basins

The Kanmantoo Fold Belt is the oldest and westernmost coast of the larger Tasman Fold Belt that contains the eastern third of Australia. The Kanmantoo Fold Belt was active between 500 and 300 Ma, forming the prominent ridges of the eastern Mount Lofty ranges and Kangaroo Island. In the southeast, the rocks were buckled during subsequent breakup from Gondwanaland (170 to 40 Ma) forming the Otway and Murray Basins. The Otway Basin extends from the Port Phillip region to Kingston SE in the west, with the western section called the Gambier Basin, which extends along the coast from the SA-Victorian border to Kingston SE. It is exposed at the coast as low Gambier limestones, capped in places by the Quaternary calcareous dunes and barriers of the Bridgewater Formation.

The Murray Basin occupies the low Coorong coastal section that lies between Kingston SE and Middleton. The modern Coorong barrier and lagoon is the most recent of a 300 km wide series of calcareous barrier ridges extending as far inland as Mildura and subaqueously out onto the inner continental shelf as submerged calcareous barrier ridges/reefs.

1.4.2   Kanmantoo Fold Belt

The positive relief of the Kanmantoo Fold Belt is evident in the Mount Lofty Ranges and Kangaroo Island. The fold belt rocks begin at Middleton and form the eastern and southern Fleurieu Peninsula, from Middleton to Cape Jervis Bay and most of Kangaroo Island. Along the southern Fleurieu coast the rocks are composed of granites (Port Elliot to Victor Harbor), with marine greywacke, siltstones and arkose metasedimentary rocks forming the remainder, including the cliffs that dominate the north and western coast of Kangaroo Island.

1.4.3   Adelaide Geosyncline

The Adelaide Geosyncline runs from the northern tip of Kangaroo Island up along the western Mount Lofty and Flinders Ranges. The heavily folded metasedimentary rocks are only exposed at the coast between Cape Jervis and Seaview, where a range of steeply dipping sandstone, limestone and glacial deposits are exposed in the readily eroded slopes.

1.4.4   St Vincent Basin

The St Vincent Basin is occupied by St Vincent Gulf and includes the low gradient, gently sloping coastal plain that extends north of Adelaide. The shoreline consists of wide, very low gradient Holocene salt flats, with occasional high tide shelly beach ridges, fronted in places by mangroves, all fronted by wide sand flats and seagrass meadows.

1.4.5   Gawler Craton: Yorke Peninsula

The Gawler Craton forms the basement rocks of the Yorke Peninsula, which are only occasionally exposed at the coast, as in the granites between Stenhouse Bay and Point Souttar, and the schist and volcanics at Port Victoria, Wardang Island and Point Pearce. Most of the coast and hinterland of the peninsula is covered by calcareous Quaternary marine and aeolian deposits.

1.4.6   Gawler Craton: Pirie Trough

The Pirie Trough occupies the northern Spencer Gulf north of Port Broughton on the east coast and Franklin Harbour on the west coast. For the most part this shoreline consists of wide, low gradient Holocene coastal plain, with salt flats, beach ridges and sand flats in the west, grading into longitudinal dunes. Along the northwest shoreline some rocks associated with the Stuart Shelf province occur between Port Augusta and Stony Point, with Quaternary sediment further south.

1.4.7   Gawler Craton: Eyre Peninsula

The ancient Gawler Craton dominates the Eyre Peninsula, forming the coastal basement from Stony Point to Cape Adieu. While much of the peninsula is blanketed by Quaternary aeolian and, at the coast, marine sediments, all coastal salients and headlands between Franklin Harbour and Cape Adieu are tied to the cratonic bedrock, usually granite and, in places, its deeply weathered remnants.

1.4.8 Gawler Craton: Eucla Basin

The Eucla Basin consists of Cainozoic shallow calcareous marine sediments. They are covered by the extensive Yalata dune fields in the east between Cape Adieu and Head of Bight. The uplifted limestone of the Eucla Basin forms the flat Nullarbor Plain. At the coast it is exposed as 60-90 m high vertical cliffs for 210 km between Head of Bight and the border at Wilson Bluff. Beyond the border the cliffs continue, at first fronted by the Roe Plain and then forming the 180 km long Baxter Cliffs, which run inland south of Point Culver, finally terminating inland from Point Malcolm, 550 km west of the border a total length (Head of Bight-Point Malcolm) of 760 km.

Table 1.5   Age of South Australian coastal features

• rocks from	2300 Ma (Eyre Peninsula)
• shape of coast	100 Ma (breakup)
• calcarenite from	2 Ma
• most beaches less than	6500 years old

1.4.9 Summary

South Australia has one of the more ancient land surfaces in the world, with all the basement west of St Vincent Gulf composed of part of the Gawler Craton, which dates back to 2300 Ma. The Gawler Craton is part of the massive Australian Craton, which extends to the Western Australian coast. In the east, buckling, followed by deep sedimentation and subsequent folding and uplift, produced the Adelaide Geosyncline, a range of mountains running from Flinders Ranges down through the Fleurieu Peninsula to Kangaroo Island. When southern Australia broke away from Gondwanaland beginning about 150 Ma, further buckling formed the Eucla, Otway and Murray basins that are now largely filled in with Cainozoic (Eucla) and Quaternary (Murray) marine sediments. These sediments deposited during 20 or more high stands of sea level over the past 2 Ma and now dominate both the exposed sections of the entire coast, forming high calcarenite bluffs, cliffs and inshore reefs, as well as all modern beach systems, that occupy 62% of the shoreline. In the gulfs they form the extensive sand flats and beach ridges.

1.5   Coastal Processes

Coastal processes are the marine, atmospheric and biological activities that interact over time with the geology and sediments to produce a particular coastal system or environment. At the shoreline, the four great spheres; the atmosphere, the hydrosphere or ocean, the lithosphere or earth's surface and the biosphere; all coexist. Consequently it is often the most dynamic part of the earth's surface; this dynamics most evident on sandy beaches, particularly those exposed to high waves. South Australian sandy beaches consist of lithospheric and biotic elements, namely terrigenous and calcareous sand, lying ultimately on bedrock geology. They are acted on by the waves, tides and currents of the hydrosphere, by the wind, rain and temperature of the atmosphere, and play host to the fauna of the biosphere. The following sections examine the atmospheric, marine and biological processes that provide the energy to build and maintain South Australian beach systems.

1.5.1   Climate processes

Climate's contribution to beaches is in two main areas. Climate interacts with the geology and biology to provide the geo- and bio-chemistry to weather the land surface, which, together with the physical forces of rain, runoff, rivers and gravity, erode and transport sediments to the coast. At the coast it is also the climate, particularly winds, that interact with the ocean to generate waves and currents that are essential to move and build this sediment into beaches and dunes.

On a global scale, beaches can be classified by their climate. The *polar beaches* of the Arctic Ocean and those surrounding Antarctica, including Australia's large Antarctic Territory, are all dark in colour and composed of coarse sand, cobbles and even boulders. They receive only low waves and have little or no surf. The coarse beaches have steep swash zones and, because of the coarse sediment, there are no dunes. All these characteristics are a product of the cold polar climate, which permits little chemical weathering, hence the dark unstable minerals in the sediments, while the dominance of cold physical weathering ensures a supply of coarse-large sediments.

Temperate middle latitude beaches typical of southern Australia have sediments composed predominantly of well-weathered quartz sand grains, with variable amounts of biotic fragments (shell, algae, coral, etc). In addition, wind and wave energy associated with the strong mid latitude westerly wind stream is relatively high. The waves produce energetic surf zones, while the winds can build massive coastal dune systems, as is typical of the open South Australian coast.

Tropical beaches of northern Australia reside in areas of lower winds of the equatorial low pressure area (the doldrums) and the great Trade winds of moderate velocity. Consequently wave energy is low to moderate at best, unless the shore is exposed to higher waves generated in the mid-latitudes. The areas of lower waves and winds tend to have steep high tide beaches with little or no surf and few dunes. Their sediments, however, are often white, being composed of well-weathered quartz sands derived from plentiful tropical rivers and bleached coral and algal fragments derived from coral and algal reefs. Areas exposed to the Trades can have more energetic beaches and, in places, extensive coastal dunes.

The South Australian coast extends over 6.5° of latitude or 720 km north to south, from 31.5°S at Head of Bight and 32.5°S at Port Augusta to 36°S on Kangaroo Island and 38°S at Port MacDonnell. This places it on the boundary between the subtropics and mid latitudes. The subtropical high pressure system dominates the summer climate, bringing warm to hot dry conditions, while during winter the cooler, more humid mid latitude cyclones impact the southern part of the state. These bring both cold fronts and associated frontal rainfall (Fig. 1.8).

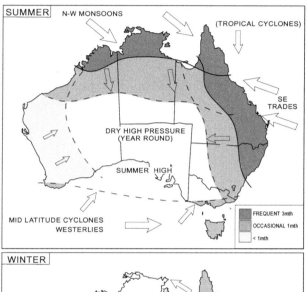

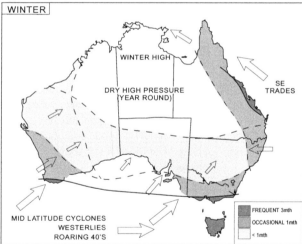

Figure 1.8   *The Australian climate is dominated year round by the dry high pressure anticyclones that sit over much of Australia. In summer, the high is centred over 36°S and brings dry conditions to the southern parts of the continent. In winter, the dry high moves north to dominate central and northern Australia, and permits mid latitude cyclones and their cold fronts to bring cool weather and rain across southern Australia. In South Australia they regularly penetrate and bring rain across the southern Eyre, Yorke and Fleurieu Peninsulas and to the South East. This figure shows the source, season and extent of influence of the major air masses around the Australian coast.*

1.5.2   Subtropical high pressure systems

The most dominant feature of the South Australian climate is the great subtropical high, which sits over the Australian continent year round, shifting north over the centre of the continent in winter and located squarely over the latitude of Adelaide (35°S) during summer. The highs bring dry, stable air conditions, with clear sky and light winds. Winds flow counterclockwise around the highs with a westerly stream along the southern half, with east through southeast (the south east Trades) across the northern half. In South Australia summers are warm to occasionally hot and generally dry.

During winter the highs shift north and are centred on 30°S. This permits the mid latitude cyclones, which track year round across the Southern Ocean, to penetrate the southern coast of the continent. The low pressure cyclones and their associated cold fronts bring periodic winter frontal rain to the southern parts of the Eyre, Yorke and Fleurieu Peninsulas and the South East, as well as cool to occasional cold weather. In between the lows, the high can reestablish itself, bringing drier conditions. Winter conditions are therefore cool to mild and wetter, with all coastal locations receiving a significant winter maximum in their rainfall (Fig. 1.9).

1.5.3   Sea breeze systems

Associated with high pressure conditions are the coastal sea breezes. Sea breeze is a result of the differential heating of the land and sea surface, particularly during summer, when the land can heat up considerably. In the absence of zonal westerly winds, the hot air over the land rises to be replaced by cooler air flowing in from the sea, hence the *sea breezes*. The breeze usually commences in mid to late morning after the land has heated and continues into the afternoon. At night the reverse may occur, with the land cooling to a lower temperature, particularly on calm clear nights, relative to the sea surface. If the land is cooler than the sea, then rising air over the sea may be replaced by cooler air from over the land, leading to an early morning land breeze. The land breezes are usually less frequent and of lower velocity than the sea breezes.

The direction of the sea breeze is dependent on the orientation of the coastline. The breezes tend to flow onshore, with a redirection in the Southern Hemisphere to the left of flow, owing to the Coriolis effect. On south facing coasts the southerly breeze comes from the southeast, on west facing coasts from the southwest and on east facing coasts from the northeast. For this reason the direction of sea breezes varies along the South Australian coast from southwest through to northeast.

South Australian Coastal Rainfall (mm)

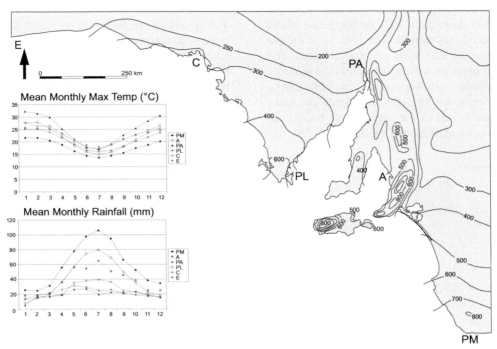

Figure 1.9   *South Australia's annual rainfall, showing the marked decrease from south to north, with local maxima due to orographic effects on Kangaroo Island and Adelaide ranges. Inserts indicate monthly rainfall at selected stations, which highlight the winter maxima, and mean maximum monthly temperature for the same locations.*

1.5.4 Mid Latitude Cyclones

Mid latitude cyclones are part of the subpolar low pressure system that extends around the entire Southern Hemisphere in a belt centred on 40° to 60°S latitude, the so called 'Roaring Forties'. The northern side of this system forms the southern boundary of the subtropical high pressure system. Like the high it shifts with the seasons, moving closer to the southern Australian continent in winter and further south in summer. The lows or cyclones that are embedded in this belt are continually moving from west to east. On average, one passes south of Australia every three to four days with between 80 and 100 passing south of the continent each year. These cyclones are responsible for both the spiraling westerly wind streams and the persistent southerly waves that arrive on average 350 days a year along the southern Australian coast.

1.5.5 South Australian coastal climate

The climate of South Australia is dependent on the subtropical highs and mid latitude lows, with moderating influence, particularly in summer, from the local sea breeze systems. In summer, when the highs dominate, hot, dry conditions prevail, with only occasional cooler changes associated with mid latitude cyclones. During winter, while the high is still dominant, cold fronts, associated with the top of the mid latitude cyclones, can penetrate the southern part of the state in between the

slowly moving highs, bringing frontal rainfall and cool to cold conditions.

The *rainfall* along the coast reflects three factors. First, as for the whole state, it is highly seasonal, with most rain falling in winter (Fig. 1.9). Second, there is a marked decrease from south to north, owing to the increasing distance from the source, the mid latitude cyclones. Third, there are some local orographic effects; that is, rainfall induced by a topographic obstacle, such as the Mount Lofty Ranges. This results in higher rainfall on the windward, westerly sides of these ranges.

1.5.6 Coastal winds

The coastal wind regime along the South Australian coast is dominated by the westerly flow associated with the sub tropical high. The westerlies dominate year round. However during summer they tend to be lighter winds, while during winter they strengthen with the addition of westerly cold fronts and associated cyclonic conditions. The only other wind system of note is the summer sea breezes which tend to occur when the centre of the high sits over the state, leading to low to calm anticyclonic winds and hot temperatures, all of which favour the development of strong sea breeze conditions.

The coastal winds are extremely important for coastal processes, as they generate local sea waves, they induce upwelling during northerly conditions and downwelling during southerly conditions and they are responsible for all aeolian (wind blown) sand transport, in other words, the formation of the massive South Australian coastal dune systems.

1.6   Ocean Processes

Oceans occupy 71% of the world's surface. They therefore influence much of what happens on the remaining land surfaces. Nowhere is this more the case than at the coast and nowhere are coasts more dynamic than on sandy beach systems. The oceans are the immediate source of most of the energy that drives coastal systems. Approximately half the energy arriving at the world's coastlines is derived from waves; much of the rest arrives as tides, with the remainder contributed by other forms of ocean and regional currents. In addition to supplying physical energy to build and reshape the coast, the ocean also influences beaches through its temperature, salinity and the rich biosphere that it hosts.

There are eight types of ocean processes that impact the South Australian coast, namely: wind and swell waves, tides, shelf waves, ocean currents, local wind driven currents, upwelling and downwelling, sea surface temperature and ocean biota (Table 1.6). Each of these and their impacts is discussed below.

1.6.1   Ocean waves

There are many forms of waves in the ocean ranging from small ripples to wind waves, swell, tidal waves, tsunamis and long waves, including standing and edge waves; the latter lesser known but very important for beaches. In this book, the term 'waves' refers to the wind waves and swell, while other forms of waves are referred to by their full name. The major waves and their impact on beaches are discussed in the following sections.

1.6.1.1   Wind waves

Wind waves, or sea, are generated by wind blowing over the ocean. They are the waves that occur in what is called the area of wave generation; as such they are called '*sea*'.

Five factors determine the size of wind waves:

- *Wind velocity* - wave height will increase exponentially as velocity increases;
- *Wind duration* - the longer the wind blows with a constant velocity and direction, the larger the waves will become until a fully arisen sea is reached; that is, the maximum size sea for a given velocity and duration;
- *Wind direction* will determine, together with the Coriolis force, the direction the waves will head;
- *Fetch* - the sea or ocean surface is also important; the longer the stretch of water the wind can blow over, called the fetch, the larger the waves;
- *Water depth* is important, as shallow seas will cause wave friction and possibly breaking. This is not a problem in the deep ocean, which averages 4.2 km in depth, but is very relevant once waves start to cross the South Australian continental shelf, which averages less than 100 m and enter the gulfs and large coastal bays which are considerably shallower.

The biggest seas occur in those parts of the world where strong winds of a constant direction and long duration blow over a long stretch of ocean. The part of the globe where these factors occur most frequently is in the southern oceans between 40° and 50°S, where the Roaring Forties and their westerly gales prevail. Satellite sensing of all the world's oceans found that the world's biggest waves, averaging 6 m and reaching up to 20 m, occurred most frequently around the Southern Ocean, centred on 40°S and including a band across the south of Australia. Waves from this source dominate the entire open South Australian coast.

1.6.1.2   Swell

Wind waves become swell when they leave the area of wave generation, by either travelling out of the area where the wind is blowing or when the wind stops blowing. Wind waves and swell are also called free waves or progressive

Table 1.6   Major ocean processes impacting the coast

Process	Area of coastal impact	Type of impact
waves – sea & swell	shallow coast & beach	wave currents, breaking waves, wave bores
tides	shoreline & inlets	sea level, currents
shelf waves	shoreline & inlets	sea level
ocean currents	continental shelf	currents
local wind currents	nearshore & shelf	currents
upwelling & downwelling	nearshore & shelf	currents & temperature
ocean temperature	entire coast	temperature
biota	entire coast	varies with environment & depth

waves. This means that once formed, they are free of their generating mechanism, the wind and they can travel long distances without any additional forcing. They are also progressive, as they can move or progress unaided over great distances.

Once swell leaves the area of wave generation, the waves undergo a transformation that permits them to travel great distances with minimum loss of energy. Whereas in a sea the waves are highly variable in height, length and direction, in swell the waves decrease in height, increase in length and become more uniform and unidirectional. As the speed of a wave is proportional to its length, they also increase in speed (Fig. 1.10).

A quick and simple way to calculate the speed of waves in deep water is to measure their period, that is, the time between two successive wave crests. The speed is equal to the wave period multiplied by 1.56 m. Therefore, a 10 second wave travels at 10 x 1.56 metres per second, which equals 15.6 m/s or 56 km per hour. In contrast, a 5 second wave in the gulfs travels at 28 km per hour and the longer ocean waves at 67 km per hour for a 12 second wave and 79 km per hour for a 14 second wave. What this means is that the long ocean swell is travelling much faster than the short gulf seas and that as sea and swell propagate across the ocean, the longest waves travel fastest and arrive first at the shore.

Swell also travels in what surfers call 'sets' or more correctly *wave groups*, that is, groups of higher waves followed by lower waves. These wave groups are a source of long, low waves (the length of the groups) that become very important in the surf zone, as discussed later.

Swell and seas will move across the ocean surface through a process called orbital motion. This means the wave particles move in an orbital path as the wave crest and trough pass overhead. This is the reason the wave form moves, while the water, or a person or boat floating at sea, simply goes up and down, or more correctly around and around. However when waves enter water where the depth is less than 25% of their wave length (wave length equals wave period squared, multiplied by 1.56; for example, a 10 second wave will be 10 x 10 sec x 1.56 = 156 m in length and a 5 second wave only 39 m long) they begin to transform into shallow water waves, a process that may ultimately end in wave breaking. Using the above calculations this will happen on the open coast when the long 12-14 sec swell reaches between 50 and 80 m depth, while in the gulfs the short 3-5 sec seas will begin to feel bottom between 5 and 10 m.

As waves move into shallow water and begin to interact with the seabed or 'feel bottom', four processes take place, affecting the wave speed, length, height, energy and ultimately the type of wave breaking (Fig. 1.11).

Ocean Wave Generation, Transformation and Breaking

Wave type	Breaking	Shoaling	Swell	Sea
Environment	Shallow water - surf zone	Inner continental shelf	Deep water >> 100m	Deep water >>100m Long fetch = sea/ocean surface Wind velocity ▲ waves▲ Wind duration ▲ waves▲ Wind direction = wave direction
Distance travelled	~100 m	1 to 100 km	100's to 1000's km	100's to 1000's km
Time required	seconds	minutes	hours to days	hours to days
Wave profile	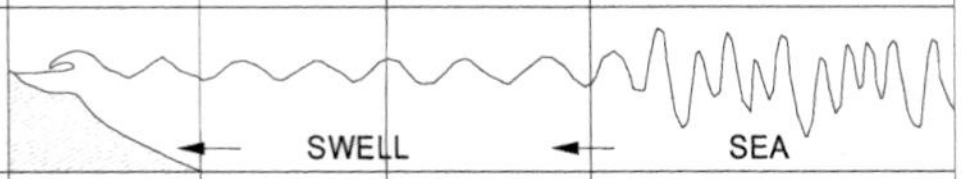			
Water depth	1.5 x wave height	< 100m	>> 100m	>> 100m
Wave character	wave breaks wave bore swash	higher shorter steeper same speed	regular lower longer flatter faster	variable height high short steep slow
Example: height (m) period (sec) length (m) speed (km/hr) distance travelled (km/day)	2.5 to 3 12 0 to 50 0 to 15 -	2 to 2.5 12 50 to 220 15 to 60 -	2 to 3 12 220 66 1600	3 to 5 6 to 8 50 to 100 33 to 45 800 to 1100

Figure 1.10 *Waves begin life as 'sea waves' produced by winds blowing over the ocean or sea surface. If they leave the area of wave generation they transform into lower, longer, faster and more regular 'swell', which can travel for hundreds to thousands of kilometres. As all waves reach shallow water, they undergo a process called 'wave shoaling' which causes them to slow, shorten, steepen and finally break. This figure provides information on the characteristics of each type of wave. The South Australian coast receives year round swell on all open coast, with shorter wind waves in the upper gulfs.*

Wave speed decreases with decreasing water depth.

Variable water depth produces variable wave speed, causing the wave crest to travel at different speeds over variable seabed topography. At the coast this leads to *wave refraction*. This is a process that bends the wave crests, as that part of the wave moving faster in deeper water overtakes that part moving slower in shallower water.

At the same time that the waves are refracting and slowing, they are interacting with the seabed, a process called *wave attenuation*. At the seafloor, some potential wave energy is transformed into kinetic energy as it interacts with the seabed, doing work such as moving sand. The loss of energy causes a decrease or attenuation in the overall wave energy and therefore lowers the height of the wave.

Finally, as the water becomes increasingly shallow, the waves shoal, which causes them to slow further, decrease in length but increase in height, as the crest attempts to catch up to the trough. The speed and distance over which this takes place determines the type of *wave breaking*.

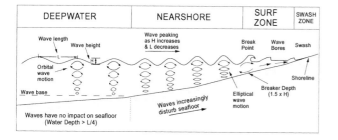

Figure 1.11 *A cross section of a beach showing the transformation of deepwater sea waves and swell as they move into shallow coastal waters. On the surface they can be seen to slow, shorten, steepen and perhaps increase in height. At the break point they break and move across the surf zone as wave bores (broken white water) and finally up the beach face as swash. Below the surface, the orbital wave currents are also interacting with the seabed, doing work by moving sand and ultimately building and forever changing the beach systems.*

Wave types: sea and swell

Waves are generated by wind blowing over water surfaces.
Large waves require very strong winds, blowing for many hours to days, over long stretches of deep ocean.
Sea waves occur in the area of wave generation, in close vicinity to the mid latitude cyclones and in the gulfs and large bays.
Swell are sea waves that have travelled out of the area of wave generation. Swell dominates the southern Australian open coast, but is filtered out of the upper gulfs and some bays.

1.6.1.3   Wave breaking

Waves break basically because the wave trough reaches shallower water (such as the sand bar) ahead of the following crest. The trough therefore slows down, while the crest in deeper water is still travelling a little faster. Depending on the slope of the bar and the speed and distance over which this occurs, the crest will attempt to 'overtake' the trough by spilling or even plunging forward and thereby breaking.

There are three basic types of breaking wave:

•   *Surging breakers,* which occur when waves run up a steep slope without appearing to break. They transform from an unbroken wave to beach swash in the process of breaking. Such waves can be commonly observed on steeper beaches when waves are low, or after larger waves have broken offshore and reformed in the surf zone. They then may reach the shore as a lower wave, which finally surges up the beach face as swash.

•   *Plunging or dumping waves,* which surfers know as a tubing or curling wave, occur when shoaling takes place rapidly, such as when the waves hit a reef or a steep bar and/or are travelling fast. As the trough slows, the following crest continues racing ahead and as it runs into the stalling trough, its forward momentum causes it to both move upward, increasing in height and throw forward, producing a curl or tube.

•   *Spilling breakers* on the other hand, occur when the seabed shoals gently and/or waves are moving slowly, resulting in the wave breaking over a wide zone. As the wave slows and steepens, only the top of the crest breaks and spills down the face of the wave. Whereas a plunging wave may rise and break in a distance of a few metres, spilling waves may break over tens or even hundreds of metres.

1.6.1.4   Broken waves

As waves break they are transformed from a progressive wave to a mass of turbulent white water and foam called a *wave bore*. It is also called a 'wave of translation', as unlike the unbroken progressive wave, the water actually moves or translates shoreward. Surfers, assisted by gravity, surf on the steep part of the breaking wave. Once the wave is broken, boards, bodies and whatever can be propelled shoreward with the leading edge of the wave bore, while the turbulence in the wave bore is well known to anyone who has dived into or under the white water.

1.6.2 Surf zone processes

Ocean waves can originate thousands of kilometres from the coast. They can travel as swell for days to reach their destination. However on reaching the coast, they can undergo wave refraction, shoaling and breaking in a matter of minutes to seconds. Once broken and heading for shore, the wave has been transformed from a progressive wave containing potential energy to a wave bore or wave of translation with kinetic energy, which can do work in the surf zone.

There are three major forms of wave energy in the surf zone - broken waves, surf zone currents and long waves (Table 1.7).

Table 1.7   Wave motions in the surf zone

Wave form	Motion	Impact
Unbroken wave	orbital	stirs sea bed, builds ripples
Breaking wave	crest moves rapidly shoreward	wave collapses
Wave bore	all bore moves shoreward	shoreward moving turbulence
Surf zone currents	water flows shoreward, longshore and seaward (rips)	moves water and sediment in surf, builds bars, erodes troughs
Long waves	slow on-off shore	determines location of bars & rips

Broken waves consist of wave bores and perhaps reformed or unbroken parts of waves. These move shoreward to finally run up and down the beach as swash uprush and backwash. Some of the backwash may reflect out to sea as a reflected wave, albeit a much smaller version of the original source.

Surf zone currents are generated by broken and unbroken waves, wave bores and swash. They include orbital wave motions under unbroken or reformed waves; shoreward moving wave bores; the uprush and backwash of the swash across the beach face; the concentrated movement of the water along the beach as a longshore current; and where two converging longshore or rip feeder currents meet, the seaward moving rip currents.

The third mechanism is a little more complex and relates to *long waves* produced by wave groups and, at times, other mechanisms. Long waves accompany sets of higher and lower waves. However, the long waves that accompany them are low (a few centimetres high), long (perhaps a few hundred metres) and invisible to the naked eye. As sets of higher and lower waves break, the accompanying long waves also go through a transformation. Like ocean waves they also become much shorter as they pile up in the surf zone, but unlike ocean waves, they do not break, but instead increase in energy and height toward the shore. Their increase in energy is due to what is called 'red shifting', a shift in wave energy to the red or lower frequency part of the wave energy field. These waves become very important in the surf zone, as their dimensions ultimately determine the number and spacing of bars and rips.

As waves arrive and break every few seconds, the energy they release at the break point diminishes shoreward, and the wave bores decrease in height toward the beach. The energy released from these bores goes into driving the surf zone currents and into building the long waves. The long wave crest attains its maximum height at the shoreline. Here it is visible to the naked eye in what is called *wave set-up* and *set-down*. These are low frequency, long wave motions, with periods in the order of several times the breaking wave period (30 sec-few minutes), that are manifest as a periodic rise (set-up) and fall (set-down) in the water level at the shoreline. If you sit and watch the swash, particularly during high waves,

you will notice that every minute or two the water level and maximum swash level rises then rapidly falls.

The height of wave set-up is a function of wave height and also increases with larger waves and on lower gradient beaches, to reach as much as one third to one half the height of the breaking waves. This means that if you have 1, 2, 5 and 10 m waves, the set-up could be as much as 0.3, 0.6, 1.5 and 3.0 m high, respectively. For this reason, wave set-up is a major hazard on high energy low gradient South Australian beaches, particularly beaches like Goolwa.

Because the waves set up and set down in one place, the crest does not progress. They are therefore also referred to as a *standing wave*, one that stands in place with the crest simply moving up and down. These standing long waves are extremely important in the surf zone as they help determine the number and spacing of bars and rips. This interaction is discussed in Section 2.2 on beach dynamics.

1.6.3 Wave measurements

While it is easy to see waves and to make an estimate of their height, period, length and direction, accurate measures of these statistics are more difficult. If we are, however, to properly design for and manage the coast we need to know just what type of waves are arriving at the shore. Traditionally, wave measurements have been made by observers on ships and at lighthouses visually estimating wave height, length and direction. This was the case in South Australian lighthouses at Cape Northumberland, Cape Willoughby, Cape Du Couedic and Cape Borda until they were automated in the 1970's. Since then there were no regular wave observations made in South Australian waters until November 2000 when the Bureau of Meteorology installed a Waverider buoy off Cape Du Couedic.

The Datawell Waverider buoy is the present state-of-the-art on-site wave recording device. It operates using an accelerometer housed in a watertight buoy, about 50 cm in diameter. The buoy is chained to the sea floor, usually in about 80 m water depth. As the waves cause the buoy to rise and fall, the vertical displacement of the buoy is recorded by the accelerometer. This information is transmitted to a shore station and then by phone line to a central computer, where it is recorded. The first Australian Waverider buoy was installed off the Gold Coast in 1968. Today, Queensland and New South Wales have a network of Waverider buoys stretching from Weipa to Eden, the most extensive in the world.

The Cape Du Couedic Waverider is the first permanent system to operate in the South Australian region. As it has only been operating for six months as this book goes to press, the recordings to date only cover the summer period and are preliminary. The data does however indicate a persistent high swell (> 2 m) generated by the mid-latitude cyclones passing south of the continent, with peaks in wave height occurring 3-4 times each month, with the maximum wave height exceeding 6 m. During the recording period significant wave height exceeded 3.5 m 10% of the time, 2 m 50% and 1 m 99%, with no calms recorded. Table 1.8 provides a summary of the mean wave characteristics. In due course this buoy will provide an excellent record of longer term wave conditions on the South Australian coast.

Table 1.8   Cape Du Couedic Waverider data (December 2000 - February 2001)

	Median Hs (m)	Mean Tz (sec)	Tp (sec)
Dec	2.5	8	12
Jan	2.0	8	12
Feb	1.8	7	12

Source: Bureau of Meteorology

The high southerly swell is supported by observations made by the US Navy satellite GEOSAT in March 1985. This satellite uses radar altimeter to measure global wind and wave conditions. Young and Holland (1996) provide an excellent presentation of the satellite results, for all oceans. Some subsequent figures have been derived from this publication.

How to estimate wave height from shore

Wave height is the vertical distance from the trough to the crest of a wave. It is easier to make a visual measurement when waves are relatively low and if there are surfers in the water to give you a reference scale. If a surfer is standing up and the wave is waist height, then it's about 1 m; if as high as the surfer, about 1.5 m; if a little higher, then it's about 2 m. For big waves, just estimate how many surfers you could 'stack' on top of each other to get a general estimate, i.e. two surfers, about 4 m, three, about 6 m and so on. Many surfers prefer to underestimate wave height by as much as 60% and a substantial number still estimate height in the old imperial measure of 'feet', as suggested below:

Actual wave height	surfer's (under)estimate	% underestimated
0.5 m	1 ft (0.3 m)	60%
1.0 m	2 ft (0.6 m)	60%
1.5 m	3 ft (0.9 m)	60%
2.0 m	5 ft (1.5 m)	80%
2.5 m	6 ft (1.8 m)	70%
3.0 m	8 ft (2.4 m)	80%

1.6.4   South Australian wave climate

Wave climate refers to the seasonal variation in the source, size and direction of waves arriving at a location. Waves on the South Australian coast originate from three possible sources - mid latitude cyclones, high pressure systems and the more localised sea breeze systems. In addition, in the upper gulfs where swell is absent, all waves are generated by local winds and arrive as short, steep sea.

Swell: Mid latitude cyclones produce year round swell that arrives right across the southern Australian coast, with a slight summer minimum and late winter maximum. The swell is long (10-14 sec), moderate to high (2-3 m) and arrives predominantly from the west (20%) and southwest (60%), being more southerly in the north of the Bight and more westerly in the South East (Table 1.9). Swell height is also higher toward the south, with Cape Northumberland receiving 0-2 m swell for 31% and 2-4 m for 62% of the year, compared to 57% and 29% respectively for Cape Du Couedic. These heights are likely to decrease into the Bight.

Figure 1.12 illustrates the global wave environment based on satellite altimetry. The figure clearly indicates that the world's highest average waves occur across the Southern Ocean, including south of South Australia. In coastal waters the waves average 4-5 m for 10% of the year, 2-3 m for 50% and 1-2 m for 90%. These global figures agree reasonably well with the lighthouse and limited Waverider observations.

While long term detailed observations and recording of waves are not yet available for the coast, the generalised satellite records provide an accurate overview of the deepwater wave climate and its source regions, while the qualitative lighthouse observations provide an approximation of the nearshore coastal wave heights in two exposed locations.

In summary, the dominant source of waves are the year-round mid latitude cyclones centred between 40 and 60°S. Waves arrive predominantly from the southwest and west. GEOSAT data indicates that the waves are rarely less 0.5 m (<10%), that they average 2.5 m and reach 3 m 40% of the year, 4 m 15% and 5 m about 5% of the time. The waves are high year round, with a slight decrease in summer when the mid latitude cyclones are located further to the south and decrease in intensity, with the lowest waves in January (1.5-2.5 m) and the highest in

Table 1.9   Lighthouse swell observations

Cape Du Couedic lighthouse observations (1967-1973, n = 1562) % month

	J	F	M	A	M	J	J	A	S	O	N	D	Total
calm	3	2	1	2	4	8	0	1	5	0	4	5	3
0-2	58	56	64	69	62	55	56	43	55	43	55	51	57
2-4	29	37	32	21	26	26	35	40	28	29	26	33	29
>4	9	5	3	8	8	11	9	16	12	28	15	11	11

Directions:
NE	=	4.9%
E	=	1.2
SE	=	5.3
S	=	4.4
SW	=	**57.0**
W	=	**18.2**
NW	=	6.7
N	=	1.1

Cape Northumberland lighthouse observations (1957-1978, n ≈ 7500)

	J	F	M	A	M	J	J	A	S	O	N	D	Total
calm		1	1	1						1	1	1	1
0-2	44	39	35	24	28	29	21	21	26	30	31	40	31
2-4	51	57	60	67	63	64	70	70	66	64	63	55	62
>4	4	3	4	8	8	7	9	9	8	5	6	4	6

All directions: W, SW, S, SE

later winter (July-October, 3.5 m) and monthly averages usually 2.5 m high and higher. At the coast the lighthouse observations agree with satellite observations. The greatest number of 0-2 m waves are observed in November-April and the greatest number of 2-4 m waves in July and August. In total, the South Australian coast faces squarely into one of the world's highest energy deepwater wave environments. Waves reach the coastal zone year round as a moderate to high swell.

Between the outer coastal zone (continental shelf) and the shore, however, considerable wave transformation can take place. Exposed beaches fronted by deeper shelf and nearshore zones and free of reef receive most of the deepwater energy, as along the Canunda coast. Many beaches are however protected by a lower gradient shelf and in particular the prevalence of calcarenite reefs, so that some like Lacepede Bay may face directly into the prevailing waves, but receive none at the shore.

The breaker wave climate, as opposed to the deepwater wave climate, is therefore a function of the deepwater wave height, less the amount of wave energy (and height) lost as the waves cross the continental shelf and nearshore zone. This loss can range from near zero on deep, steep shelves, to 100% on wide, shallow shelves and in the upper gulfs. As a consequence of the prevalence of calcarenite reefs and islands, and the variable coastal orientations along the South Australian coast, breaker wave height, derived from deep ocean waves, ranges from an estimated maximum average of 2.5 m to zero.

Wave period: Still less is known of wave periods along the coast. As an approximation, deepwater and breaker waves average 12-14 seconds, while in the gulfs and large bays wind wave period is between 3-5 seconds.

Sea waves: While the mid latitude cyclones pass often well south of the continent (40-50°S), the South Australian coast is usually under the direct influence of high pressure systems and their westerly wind stream. These winds generate westerly sea conditions, which are superimposed on top of the swell. The occurrence and size of the seas is entirely dependent on local wind conditions. Under strong westerly winds they may reach a height of a few metres with periods of several seconds. In the gulfs they produce predominantly westerly waves that dominate more on the eastern shores. The size of these waves is influenced by the fetch, which narrows northward in both gulfs and water depth, an important factor in the relatively shallow gulfs. As a consequence, gulf wind waves are usually low to moderate in height (1-2 m), short (a few seconds) and steep.

Sea breezes generate a low to moderate velocity onshore wind of limited duration and fetch. As a consequence they generate only low (<1 m), short, steep seas, usually from late morning to afternoon, with direction varying around the coast in response to changing sea breeze direction.

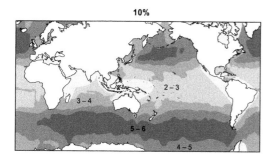

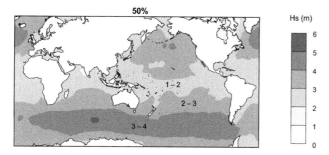

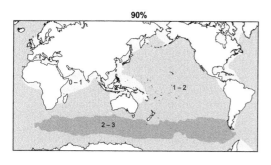

Figure 1.12 *Average global wave heights. Note the dominance of high waves south of Australia year round, with deepwater waves south of South Australia exceeding 5-6 m for 10%, 3-4 m for 5% and 2-3 m for 90% of the year (modified from Young and Holland, 1995).*

Freak waves, king waves, rogue waves and tidal waves

Freak waves do not exist.

All waves travel in wave groups or sets. A so-called 'freak', 'king' or 'rogue' wave is simply the largest wave or waves in a wave group. The fact that these waves can also break in deep water, due to oversteepening, also adds to their demeanour.

Unusually high waves are more likely in a sea than a swell. For this reason they are more likely to be encountered by yachtsmen than surfers or rock fishermen.

Tidal waves arrive on the South Australian coast twice each day. These are related to the predictable movement of the tides and not the damaging tsunamis with which they are commonly confused. Tidal waves are discussed in Section 1.6.5.

1.6.5 Tides

Tides are the periodic rise and fall in the ocean surface, due to the gravitational force of the moon and the sun acting on a rotating earth. The amount of force is a function of the size of each and their distance from the earth. While the sun is much larger than the moon, the moon exerts 2.16 times the force of the sun because it is much closer to earth. Therefore, approximately 2/3 of the tidal forces are due to the moon and are called the *lunar* tides. The other 1/3 are due to the sun and these are called *solar* tides.

Because the rotation and orbit of the earth and the orbit of the moon and sun are all rigidly fixed, the *lunar* tidal period, or time between successive high or low tides, is an exact 12.4 hours; while the *solar* period is 24.07 hours. As these periods are not in phase, they progressively go in and out of phase. When they are in phase, their combined force acts together to produce higher than average tides, called *spring tides*. Fourteen days later when they are 90^0 out of phase, they counteract each other to produce lower than average tides, called *neap tides*. The whole cycle takes 28 days and is called the lunar cycle, over a lunar month.

Note: Spring tides are also called 'king' tides and are highest around New Year and Christmas.

Spring or *king tides* are not responsible for beach erosion, unless they happen to coincide with large waves.

The actual tide is in fact a wave, correctly called a *tidal wave,* not to be confused with tsunamis. They consist of a crest and trough, but are hundreds of kilometres in length. When the crest arrives it is called *high tide* and the trough *low tide*. Ideally, the tidal waves would like to travel around the globe. However the varying size, shape and depth of the oceans, plus the presence of islands, continents, continental shelves and small seas complicate matters. The result is that the tide breaks down into a series of smaller tidal waves that rotate around an area of zero tide called an *amphidromic point*. In the Southern Hemisphere, the Coriolis force causes the tidal waves to rotate in a clockwise direction and anticlockwise in the Northern Hemisphere.

Tides in the deep ocean are zero at the amphidromic point and average less than 20 cm over much of the ocean. However three processes cause them to be amplified in shallow water and at the shore. The first is due to shoaling of the tidal waves across the relatively shallow (< 150 m deep) continental shelf. Like breaking waves they are amplified due to wave shoaling processes and increase in height (tide range) up to 1 to 3 m, as in the upper gulfs and in some locations (not in South Australia) even break as a tidal bore. Secondly, when two tidal waves arriving

Table 1.10   Tidal characteristics of South Australian ports

Location	Mean spring high tide (m)	Mean spring low tide (m)	Mean spring tide range (m)	Relative time of arrival 0 hr = Adelaide - = before, subtract + = after, add hours
South East				
Port MacDonnell	1.1	0.2	0.9	-1.02
Beachport	1.1	0.3	0.8	-1.33
Robe	1.1	0.2	0.9	-1.24
Kingston SE	1.2	0.3	0.9	-1.16
Victor Harbor	1.2	0.2	1.0	-1.05
Cape Jervis	1.3	0.1	1.2	-0.26
Kangaroo Island				
Hog Bay	*1.4*	*0.2*	*1.2*	*-0.53*
Kingscote	*1.4*	*0.1*	*1.3*	*-1.01*
Vivonne Bay	*1.2*	*0.2*	*1.0*	*-1.40*
Gulfs				
Port Noarlunga	1.9	0.3	1.6	-0.11
Adelaide	2.4	0.3	2.1	0.0
Ardrossan	2.9	0.4	2.5	-0.03
Wool Bay	2.7	1.1	1.6	+0.07
Edithburgh	2.0	1.1	0.9	-0.31
Stenhouse Bay	0.9	0.3	0.6	+0.10
Wedge Island	1.1	0.4	0.7	-0.13
Pondalowie Bay	1.0	0.4	0.6	-0.01
Point Turton	1.3	0.4	0.9	+1.06
Port Victoria	1.4	0.6	0.8	+1.04
Wallaroo	1.7	0.3	1.4	
Port Broughton	1.6	0.4	1.2	+3.09
Port Pirie	2.9	0.5	2.4	
Port Augusta	3.2	0.5	2.7	+0.25
Whyalla	2.6	0.4	2.2	
Eyre Peninsula				
Arno Bay	1.6	0.1	1.5	-2.05
Port Neill	1.8	0.2	1.6	-4.08
Port Lincoln	1.5	0.2	1.3	
Taylors Landing	1.3	0.2	1.1	-0.29
Thistle Island	1.1	0.2	0.9	-0.40
Elliston	1.4	0.4	1.0	-0.17
Blanche Port	1.4	0.4	1.0	+0.10
Thevenard	1.7	0.3	1.4	
Port Eyre	1.5	0.8	0.7	-0.31
Eucla	1.3	0.3	1.0	-1.30

from different directions converge, they may be amplified. Finally, in certain large embayments the tidal wave can be amplified by a process of wave resonance, which causes the tide to reach heights of several metres, as occurs in parts of northwest Australia.

Tides are classified as being micro-tidal when their range is less than 2 m, meso-tidal when between 2 and 4 m, macro-tidal when greater than 4 m and mega-tides greater than 8 m.

1.6.5.1   South Australian tides

There is a systematic variation in both the height of the tide along the South Australian coast and its time of arrival (Fig. 1.14). This is due to four factors:

- the tidal wave approaches from the southwest, reaching open coast first and the upper gulfs last;
- it slows down when moving through Investigator Strait and Backstairs Passage and up the shallow gulfs;
- it undergoes little amplification in crossing the relatively deep continental shelf;
- it is however amplified two to three-fold in the shallow gulfs.

As a result, the South Australian coast has a micro-tide range along the entire open coast, with a spring range that varies from between 0.6 and 1.4 m. In both gulfs the tides are amplified twofold, to 2.5 m at Ardrossan in Gulf St Vincent and 3 m at Port Augusta at the top of Spencer Gulf (Figs. 1.14, 1.15 and Table 1.10).

In areas of higher tide range, tides can influence wave height, as high tide and deeper nearshore waters result in less wave shoaling and higher waves at the shore, while shallower water at low tide induces greater wave shoaling and lower waves at the shore.

1.6.5.1.1 Time of tide

The southern Australian tidal wave is part of an amphidromic system located south of Tasmania. The tidal wave approaches from the south, first reaching the South East ports of Beachport and Robe, together with the far western port of Eucla, 1.5 hours ahead of Adelaide. Within the next hour and a half it reaches all the open coast ports and all of Gulf St Vincent, with Ardrossan, Wool Bay and Stenhouse Bay just a few minutes behind Adelaide. In Spencer Gulf it is slowed by the shallow waters, arriving at Port Victoria 1 hour after Adelaide and

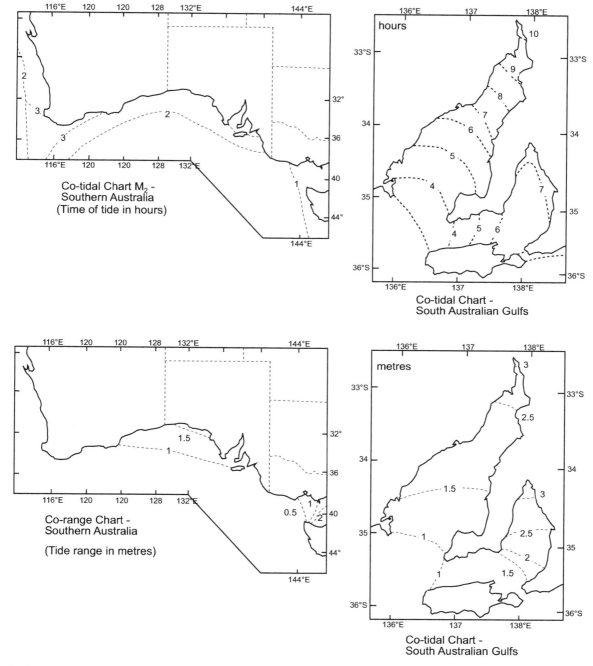

Figure 1.14 *Co-range and co-tidal lines for the South Australian coast. The tidal waves approach from the south and slow and amplify as they enter the gulfs (modified from Easton, 1970).*

Australian Beach Safety and Management Program

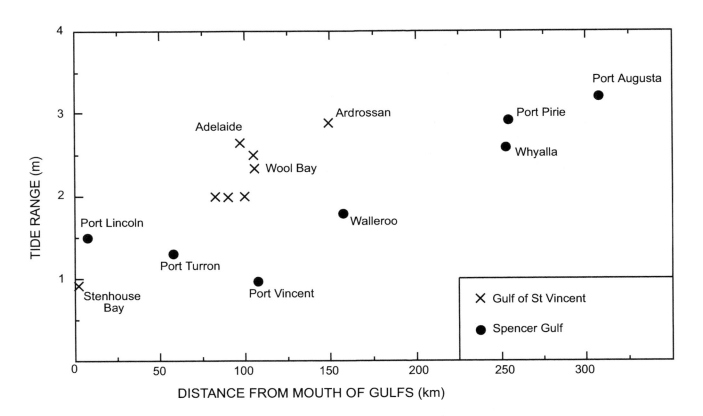

Figure 1.15 *Tidal amplification into the gulfs. Port Augusta has a tide range three times that on the open coast.*

Port Broughton 3 hours later, finally reaching Port Augusta 4 hours after Adelaide.

1.6.5.1.2 Tidal currents

Tidal currents are generally weak on the open coast, but strengthen considerably into the shallow waters of the gulfs and in confined passages, such as Backstairs Passage and in all constricted bays. The stronger currents are capable of transporting sediment and building substantial flood tide deltas inside the entrances to some bays such as Murray Mouth, Venus Bay, Baird Bay and Davenport Creek. In addition, all tidal creeks and lagoons are dominated by tidal flows.

1.6.6 Shelf waves

Shelf waves are periodic oscillations in sea level that occur along the coast. Across the southern Australian coast there are two additional meteorologically driven sources of sea level oscillations, one a progressive wave, the other a standing wave. The progressive shelf waves move from west to east across the southern coast, with a period between waves of 5 to 20 days, that is, the time between crest and trough. They arrive at Western Australian stations and move across the southern coast and on to eastern Australia. These waves raise sea level, for a few days at a time, by as much as 1.5 m, but more

normally 50 cm or less.

The second type is more seasonal, with periods of several months and which affect the entire coast simultaneously, that is, they stand against the coast. These can also result in sea level oscillations, as much as 0.5 m. As tides on the open coast are normally of the order of a few decimetres, these waves can cause abnormally high or low tides, depending on whether the crest or trough is at the coast. They also make it difficult for the casual observer to determine the actual height of the tide.

1.6.7 Ocean currents

Ocean currents refer to the wind driven movement of the upper 100 to 200 m of the ocean. The major wind systems blowing over the ocean surface drive currents that move in large ocean gyres, spanning millions of square kilometres. The dominant current south of Australia is the great Antarctic Circumpolar Current, which runs westerly at relatively high speed (1-3 knots), driven by the strong westerly wind stream. This current runs continuous around the southern oceans, centred between 50-60°S, well south of the continent. Between the main current and the southern Australian coast, the currents, while still predominantly westerly, decrease in velocity owing to increasing distance from the main westerly wind-stream, more variable wind direction and the impact of the relatively shallow continental shelf. Closer to the coast the currents become increasingly dependent on local winds, which still however are predominantly westerly. In the South East a seasonal reversal in the coastal current can occur, with an easterly flow of water from Bass Strait flowing up along the South East coast.

1.6.8 Other currents

There are several other forms of ocean currents driven by winds, density, tides, shelf waves and ocean waves. It is not uncommon to have several operating simultaneously. Each will have a measurable impact on the overall current structure and must be taken into account if one needs to know the finer detail of the coastal currents, their direction, velocity and temperature. Along the South Australian coast, in addition to the ocean and tidal currents, the next most important are those associated with the shelf waves and wind driven upwelling and downwelling. The latter are associated with regional winds. Strong west through southwest winds push the surface waters to the left (north) causing a piling up of warmer ocean water at the coast which downwells to return seaward. When strong (and often hot) northerly winds prevail, the water is also pushed to the left (west) pushing the warmer surface away from the shore, particularly in the South East, which is replaced at the shore by the upwelling of cooler, deeper bottom water. The combination of this cooler water and warm air can lead to condensation in the overlying warmer air and the formation of a sea fog or mist along the coast, not uncommon along the South East.

1.6.9 Sea surface temperature

The sea temperature along the South Australian coast is a product of three main processes: firstly, the latitudinal location of the coast between 31.5-38°S determines the overhead position of the sun and the amount of solar radiation available to warm the ocean water; secondly, the coastal water depth and turnover with the shallower gulf waters able to heat more than the adjacent deeper ocean waters; and third, the intrusion of either warmer gulf water or cooler ocean water into the coastal zone, including the impact of upwelling and downwelling.

On a seasonal basis, the summer water temperature ranges from the low 20's in the upper Bight to high teens (17-18°C) in the South East, with higher temperatures in the gulfs and protected bays. In winter, temperatures right along the coast drop to the mid to low teens (13-15°C). Northerly winds induce an offshore movement of the ocean surface waters, which are replaced by cooler bottom waters (upwelling).

1.6.10 Salinity

All oceans and seas contain dissolved salts derived from the erosion of land surfaces over hundreds of millions of years. Chlorine and sodium dominate and, together with several other minerals, account for the dissolved 'salt'. The salts are well mixed and globally average 35 parts per thousand, increasing slightly into the dry sub-tropics and decreasing slightly in the wetter latitudes. Along the open coast salinity maintains this average. However in the Coorong lagoon, the gulfs and some bays, evaporation combined with limited circulation permits a build up in salinity. In the Coorong lagoon salinity increases from 35-55 $^o/_{oo}$ in the northern lagoon, to 60-110 $^o/_{oo}$ in the southern lagoon. Salinities also increase in the shallow upper reaches of the gulfs and in confined bays such as Baird Bay, particularly in summer.

1.7   Biological Processes

Biological processes are extremely important along the entire southern Australian coast and continental shelf, rivaling the more publicised northern Australian tropical systems. The main biological systems that directly impact the coast and many beaches are the coastal dune vegetation, the samphire vegetation of the supratidal and salinity-affected coastal flats, the mangrove woodlands, the inter- to subtidal seagrass meadows, the in places rich beach ecology and in particular the extensive shelf temperate carbonate province.

1.7.1 Coastal dune vegetation

Coastal dunes in South Australia are vegetated by a succession of plants beginning with herbs and grasses on the foredunes and grading landward into a combination of sedgelands and shrublands, with some mallee growing in more protected locations. See Oppermann (1999) for a detailed survey of the South Australian coastal vegetation.

The incipient foredune and outer foredune have a typical succession of the herb *Cakile maritima*, backed by the prolific *Spinifex sericeus*, often accompanied by *Euphorbia paralias*, particularly in the east. On the western Eyre Peninsula *Triodia compacta* is more dominant on the dunes.

The main foredunes and hind dunes are dominated by a range of shrublands, with the more important including *Atriplex cinerea, Atriplex vesicaria, Atriplex paludosa, Leucophyta brownii, Melaleuca lanceolata, Tetragonia implexicoma, Nitraria billardierei and Olearia axillaris* occurring along many of the gulf and western dune fields, and *Leucopogon parviflorous* and *Olearia axillaris* in the centre and east.

1.7.2 Samphire vegetation

Samphire vegetation in association with algae grows in the lower swales between beach ridges and in the saline back barrier depressions and dry lagoons. The samphire vegetation is low (<1 m) and scrubby and dominated by the family *Chenopodiaceac*.

1.7.3 Mangroves

Mangroves grow between mean sea level and neap high tide and require some degree of protection from wave attack. The wide tidal flats of the gulfs and protected Eyre Peninsula bays provide an ideal habitat for the single species *Avicennia maritima*. They extend along the coast from just north of Adelaide to Davenport Creek in Tourville Bay; the westernmost mangroves on the southern Australian coast.

1.7.4 Seagrass meadows

Seagrass grows in the lower intertidal and particularly the shallow subtidal zone. Two species dominate, with the grassy *Zostra* growing between mean sea level and low tide, while the taller *Posidonia australis* grows generally below low tide to a few metres depth.

Besides helping to stabilise the nearshore sands, the meadows support a rich epibiota and contribute a high proportion of red algae, foraminifera and bivalve fragments to the beach sediments, as well as seagrass roots and detritus. This material has been reworked by the waves and tides to build the extensive tidal shoals and sand flats in the upper gulfs and some bays, as well as the backing beaches.

1.7.5 Beach ecology

The basis of the beach ecosystem are the microscopic diatoms that live in the water column (called phytoplankton) and microscopic meiofauna that live on and between sand grains; including bacteria, fungi, algae, protozoa and metazoa. Feeding on these are larger organisms that live in the sand, including meiofauna such as small worms and shrimp; and filter feeding benthos such as molluscs and worms. In the water column they are also preyed upon by zooplankton such as amphipods, isopods, mysids, prawns and crabs. At the top of the food chain are the fish and sea birds, and the occasional mammals such as dolphins, dugongs and whales.

A number of hard-bodied organisms also contribute their skeletal material to the beach in the form of sediment.

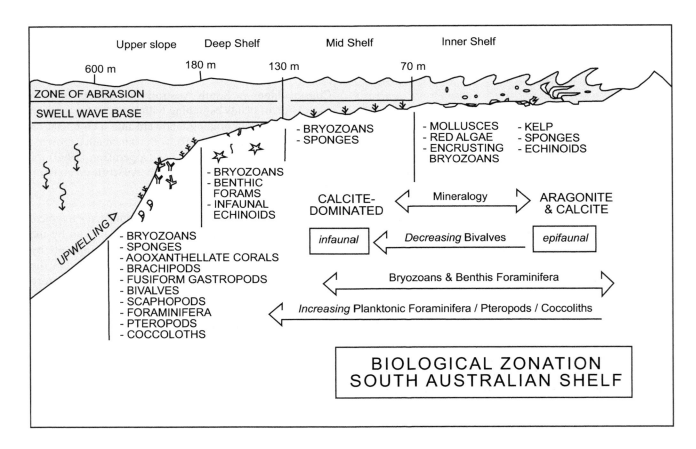

Figure 1.16 *Typical shelf biota and carbonate environments on the southeast South Australian-western Victorian continental shelf (from Boreen, et al. 1993).*

When all organisms die, their internal and/or external skeletons can be broken up, abraded and washed onto beaches by waves.

Biotic debris found in Holocene beach ridges and swales along the Eyre Peninsula include the following species: *Katelysia peroni, Katelysia scalarina, Amesodesma cuneata, Spisula trigonella, Anadara trapesia, Brachydontes erosus, Batillariella estuarina, Zeacumantus diemenensis, Salinator fragilis, Thalotia conica, Prothalotia* sp.*, Laemodonta ciliata, Bembicium melanostoma, Coxiella* sp.*, Phasianella australia, Bulla botanica* and *Diala.*

Rocky shore species also found in the beach ridge deposits include*: Tellina deltoidalis, Trichomya hirstua, Cominella eburnea, Austrocholea constricta, Phasisanotrochus rutilus, Sanguinolaria biradiata, Mactra pura, Fulvia tenuoicostata, Ostrea angasi* and *Bemicium auratum.*

1.7.6 Shelf biota

The biota of the southern Australian continental shelf makes a major contribution to the entire coast. The shelf contains a range of distinctive habitat assemblages grading from the inner, mid to outer shelf and including the upper continental slope (Fig. 1.16).

The shallow (30-70 m) **inner shelf** is dominated by wave action and contains a range of hard-bodied organisms, particularly molluscs, red algae, encrusting bryozoans, echinoids and soft kelps and sponges. Wave abrasion occurs across the inner shelf permitting erosion and transport of the carbonate detritus. The mid to outer shelf and slopes are dominated by a range of *Bryozoans* species. The **mid shelf** (70 to 130 m depth) is reworked by shoaling waves and is a zone of carbonate production and accumulation. The **outer shelf** (130-180 m depth) is only reworked during storm wave conditions and contains finer bioclastic sands. The top of the continental slope (180-350 m depth) has extensive bryozoan/sponge/coral communities which lead to the accumulation of muddy skeletal sands.

This massive area of carbonate production (shelf carbonate factory) is all the more important because during successive sea level regressions (falling sea level) and transgressions (rising sea level) a massive volume of carbonate detritus is reworked shoreward by waves during the rising sea level and deposited as the carbonate rich beaches and dunes that dominate the entire coast. On average, approximately 800 m³ of carbonate sand have been moved onshore for every metre of beach in South Australia. It has been estimated that on exposed high energy beaches backed by massive dune systems, this figure can reach up to 100 000 m³/m. In addition, at present sea level sediment can continue to be moved shoreward from the shallow inner shelf region, while in the gulfs and protected bays it accumulates in vast sand shoals and intertidal sand flats.

2. BEACH SYSTEMS

Beaches throughout the world consist of wave deposited sediment and lie between the base of wave activity and the limit of wave run-up or swash. They therefore include the dry (subaerial) beach, the swash or intertidal zone, the surf zone and, beyond the breakers, the nearshore zone (Fig. 2.1). Usually only the dry beach and swash zone are clearly visible, while bars and channels are often present in the surf zone but are obscured below waves and surf, while the nearshore is always submerged. The shape of any surface is called its morphology, hence beach morphology refers to the shape of the beach, surf and nearshore zone (Fig. 2.2).

Figure 2.1   *Goolwa Beach is a popular wide, low gradient beach, backed by large dunes, with a 500 m wide, usually high energy surf zone.*

2.1   Beach Morphology

As all beaches are composed of sediment deposited by waves, beach morphology reflects the interaction between waves of a certain height, length and direction and the available sediment; whether it be sand, cobbles or boulders, together with any other structures such as headlands, reefs and inlets.

South Australian beaches can be very generally divided into three groups: wave dominated, tide modified and tide dominated. The *wave dominated* beaches occur along the open ocean coasts. They are exposed to persistent ocean swell and waves and low tides (<2 m). The *tide modified* beaches occur in the gulfs and some large bays, where waves are low and (in the gulfs) tide range higher. The *tide dominated* beaches are those beaches in the area of tide modification which receive such low waves, owing to some form of protection, that they become increasingly dominated by the tides and are a mix of beach and tidal flats. The remainder of this chapter is devoted to a description of first the higher energy wave dominated beaches (section 2.3), followed by the tide modified (section 2.4) and then the tide dominated beaches (section 2.5).

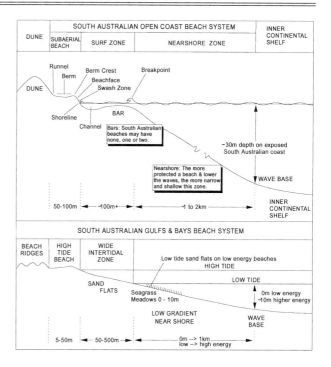

Figure 2.2   *Examples of two typical South Australian beach systems. The higher energy open coast beaches (upper) consist of the dry subaerial beach above the shoreline, the surf zone containing bars, troughs and breaking waves, and the nearshore zone which extends seaward of the breaker zone out to modal wave base. Wave base is the depth to which ocean waves can move beach sands. Seaward lies the inner continental shelf. The approximate width and depth of each zone are indicated. In the more protected gulfs and some larger bays (lower) the lower waves and higher tide result in a lower gradient and shallower beach, intertidal and nearshore zone, with wave base as shallow as low tide.*

2.1.1   Wave dominated beaches

The simplest way to describe a beach is in two dimensions, as shown in Figure 2.2. The beach consists of three zones: the subaerial beach, the surf zone and the nearshore zone.

2.1.2   Subaerial Beach

The *subaerial beach* is that part of the beach above sea level, which is shaped by wave run-up or swash. It starts at the shoreline and extends up the relatively steep swash zone or beach face. This may be backed by a flatter berm or cusps, which in turn may be backed by a runnel, where the swash reaches at high tide. Behind the upper limit of spring tide and/or storm swash usually lies the edge of dune vegetation. The subaerial, or dry, beach is that part which most people go to and consider 'the beach'. However, the real beach is far more extensive, in places extending several kilometres seaward, with the subaerial beach forming the figurative tip of the iceberg.

2.1.3 Swash or intertidal zone

On wave dominated beaches the swash zone connects the dry beach with the surf. The swash zone is the steeper part of the shoreline across which the broken waves run up and down. As wave height decreases and tide range increases, this zone tends to become flatter and considerably wider, and is termed the intertidal zone on tide modified to tide dominated beaches.

2.1.4 Surf Zone

The *surf zone* extends seaward of the shoreline and out to the area of wave breaking. This is one of the most dynamic places on earth. It is the zone where waves are continuously expending their energy and reshaping the seabed. It can be divided into the area of wave breaking, often underlain by a bar, and, immediately shoreward, the area of wave translation where the wave bore (white water) moves toward the shoreline, transforming along the way into surf zone currents and, at the shoreline, into swash. Surf zones are up to 500 m wide on exposed South Australian beaches, where waves average over 2 m and break across a double bar system. They decrease in width and bar number as wave height decreases. On tide modified beaches the usually narrow surf zone is transient with the tide, while it is usually nonexistent on tide dominated beaches.

2.1.5 Nearshore Zone

On wave dominated beaches the *nearshore zone* is the most extensive part of the beach. It extends seaward from the outer breakers to the maximum depth at which average waves can mobilise beach sediment and move it shoreward. The point is called the *wave base*, referring to the base of wave activity. On the high energy open South Australian coast, where waves average over 2 m and commonly reach several metres, it usually lies at a depth between 30 and 50 m and may extend 1, 2 or even 3 km out to sea. It decreases in depth and width as wave height decreases. On tide modified beaches it is usually at most a few metres deep, while on tide dominated beaches it may terminate at low tide. The fringe of the seagrass meadows, usually a metre or less below low tide, is a good indicator of its outer limit on lower energy gulf and bay shores.

2.1.6 Three dimensional beach morphology

In three dimensions, wave dominated beaches become more complex. This is because most beaches are not uniform alongshore, but vary in a predictable manner.

Beaches vary longshore on two scales. First, they are usually swash-aligned, meaning they are aligned parallel to the crest of the dominant wave. If the wave crest is bent or deformed as it approaches the shoreline, owing to the presence of obstacles such as reefs, rocks and headlands, then the beach will also be shaped to fit the wave crest. This is because the waves to begin to 'feel' the shallower seabed (reefs, rocks etc.) and slow down, while the part of the wave crest that is still in deeper water moves more rapidly. As a result, the waves bend or refract toward the shore, as well as decreasing in height in lee of the obstacle, while where unimpeded they approach the beach straighter and higher. The overall effect of the bending wave crests is to cause a spiral or curvature in the shape of the beach, so that the curvature increases toward the more protected end. This bending of the wave crests is called *wave refraction* (Fig. 2.3), while the loss of wave energy and height is called *wave attenuation*. The decrease in wave height along the beach results in lower breakers, a narrower and shallower surf zone, a narrower and often steeper swash zone and a lower and narrower subaerial beach. The average South Australian beach is only 1.3 km in length and usually bordered by headlands and reefs. As a result there is considerable opportunity for these structures to influence the size and direction of waves arriving at many beaches.

Figure 2.3 *Wave refraction around Point James. The high wave energy beach contrasts with the protected beach in lee of the point.*

The second scale of longshore beach variation relates to any and all undulations on the beach and in the surf zone, usually with scales in the order of tens of metres to as much as 500 m or more on very high energy beaches. Variable longshore forms produced at this scale include regular beach cusps located in the high tide swash zone and spaced between 20 and 40 m, and all variation in rips, bars, troughs and any undulations along the beach, usually with spacing of between 200 and 500 m. These features are associated with rip circulation and are known as rip channels, crescentic and transverse bars, and megacusp horns and embayments. Each of these and their associated beach type is described in section 2.3.

2.2　Beach Dynamics

Beach dynamics refers to the dynamic interaction between the breaking waves and currents and the sediments that compose the beach. In the long term (tens to thousands of years) this interaction builds all beaches and contributes sand to form backing dune systems. It can also ultimately erode these beaches. In the shorter term (days to months), changing wave conditions produce continual changes in beach response and shape, as sand is moved onshore (beach accretion) and offshore (beach erosion), together with the associated movement of the shoreline, bars and channels.

Five factors determine the character of a beach. These are the size of the sediment, the height and length of the waves, the characteristics of any long waves present in the surf zone and the tide range. The impact of each is briefly discussed below.

2.2.1　Beach sediment

The size of beach sediment determines its contribution to beach dynamics. Unlike in air where all objects fall at the same speed, sediment falls through water at a speed proportional to its size. Very fine sediment, like clay, will simply not sink but stay in suspension for days or weeks, causing turbid, muddy water. Silt, sediment coarser than clay but finer than sand, takes up to two hours to settle in a laboratory cylinder (Table 2.1). As a consequence fine sediments usually stay in suspension through the energetic surf zone and tidal inlets and are carried out to sea to settle in deep water on the continental shelf.

Table 2.1　Sediment size and settling rates

Material	Size – diameter	Time to settle 1 m
Clay	0.001-0.008 mm	hours to days
Silt	0.008-0.063 mm	5 min to 2 hours
Sand	0.063-2 mm	5 sec to 5 min
Cobble	2 mm-6.4 cm	1 to 5 sec
Boulder	> 6.4 cm	< 1 sec

Sand takes between a few seconds for coarse sand, to five minutes for very fine sand, to settle through 1 m of water. For this reason it is fine enough to be put into suspension by waves, yet coarse enough to settle quickly to the seabed as soon as waves stop breaking. In the energetic breaking wave environment, anything as fine or finer than very fine sand stays in continual suspension and is flushed out of the beach system into deeper, quieter water. This is why ocean beaches never consist of silts or mud. Most beaches consist of sand because it can be transported in large quantities to the coast and settle fast enough to remain in the surf zone.

Fifty South Australian beaches (3.5%) consist of cobbles and boulders. Cobbles and boulders require substantial wave or current energy to be lifted or moved and then settle immediately. They are therefore only rarely moved, such as during extreme storms, and then only very slowly and over short distances. Consequently, beaches containing such coarse sediment always have a nearby source, usually an eroding cliff or bluff.

Depending on the nature of its sediment, each beach will inherit a number of characteristics. Firstly, the sediment will determine the mineralogy or composition of the beach. Sediment derived from the land via rivers and creeks is usually quartz sand or silica. In South Australia, however, rivers and streams are few and most sand has been derived from the biotic detritus of the seagrass meadows and inner continental shelf and is carbonate (algal, shell detritus etc.) as discussed in section 1.7. Secondly, the size of the sediment will, along with waves, determine beach shape and dynamics. Fine sand produces a low gradient (1 to 3°) swash zone, wide surf zone and potentially highly mobile sand. Medium to coarse sand beaches have a steeper gradient (4 to 10°), a narrower surf zone and less mobile sand. Cobble and boulder beaches are not only very steep (> 8°), but they have no surf zone and are usually immobile. Therefore, identical waves arriving at adjacent fine, medium and coarse sand beaches will interact to produce three distinctly different beaches.

Likewise, three beaches having identical sand size, but exposed to low, medium and high waves, will have three very different beach systems. Therefore, it is not just the sand or the waves, but the interaction of both, along with long waves and tides, that determine the nature of our beaches.

2.2.2　Wave energy - long term

Waves are the major source of energy to build and change beaches. Seaward of the breaker zone, waves interact with the sandy seabed to stir sand into suspension and, under normal conditions, slowly move it shoreward. The wave-by-wave stirring of sand across the nearshore zone and its shoreward transport has been responsible for the delivery of all the sand that presently composes the beaches and coastal sand dunes of South Australia.

Waves are therefore responsible for supplying the sand to build beaches. The higher the waves, the greater the depth from which they can transport sand and the faster they can transport it. Consequently, the largest volumes of sand that build the biggest beaches and dunes are all in areas of very high waves. Along parts of the South East and Eyre Peninsula, massive amounts of up to 100 000 m³ of sand have been transported onshore for every metre of beach, amounting to 1 000 000 m³ for every few metres of beach. However, these same large waves can just as rapidly erode the very dynamic beaches they initially built. For this reason many high energy South

Australian beaches have left remnants of dunes on top of cliffs and bluffs, while the beaches in many cases have been completely eroded and removed.

Lower waves can only transport sand from shallow depths and at slower rates. Consequently, they build smaller barrier systems, usually delivering less than 10 000 m³ for every metre of beach, as is typical of more sheltered parts of the open coast. However, these beaches are less dynamic, more stable and are less likely to be eroded.

2.2.3   Wave energy - short term

Waves are not only responsible for the long term evolution of beaches, but also the continual changes and adjustments that take place as wave conditions vary from day to day. As noted above, a wave's first impact on a beach is felt as soon as the water is shallow enough for wave shoaling to commence, usually in less than 30 m water depth. As waves shoal and approach the break point, they undergo a rapid transformation which results in the waves becoming slower and shorter, but higher, and ultimately breaking, as the wave crest overtakes the trough.

As waves break, they release kinetic energy, energy that may have been derived from the wind some hundreds or even thousands of kilometres away. This energy is released as turbulence, sound (the roar of the surf) and even heat. The turbulence stirs sand into suspension and carries it shoreward with the wave bore. The wave bore decreases in height shoreward, eventually turning into swash as it reaches the shoreline.

Breaking waves, wave bores and swash, together with unbroken and reformed waves, all contribute to a shoreward momentum in the surf zone. As these waves and currents move shoreward, much of their energy is transferred into other forms of surf zone currents, namely longshore, rip feeder and rip currents, and long waves and associated currents. The rip currents are responsible for returning the water seaward, while the long wave currents play a major role in shaping the surf zone.

It is therefore the variation in waves and sediment that produces the seemingly wide range of beaches present along the coast, ranging from the steep, narrow, protected beaches to the broad, low gradient beaches with wide surf zones, large rips and massive breakers. Yet every beach follows a predictable pattern of response, largely governed by its sediment size and prevailing wave height and length. The types of beaches that can be produced by waves and sand are discussed in section 2.3.

2.3   Beach Types

Beach type refers to the prevailing nature of a beach, including the waves and currents, the extent of the nearshore zone, the width and shape of the surf zone,

including its bars and troughs, and the dry or subaerial beach. *Beach change* refers to the changing nature of a beach or beaches along a coast as wave, tide and sediment conditions change.

The first comprehensive classification of wave dominated beaches was developed by the Coastal Studies Unit (CSU) at the University of Sydney in the late 1970's. These investigations included instrumenting the wide, energetic surf zone at Goolwa Beach (Fig. 2.1). This classification is now used internationally, wherever tide range is less than 2 m. In Australia it applies to most of the southern coast, from Fraser Island in the east around to Exmouth Peninsula in the west. In the upper gulfs and across northern Australia, waves are lower to nonexistent and tides are generally higher, producing a different range of beach types. In the early 1990's the CSU undertook research along the central Queensland coast which resulted in the identification of a tide modified and tide dominated range of beach types. Based on this work the full range of South Australian beach types is summarised in Table 2.2. In the following sections each of the beach types is described, together with examples and photographs of the ten beach types that occur along the South Australian open coast and gulfs.

2.3.1   Wave dominated beach types

Wave dominated beaches consist of three types: reflective, intermediate or dissipative, with the intermediate having four states as indicated in Figures 2.4 and 2.5. In South Australia all wave dominated beach types occur on the open coast.

2.3.2   Dissipative beaches (D)

There are 14 dissipative beaches in South Australia, more than in any other part of Australia. This is because the open coast receives persistent high waves and a number of these exposed beaches are composed of fine sand. The beaches average 19.3 km in length and occupy 270 km or 8% of the entire coast.

When *dissipative* beaches occur, the combination of high waves and fine sand ensures that they have wide surf zones and usually two to occasionally three shore parallel bars, separated by subdued troughs. The beach face is composed of fine sand and is always wide, low and firm, firm enough to support a 2WD drive car. The popular Goolwa Beach, south of Adelaide, is one of Australia's more accessible dissipative beaches (Fig. 2.1) .

Wave breaking begins as high spilling breakers on the outer bar, which reform to break again and perhaps again, on the inner bar or bars. In this way they dissipate their energy across the surf zone, which may be up to 300-500 m wide (Fig. 2.6). This is the origin of the name 'dissipative'.

Table 2.2   South Australian beach types by number and length

Beach Type	Symbol	No. of beaches		% of total beaches		Mean length (km)		Total length (km)		Proportion of length of all beaches (%)		
		SA	KI	SA	KI	SA	KI	SA	KI	SA	KI	Total
Wave dominated												
Dissipative	D	14	0	1	0	19.3	-	270	0	13	0	12
Longshore bar trough	LBT	7	0	<1	0	2.2	-	15	0	<1	0	<1
Rhythmic bar beach	RBB	33	20	2	9	2.8	1.0	91	21	5	13	5
Transverse bar rip	TBR	159	26	11	12	1.8	1.0	288	27	14	17	14
Low tide terrace	LTT	164	36	11	17	1.2	0.9	203	34	10	22	11
Reflective	R	571	109	39	50	0.6	0.3	354	28	16	18	18
Tide modified												
Reflective + LTT	R+LTT	42	0	3	0	1.1	-	45	0	2	0	2
Reflective + TBR	R+BR	2	0	<1	0	8.0	-	16	0	<1	0	<1
Ultradissipative	UD	0	0	0	0	-	-	0	0	0	0	0
Tide dominated												
R + sand ridges	R+SR	29	15	2	7	1.5	1.4	44	21	2	13	3
R + sand flats	R+SF	364	12	25	6	1.4	2.2	517	27	26	17	25
R + tidal sand flats	R+TSF	69	0	5	0	2.6	-	177	0	9	0	8
TOTAL		1454	218	100	100	1.4	0.7	2020	158	100	100	100

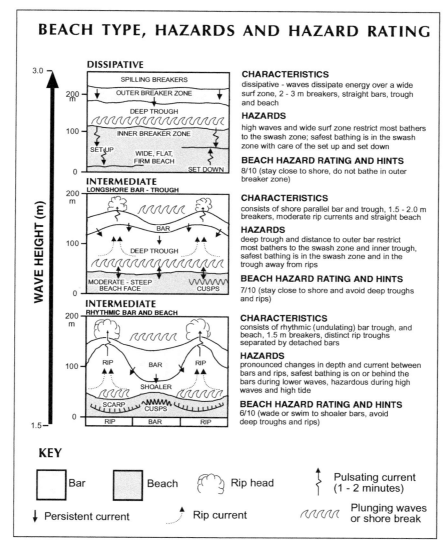

Figure 2.4   *A plan view of the rhythmic bar and beach, longshore bar and trough and dissipative beach types. As wave height increases between the rhythmic and dissipative beaches, the surf zone increases in width, rips initially increase and then are replaced by other currents, and the shoreline becomes straighter. The physical characteristics and beach and surf hazards associated with each type are indicated.*

Australian Beach Safety and Management Program

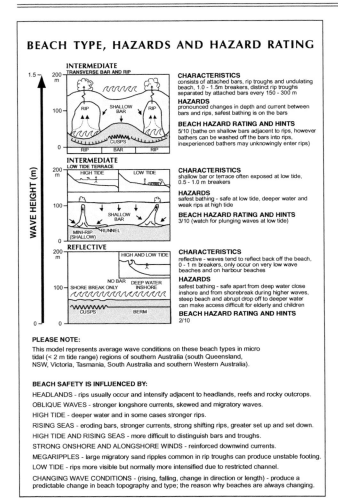

BEACH TYPE, HAZARDS AND HAZARD RATING

INTERMEDIATE
TRANSVERSE BAR AND RIP

CHARACTERISTICS
consists of attached bars, rip troughs and undulating beach, 1.0 - 1.5m breakers, distinct rip troughs separated by attached bars every 150 - 300 m

HAZARDS
pronounced changes in depth and current between bars and rips, safest bathing is on the bars

BEACH HAZARD RATING AND HINTS
5/10 (bathe on shallow bars adjacent to rips, however bathers can be washed off the bars into rips, inexperienced bathers may unknowingly enter rips)

INTERMEDIATE
LOW TIDE TERRACE

CHARACTERISTICS
shallow bar or terrace often exposed at low tide, 0.5 - 1.0 m breakers

HAZARDS
safest bathing - safe at low tide, deeper water and weak rips at high tide

BEACH HAZARD RATING AND HINTS
3/10 (watch for plunging waves at low tide)

REFLECTIVE

CHARACTERISTICS
reflective - waves tend to reflect back off the beach, 0 - 1 m breakers, only occur on very low wave beaches and on harbour beaches

HAZARDS
safest bathing - safe apart from deep water close inshore and from shorebreak during higher waves, steep beach and abrupt drop off to deeper water can make access difficult for elderly and children

BEACH HAZARD RATING AND HINTS
2/10

PLEASE NOTE:
This model represents average wave conditions on these beach types in micro tidal (< 2 m tide range) regions of southern Australia (south Queensland, NSW, Victoria, Tasmania, South Australia and southern Western Australia).

BEACH SAFETY IS INFLUENCED BY:
HEADLANDS - rips usually occur and intensify adjacent to headlands, reefs and rocky outcrops.
OBLIQUE WAVES - stronger longshore currents, skewed and migratory waves.
HIGH TIDE - deeper water and in some cases stronger rips.
RISING SEAS - eroding bars, stronger currents, strong shifting rips, greater set up and set down.
HIGH TIDE AND RISING SEAS - more difficult to distinguish bars and troughs.
STRONG ONSHORE AND ALONGSHORE WINDS - reinforced downwind currents.
MEGARIPPLES - large migratory sand ripples common in rip troughs can produce unstable footing.
LOW TIDE - rips more visible but normally more intensified due to restricted channel.
CHANGING WAVE CONDITIONS - (rising, falling, change in direction or length) - produce a predictable change in beach topography and type; the reason why beaches are always changing.

Figure 2.5 *A plan view of the reflective, low tide terrace and transverse bar and rip beach types. As wave height increases between the reflective and transverse beaches, the surf zone and bar increase in width, rips form and increase in size, and the shoreline becomes crenulate. The physical characteristics and beach and surf hazards associated with each type are indicated.*

In the process of continual breaking and re-breaking across the wide surf zone, the incident or regular waves decrease in height and may be indiscernible at the shoreline. The energy and water that commenced breaking in the original wave is gradually transferred in crossing the surf zone to a lower frequency movement of water, called a standing wave. This is known as red shifting, where energy shifts to the lower frequency, or red end, of the energy spectrum.

At the shoreline, the standing wave is manifest as a periodic (every 60 to 120 seconds) rise in the water level (set-up), followed by a more rapid fall in the water level (set-down). As a rule of thumb, the height of the set-up is 0.3 to 0.5 times the height of the breaking waves (i.e. 1 to 1.5 m for a 3 m wave). Because the wave is standing, the water moves with the wave in a seaward direction

during set-down, with velocities between 1 and 2 m/sec closer to the seabed. As the water continues to set down, the next wave is building up in the inner surf zone, often to a substantial wave bore, 1 m+ high. The bore then flows across the low beach face and continues to rise, as more water moves shoreward and sets up. This process continuously repeats itself every one to two minutes.

Because of the fine sand and the large, low frequency standing wave, the beach is planed down to a wide, low gradient, with the high tide swash reaching to the back of the beach, often leaving no dry sand to sit on at high tide.

Figure 2.6 *Dog Fence beach in the western Eyre Peninsula is a typical high energy dissipative beach with a 500 m wide double bar surf zone and backing massive dune fields.*

2.3.2.1   Dissipative beach hazards

The wide surf zones and high waves associated with dissipative beaches keep most bathers to the inner swash and surf zone. They are relatively safe close inshore, though not without some surprises, while the mid to outer surf zone is only for the fittest and most experienced surfers.

Dissipative beach hazards

Most people do not venture far into dissipative surf zones as they are put off by their extremely wide surf and high outer breakers. However, if you do, this is what to watch out for:

- Outer surf zone - spilling breakers. Bigger sets break well seaward and catch surfers inside.

- Troughs - usually on/offshore currents, but chance of longshore and even rip currents, particularly under lower (< 1.5 m) wave conditions.

- Inner surf zone - watch for standing wave bores that can knock you over, fortunately shoreward. Set-down

---

produces an often strong seaward flow, particularly closer to the sea bed, which may also drag children off their feet.

- Swash zone/beach face - this is where most bathers stay and where most get into trouble, owing to the set-up and set-down. Be aware that water level will vary considerably between set-up and set-down, and currents will reverse from onshore to offshore. At best you will be knocked over by the incoming bore, at worst you might be dragged seaward by the set-down. Children in particular are most at risk. Some young children, even babies in prams and parked cars, have been left on a seemingly safe part of the beach face, only to have a higher than usual set-up engulf them in water.

- **Summary:** Dissipative beaches are dangerous, however in South Australia they only occur in a few locations and when the seas are very big, so most people don't consider swimming, or at least not beyond the swash zone. Definitely for experienced surfers only.

2.3.3 Intermediate beaches

Intermediate beaches refer to those beach types that are intermediate between the lower energy reflective beaches and the highest energy dissipative beaches. The most obvious characteristic of intermediate beaches is the presence of a surf zone with bars and rips. On the South Australian coast 445 (40%) of the 1164 wave dominated beaches are intermediate. They are the most common beach on the open coast, but do not extend into the gulfs and protected bays. The open coast dominance is a result of the persistent high deepwater waves and intermediate types will form on the most exposed beaches, where waves average over 2 m and the sand is medium, through to more sheltered beaches with waves down to 0.5 m high.

However, between those beaches produced by 0.5 m waves and 2 m plus waves, there is quite a range in the shape and character of the beach. For this reason, intermediate beaches are classified into four beach states. The lowest energy state is called *low tide terrace*, then as waves increase, the *transverse bar and rip*, then the *rhythmic bar and beach*, and finally the *longshore bar and trough*. Each of these beaches is described below.

2.3.4 Longshore bar and trough (LBT)

The *longshore bar and trough* beach type is rare on the South Australian coast, usually occurring as the outer bars on the most exposed section of the south-east sand islands (Fig. 2.7). There are only seven LBT beaches in the state, totalling 15 km in length. They very rarely occur as the inner bar, where people swim, and then only following periods of high seas.

Longshore bar and trough beaches are characterised by waves averaging 1.5 m or more, which break over a near continuous longshore bar located between 100 and 150 m seaward of the beach, with a 50 to 100 m wide, 2 to 3 m deep longshore trough separating it from the beach. The beach face is straight alongshore and usually has a low gradient. While the bar, trough and beach may look straight and devoid of rips, the bar is usually crossed by rips every 250 to 500 m. The wide deep trough and the presence of less obvious rips make this a particularly hazardous swimming beach.

Figure 2.7 *Gunyah Beach on the lower Eyre Peninsula is one of the highest energy beaches in South Australia. Because of the medium sand, it has a longshore bar and trough system during above average waves, as seen in this view. Waves break across the outer bar and reform in the shore parallel trough. Widely spaced rips cut across the bar.*

Higher waves tend to break continuously along the bar, with lower waves not breaking in the vicinity of the rip gaps. The wave bores cross the bar and enter the deeper longshore trough, where they quickly reform and continue shoreward as a lower wave to break or surge up the beach face. The water moving shoreward into the trough returns seaward using two mechanisms. Firstly, the water piles up along the beach face as wave set-up. As the water sets down, it moves both seaward as a standing wave and longshore to feed the rip current. Secondly, as the converging feeder currents approach the rip, they accelerate, causing additional set-up in lee of the rip. As this set-up sets down, the rip pulses seaward.

2.3.4.1 Longshore bar and trough hazards

Longshore bar and trough beaches are hazardous. They require large waves to form and they have wide deep channels and troughs running continuously along the beach, which contain longshore moving feeder currents and periodic rip currents.

Longshore bar and trough beach hazards

- Bar - it's usually a long swim across a deep trough containing longshore and rip currents to reach the bar. At low tide, water pouring off the bar into the trough may make it difficult to get up onto the bar, or may wash you off the bar. Waves break more heavily on the bar at low tide.

- Trough - a wide, 2 to 3 m deep trough runs between the bar and beach and is occupied by longshore currents, which intensify toward the rips and seaward moving rip currents.

- High tide - deeper bar and troughs, weaker longshore currents and rips.

- Low tide - shallower channel, but often still greater than 2 m deep and stronger longshore and rip currents, shallower bar with plunging waves.

- Higher waves - heavier waves breaking, stronger longshore and rip currents.

- Oblique waves - bias in longshore current in direction of waves, increase in velocity of current, plus skewed rip currents.

- Summary: Unless you are an experienced surf swimmer stay in the swash zone landward of the trough. As waves usually break first on the outer bar, they are lower at the beach face and can result in moderately safe conditions at the shore.

2.3.5   Rhythmic bar and beach (RBB)

The *rhythmic bar and beach* type is the highest energy beach type that commonly occurs on the open coast. In all, 33 beaches, averaging 2.8 km in length, are of this type. These energetic beaches require two primary ingredients for their formation, relatively fine-medium sand and exposure to the high deepwater waves. They occur where waves average at least 1.5 m and sand is fine to medium (Fig. 2.8).

Rhythmic beaches consist of a rhythmic longshore bar that narrows and deepens where the rips cross the breakers, and in between broadens, shoals and approaches the shore. It does not, however, reach the shore, with a continuous rip feeder channel feeding the rips to either side of the bar. The shoreline is usually rhythmic with protruding megacusp horns in lee of the detached bars and commonly scarped megacusp embayments behind the rips. The surf zone may be up to 100 to 250 m wide and the bars and rips are spaced every 250 to 1000 m alongshore.

The shallower rhythmic bar causes waves to break more heavily, with the white water flowing shoreward as a wave

bore. The wave bore flows across the bar and into the backing rip feeder channel. The water from both the wave bore and the swash piles up in the rip feeder channel and starts moving sideways toward the adjacent rip embayment, which may be several metres to more than 100 m alongshore. The feeder currents are weakest where they diverge behind the centre of the bar, but pick up in speed and intensity toward the rip. In addition, the rip feeder channels deepen toward the rip.

Figure 2.8   *Waitpinga Beach south of Adelaide usually has a well developed rhythmic bar and beach system, with pronounced shoreline rhythms and well developed rips cutting across the bar.*

In the adjacent rip channels, waves break less or often not at all. They may move unbroken across the rip to finally break or surge up the steeper rip embayment swash zone. The strong swash often causes slight erosion of the beach face and cuts an erosion scarp.

In the rip embayment, the backwash returning down the beach face combines with flow from the adjacent rip feeder channels. This water builds up close to shore (called wave set-up), then pulses seaward as a strong, narrow rip current. The currents pulse every 30 to 90 seconds, depending on wave conditions. The rip current accelerates with each pulse and persists with lower velocities between pulses. Rip velocities are usually less than 1 m per second (3.5 km/h), but will increase up to 2 m per second in confined channels and under higher waves.

To identify this beach type, looks for the pronounced longshore beach rhythms, i.e. the shoreline is very sinuous. The shallowest, widest bars and heaviest surf lie off the protruding parts of the shore (the megacusps). Water flows off the bars, into the feeder channel, along the beach to the deeper rip embayment, then seaward in the rip current.

2.3.5.1   Rhythmic bar and beach hazards

This is the most hazardous beach type commonly occurring along the southeast coast. Most people are put off entering the surf by the deep longshore trough containing rips and their feeder currents. If you are swimming or surfing on a rhythmic beach, the following highlights some common hazards.

Rhythmic bar and beach hazards

- Bar - just to reach the bar requires crossing the rip feeder channel. This may be an easy wade at low tide or a difficult swim at high tide. Be very careful once the water exceeds waist depth, particularly if a current is flowing. Also, as you reach the bar, water pouring off the bar may wash you back into the channel.
- The centre of the bar is relatively safe at low tide, but at high tide you run the risk of being washed into the rip feeder or rip channel.
- Rip feeder channel - depth varies with position and tide, both depth and velocity increase toward the rip.
- Rip - the rip channel is usually 2 to 3 m deep, with a continuous, but pulsating, rip current.
- High tide - deeper bar and channels, but weaker currents and rips.
- Low tide - waves break more heavily and may plunge dangerously, shallower bar and channels, but stronger currents and rips.
- Oblique waves - skew bar and rips alongshore.
- Higher waves - intensify wave breaking and strength of all currents.
- Summary: Caution is required by the young and inexperienced on rhythmic beaches, as the bar is separated from the beach by often deep channels and strong currents.

2.3.6 Transverse bar and rip (TBR)

The *transverse bar and rip* is the most common intermediate beach type. There are 159 TBR beaches, with a total length of 288 km (9% of coast). They occur on beaches composed of fine to medium sand and exposed to waves averaging 1-1.5 m. TBR beaches receive their name from the fact that as you walk along the beach, you will see bars transverse or perpendicular to and attached to the beach, separated by deeper rip channels and currents (Fig. 2.9). The bars and rips are usually regularly spaced every 150 to 250 m, but can reach spacings of 500 m. Their surf zones range from 50 to 100 m wide.

Rip currents

Beach rips: Rip currents are a relatively narrow, seaward moving stream of water. They represent a mechanism for returning water back out to sea that has been brought onshore by breaking waves. They originate close to shore as broken waves (wave bores) flow into longshore rip feeder troughs. This water moves along the base of the beach as rip feeder currents. On normal beaches, two currents arriving from opposite directions usually converge in the rip embayment, turn and flow seaward through the surf zone. The currents usually

---

maintain a deeper rip feeder trough close to shore and a deeper rip channel through the surf zone.

The converging currents turn, accelerate and flow seaward through the surf zone, either directly or at an angle at speeds up to 1.5 m/sec. As the confined rip current exits the surf zone and flows seaward of the outer breakers, it expands and may meander as a larger rip head. Its speed decreases and it will usually dissipate within a distance of two to three times the width of the surf zone.

Rip currents will exist in some form on ALL beaches where there is a surf zone. Their spacing is usually regular and ranges from as close as 50-100 m on some of the higher energy, wind and wave dominated gulf beaches, up to 500 m and more on the swell dominated open coast.

Under typical wave conditions there are about 1500 beach rips along the South Australian coast with an average spacing of 300 m, thereby occupying 450 km of the coast (14%).

Topographic rips: Headlands and reefs in the surf will induce additional rips, called topographically controlled rips, with the rip usually flowing out beside the obstacle. There are approximately 800 topographic rips located permanently along the South Australian coast, occurring on all beaches with surf and headlands and/or surf zone reefs.

In total, South Australia has about 2,300 beach and topographic rip systems operating on a typical day.

Rip current spacing

- spacing approximately = surf zone width x 4
- on open coast, rip spacing is commonly 250 to 350 m
- on highest energy beach, spacing can reach 400 to 600 m
- also a function of beach slope, the lower the slope (hence wider the surf zone), the wider the rip spacing

---

Transverse bars and rip beaches are discontinuous alongshore, being cut by prominent rips. The alternation of shallow bars and deeper rip channels causes a longshore variation in the way waves break across the surf zone. On the shallower bars waves break heavily, losing much of their energy. In the deeper rip channels they will break less and possibly not at all, leaving more energy to be expended as a shorebreak at the beach face. Consequently, across the inner surf zone and at the beach face, there is an alternation of lower energy swash in lee of the bars and higher energy swash/shorebreak in lee of the rips.

Figure 2.9 *Well developed transverse bars (waves breaking) and rip channels (no waves breaking) along Waitpinga Beach following a period of moderate waves.*

This longshore variation in wave breaking and swash causes the shoreline to be reworked, such that slight erosion usually occurs in lee of the rips and slight deposition in lee of the bars. This results in a rhythmic shoreline, building a few metres seaward behind the attached bars as deposition occurs and being scoured out and often scarped in lee of the rips. The rhythmic undulations are called megacusp horns (behind the bars) and embayments (behind the rips). Whenever you see such rhythmic features, which have a spacing identical to the bar and rips (100 to 500+ m), you know rips are present.

The transverse surf zone has a cellular circulation pattern. Waves tend to break more on the bars and move shoreward as wave bores. This water flows both directly into the adjacent rip channel and, closer to the beach, to the rip feeder channels located at the base of the beach.

The water in the rip feeder and rip channel then returns seaward in two stages. Firstly, water collects in the rip feeder channels and the inner part of the rip channel, building up an hydraulic head against the lower beach face. Once high enough, it pulses seaward as a relatively narrow accelerated flow, the rip. The water usually moves through the rip channel, out through the breakers and seaward for a distance usually less than the width of the surf zone, that is, a few tens of metres.

The velocity of the rip currents varies tremendously. However, on a typical beach with waves less than 1.5 m, they peak at about 1 m per second, or 3.5 km per hour, about walking pace. However, under high waves they may double that speed. What this means is that under average conditions, a rip may carry someone out from the shore to beyond the breakers in 20 to 30 seconds. Even an Olympic swimmer would only be able to maintain their position, at best, when swimming against a strong rip.

Two other problems with rips and rip channels are their depth and their rippled seabed. They are usually 0.5 to 1 m deeper than the adjacent bar, reaching maximum depths of 3 m. Furthermore, the faster seaward flowing water forms megaripples on the floor of the rip channel. These are sand ripples 1 to 2 m in length and 0.1 to 0.3 m high, that slowly migrate seaward. The effect of the depth and ripples on bathers is to provide both variable water depth in the rip channels and a soft sand bottom, compared to the more compact bar. As a result, it is more difficult to maintain your footing in the rip channel for three reasons: the water is deeper, the current is stronger and the channel floor is less compact. Also, someone standing on a megaripple crest that is suddenly washed or walks into the deeper trough, may think the bottom has 'collapsed'. This may be one source of the 'collapsing sand bar' myth, an event that can not and does not occur.

2.3.6.1 Transverse bar and rip beach hazards

Transverse beaches are one of the main reasons many South Australian beach breaks are so good for surfing. However, the good surf is also a hazard to the unwary swimmer and most drownings and rescues occur with this beach type. The shallow bars tempt people into the surf, while lying to either side are the deeper, more treacherous rip channels and currents.

Transverse bar and rip beach hazards

- Bars - the centres of the attached bars are the safest place to swim. They are shallow, furthest from the rip channels and the wave bores move toward the shore.
- Rips - the rips are the cause of most rescues and drownings, so they are best avoided unless you are a very experienced surfer.
- Rip feeder channels - any channel close to shore has been carved by currents and is part of the surf zone circulation. It will be carrying water that is ultimately heading out to sea. Rip feeder channels usually run along behind and to the sides of the bar, adjacent to the base of the beach.
- In the rip embayment, the feeder currents converge and head out to sea. If you are not experienced, stay away from any channels, particularly if water is greater than waist depth and is moving.
- Children on floats must be very wary of feeder channels as they can drift from a seemingly calm, shallow, inner feeder channel located right next to the beach, rapidly out into a strong rip current.
- Breakers - waves will break more heavily on the bar at low tide, often as dangerous plunging waves or dumpers. In the rip embayment, the shorebreak will be stronger at high tide.
- Higher waves - when waves exceed 1 to 1.5 m, both wave breaking and rip currents will intensify.
- Oblique waves - these will skew both the bars and rips alongshore and may make the rips more difficult to spot.

- Low tide - rip currents intensify at low tide, but are more confined to the rip channel.
- High tide - rip currents are weaker and may be partially replaced by a longshore current, even across the bar.
- Summary: It is relatively safe on the bars during low to moderate waves, but beware, as many hazards, particularly rips, lurk for the young and inexperienced. Stay on the bar/s and well away from the rips and their side feeder currents.

2.3.7   Low tide terrace (LTT)

Low tide terrace beaches are the lowest energy intermediate beach type. They occur on the open coast where sand is fine to medium and wave height averages between 0.5 and 1 m. In South Australia there are 164 low tide terrace beaches, representing 11% of the mainland beaches and occupying 203 km of coast. They tend to occur toward the lower energy, more protected end of long beaches (Fig. 2.10).

Low tide terrace beaches are characterised by a moderately steep beach face, which is joined at the low tide level to an attached bar or terrace, hence the name - low tide terrace. The bar usually extends between 20 and 50 m seaward and continues alongshore, attached to the beach. It may be flat and featureless, have a slight central crest, called a ridge, and may be cut every several tens of metres by small shallow rip channels, called *mini rips*.

Figure 2.10   *Chiton Rocks is a typical low tide terrace beach, with a usually attached and continuous low tide bar, shown here with a few rocks also on the bar.*

At high tide when waves are less than 1 m, they may pass right over the bar and not break until the beach face, behaving much like a reflective beach. However, at spring low tide, the entire bar is usually exposed as a ridge or terrace running parallel to the beach. At this time, waves break by plunging heavily on the outer edge of the bar. At mid tide, waves usually break right across the shallow bar.

Under typical mid tide conditions, with waves breaking across the bar, a low 'friendly' surf zone is produced. Waves are less than 1 m and most water appears to head toward the shore. In fact it is also returned seaward, both by reflection off the beach face and via the mini rips, even if no rip channels are present. The rips, however, are usually weak, ephemeral and shallow.

2.3.7.1   Low tide terrace hazards

Low tide terrace beaches are the safest of the intermediate beaches, because of their characteristically low waves and shallow terrace. However, changing wave and tide conditions do produce a number of hazards to swimmers and surfers.

Low tide terrace beach hazards
- High tide - deep water close to shore; behaves like a reflective beach.
- Low tide - waves may plunge heavily on the outer edge of the bar, with deep water beyond. Take extreme care if body surfing or body boarding in plunging waves, as spinal injuries can result.
- Mid tide - more gently breaking waves and waist deep water, however weak mini rips return some water seaward.
- Higher waves - mini rips increase in strength and frequency and may be variable in location.
- Oblique waves - rips and currents are skewed and may shift along the beach, causing a longshore and seaward drag.
- Most hazardous at mid to high tide when waves exceed 1 m and are oblique to shore, such as during a strong summer northeast sea breeze.
- Summary: One of the safer beach types when waves are below 1 m high, at mid to high tide. Higher waves, however, generate dumping waves, strong currents and ephemeral rips, called 'side drag', 'side sweep' and 'flash' rips by lifesavers.

2.3.8   Reflective beaches (R)

Reflective sandy beaches lie at the lower energy end of the wave dominated beach spectrum. They are characterised by steep, narrow beaches usually composed of coarser sand with low waves. On the South Australian open coast, sandy beaches require waves to be less than 0.5 m to be reflective. For this reason they are also found inside the entrance to bays, at the lower energy end of some ocean beaches and in lee of many of the calcarenite reefs and rock platforms that front many South East and Eyre Peninsula beaches.

In South Australia there are 571 reflective beaches. They are the most common beach type, making up 39% of the mainland beaches and 50% of Kangaroo Island beaches. They are however also the shortest of the beaches as they tend to form in protected pockets in lee of reefs and headlands, even along high energy sections of coast. As a result they have an average length of 600 m and a total shoreline length of 354 km, which represents only 16% of the sandy beach coast.

Reflective beaches are a product of both coarser sand and lower waves. Consequently, all 50 South Australian beaches composed of gravel, cobble and boulders are always reflective, no matter what the wave height.

Reflective beaches always have a steep, narrow beach and swash zone. Beach cusps are commonly present in the upper high tide swash zone. They also have no bar or surf zone as waves move unbroken to the shore, where they collapse or surge up the beach face (Fig. 2.11).

Reflective beach morphology is a product of four factors. First, low waves will not break until they reach relatively shallow water (< 1 m); second, the coarser sand results in a steep gradient beach and relatively deep nearshore zone (> 1 m); third, because of the low waves and deep water, the waves do not break until they reach the base of the beach face; and finally, because the waves break at the beach face, they must expend all their remaining energy over a very short distance. Much of the energy goes into the wave swash and backwash, the rest is reflected back out to sea as a reflected wave, hence the name reflective.

Figure 2.11 *Port Elliot beach is well protected by headlands, rocks and reefs which lower waves to less than 1 m and maintain a typical steep, cusped reflective beach.*

The strong swash, in conjunction with the usually coarse sediment, builds a high, steep beach face. The *cusps* which often reside on the upper part of the beach face are a product of sub-harmonic edge waves, meaning the waves have a period twice that of the incoming wave. The edge wave period and the beach slope determine the edge wave length, which in turn determines the cusp spacing. On the South Australian coast, cusp spacing can range from 20 to 40 m.

Another interesting phenomenon of most reflective beaches is that all those containing a range of sand sizes have what is called a *beach step*. The step is always located at the base of the beach face, around the low water mark. It consists of a band containing the coarsest material available, including rocks, cobbles, even boulders and often numerous shells. Because it contains coarser material, its slope is very steep, hence the step-like shape. They are usually a few decimetres in height, reaching a maximum of perhaps a metre. Immediately seaward of the step, the sediments usually fine markedly and assume a lower slope.

The reason for the step is twofold. The unbroken waves sweep the coarsest sediment continuously toward the beach and the step. The same waves break by surging over the step and up the beach face. However, the swash deposits the coarsest, heaviest material first, only carrying finer sand up onto the beach, then the backwash rolls any coarse material back down the beach. The coarsest material is therefore trapped at the base of the beach face by both the incoming wave and the swash and backwash.

2.3.8.1   Reflective beach hazards

The low waves and protected locations that characterise reflective beaches usually lead to relatively safe swimming locations. However, as with any water body, particularly one with waves and currents, there are hazards present that can produce problems for swimmers and surfers.

Reflective beach hazards
- Steep, soft beach face - may be a problem for toddlers, the elderly and disabled people.
- Relatively strong swash and backwash - may knock people off their feet.
- Step - causes a sudden drop off from shallow into deeper water.
- Deep water - absence of bar means deeper water close into shore, which can be a problem for non-swimmers and children.
- Surging waves and shorebreak - when waves exceed 0.5 m, they break increasingly heavily over the step and lower beach face. They can knock unsuspecting swimmers over. If swimming seaward of the break, swimmers may experience problems returning to shore through a high shorebreak.
- Most hazardous when waves exceed 1 m and shorebreak becomes increasingly powerful.
- Where fronted by a rock platform or reef, additional hazards are associated with the presence of the rock/reef.
- Summary: Relatively safe under low wave conditions, so long as you can swim. Watch children as deep water is close to shore. Hazardous shorebreak and strong surging swash under high waves (> 1 m).

2.3.9 Determining wave dominated beach type

The type of wave dominated beach that occurs along the southeast coast is a function of the modal wave height, which has a maximum of 2 m+, the wave period, which averages 12 seconds and finally the sand size. While the wave period is essentially constant for the open coast, the wave height is at a maximum on exposed locations, decreasing into more protected environments and the gulfs and sand size can range from fine to coarse sand.

Figure 2.12 provides a method for determining the predicted beach type based on these three parameters. Only the highest energy beaches, composed of fine sand and with waves averaging over 2 m, produce dissipative beaches in South Australia. The same waves on beaches composed of slightly coarser (fine to medium) sand produce high energy intermediate beach systems, in particular longshore bar and trough and rhythmic bar and beach. Where waves are reduced slightly to average between 1 and 1.5 m, they form transverse bar and rip beaches, while low tide terrace beaches occur where waves have been reduced to around 1 m and reflective beaches where waves are lowered still further to average about 0.5 m.

2.4 Tide modified beaches

While the open coast is dominated by high energy waves, substantial sections of the South Australian coast are well protected from waves, permitting tides to become increasingly dominant. In both St Vincent and Spencer gulfs, in lee of Kangaroo Island and in some of the larger, more sheltered bays of the Eyre Peninsula, ocean wave energy often decreases to near zero, leaving local wind waves and tides to form the beaches. As a consequence the beaches become increasingly tide modified and even tide dominated. Of the South Australian sand-beach coastline, 3% is tide modified and 37% is tide dominated (Table 2.2).

By definition, tide modified beaches occur when the tide range is between 3 and 15 times the wave height. Where tides remain low (i.e. 1 m or less), as along all the open coast, the waves must be less than 0.3 m to be considered tide modified. These conditions can produce three beach types - ultradissipative, reflective plus bar and rips and reflective plus low tide terrace.

In the gulfs the impact of the lower and shorter wind waves is to produce a smaller version of the three low energy wave dominated beaches, that is: the TBR, LTT and R. The wave period is shorter, averaging 5 seconds compared to 12 seconds on the open coast. As a result when rips do form, their spacing is usually less than 100 m, down to as close as 50 m. At the same time the higher tide range acts to essentially dislocate the surf zone from the swash zone. What this means is that all tide modified beaches

have a relatively steep high tide beach, a wide, low to very low gradient intertidal zone and only at low tide a surf zone that can resemble the bars and rips of the open coast. The wide intertidal zone therefore separates the upper 'beach' from the lower 'surf zone' by as much as a few hundred metres. Basically three types of tide modified beaches occur , generally in areas moderately to well exposed to the westerly winds and their associated waves. These are illustrated in Figure 2.13 and described below.

2.4.1 Ultradissipative (UD)

Ultradissipative beaches occur in higher energy tide-modified locations, where the beaches are also composed of fine to very fine sand. They are characterised by a very wide (200-400 m) intertidal zone, with a low gradient high tide beach and a very low gradient to almost horizontal low tide beach. Because of the low gradient right across the beach, waves break across a relatively wide, shallow surf zone as a series of spilling breakers. This wide, spilling surf zone dissipates the waves to the extent that they are known as 'ultradissipative' beaches. During periods of higher waves (>1 m), the surf zone can be well over 100 m wide, though still relatively shallow. There are, however, no ultradissipative beaches in South Australia.

Basically the fine sand induces the low gradient, while the tide range moves the higher waves backwards and forwards across the wide intertidal zone every six hours. The two act to plane down the beach, while the lack of stationarity or stability of the shoreline precludes the formation of bars and rips.

Hazards

The major hazards associated with ultradissipative beaches are their usually higher waves, the relatively deep water off the high tide beach, the long distance from the shore to the low tide surf and the often considerable distance from the shoreline out to beyond the breakers. Currents run along the beach when waves arrive at angles, however strong rip currents are generally absent. Seaward of the breakers however, shore parallel tide currents also increase in strength.

2.4.2 Reflective plus bar & rips (R+BR)

Reflective beaches fronted by low tide bars and rips occur in similar environments to the ultradissipative, only on beaches with fine to medium sand. The slightly coarser sand results in more moderate beach gradients. These beaches are consequently characterised by a relatively steep high tide beach, a moderately wide, lower gradient intertidal beach, which may be featureless or with low sand shoals and a low gradient low tide surf zone containing either transverse bars and rips, or occasionally rhythmic bars and rips. There are only a few of these

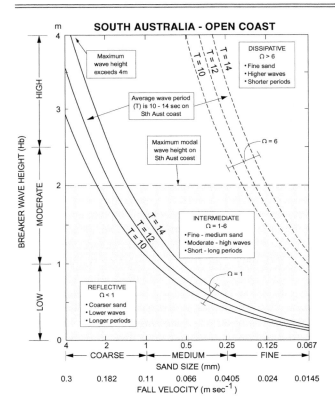

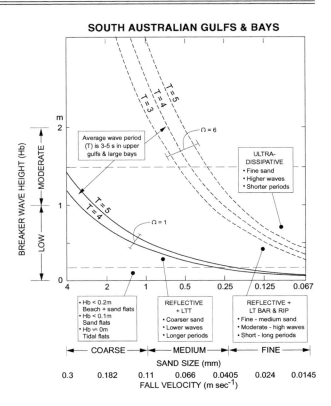

Figure 2.12 *A plot of breaker wave height versus sediment size, together with wave period, that can be used to determine approximate Ω and beach type. To use the chart, determine the breaker wave height, period and grain size/fall velocity (mm or cm/sec). Read off the wave height and grain size, then use the period to determine where the boundary of reflective/intermediate, or intermediate/dissipative beaches lies. Ω = 1 along solid T lines and 6 along dashed T lines. Below the solid lines Ω < 1 and the beach is reflective, above the dashed lines Ω > 6 and the beach is dissipative, between the solid and dashed lines Ω is between 1 and 6 and the beach is intermediate.*

beaches in South Australia. They tend to occur in the lower gulfs where occasional swell waves also impact the beaches.

The morphology of these beaches is in part due to the steeper gradients, which produce the steep high tide beach. At spring high tide, waves usually surge against the beach face with no surf. During mid tide, the waves spill across the lower gradient intertidal zone, while at low tide the cellular circular flow associated with the rips dominates the surf. The surf and rips are able to form on this beach type owing to the period of time (2-4 hr) that the surf zone is roughly located over the low tide bar. This provides the waves with sufficient time to generate a surf zone similar to the intermediate beaches of the wave dominated coast.

Hazards

As with all tide modified beaches, the nature of the beach, surf zone and associated hazards changes with the state of the tide as well as the height of the waves. Under normal moderate wave conditions (Hb=1 m) the high tide beach

is relatively safe, apart from the surging waves and deeper water off the beach, the mid tide beach has a lower gradient and wider, more dissipative surf zone, while the low tide beach is the most hazardous with more heavily breaking waves, rip feeder and rip currents, as well as being up to 300 m seaward of the high tide shoreline. The rip currents are, however, only active at low tide, while at mid to high tide the rip morphology is simply submerged below the zone of wave shoaling, seaward of the breaker zone.

2.4.3   Reflective plus low tide terrace (R+LTT)

The lowest energy of the tide modified beaches is the reflective high tide beach fronted by a low tide terrace. This is the most common of the tide modified beaches, with 42 occurring on the more exposed eastern and south facing shores in mid Spencer Gulf. The beaches are characterised by a steeper high tide beach, at the base of which is an abrupt change in slope and a wider, low gradient low tide terrace, which may occasionally contain rips (Fig. 2.14). They average 1.1 km in length and occupy 2% of the sandy-beach coastline.

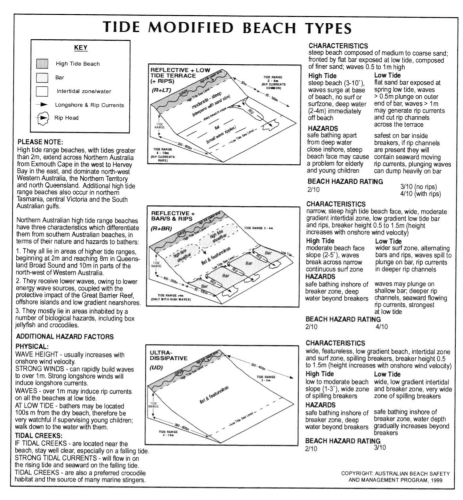

Figure 2.13 *A three dimensional sketch of the three tide modified South Australian beaches: beach and low tide terrace, beach with low tide bars and rips, and ultradissipative.*

At high tide, waves surge up the steep beach face. This continues as the tide falls, until the shallower water of the terrace induces wave breaking increasingly across the terrace. At low tide, waves spill over the outer edge of the terrace, with the inner portion remaining dry during spring low tide. If rips are present, they will cut a channel across the terrace. Like the bar and rip beaches, the rips are only active at low tide.

Hazards

This beach undergoes a marked change in morphology between high and low tide. At high tide hazards are associated with the waves surging against the steep high tide beach, together with the deeper water off the beach, while at low tide, waves spill across the broad, shallow terrace, with hazards associated with the deeper water off the terrace and the rips, if present.

2.5   Tide dominated beaches

When the tide range begins to exceed the wave height by between 10 and 15 times, then the tide becomes increasingly important in the beach dynamics, at the expense of the waves. In South Australia these conditions only occur in the upper gulfs where waves are very small and particularly in Spencer Gulf where tide range increases to 3 m. Basically there are three beach types - two with a high tide beach and sand flats, while the third is essentially a transition from sand flat to a tidal flat. These 'beaches' represent a transition from the high tide wave dominated beach to the mid to low tide, more tide dominated tidal sand flats (Fig. 2.15). They still receive sufficient wave energy to build the high tide beach, remove any fine silts and mud to maintain sand, rather than mud flats, and on some beaches to build multiple sand ridges.

2.5.1   Beach plus sand ridges (R+SR)

The high tide beach fronted by multiple sand ridges occurs in 29 locations occupying 44 km of the coast in the gulfs and some of the larger bays on the Eyre Peninsula. These systems usually have a moderate to steep high tide beach, with shore parallel, sinuous, low amplitude, evenly spaced sand ridges extending out across the inter- to sub-tidal

a)

b)

c)

Figure 2.14 *Many of the less exposed gulf beaches consist of a relatively steep high tide beach, with a sharp break in slope and a low gradient intertidal tide terrace, as seen here at (a) Somerton, with a low tide terrace and poorly developed rips; (b) Brighton, with a flat low tide terrace; and (c) Semaphore, with three intertidal bars and troughs.*

sand flats (Fig. 2.16). These are not to be confused with the higher relief (sand) bars of the higher energy beaches. The beach is only active at high tide when either very low waves or calms prevail. The intertidal zone and ridges are usually inactive. The exact mode of formation is unknown, though it is suspected that the ridges are active and formed during infrequent periods of higher waves acting across the intertidal ridged zone.

The number of ridges averages seven (sd=8), but can range up to 40. The ridges are very low amplitude, no more than a few centimetres to a decimetre from trough to crest and average about 50 m in spacing. They tend to parallel the coast, but can at times lie obliquely to the shore, while at other places they merge into more complex patterns.

Hazards

The major hazard with these low energy beaches is the relatively deep water off the high tide beach and, in places, associated tidal currents. Low tide is dominated by the wide, shallow to exposed ridges and sand flats.

2.5.2 Beach plus sand flats (R+SF)

The most common tide dominated beach is the high tide beach fronted by very low gradient, flat, featureless sand flats, which average 430 m in width (sd=550 m) (Figs. 2.17 & 2.18). There are 364 such beaches in the gulfs and larger bays, averaging 1.4 km in length and occupying 517 km of coast, or 26% of the sand coast. They are the second most dominant beach type, after reflective beaches.

Hazards

The only hazards associated with these beaches are the deeper water off the high tide beach and the slight chance of tidal currents off the beach. Low tide reveals a wide, shallow to exposed tidal flat.

2.5.3 Beach plus tidal sand flats (R+TSF)

Tidal flats are not by definition 'beaches', however they are included here for two reasons. First, these tidal flats represent a gradation from the true high tide beaches of the above two beach types, to the mangrove vegetated tidal flats that dominate the tide dominated sections of the coast. Secondly, a number of sand flats are labelled 'beaches' and known locally as such. The main difference between the beach+sand flats and beach+tidal flats, is that the tidal flats may be vegetated with seagrass and even scattered mangroves. There are 69 beaches of this type, all occurring in the very low energy sections of the upper St Vincent and Spencer gulfs and in parts of Denial Bay.

Hazards

The only physical hazards associated with tidal sand flats are the deeper water over the flats at high tide, the increased likelihood of tidal currents moving across or parallel to the flats and their wide distance at low tide, which increases the chance of being caught in the intertidal zone by the rising tide.

2.6   Calcarenite

Calcarenite is a lithified (cemented) sand composed predominantly of calcareous material. The 'calc' stands for the calcareous and 'arenite' means sand. The South Australian coast is dominated by two types of calcarenite – aeolian or dune calcarenite, when sand dunes have been cemented, commonly called *dunerock*, and beach calcarenite, where the beach has been cemented, commonly called *beachrock* (Fig. 2.19).

Along the entire South Australian coast and particularly on the exposed open coast, massive deposits of calcareous sand have been deposited since the beginning of the Quaternary (2 million years ago). When exposed to long periods (thousands of years) of pedogenesis (soil forming processes), the sand can undergo transformation into massive calcrete, normally near the top of the surface. When this occurs, the original sand grains and structures are no longer present. Below this surface the sand undergoes partial cementation, which leaves the individual grains and structures visible, though cemented. In addition, towards the surface the vegetation interacts with the sand to produce a range of lithocasts (limestone cast around a trunk or root) and lithoskels (limestone cast inside a trunk or root). What all this means is that if the lithified beach and dune is subsequently exposed to wave attack, much will remain as a resilient rock or reef structure, which exerts considerable influence on the present coast (Fig. 2.19).

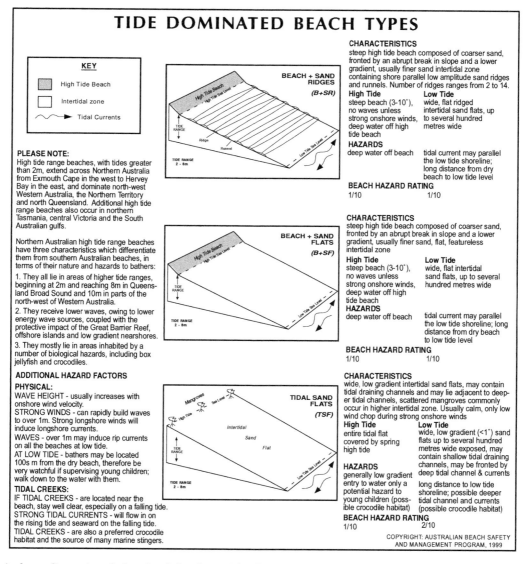

Figure 2.15 *A three dimensional sketch of the three tide dominated South Australian beaches: beach+ridged sand flats, beach+sand flats and tidal sand flats.*

2.7   Larger scale beach systems

The beach systems described above are all part of larger scale beach and barrier systems, the barriers including backing beach and foredune ridges and, in places, sand dunes, as well as adjoining tidal creeks and inlets and headlands and reefs. Figure 2.20 provides a schematic overview of the typical arrangement of some of these beach and associated barrier systems. The lowest energy systems receive low waves and are backed by low beach ridge plains, as in Lacepede Bay and in the gulfs. Moderate energy systems are backed by prograding foredune ridge plains as at Guichen and Rivoli bays. Moderate to higher energy beaches tend to have backing blowouts and parabolic dunes (common along much of the coast), while the highest energy systems are backed by massive dune transgression, as along the northern Coorong.

a)

a)

b)

Figure 2.17   *a) A low energy high tide beach, intertidal sand flats and seagrass meadows off The Pines beach, Cape Yorke. b) Ground view of a low energy, seagrass littered high tide beach and intertidal sand flats at Port Clinton, Gulf St Vincent.*

b)

Figure 2.16   *Multiple sand ridges in (a) Hardwicke Bay, Yorke Peninsula; and (b) Bosanquet Bay near Ceduna.*

Figure 2.18   *Very low energy high tide beach and partly vegetated tidal sand flats, near Point Hughes in upper Spencer Gulf.*

a)

b)

c)

Figure 2.19   *Pleistocene beach and dune calcarenite may occur as intertidal platforms (a), as shore parallel beachrock reefs (b), both at Coymbra; and as massive layered sea cliffs (c), shown here at Groper Bay and common along the entire coast.*

Figure 2.20   *Schematic cross-section of typical low, moderate and high energy South Australia beach-dune-barrier systems. As wave height increases the width and depth of the surf and nearshore zone increases, as does the size and instability of the backing dune systems.*

3.   SOUTH AUSTRALIAN BEACHES AND HAZARDS

3.1   Beach usage and surf lifesaving

The Australian coast remained untouched until the arrival of the first Aborigines some tens of thousands of years ago. The coast and beaches that they found, however, no longer exist. Probably crossing to Australia during one of the glacial low sea level periods, they not only reached a far larger, cooler, drier and windier continent, but one where the shoreline was some tens of metres below present sea level. Hence the coast they walked and fished now lies out below sea level on the inner continental shelf.

The present Australian coast only obtained its position some 6500 years ago, when sea level rose to approximately its present position. As it was rising, at about 1 m every 100 years, the Aborigines no doubt followed its progress by slowly moving inland and to higher ground. Therefore, we can assume that usage of the present South Australian beaches began as soon as they began forming some 6000 years ago and continued in the traditional way until the 1800's.

The first European settlers in South Australian waters were the sealers and whalers who established camps at suitable protected beach locations on Kangaroo Island (such as American Bay and Hog Bay) and various mainland sites, including Fishery Bay and Fowlers Bay. Following formal settlement and the spread of farming along the coast, South Australia became renowned for its initial beach 'ports', where produce was loaded from the shore directly onto small ships. These were slowly replaced during the latter half of the 19th century by long jetties built out across the sand flats to reach deeper water. Jetties were built across many beaches in the South East, in the gulfs and around the Eyre Peninsula, with the westernmost jetty at Fowlers Bay. These jetties continued to service the wheat ships into the 1950's and today many remain as popular fishing sites.

Because of the prevalence of many long beaches, including Australia's longest, coupled with the general absence of inlets, the beaches also served as early travel routes. John Eyre followed them right across the Great Australian Bight, becoming the first European to cross the southern part of the continent. Even today some Adelaide and South East beaches are popular 'roads', including traffic signs and parking areas, while 4WD vehicles use most beaches in the state as offroad 'highways'. South Australia, perhaps more than any other state, has made use of its beaches for shipping and transport purposes.

Only in the early 1900's did beach and surf swimming become a popular recreational pastime. Along the Adelaide coast the public used the relatively safe gulf beaches from Seacliff up to Largs Bay, while in country areas, places like Beachport, Robe, Port Elliot and Ceduna offered generally safe swimming. However, the gradual use of the more exposed beaches in the South East, Fleurieu, Yorke and Eyre Peninsulas exposed swimmers to hazardous conditions, resulting in an increase in rescues and drownings.

3.1.1   The Surf Lifesaving Movement

To understand the formation of the surf lifesaving clubs and the broader Surf Life Saving Association is to realise two things. Firstly, in their rush to the surf, most beachgoers could not swim and had little or no knowledge of the surf and its dangers. Secondly, the open coast of southern Australia is as dangerous as it is inviting to the unprepared.

All beaches patrolled by surf lifesavers and lifeguards are listed in Table 3.1 by year of establishment and in Table 3.2 from east to west.

The first surf lifesaving clubs and their national association, formed in 1907, had embarked upon the establishment and growth of an organisation that has become an integral part of Australian beach usage and culture, and through which it is so readily identified internationally.

Now, nearly a century after the initial rush to the beaches and the foundation of the early surf lifesaving clubs, both beach usage and the 274 surf lifesaving clubs around the coast are accepted as part of Australian beaches and beach life. However, at the beginning of the 21st century, beach usage is undergoing yet another surge as the Australian population and visiting tourists increasingly concentrate on the coast. This is resulting in more beaches being used, more of the time, by more people, many of whom are unfamiliar with beaches and their dangers. In addition, most of the newly used beaches have no surf lifesaving clubs or patrols.

There is now a greater need than ever to maintain public safety on these beaches, a service that is provided on patrolled beaches by volunteer lifesavers and professional lifeguards. This book is the result of a joint Surf Life Saving Australia, Surf Life Saving South Australia and the Coastal Studies Unit project that is addressing this problem. The book is designed to provide information on each and every beach in South Australia, including Kangaroo Island and beaches on five smaller islands, in all, 1755 beaches. It contains information on each beach's general characteristics and suitability for swimming, surfing and fishing. In this way, swimmers may be forewarned of potential hazards before they get to the beach and consequently swim more safely.

Table 3.1   Formation of South Australian Surf Lifesaving Clubs (first to last)

Year established	Surf Lifesaving Club
Henley	1925
Port Noarlunga	1933 (Royal) 1953 (Surf)
Port Elliot	1935
Moana	1938 (Royal) 1952 (Surf)
Glenelg	1931 (Royal) 1952 (Surf)
Brighton	1951
Seacliff	1952 (formed as 'Seacliff Club' in 1929)
Semaphore	1953
Christies Beach	1954
Grange	1956
Chiton Rocks	1957
Whyalla	1957
South Port	1959
West Beach	1959
Somerton	1960
North Haven	1966
Hallett Cove	1968-69
Aldinga Bay	1978
Normanville	1999

Oldest and Youngest Clubs

The oldest surf lifesaving club in South Australia is Henley SLSC, formed in 1925, while the youngest is Normanville SLSC, formed 74 years later in 1999.

Table 3.2   South Australian patrolled beaches (east to west)

Beach	SLSC	Patrol Season	Lifeguard
Port Elliot	x	Nov-Mar	
Chiton Rocks	x	Nov-Mar	
Normanville	x	Dec-Mar	January
Aldinga (3)	x	Nov-Mar	
Moana	x	Nov-Mar	
South Port	x	Nov-Mar	
Port Noarlunga	x	Nov-Mar	
Christies Beach	x	Nov-Mar	
O'Sullivan	x	Nov-Mar	
Seacliff	x	Nov-Mar	
Brighton	x	Nov-Mar	Dec-Mar
Somerton	x	Nov-Mar	Dec-Mar
Glenelg	x	Oct-Apr	Dec-Mar
West Beach	x	Nov-Mar	
Henley	x	Nov-Mar	
Grange	x	Nov-Mar	
Semaphore	x	Nov-Mar	
North Haven	x	Nov-Mar	
Whyalla	x	Nov-Mar	
Total			

If you are interested in joining a surf lifesaving club or learning more about surf lifesaving, contact Surf Life Saving Australia, your state centre (listed below) or your nearest surf lifesaving club.

Surf Life Saving Australia 128 The Grand Parade Brighton-Le-Sands NSW 2216 Phone: (02) 9597 5588 Fax: (02) 9599 4809 Email: info@slsa.asn.au	**Surf Life Saving South Australia** PO Box 82 Henley Beach SA 5022 Phone: (08) 8356 5544 Fax: (08) 8235 0910 Email: surflifesaving@surfrescue.com.au
Surf Life Saving New South Wales PO Box 430 Narrabeen NSW 2101 Phone: (02) 9984 7188 Fax: (02) 9984 7199 Email: experts@surflifesaving.com.au	**Surf Life Saving Victoria** A W Walker House Beaconsfield Parade St Kilda VIC 3182 Phone: (03) 9534 8201 Fax: (03) 9534 0311 Email: slsv@slsv.asn.au
Surf Life Saving Queensland PO Box 3747 South Brisbane QLD 4101 Phone: (07) 3846 8000 Fax: (07) 3846 8008 Email: slsq@lifesaving.com.au	**Surf Life Saving Tasmania** GPO Box 1745 Hobart TAS 7001 Phone: (03) 6231 5380 Fax: (03) 6231 5451 Email: slst@slst.asn.au
Surf Life Saving Western Australia PO Box 1048 Osborne Park Business Centre WA 6017 Phone: (08) 9244 1222 Fax: (08) 9244 1225 Email: slswa@slswa.asn.au	**Surf Life Saving Northern Territory** PO Box 43096 Casuarina NT 0811 Phone: (08) 8941 3501 Fax: (08) 8981 3890 Email: slsnt@topend.com.au

3.1.2 Physical beach hazards

Beach hazards are elements of the beach-surf environment that expose the public to danger or harm. *Beach safety* is the mitigation of such hazards and requires a combination of common sense, swimming ability and beach-surf knowledge and experience. The following section highlights the major physical hazards encountered in the surf, with hints on how to spot, avoid or escape from such hazards. This is followed by the biological hazards.

There are seven major physical hazards on South Australian beaches (Fig. 3.1):

1. water depth (deep and shallow)
2. breaking waves
3. surf zone currents (particularly rip currents)
4. tidal currents
5. strong winds
6. rocks, reefs and headlands
7. water temperature

In the surf zone, three or four hazards, particularly water depth, breaking waves and currents, usually occur together. In order to swim safely, it is simply a matter of avoiding or being able to handle the above when they constitute a hazard to you, your friends or children.

3.1.2.1 Water depth

Any depth of water is potentially a hazard.
- *Shallow water* is a hazard when people are diving in the surf or catching waves. Both can result in spinal injury if people hit the sand head first.
- *Knee depth* water can be a problem for a toddler or young child.
- *Chest depth* is hazardous to non-swimmers, as well as to panicking swimmers. In the presence of a current, it is only possible to wade against the current when water is below chest depth. Be very careful when water begins to exceed waist depth, particularly if younger or smaller people are present and if in the vicinity of rip or tidal currents.

a)

b)

c)   d)

Figure 3.1   *South Australian physical beach hazards include deep water, breaking waves, rips and tidal currents, rocks and reefs, as illustrated in this series of photographs. (a) Highly irregular rock and reef dominated coast at Carpenters Rocks; (b) high waves breaking close to shore at Cape Tournefort; (c) shore parallel beachrock reef inducing a strong permanent rip near Head of Bight; and (d) deep tidal channel at Venus Bay.*

Water Depth
- Safest: knee deep - can walk against a strong rip current
- Moderately safe: waist deep - can maintain footing in rip current
- Unsafe: chest deep - unable to maintain footing in rip current

Remember: what is shallow and safe for an adult can be deep and distressing for a child.

Shallow water hazards

Spinal injuries are usually caused by people surfing dumping waves in shallow water, or by people running and diving into shallow water.

To avoid these:
- Always check the water depth prior to entering the surf.
- If unsure, WALK, do not run and dive into the surf.
- Only dive under waves when water is at least waist deep.
- Always dive with your hands outstretched.

Also
- Do not surf dumping waves.
- Do not surf in shallow water.

3.1.2.2 Breaking waves

As waves break, they generate turbulence and currents which can knock people over, drag and hold them under water, and dump them on the sand bar or shore. If you do not know how to handle breaking waves (as most people don't), stay away from them, stay close to shore and on the inner part of the bar.

If you are knocked over by a wave, remember two points – the wave usually holds you under for only two to three seconds (though it may seem like much longer), therefore do not fight the wave, you will only waste energy. Rather, let the wave turbulence subside, then return to the surface. The best place to be when a big wave is breaking on you is as close to the seafloor as possible. Experienced surfers will actually 'hold on' by digging their hands into the sand as the wave passes overhead, then kick off the seabed to speed their return to the surface.

If a wave does happen to gather you up in its wave bore (white water) it will only take you towards the shore and will quickly weaken, allowing you to reach the surface after two to three seconds and usually leave you in a safer location than where you started.

Breaking waves and wave energy

Surging waves - safe when low
- break by surging up beach face - usually less than 50 cm high
- can be a problem for children and elderly, who are more easily knocked over
- become increasingly strong and dangerous when over 50 cm high

Spilling waves - relatively safe
- break slowly over a wide surf zone
- are good body surfing waves

Plunging (dumping) waves - the most dangerous waves
- break by plunging heavily onto sand bar
- strong wave bore (white water) can knock swimmers over
- very dangerous at low tide or where water is shallow
- waves can dump surfers onto sand bar, causing injury
- most spinal injuries are caused by people body surfing or body boarding on dumping waves
- to avoid spinal injury, do not surf dumping waves or in shallow water. If caught by a wave do not let it dump you head first, turn sideways and cover your head with your arms
- only very experienced surfers should attempt to catch plunging waves

Wave energy ≈ square of the wave height
Wave energy represents the amount of power in a wave of a particular height.
     0.3 m wave = 1 unit wave energy/power
     1.0 m wave = 11 units
     1.5 m wave = 25 units
     2.0 m wave = 44 units
     2.5 m wave = 70 units
     3.0 m wave = 100 units
Therefore, a 3 m wave is 10 times more powerful than a 1 m wave and 100 times more powerful than a 0.3 m wave.

Rip and surf zone current velocity
Breaking waves   travel at 3-4 m/sec (10-15 km/hr)
Wave bores (white water)   travel at 1-2 m/sec (3-7 km/hr)
Rip feeder and longshore currents travel at 0.5 - 1.5 m/sec (2-5 km/hr)
Rip currents under average wave conditions (< 1.5 m high) attain maximum velocities of 1.5 m/sec = 5.4 km/hr
     (Note: Olympic swimmers swim at about 7 km/hr)
     An average rip in a surf zone 50 m wide can carry you outside the breakers in as little as 30 seconds.

3.1.2.3 Surf zone currents and rip currents

Surf zone currents and particularly *rip currents* are the biggest hazards to most swimmers. They are the hardest for the inexperienced to spot and can generate panic when swimmers are caught by them. See the following section on rips, on how to identify and escape from rip currents.

The problem with currents, particularly rip currents, is that they can move you unwillingly around the surf zone and ultimately seaward (Fig. 3.2). In moving seaward, they will also take you into deeper water and possibly toward and beyond the breakers. As mentioned earlier, currents are manageable when the water is below waist level, but as water depth reaches chest height they will sweep you off your feet.

3.1.2.4 Tidal currents

Tidal currents are generally weak on the open coast and have little impact on beach processes. However, in the gulfs and confined straits, bays and inlets, the combination of shallower, constricted water and in places increasing tide range can generate strong tidal currents. Where such constrictions and stronger currents occur, tides must flow in and out twice a day and in so doing generate strong reversing currents, which also maintain deeper tidal channels (Fig. 3.1d). As some settlements are located on bays and near inlets, and some beaches are located on or adjacent to inlets and bay mouths, these strong currents and their deep channels are a very real hazard on all such beaches. They are particularly hazardous on a falling tide as the currents flow seaward.

When swimming or even boating in tidal creeks, always check the state and direction of the tide and be prepared for strong currents. Swimmers should not venture beyond waist deep water.

3.1.2.5 Strong winds

The South Australian coast is exposed to predominantly westerly winds associated with both the subtropical high and the passage of cyclonic cold fronts. The strongest winds are associated with the cold fronts which tend to hit from the west and gradually swing more to the south. As a consequence wind strength and direction must be factored into the level of hazard on South Australian beaches, in particular the higher energy open coast beaches (Table 3.3a). In the gulf, these winds are responsible for most, and in places all, of the waves, with wind strength relating directly to wave height (Table 3.3b) and hence the level of beach hazard. Whenever strong winds occur, their direction determines the impact they will have on beach and surf conditions, as follows:

- *Onshore winds* will help pile more water onto the beach and increase the water level at the shore. They also produce more irregular surf, which makes it more difficult to spot rips and currents.

- *Longshore winds,* particularly strong westerly winds, will cause wind waves to run along the beach, with accompanying longshore and rip currents also running along the beach. The waves and currents can very quickly sweep a person along the beach and into deeper rip channels and stronger currents.

- *Offshore winds* tend to clean up the surf. They are generally restricted to hot summer northerly conditions. However, if you are floating on a surfboard, bodyboard, ski or wind surfer, it also means it will blow the board offshore. In very strong offshore winds, it may be difficult or impossible for some people to paddle against this wind.

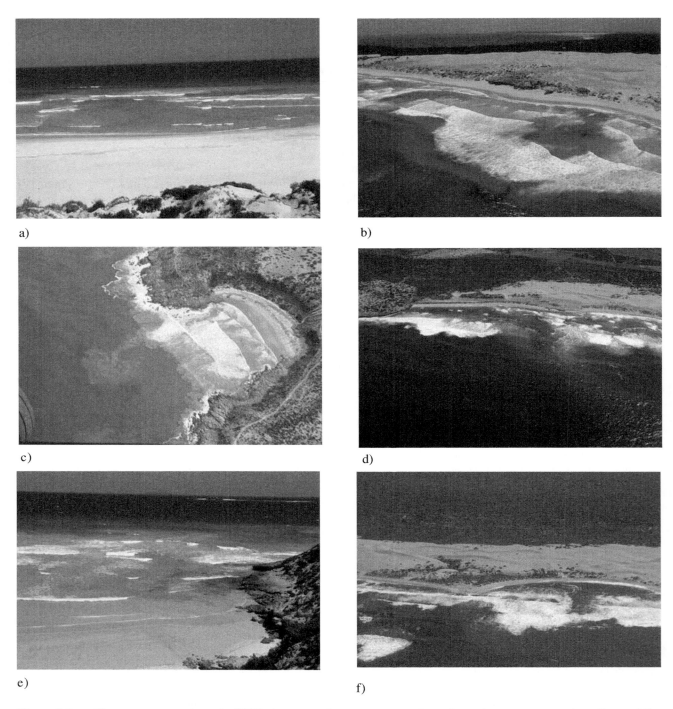

a)
b)
c)
d)
e)
f)

Figure 3.2   *There are approximately 3000 rips operating at any one time along the more exposed sections of the South Australian coast. These photographs illustrate a number of typical rips. (a & b) Well developed beach rips at Gunyah Beach; headland rips at Point James (c), Black Rocks (d), Cabbotts Well (e); and reef controlled rips at Coymbra (f).*

Table 3.3   (a) Wind hazard rating for South Australian beaches, based on wind direction and strength. Winds blowing on and alongshore will intensify wave breaking and surf zone currents, with strong longshore winds capable of producing a strong longshore drag. Their impact on surf zone hazards and beach safety is indicated by the relevant hazard number, which should be added to the prevailing beach hazard rating. (b) Gulf waters sea wave height based on strength of onshore wind. Maximum sea height in the gulf is about 3 m. In addition, gulf waves are short and steep, compared to ocean waves of the same height.

(a) South Australian ocean coast

Wind Hazards (Add to Beach Hazard Rating)				
Direction	Light	Mod	Strong	Gale
Longshore	0	1	3	4
Onshore	0	1	2	3
Offshore	0	1	1	2

(b) The Gulfs

Onshore Wind Strength	Max Sea Wave Height (m)
Light	0.3
Moderate	0.5
Strong	1.0
Gale	3.0 - 4.0

Period 3 - 5 seconds

3.1.2.6   Wind generated waves and currents

Winds in the Southern Ocean are responsible for the high swell that arrives at the coast, while the same winds also drive the great circumpolar ocean current. Closer to shore and at the coast winds generate seas, rather than swell and can generate local wind driven currents. As Table 3.3b indicates, local winds are responsible for all wind wave (sea) generation in the gulfs. In addition these same winds along the open coast will generate a sea on top of any existing swell, hence the weather forecast for sea and swell conditions. The locally generated seas are by definition short, steep and more irregular (some say confused) compared to swell.

If winds are of sufficient strength and duration they can generate locally wind driven currents at the coast. In South Australian open waters, most wind driven currents set to the east following the larger ocean currents to the south. Changes to this pattern occur during strong northerly winds which push water offshore, causing upwelling, while strong onshore winds push water onshore, causing downwelling, as previously discussed in section 1.6.8.

3.1.2.7   Rocks, reefs and headlands

Most open coast South Australian beaches have some rocks, calcarenite reefs and headlands. These pose problems on higher energy beaches because they cause additional wave breaking, generate more (and stronger) rips (topographic rips), and have hard and often dangerous surfaces. When they occur in shallow water and/or close to shore, they are also a danger to people walking, swimming or diving because of the hard seabed and the fact that they may not be visible from the surface.

Rocks, reefs and headlands

- if there is surf against rocks or a headland, there will usually be a rip channel and current (topographic rip) next to the rocks
- rocks and reefs can be hidden by waves and tides, so be wary
- most calcarenite reefs will be submerged at high tide
- do not dive or surf near rocks, as they generate greater wave turbulence and stronger currents
- rocks often have sharp, shelled organisms growing on their surface which can inflict additional injury
- if walking or fishing from rocks, be wary of being washed off by sets of larger waves

---

3.1.2.8   Safe swimming

The safest place to swim is on a patrolled beach between the red and yellow flags, as these indicate the safest area of the beach and the area under the surveillance of the lifesavers. If there are no flags then stay in the shallow inshore or toward the centre of attached bars, or close to shore if water is deep. However, remember that rip feeder currents are strongest close to shore and rip currents depart from the shore. The most hazardous parts of a beach are in or near rips and/or rocks, outside the flags or on unpatrolled beaches and when you swim alone.

Remember these points:
- DO swim on patrolled beaches.
- DO swim between the red and yellow flags.
- DO swim in the net enclosure (where present).
- DO observe and obey the instructions of the lifesavers or lifeguards.
- DO swim close to shore, on the shallow inshore and/or on sand bars.
- ALWAYS have at least one experienced surf swimmer in your group.
- NEVER swim alone.
- DO swim under supervision if uncertain of conditions.
- DO NOT enter the surf if you cannot swim or are a poor swimmer.
- DO NOT swim or surf in rips, troughs, channels or near rocks.
- DO NOT enter the surf if you are at all unsure where to swim or where the rips are.
- BE AWARE of hypothermia caused by exposure to cold air and water, particularly on bare skin and with small children. Wind will add to the chill factor.

Patrolled beaches
- swim between the red and yellow flags
- obey the signs and instructions of the lifesavers or lifeguards
- still keep a check on all the above, as over 100 people are rescued from patrolled beaches in South Australia each year

Unpatrolled beaches
- always look first and check out the surf, bars and rips
- select the safest place to swim, do not just go to the point in front of your car or the beach access track
- try to identify any rips that may be present
- select a spot behind a bar and away from rips and rocks
- on entering the surf, check for any side currents (these are rip feeder currents) or seaward moving currents (rip currents)
- if these currents are present, look for a safer spot
- it's generally safer to swim at low tide, if you avoid the rips

Children
- NEVER let them out of your sight
- ADVISE them on where to swim and why
- ALWAYS accompany young children or inexperienced children and teenagers into the surf
- REMEMBER they are small, inexperienced and usually poor swimmers and can get into difficulty at a much faster rate than an adult

3.1.3 Beach Hazard Rating

The *beach hazard rating* refers to the scaling of a beach according to the physical hazards associated with its beach type, together with any local physical hazards. It ranges from the low, least hazardous rating of 1 to a high, most hazardous rating of 10. It does not include biological hazards, these are discussed later in this chapter. The beach characteristics and hazard rating for wave dominated, tide modified and tide dominated beaches are shown in Figures 2.4, 2.5, 2.13 and 2.15.

Each beach, depending on its beach type and typical wave conditions, will have a *modal beach hazard rating* that typifies that beach. Figure 3.3 plots both the wave height required to generate the ten South Australian beach types, as well as the hazard rating associated with those beach type-wave conditions. The hazard rating ranges from a high of 6 on rip-dominated RBB and TBR beaches exposed to waves averaging 1.5 m, to a low of 1 on tide dominated sand flats. It can vary rapidly between adjoining beaches as wave and beach conditions change, as illustrated in Figure 3.4.

BEACH HAZARD RATING GUIDE
Impact of changing breaker wave height on hazard rating for each beach type

Wave Dominated Beaches

BEACH TYPE / WAVE HEIGHT	< 0.5 (m)	0.5 (m)	1.0 (m)	1.5 (m)	2.0 (m)	2.5 (m)	3.0 (m)	> 3.0 (m)
Dissipative	4	5	6	7	8	9	10	10
Long Shore Bar Trough	4	5	6	7	7	8	9	10
Rhythmic Bar Beach	4	5	6	6	7	8	9	10
Transverse Bar Rip	4	4	5	6	7	8	9	10
Low Tide Terrace	3	3	4	5	6	7	8	10
Reflective	2	3	4	5	6	7	8	10

Tide Modified Beaches
(at high tide - at low tide add 1)

Ultradissipative	1	2	4	6	8	10	10	
Reflective + Bar & Rips	1	2	3	5	7	9	10	
Reflective + LTT	1	1	2	4	6	8	10	

Tide Dominated Beaches
(at high tide - at low tide add 1)

Beach + Sand Ridges	1	1	2	Waves unlikely to exceed 0.5 - 1m
Beach + Sand Flats	1	1		Note: if adjacent to tidal channel, beware of deep water and strong tidal currents.
Tidal Sand Flats	1			

BEACH HAZARD RATING	KEY TO HAZARDS
Least hazardous: 1 - 3	Water depth and/or tidal currents
Moderately hazardous: 4 - 6	Shorebreak
Highly hazardous: 7 - 8	Rips and surfzone currents
Extremely hazardous: 9 - 10	Rips, currents and large breakers

NOTE: All hazard level ratings are based on a swimmer being in the surf zone and will increase with increasing wave height or with the presence of features such as inlet, headland or reef induced rips and currents. Rips also become stronger with falling tide.

BOLD gradings indicate the average wave height usually required to produce the beach type and its average hazard rating.

Figure 3.3 *Matrix for calculating the prevailing beach hazard rating for wave dominated, tide modified and tide dominated beaches, based on beach type and prevailing wave height and, on tide modified beaches, state of tide.*

As the modal hazard rating refers to average wave conditions, any change in wave height, length or direction will change the surf conditions and accompanying hazards. This will in turn change the beach's hazard rating. Figure 3.3 provides a matrix for calculating the actual or *prevailing beach hazard rating* for wave dominated, tide-modified and tide-dominated South Australian beaches. It assumes you know the beach type and can estimate the breaker wave height. With these two factors, the prevailing beach hazard rating can be read off the chart. This figure also describes the general hazards associated with each beach type as wave conditions rise and fall.

Local factors must also be considered in calculating the prevailing beach hazard rating, particularly hazards such as rocks, reefs, inlets and headlands, and the presence of

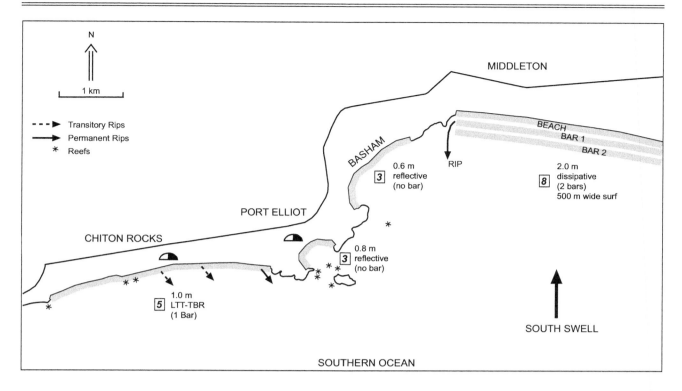

Figure 3.4   *Breaker wave conditions, beach type and the hazard level can change considerably between adjoining beaches. For example the few kilometres of coast between Goolwa and Parsons Beach have a considerable range in wave and beach conditions owing to changes in wave height and sand size. Goolwa receives waves averaging over 2 m and has a hazardous 500 m wide double bar system. Port Elliot beach is protected and has low energy reflective beaches with usually no surf, while Waitpinga and Parsons are again exposed, but with coarser sand than Goolwa, have very well developed rip systems. Beach hazard rating ranges from a high 8 on Middleton, to 7 at Waitpinga and Parsons, to a low 3 at Port Elliot.*

strong winds. These will generally increase the rating by 1, with Table 3.3 providing a matrix for calculating wind hazard.

Beach Hazard Ratings

   1 - least hazardous beach

   10 - most hazardous beach

Beach hazard rating is the scaling of a beach according to the physical hazards associated with its beach type and local beach and surf environment.

Modal beach hazard rating is based on the beach type prevailing under average or modal wave conditions, for a particular beach type (Figs 2.4, 2.5, 2.13, 2.15) or beach.

Prevailing beach hazard rating refers to the level of beach hazard associated with the prevailing wave and beach conditions on a particular day or time.

Table 3.4 summarises the rating for all South Australian beaches.

The hazard rating of South Australian beaches reflects the highly variable nature of the coast. The greatest number

of beaches (320) have a low rating of 1, indicative of the low energy, low gradient conditions in the gulfs and protected bays. At the other extreme, 11 exposed, rocky, high energy beaches have a maximum rating of 10. In between, the 797 beaches rating 3 or less are located in the gulfs and protected bays and include some of the protected, reflective, open coast beaches. Two hundred and seventy nine beaches rate a moderately hazardous 4 to 5, while 378 (26%) rate a high 6 to 10. The higher rating beaches are all dominated by rip currents, including topographic rips and additional hazards such as rocks and reefs.

Compared to other southeastern states, South Australia has the highest number of hazardous beaches, particularly those rating 8 to 10 (Table 3.5). Overall South Australia has 503 beaches rating 5 or above, compared to 413 in Victoria, 439 in NSW and 46 in Queensland. Clearly the high waves of the Southern Ocean produce more hazardous beaches than those along the east coast. South Australia also has a high number of beaches with low hazard ratings, a function of the protected gulfs and larger bays. These are comparable to the protected beaches of Port Phillip Bay and the Queensland beaches in lee of the Great Barrier Reef. Such protected beach types are, however, absent from the open Victorian and NSW coasts.

Table 3.4   South Australian beach hazard rating, by number of beaches and their length (excluding wind hazard) for the mainland coast (SA) and for Kangaroo Island (KI).

Beach Hazard Rating	No. of beaches		% of total beaches		Mean length (km)		Total length (km)		Proportion of length of beaches (%)		
	SA	KI	SA	KI	SA	KI	SA	KI	SA	KI	Total
1	320	14	22	6	1.8	2.7	581	38	29	24	28
2	271	15	19	7	1.3	0.6	349	9	17	6	16
3	206	22	14	10	1.0	0.8	202	18	10	12	10
4	154	19	11	9	0.6	0.7	97	13	5	8	5
5	125	29	9	13	0.7	0.2	88	6	4	4	4
6	93	39	6	18	1.0	0.4	91	14	5	9	5
7	137	6	9	3	1.6	3.4	215	21	11	13	11
8	117	23	8	11	2.8	0.7	333	16	16	10	16
9	20	14	1	6	3.1	0.8	62	11	3	7	3
10	11	37	1	17	0.3	0.3	3	9	<1	6	<1
TOTAL	1454	218	100	100	1.4	0.7	2020	156	100	100	100

Table 3.5   Beach hazard rating of South Australian, Victorian, NSW and Queensland beaches (Source: Short, 1993, 1996, 2000)

Beach Hazard Rating	South Australia	Victoria	Victoria: Port Phillip Bay	NSW	Queensland: Southeast Coast	Queensland: in lee of Great Barrier Reef*
10	11	2	0	3	0	0
9	20	11	0	0	0	0
8	117	77	0	7	1	0
7	137	148	0	112	9	0
6	93	109	0	232	23	0
5	125	66	0	85	12	1
4	154	85	7	134	7	38
3	206	53	37	103	17	216
2	271	9	27	45	33	352
1	320	0	61	0	20	280
Total beaches	1454	560	132	721	122	887

* south of Cooktown.

3.1.4   Rip identification - how to spot rips

To the experienced surfer rips are not only easy to spot, but they are the surfer's friend, providing a quick way (and at times the only way) to get 'out the back' beyond the breakers, as well as carving channels to produce better waves. To the inexperienced however, rips are not only unknown or 'invisible' to them, but if caught in one it can be a terrifying and even fatal experience. Most recreational swimmers and visitors do not have the time or desire to become experienced swimmers and surfers. In order to assist them, a check list of features that indicate a rip or rips are present on the beach is noted below:

Rip Current Spotting Check List

Note: any one of these features indicates a rip, but not all will necessarily be present.
   * indicates always present
   + indicates may be present

---

* A seaward movement of water (Fig. 3.2) either at right angles to or diagonally across the surf. To check on currents, watch the movement of the water or throw a piece of driftwood or seaweed into the surf and follow its movement.

* Rips only occur when there are breaking waves seaward of the beach. If water is moving shoreward, it must return seaward somewhere.

* Disturbed water surface (ripples or chop) above the rip, caused by the rip current as it pushes against incoming waves and water. May be difficult to spot.

+ Longshore rip feeder channels and/or currents running alongshore, hard against the base of the sloping beach face. Rips are usually supplied by two rip feeder channels, one on either side of the rip.

+ Rhythmic or undulating beach topography, with the rips located in the indented rip embayments.

+ Areas where waves are not breaking, or are breaking less across a surf zone, owing to the deeper rip channel.

+ A deep channel or trough, usually located between

two bars or against rocks. The channel may contain inviting, clear, calmer water compared to the adjacent turbulent surf on the bars. However, do not be fooled. In the surf, calm water usually means it's deeper and contains currents.

+ A low point in the bar where waves are not breaking, or break less. This is where the rip channel exits the surf zone.

+ Turbid, sandy water moving seaward, either across the surf zone and/or out past the breakers.

+ In the rip feeder and rip channel, the stronger currents produce a rippled seabed. These ripples are called megaripples and are sandy undulations up to 30 cm high and 1.5 m long. If you see or can feel large ripples on the seabed, then strong currents are present, so stay clear.

+ If you see one rip on a long beach, there will be more if wave height remains the same along the beach. Rip spacing can vary from 150 to 500 m, depending on wave conditions.

+ If there is surf and rocks, reef or a headland, a rip will always flow out close to or against the rocks. These are called **topographic rips**, in that they are controlled by the topography of the headland or reef. These rips are often permanent and have local names like *the express, the accelerator, the garbage bowl and the alley.*

3.1.4.1 Rips - how to escape if caught in one

• If the water is less than chest deep, adults should be able to walk out of a rip. Conversely, avoid going into deeper water. So if you are in any surf current, become very careful once the water exceeds waist depth. Get out while the water is shallow.

• Most people rescued in rips are children. Never let them out of your sight and if they get into difficulties, go to them immediately while the water is still relatively shallow.

• As long as you can swim or float, the rip will not drown you. There is no such thing as an undertow associated with rips, or for that matter, with surf zone currents. Only breaking waves can drive you under water. Most swimmers who drown when caught in rips do so because they panic. So stay calm, tread water, float and conserve your energy.

• If there are people/lifesavers on the beach, raise one arm to signal for assistance.

• Do not try to swim or wade in deep water directly against the rip, as you are fighting the strongest current. There are easier ways out.

• Where possible, wade rather than swim, as your feet act as an anchor and help you fight the current.

• If it is a relatively weak and/or shallow rip, swim or wade sidewards to the nearest bar. Once on the bar, walk or let the waves or wave bores return you to shore.

• If it is a strong and/or deep rip, go with it through the breakers. Do not panic. When beyond the

breakers, slowly and calmly swim alongshore in the direction of the nearest bar, indicated by heavier wave breaking. If you are not a surfer, simply wait for a lull in the waves, swim into the break and allow the waves to wash you to shore. Stay near the surface so the broken wave can wash you shoreward. Do not dive under the waves as they will wash over you.

• To summarise: stay calm, swim sideways toward breakers or the bar and let the broken waves return you to shore. Raise an arm to signal for help if people are on the beach.

• If rescued, thank the rescuer.

3.2   Surf Safety

Surfing, as opposed to swimming, requires the surfer to go out to and beyond the breakers, so he or she can catch and ride the waves, in other words, go surfing. This can be done using just your body (body surfing) or a range of surfboards, bodyboards and skis.

Surfing safely requires a substantially greater knowledge of the surf, compared to swimming on the bar or close to shore. The following points should be observed before you begin to surf:

1. You must be a strong swimmer.
2. You must also be experienced at swimming in the surf.
3. You must be able to tell if and when a wave will break.
4. You must know the basics of how breaking waves and the surf zone operate. You should be able to spot rip currents.
5. Equally, you should know what hazards are associated with the surf, including breaking waves, rips, reefs, rocks and so on.
6. You must only use equipment that is suitable for you, i.e. the right size and level.
7. You must know how to use your equipment, whether it be flippers, bodyboard, surfboard or wave ski.
8. You should use safety equipment as appropriate, including a legrope or handrope, wetsuit, flippers and in some cases a helmet.
9. You should ensure your equipment is in good condition, with no broken fibreglass, frayed legrope, etc.

3.2.1   Some tips on safe surfing:

• Remember surfing is fun, but it is also hazardous.
• Never surf alone. If you get into trouble, who will help you?
• Before you enter the water, always look at the surf for at least five minutes. This will enable you to first, gauge the true size of the sets, which may come only every few minutes; and secondly, besides picking out the best spot to surf, you can also check out the breaker pattern, channels, currents and rips;

Australian Beach Safety and Management Program

in other words, the circulation pattern in the surf. This is important as you can use this to your advantage.

- On patrolled beaches, observe the flags, surfboard signs and directions of the lifesavers. Do not surf between the red and yellow flags or near a group of swimmers.
- If you are surfing out the back, the safest, quickest and usually the easiest way to get out is via a rip. This is because the water is moving seaward, making it easier to paddle; the rip flows in a deeper channel, resulting in lower waves; and the rip will keep you away from the bar or rocks where waves break more heavily.
- Once out past the breakers, particularly if paddling out in a rip, move sidewards and position yourself behind the break.
- Buy and read the Surf Survival Guide, published by Surf Life Saving Australia and available at all newsagents.
- Obtain an SLSA Surf Survival Certificate from your school.

3.2.2   Some general tips:

Surfers conduct many rescues around the Australian coast, so be prepared to assist if required. Remember, if you are on a surfboard, bodyboard or wave ski and have a legrope and wetsuit, you are already kitted out to perform rescues. The board is a good flotation device that can be used to support someone in difficulty. The wetsuit will keep you warm and buoyant and thereby give you more energy and flotation to assist someone in distress; and the legrope or board can be used to tow someone in difficulty, while you paddle them toward safety.

The simplest way to get someone back to shore is to lay them on your board while you swim at the side or rear of the board and let the waves wash the board, patient and you back to shore.

Some surfing hazards to watch out for when paddling out:

---

* Heavily breaking/plunging waves, particularly the lip of breaking waves.
* Rocks and reefs.
* Strong currents, particularly in big surf.
* Other surfers and their equipment.

...when you are surfing:

---

* Other surfers - the surfer on the wave has more control and is responsible for avoiding surfers paddling out, or in the way.
* Heavily breaking waves.
* Your own and other surfboards. They can and do hit you and can knock you out.
* Shallow sand bars.
* Rocks and reefs.
* Close-out sets and big surf.

...when returning to shore:

---

* Heavy shorebreaks.
* Rocks and reefs.
* Strong longshore/rip feeder currents.

Remember: The greatest danger to surfers is to be knocked out and drown. Most surfers are knocked out by their own boards or by hitting shallow sand or rocks. This can be avoided by always covering your head with your arms when wiped out, by wearing a wetsuit for flotation, by wearing a helmet and by surfing with other surfers who can render assistance if required.

Sharks

This book does not deal with biological hazards. However some mention must be made of sharks, as there were two fatal attacks on surfers by white pointer sharks on consecutive days, in 2000, on the western Eyre Peninsula at Cactus and Blackfellows near Elliston. A month later another fatal attack followed at Cottesloe, a small Perth beach patrolled by two surf clubs and there have since been attacks in Queensland and New South Wales, totalling five attacks over a six month period.

There is no way of avoiding sharks once you enter their territory. If you are concerned about sharks then it is best to stay out of their domain. However all surfers and divers and many swimmers are prepared to spend some time in the ocean with the knowledge that the chances of being attacked are extremely small. On average only one person is taken in all of Australia each year. Hopefully 2000-2001 was an unfortunate aberration to this statistic.

The best reference for biological hazards is:
Venomous and Poisonous Marine Animals - a Medical and Biological Handbook, 504 pp.
　　　edited by J A Williamson, P J Fenner, J W Burnett and J F Rifkin published by University of New South Wales Press, Sydney, 1996
　　　ISBN 0 86840 279 6
Available through UNSW Press or the Medical Journal of Australia.
This is an excellent and authoritative text, which provides the most extensive and up-to-date description and illustrations of these marine animals and the treatment for their envenomation.

3.3   Rock fishing safety

The rocky sections are the most hazardous part of the South Australian coast, with most fatalities due to fishermen drowning after being washed off the rocks.

Rock fishing is hazardous because:

- Deep water lies immediately off the rocks, often containing submerged reefs and rocks and heavily breaking waves.
- Occasional higher sets of waves can wash unwary fishermen off the rocks.
- Most fishermen are not prepared or dressed for swimming, as they are often wearing heavy waterproof clothing, shoes and tackle.
- Many fishermen are not experienced surf swimmers and many cannot even swim.

To minimise your chances of joining this distressing statistic, two points must be heeded. Firstly, avoid being washed off and secondly, if you are washed off, make sure you know how to handle yourself in the waves until you can return to the rocks or await rescue.

The biggest problems usually occur when inexperienced fishermen are washed off rock platforms. To compound the problem, they either cannot swim or are not prepared for a swim. You only need to watch experienced board and body surfers surfing rocky point and reef breaks, to realise rocks are not a serious hazard to the experienced and the properly equipped.

So the rules are:-

1.   Before you leave home:

- Check the weather forecast. Avoid rock fishing in strong winds and rain.
- Phone the boat or surf forecast and check the wave height. Avoid waves greater than 1 m.
- Check the tide state and time. Avoid high and spring tides.
- Are you suitably attired for rock fishing, particularly footwear?
- Are you suitably attired in case of being washed off the rocks?
- A loose fitting wetsuit is both comfortable and warm, and it will keep you afloat and protected if washed off the rock platform.

2.   Before you start fishing:

- Check the waves for ten minutes, particularly watching for bigger sets.
- Choose a spot where you consider you will be safe.
- When choosing a spot to fish: if the rocks are wet, then waves are reaching that spot, if the rocks are dry, waves are not reaching them, but may if the tide is rising or wave height is increasing.
- Ensure you have somewhere to easily and quickly retreat to, if threatened by larger waves.
- Place your tackle box and equipment high and dry.

3.   When you are fishing:

- Never turn your back on the sea, unless it is a safe location.
- Watch every wave.
- Be aware of the tide, if it is rising, the rocks will become increasingly awash.
- Watch the waves, to check for:
    - increasing wave height, leading to more hazardous conditions;
    - the general pattern of wave sets, it is the sets of higher waves that usually wash people off rocks.
- Remember, 'freak waves' exist only in media reports. No waves are freak, all that happens is that a set of larger waves arrives, as any experienced fisherman or surfer can tell you. These larger sets are likely to arrive every several minutes.
- Do not fish alone, two can watch and assist better than one.
- If you see a larger set of waves approaching - retreat. If you cannot retreat, lie flat and attach all your limbs to the rock. Forget your gear, you are more valuable. As soon as the wave has passed, get up and retreat.
- Wear sensible clothing. A wetsuit provides warmth, protection and safety, particularly if you are washed off or knocked over. Sandals with cleats to prevent slipping are also popular.

4.   If you are washed off, here are some hints:

- If you have sensible clothing, that is, clothing that will keep you buoyant, such as a wetsuit or life jacket, then you should do the following:
- Head out to sea away from the rocks, as they are your greatest danger.
- Abandon your gear, it will not keep you afloat.
- Take off any shoes or boots and you will be able to swim better.
- Tread water and await rescue, assuming there is someone who can raise the alarm.
- If you are alone, or can only be saved by returning to the rocks, try the following:
- Move seaward of the rocks and watch the waves breaking over the rocks in the general area, then:

Choose a spot where there is either:

a channel - this may offer a safer, more protected route;

a gradually sloping rock - if waves are surging up the slope, you can ride one up the slope, feet and bottom first, then grab hold of the rocks as the swash returns;

or a steep vertical face with a flat top reached by the waves - swim in close to the rocks, wait for a high wave that will surge up to the top of the rocks, float up with the wave, then grab the top of the rocks and crawl onto the rock as the wave peaks. As the wave drops, you can stand or crawl to a safer location.

4.   SOUTH AUSTRALIAN BEACHES

This chapter contains a description of every beach on the South Australian coast arranged in sequential order from east to west, followed by Kangaroo Island in chapter 5. All beaches are located by number on thirteen regional maps (Fig. 4.1a), while individual beach maps and photographs are provided of all beaches patrolled by surf clubs and/or lifeguards, as well as several popular beaches. The beaches are also arranged into the state's seven Coast Protection Districts, illustrated in Figure 4.1b and listed below, together with page numbers for each district:

Coast Protection District	Pages
1. South East	65-95
2. Fleurieu Peninsula	97-110
3. Metropolitan (Adelaide)	111-126
4. Yorke Peninsula	127-167
5. Upper Spencer Gulf	169-176
6. Eyre Peninsula	177-285
7. Kangaroo Island	287-321

To find a beach, you can use four systems:

1.   The alphabetical **BEACH INDEX** at the rear of the book lists all beaches.
2.   Use the **REGIONAL MAPS** or **COAST PROTECTION DISTRICT** of the particular section of coast to locate a beach, or follow the beaches along the coast until you find the beach.
3.   By name of the **SURF LIFESAVING CLUB**. If the beach has a surf club it will be listed in **BOLD** in the BEACH INDEX. If it differs from the beach name, then both are listed in the BEACH INDEX.
4.   By name of a surfing break. If a popular surfing break has a name, which may differ from the beach name, then use the **SURF INDEX**.

Note that most South Australian beaches have **no official name**. Where this is the case the beach has been assigned its beach number or given a name associated with the nearest named location. If you see any such beaches in the book that have a local name, please convey this information to the author, so that subsequent publications can be brought up to date. Contact details are provided on page 48.

The description of each beach contains the following information:

- **Patrolled** or **unpatrolled** by lifesavers or lifeguards
- **Patrol months** on patrolled beaches
- **No.**   Beach number: 1 to 1454 on the mainland coast, plus 218 beaches on Kangaroo Island and 83 beaches on five major islands.
- **Beach** name or names (or Beach number if no known name)
- **Surf Lifesaving Club:** name of surf lifesaving club (if present)
- **Lifeguards:** lifeguard patrol hours (if present)
- **Rating** (beach hazard rating): 1 to 10
- **Type**: see pages 30 to 43 for full explanations of beach types
  wave dominated   (D, LBT, RBB, TBR, LTT, R) including some double bar systems with **inner & outer bar**
  tide modified   (UD, R+BR, R+LTT)
  tide dominated   (R+SR, R+SF, R+TSF)
- **Length**   length of beach

Beach Types

Wave dominated beach types
D	dissipative
LBT	longshore bar & trough (also as outer bar)
RBB	rhythmic bar & beach (also as outer bar)
TBR	transverse bar & rip (also as outer bar)
LTT	low tide terrace
R	reflective

Tide modified beach types
UD	ultradissipative (none in South Australia)
R+BR	reflective+bar & rips (none in South Australia)
R+LTT	reflective+low tide terrace

Tide dominated beach types
R+SR	(HT)	beach+sand ridges
R+SF	(HT)	beach+sand flats
R+TSF	(HT)	beach +tidal sand flats

Other comments
   sediment size (cobbles, boulders)
      +rocks (scattered rocks on beach & in surf)
      +rock flats (intertidal rock/calcarenite flats)
      +rock reef (submerged rocks/calcarenite in surf and/ or nearshore)

South Australia Regional Maps

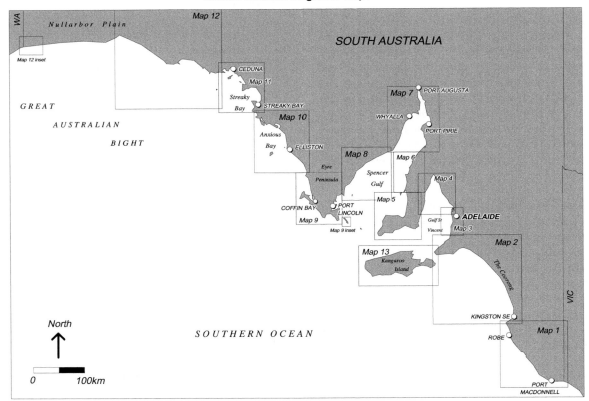

South Australian Coast Protection Districts

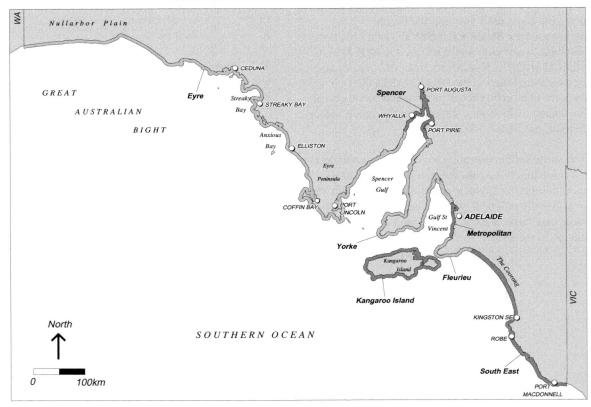

Figure 4.1a) *Map of South Australia showing the coverage of the 13 regional maps; and b) location of South Australia's seven Coast Protection Districts: South East, Fleurieu, Metropolitan, Yorke Peninsula, Spencer Gulf, Eyre Peninsula and Kangaroo Island.*

Australian Beach Safety and Management Program

- **Inner bar:** refers to inner sand bar, close to or attached to the beach
- **Outer bar:** refers to the outermost sand bar on double bar beaches, lying seaward of the inner bar and usually separated by a trough.

Comments on beach location, access, length, beach and surf conditions, including beach type and presence and location of rips.

Specific comments are made for most beaches on:
- **Swimming** - suitability and safety
- **Surfing** - good surfing spots
- **Fishing** - presence and location of gutters and holes
- **Summary** - general overview of beach

Surf lifesaving club patrol dates and hours

In South Australia, all 19 surf lifesaving clubs voluntarily patrol the beach on weekends and public holidays over summer (see Table 3.2).

Patrols generally commence in early November and run each weekend until the end of March. Most clubs commence patrols between 12 noon and 1 pm each day and finish between 5 pm and 7 pm.

Lifeguards and Contract Lifesavers

Surf Life Saving South Australia lifeguards at this time patrol Glenelg (from Glenelg to Brighton via 4WD) from December through to the end of March and at Normanville for the month of January (see Table 3.2). The patrols work on a Monday to Friday basis over the peak summer holiday period.

Most lifeguards patrol all weekdays between the hours of 10 am and 6 pm.

SOUTH EAST DISTRICT
GLENELG RIVER TO MURRAY MOUTH
(Figs. 4.2 and 4.22)

Length of Coast:	410 km
Beaches:	148 beaches (1-148)
Surf Lifesaving Clubs:	none
Major towns:	Port MacDonnell, Southend, Beachport, Robe, Kingston SE

The South East District of South Australia contains 410 km of some of the most exposed, high energy coast in the country. Most of this section faces into the prevailing high southwest waves and persistent westerly winds. Exposed beaches are all dominated by wide, energetic surf zones and are backed by massive dune systems. In addition, limestone in the far southeast dominates the coast between the border and Cape Banks, while calcarenite reefs and cliffs influence much of the remainder. Even the 194 km long Coorong beach between Cape Jaffa and Murray Mouth is modified by calcarenite reefs, which substantially lower waves along more than half its length.

Regional Map 1: Victorian Border to Cape Jaffa

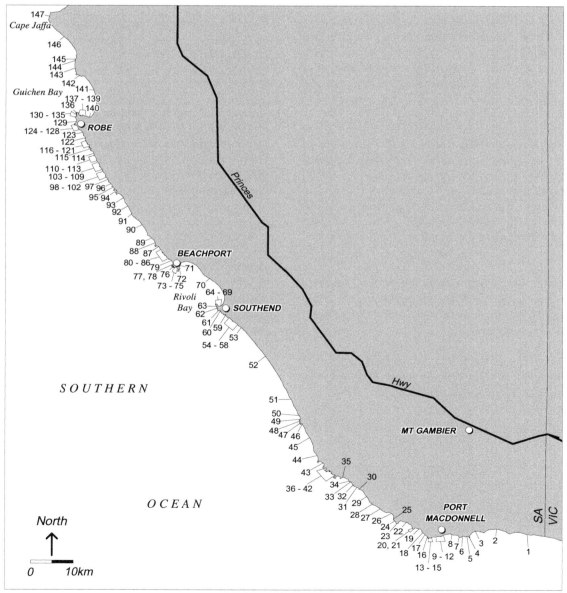

Figure 4.2   *Regional map 1 - Glenelg River (Victorian border) to Cape Jaffa*

Australian Beach Safety and Management Program

1   PICCANINNIE PONDS

No.	Beach	Unpatrolled Rating	Type		Length
1	Piccaninnie Ponds	7	Inner LTT/TBR	Outer Bar RBB/LBT	8.2 km

The first beach in South Australia is part of the large Discovery Bay beach system, which extends from Cape Bridgewater in Victoria for 43 km to the border, then another 8 km through to Green Point in South Australia. It has a total length of 51 km, making it one of Australia's longer beaches. The Glenelg River mouth is located 2 km east of the border. The South Australian section of the beach, **Piccaninnie Ponds (1)**, therefore extends from the border for 8.2 km to the low, sandy Green Point.

The beach is readily accessible via the Piccaninnie Ponds Road, with basic camping facilities at the ponds, located 500 m behind the beach, with a small car park at the beach. The freshwater ponds are up to 85 m deep and are part of the Piccaninnie Ponds Conservation Park. Their overflow drains across the beach 200 m west of the car park.

The beach faces south, with some longshore crenulations in response to offshore reefs, with Green Point protruding up to 1 km in lee of the most substantial reefs. It is composed of fine white sand and receives waves averaging 1.5 to 2 m in the east, decreasing to the west in lee of the reefs off Green Point. The high waves and fine sand maintain a 200 m wide surf zone containing an inner bar cut by several widely spaced rips, a longshore trough then a wider, rip-dominated outer bar.

Swimming: This beach usually has spilling breakers rolling across a low gradient inner surf zone, combined with widely spaced rips. Swim well clear of the rips, but beware as it is unpatrolled.

Surfing: One of the few accessible surfing beaches in Discovery Bay, usually offering low spilling waves in close and some larger waves out the back.

Fishing: There are usually a few rip holes along the beach.

Summary: A relatively popular spot for beach fishing, with good access but no facilities. It's a lovely natural beach but beware of the rips in the surf.

Piccaninnie Ponds Conservation Park
Established: 1969
Area: 397 ha
Beach: in lee of Beach 1

An attractive series of freshwater ponds underlain by subterranean caves, lying in a low, sandy coastal plain just behind the beach and dunes. Basic camping facilities provided. Snorkel and dive permits are required to enter the ponds.

Further information:
Department of Environment and Natural Resources
Mount Gambier office

2,3   BROWN BAY, RIDDOCH BAY

No.	Beach	Unpatrolled Rating	Type	Length
2	Brown Bay	6	LTT/TBR	6.3 km
3	Riddoch Bay	4	LTT	4.0 km

Brown Bay and Riddoch Bay are two gently curving embayments facing due south into the Southern Ocean, but partly protected by extensive shallow limestone reefs. Brown Bay is 6.3 km long and bordered by the low Green and Danger Points, with 4 km long Riddoch Bay extending from Danger Point to Stony Point. The Point Road runs 200 to 300 m behind the beach providing several unsealed access points and beachfront parking. There are no facilities at either beach.

Brown Bay beach (**2**) consists of fine white sand which produces a wide, low gradient beach and surf zone. Waves are lowered by the offshore reefs and average 1 to 1.5 m, decreasing in height toward either end. In the central section they maintain a 200 m wide continuous bar, cut by rips every few hundred metres.

Riddoch Bay (**3**) has more continuous offshore reefs and consequently lower waves at the shore, averaging about 0.5 m, which maintain a continuous bar and usually no rips. Eight Mile Drain cuts the eastern end of the beach, with the smaller Deep Creek toward the western end. The track adjacent to the drain provides beach access and the area is used as a camp site. The low gradient sandy beach grades into a steep cobble beach at the aptly named Stony Point.

Swimming: Be careful in Brown Bay as rips frequently occur along the beach, while Riddoch Bay is usually free of rips with lower waves, however there is often a deep hole at the mouth of Eight Mile Creek.

Surfing: Best in Brown Bay with usually a reduced swell breaking across the bar.

Fishing: The Brown Bay rip holes and the Eight Mile and Deep Creek mouths are the most popular spots.

Summary: Two readily accessible beaches offering clean white sand but no facilities.

4-7 RACECOURSE BAY

Unpatrolled			
No. Beach	Rating	Type	Length
4 Stony Point (W)	2	LTT	0.3 km
5 Racecourse Bay	3	LTT+rock flats	0.8 km
6 French Point (E)	3	LTT+rock flats	1.1 km
7 French Point (W)	3	LTT+rock flats	0.7 km

Racecourse Bay is a 2 km wide, south facing bay bordered by the low Stony and French Points. It is fronted by continuous shallow limestone reefs and substrate, which both lower waves and at low tide are exposed as extensive intertidal rock flats, as well as supplying the limestone cobble to build the high tide cobble beach ridges that run the length of the crenulate bay shoreline. Creek Road parallels the back of the beaches, with unsealed access points and informal car parking at each beach and a larger car park on the western side of French Point.

The crenulations divide the bay into three beach systems. Immediately west of **Stony Point** is a 300 m long, southwest facing high tide sand beach (**4**), fronted by a low sandy bar then rock flats and reefs. Immediately west is the 800 m long central section of **Racecourse Bay**, which has a high tide cobble beach (**5**) fronted by a narrow strip of sand and rock flats. The western beach (**6**) runs in a similar fashion for 1.1 km to the cobble ridges of **French Point**, with rock flats extending 200 m off the point. A cobble-sand beach (**7**) continues on the western side of the point for another 700 m. The access road just east of the point is used to launch boats across the beach, which are also moored off the beach in lee of the reefs.

Swimming: Three relatively protected beaches with waves usually less than 0.5 m and no rips. However, beware of the extensive intertidal to submerged rock reefs.

Surfing: None on the beaches, but a chance of some waves over the reefs off the points.

Fishing: These are shallow beaches with most fishing done from boats over the outer reefs.

Summary: Three undeveloped but easily accessible beaches best suited for a quiet picnic and wade.

8 ORWELL ROCKS

Unpatrolled			
No. Beach	Rating	Type	Length
8 Orwell Rocks	4	LTT+rips	2.3 km

Orwell Rocks are a series of 5 m high eroding limestone bluffs, which have been undercut to form small caves and rocks and reefs off the shore. They form the eastern boundary of a 2.3 km long, south facing beach (**8**) that extends west to the first of the Port MacDonnell groynes. Creek Road parallels the back of the beach, with the Port MacDonnell caravan park and sailing club located behind the centre of the beach. Two small drains, Jerusalem and Cress Creeks, also cross the beach.

The beach is a low gradient sand beach fronted by a 50 m wide bar, which can be cut by rips toward the centre of the beach. Waves average 0.5 m in the centre, increasing slightly to the east, while they decrease toward the harbour.

Swimming: The safest swimming is between the harbour and caravan park, with high waves and occasional rips toward the eastern end.

Surfing: Usually only a low beach break in the Orwell Rocks area.

Fishing: Best around the rocks at high tide.

Summary: The main swimming beach for Port MacDonnell. A readily accessible, slightly crenulate beach; very popular in summer with holiday-makers at the caravan park and used to launch boats at the sailing club.

9-11 PORT MACDONNELL

Unpatrolled			
No. Beach	Rating	Type	Length
9 Port MacDonnell (E)	1	LTT	250 m
10 Port MacDonnell (Hbr)	1	R + tidal flats	250 m
11 Port MacDonnell (W)	2	R+rock flats	300 m

Port MacDonnell (Fig. 4.3) was called a port long before the present harbour wall was constructed, thereby affording the harbour and boats additional protection from southern storms. The harbour was constructed in the early 1970's, cutting the beach in two. As sand has built out in lee of the harbour wall, two additional groynes have been built which further subdivide the beach. Today, the once continuous beach consists of four man-made beach sections (beaches 8, 9, 10 and 11).

Figure 4.3   *Port MacDonnell breakwater and outer reefs.*

To the east of the harbour is a 250 m long, low energy beach (**9**), wedged between two rock groynes. This beach has been building seaward since the harbour was constructed and the backing seawall built to protect the road is now stranded 200 m from the shoreline. Most of the new land has been turned into a large grassy park dominated by an old fishing boat. At the western end of the beach is a 100 m long groyne and a 200 m long jetty. The groyne has been built to prevent the boat ramp from being covered in sand. On the other side of the boat ramp is the very protected 250 m long harbour beach (**10**) which extends to the main harbour wall. This beach is essentially calm and often covered by seagrass debris. Small boats are moored off this beach, with larger boats out in the harbour.

On the western side of the harbour wall is another new beach (**11**) that has been building out as longshore movement has piled up to 200 m of sand against the wall. This low energy, 300 m long beach faces southwest and is fronted by extensive rock flats.

Swimming: The only harbour beach suitable for swimming is the eastern beach (9) backed by the large park. The other two are usually covered in seagrass and are too shallow.

Surfing: None.

Fishing: Most shore fishing is done off the harbour wall and jetty, or outside in boats.

Summary: These three beaches occupy the Port MacDonnell waterfront. All three have been heavily modified by the various harbour walls, groynes, seawalls and jetty and provide a variety of facilities for swimming, picnics, boat launching and jetty fishing.

12   SPRINGS ROAD

No.	Beach	Rating	Type	Length
	Unpatrolled			
12	Springs Road	2	R + rock flats	600 m

Between Port MacDonnell Harbour and Cape Northumberland is a 600 m long sandy beach (**12**), part of which has built out up to 100 m onto the fronting rock flats. Creek Road backs part of the beach, with vehicle tracks running through the scrub across the wider section to the usually calm shoreline. Owing to the rock flats, it is unsuitable for swimming or surfing and can only be fished at high tide.

13-15 CAPE NORTHUMBERLAND (E)

No.	Beach	Rating	Type	Length
	Unpatrolled			
13	Smiths Road	3	R+rock flats	120 m
14	Frog Rock	3	R+rock flats	220 m
15	Cape Northumberland (E)	3	R+rock flats	90 m

Immediately west of Port MacDonnell is the prominent 20 m high Cape Northumberland and its lighthouse. The original lighthouse was constructed in 1859 and rebuilt in 1882. The cape is composed of lithified Pleistocene dunes, which have been cliffed by wave attack. Along the eastern base of the cliffs are three small pocket beaches, each lying below 10 to 20 m high craggy cliffs and bordered (and in places fronted) by bluffs, sea stacks and reefs. The three are all accessible off the Lighthouse Road, with car parking above each beach. The most popular and accessible is Frog Rock.

Smiths Road beach (**13**) is 120 m long, faces south and is fronted by 50 m wide rock flats that are exposed at low tide. Immediately to the west is Frog Rock, a sea stack which lies just off a 220 m long beach (**14**), cutting it in two at low tide. The steep, narrow, sandy beach has exposed rocks along its base and patchy rocks and reef offshore. **Cape Northumberland** beach (**15**) is a 90 m long, east facing pocket of sand fronted by a mixture of rock flats and some sand patches. A small penguin colony is located on the eastern side of the cape.

Swimming: While all are suitable for sunbathing, swimming can be a problem owing to numerous rocks. Use caution and stay close inshore on the sandy patches.

Surfing: None on the beach, but there are some reef breaks off the beaches and cape.

Fishing: Three popular locations providing easy access to the reef flats, holes and gutters.

Summary: The cape is a scenic area and these three beaches are highly visible and accessible.

16-18 CAPE NORTHUMBERLAND (W)

Unpatrolled			
No. Beach	Rating	Type	Length
16 Cape Northumberland (W 1)	3	R+rock flats	400 m
17 Cape Northumberland (W 2)	3	R+rock flats	900 m
18 Finger Point	3	R+rock flats	650 m

On the west side of **Cape Northumberland** the craggy cliffs continue for 1 km, then decrease in height as dunes back the shoreline. Between the cape and Finger Point are three southwest facing beaches, each fronted by extensive rocks and reefs resulting in only low waves at the shore. The beaches are all backed by active dunes, which extend up over the backing slope and inland for up to 2.5 km. A sealed road runs along the back of the first beaches, with five sealed car parks providing excellent access, but no other facilities. The Mount Gambier sewerage treatment works is located in lee of Finger Point and discharges treated sewage into the surf off the point.

Cape Northumberland (W 1) beach (**16**) is 400 m long and backed by steep, 20 m high bluffs. There is a large car park above the centre of the beach and a steep, sandy track down to the steep, narrow beach. The beach is fronted by rocks and reef flats. The second beach (**17**) is 900 m long with four parking areas behind the beach. The beach is up to 30 m wide, backed by low sand dunes and fronted by rock flats and some sand patches for swimming. **Finger Point** beach (**18**) is 650 m long, with a car park at the eastern end and the sewer outlet at the western end. Rock flats front the entire beach, with waves increasing toward the western end.

Swimming: Best at high tide in the sandy patches.

Surfing: None at the beach, with the reef surf off Finger Point the best in the area.

Fishing: Excellent access to the beaches and the holes and gutters in the rocks and reefs.

Summary: Three very accessible beaches popular with fishers and daytrippers.

19   BLANCHE BAY

Unpatrolled			
No. Beach	Rating	Type	Length
19 Blanche Bay	2	R+rock flats	2 km

Blanche Bay is a 2 km wide, south facing bay lying between the low Finger and Middle Points. The beach (**19**) is backed by low, well vegetated foredune ridges and fronted by extensive patchy rock reefs, both at the beach and out in the bay, which result in lower waves and a crenulate sandy shoreline. At the shore, waves are low to calm with seagrass commonly washed up onto the beach. Access to both the bay and the point is restricted to sandy 4WD tracks. Consequently the beach is little used.

Swimming: Relatively safe in the bay owing to the mostly low waves at the shore. Best toward the centre and east where there is more sand than reef along the shore.

Surfing: None in the bay.

Fishing: Most fishing is done from boats out over the reef.

Summary: A relatively remote bay and point, mainly used by local fishers and surfers.

20, 21 MIDDLE POINT

Unpatrolled			
No. Beach	Rating	Type	Length
20 Middle Pt (E)	5	R+rock flats	500 m
21 Middle Pt (W)	5	R+rock flats	200 m

Middle Point is a low, round limestone point which protrudes to the south for 600 m and forms the boundary between Blanche and Umpherstone Bays. It is capped by both active and deflated dunes. The eroding point is fronted by rock flats up to 200 m wide, with eroded limestone cobbles forming two steep, narrow, crenulate beaches on either side of the point (**20, 21**). Intertidal rock flats front the beaches, with heavy surf on the outer rocks. Access is via 4WD tracks from both bays that lead to the rear of both cobble beaches.

Swimming: These beaches are usually unsuitable for swimming because of the cobbles, rock flats and the exposed nature of the point.

Surfing: There are a variety of reef breaks off the point.

Fishing: A popular spot with locals who fish the rock holes and gutters around the point.

Summary: An eroding, denuded point providing a good view of the Southern Ocean wave action.

22   UMPHERSTONE BAY

Unpatrolled			
No.	Beach	Rating	Type	Length
22	Umpherstone Bay	2	R+rock flats	3.5 km

Umpherstone Bay is an open, south facing, 3 km long bay containing continuous rock reefs across its mouth. The reefs extend up to 1 km off the beach and lower waves to near zero at the shoreline, producing a low and usually calm beach (**22**) covered in seagrass. The Point Douglas road enters the centre of the beach and continues on to Point Douglas, with 4WD tracks leading off the gravel road to Middle Point and west of Point Douglas. There are approximately 50 shacks lining the central and western shores of the bay, with their small fishing boats commonly lying at anchor off the beach.

Swimming: A safe and usually calm beach, but very shallow, with the added problems of rock reefs and seagrass.

Surfing: None along the beach.

Fishing: Most fishers head out to the reefs in their boats.

Summary: A quiet bay with good access, shacks and a backdrop of agricultural land.

23, 24 DOUGLAS POINT

Unpatrolled			
No.	Beach	Rating	Type	Length
23	Douglas Pt (E)	5	cobble+rock flats	150 m
24	Douglas Pt (W 1)	5	cobble+rock flats	150 m

Douglas Point is a 2 km long, 10 to 20 m high, calcarenite capped limestone point which forms the western boundary of Umpherstone Bay. The calcarenite gives it an irregular, jagged shoreline, mostly fronted by cliffs and bluffs, with two small beaches. The eastern beach (**23**) is located at the end of the Point Douglas road and consists of a mixed cobble and sand high tide beach fronted by up to 200 m of rock flats. The western beach (**24**) is a protruding cobble beach located in lee of a more substantial reef and exposed rocks.

Swimming: Neither beach is suitable for swimming owing to the rock flats and exposed location.

Surfing: There are breaks on the reefs off both beaches.

Fishing: Very popular spot for the few locals, with the bluffs surrounding the beaches providing good access to the deeper holes and gutters.

Summary: A denuded headland backed by more stable dunes extending up to 700 m inland.

25, 26 DOUGLAS POINT (W)

Unpatrolled			
No.	Beach	Rating	Type	Length
25	Douglas Pt (W 2)	2	R+rock flats	300 m
26	Douglas Pt 'bay'	2	LTT	2 km

Douglas Point forms the eastern boundary of a 2 km wide bay that faces southwest directly into strong winds and high waves. However the entire bay floor consists of limestone reefs, which, together with more prominent reefs off either end, substantially lower waves at the shoreline. The first beach (**25**) lies on the western side of the point and faces due west. It is 300 m long and protected by the point and reefs, resulting in waves averaging less than 0.5 m at the shore. These maintain a narrow, low gradient beach fronted by shallow, patchy sand and reef, and backed by vegetated bluffs, which increase in height to the north. Vehicle access and beachside parking are available at the southern end.

The main bay beach (**26**) occupies most of the bay, extending in a semicircle for 2 km to a prominent sandy, cuspate foreland. It is backed by patchy stable and unstable dune systems up to 300 m wide, while the beach has a low gradient and a 30 to 50 m wide continuous bar, which is cut by a few rips during and following higher than average waves. Usually waves average between 0.5 to 1 m. Vehicle tracks run through the dunes, with several beach access points.

Swimming: Relatively safe under normal conditions, when waves are less than 1 m. However, watch for rips along the main beach.

Surfing: Only a low shorebreak along the main beach.

Fishing: Best off the point or out over the reefs from a boat.

Summary: Two fairly remote beaches requiring local

knowledge to find the access tracks and possibly a 4WD to navigate them.

27, 28 NENE VALLEY

Unpatrolled				
No.	Beach	Rating	Type	Length
27	Nene Valley (E)	3	LTT	4 km
28	Nene Valley	3	LTT	400 m

Nene Valley is a small shack settlement (Fig. 4.4) that in the past decade has also seen subdivision for freehold houses, with a total of about 50 dwellings in 1998. It has good road access and lies 4 km off the Glyn Dale sealed road. There is a foreshore park, toilets and shower at the beach and beach boat launching, but no other facilities. Boats are usually moored off the protected beaches.

Figure 4.4   *Nene Valley is a settlement largely surrounded by the natural vegetation of the Nene Valley Conservation Park.*

Two to three metre high limestone bluffs, called the Nene Valley Rock, outcrop in front of the settlement, forming the western boundary of the 4 km long eastern beach and at either end of the 400 m long beach that fronts the western part of the settlement. There is good beach access at the settlement, otherwise a 4WD is required to drive the beach or tracks through the dunes. The eastern beach (**27**) extends from the low, sandy, cuspate foreland to the rocks. Limestone reefs extend up to 500 m seaward of the beach, reducing waves to about 0.5 m. The beach consists of a low gradient beach and narrow continuous bar, which is occasionally cut by rips, some permanently located in lee of inshore reefs.

The settlement beach (**28**) is afforded even greater protection by the reefs and conditions are usually calm at the shore. A road parallels the back of the beach, providing excellent access.

There is no 'valley' at Nene Valley, rather the location was named after the ship *Nene Valley* wrecked on reefs off the beach in 1854. The survivors had to walk 50 km to the Glenelg River, where horses were obtained to ride to Portland to report the wreck.

Swimming: Relatively safe, particularly at the Nene Valley settlement. Watch for rips when waves are breaking east of the shacks.

Surfing: Usually only a low shorebreak.

Fishing: Most fishers head out in boats to fish the reefs.

Summary: A quiet, out of the way settlement.

Nene Valley Conservation Park	
Established:	1972
Area:	373 ha
Beach:	29
Coast length:	5 km

This is a small park that consists of low and densely vegetated dunes. The park surrounds the northern and western sides of the Nene Valley settlement (Fig. 4.4) and extends for 5 km along the coast and up to 2 km inland.

29, 30 NENE VALLEY (W)

Unpatrolled				
No.	Beach	Rating	Type	Length
29	Nene Valley (W)	5	TBR	4.2 km
30	Beach 30	5	LTT+rock reefs	80 m

The western end of the Nene Valley settlement terminates at low limestone bluffs, beyond which is a 4.2 km long, low gradient, sandy beach that extends most of the way toward Blackfellows Caves. A vehicle access track at the end of the Nene Valley road provides the best access to this natural beach (**29**). It is the first beach since Piccaninnie Ponds to receive significant waves at the shore. They average 1.5 m and maintain a 200-300 m wide surf zone cut by rips every 400 m, together with permanent rips against reefs at the eastern, central and western ends. The beach is backed by stabilising Holocene dunes and crossed by the small, solitary Nene Valley Creek. It terminates at low limestone bluffs, which separate it from Beach 30.

Beach **30** is an 80 m long pocket of sand dominated by the 5 m high limestone bluffs, which form its boundaries and protrude up to 100 m off the western end. In addition the limestone forms shallow reefs off the beach. During higher waves a permanent rip drains the small surf zone.

Swimming: Be careful at both beaches as rips and energetic surf usually prevail. Stay close inshore, on the bars and clear of the rips.

Surfing: Usually a wide, spilling beach break along the main beach.

Fishing: The rocks and reefs of beach 29 and the rip holes of the longer beach are the best locations.

Summary: Two beaches mainly used by the local fishers.

31, 32 BLACKFELLOWS CAVES

Unpatrolled			
No.	Beach	Rating	Type	Length
31	Blackfellows Caves (E)	4	LTT	1.4 km
32	Blackfellows Caves (W 1)	1	cobble	400 m

Blackfellows Caves is a small fishing settlement named after the sea caves eroded into the low limestone bluffs that outcrop on the shore right at the end of the main road (Fig. 4.5). There are two beaches on either side of the outcrop. To the east is a 1.4 km long beach (**31**), which is sufficiently protected by the offshore reefs to act as a boat anchorage just off the bluffs. The reefs protect the entire beach, though wave height does increase from less than 0.5 m to up to 1 m at the eastern end, where the beach also narrows and is backed by a high tide cobble beach. A narrow continuous bar parallels the beach, with permanent reef-induced rip circulation dominating the eastern end.

To the west of the bluffs is a steep high tide cobble beach (**32**), fronted by an eroded limestone rock flat, then the reefs offshore. The reefs absorb most of the wave energy and calm conditions usually prevail at the shore. A good boat ramp and car park are located on the western side of the bluffs. Adjacent to the boat ramp are a playground, toilets and a phone, but no other facilities. A grassy foreshore reserve backs the beach, with a second large car park towards the western end.

Swimming: The best swimming is on the eastern side of the bluffs.

Surfing: None.

Fishing: The bluffs are very popular, however most fishers head further out in their boats.

Summary: A very accessible and interesting spot to stop and view the caves.

Figure 4.5 *'Blackfellows Caves' refers to sea caves in the low, protruding limestone headland (centre), with the small settlement shown behind. Shallow reefs lower waves, with calms predominating, permitting boats to moor off the beach.*

33-35 BLACKFELLOWS CAVES (W), PELICAN POINT

Unpatrolled			
No.	Beach	Rating	Type	Length
33	Blackfellows Caves (W 2)	2	R+rock flats	400 m
34	Beach 34	3	Cobbles+rock flats	450 m
35	Pelican Point	2	Cobbles+rock flats	3 km

West of Blackfellows Caves the coast is dominated by the low limestone bluffs and reefs, the eroding bluffs supplying cobbles to form many of the beaches, the reefs reducing the high offshore waves to calm at the shoreline. The Blackfellows Caves-Carpenters Rocks road parallels the back of most of the beaches, providing easy access.

Nine hundred metres west of Blackfellows Caves is a 400 m long, narrow sand beach (**33**) located below 5 m high overhanging limestone bluffs. The road runs along the top of the bluffs. The beach merges in the west with a protruding cobble foreland consisting of a series of distinct cobble ridges, (**34**) which sit atop a 50 m wide intertidal limestone rock flat that receives moderate waves on its outer edge.

A small drain separates this beach from the 3 km long crenulate cobble beach (**35**) that extends west to the low Pelican Point, another distinctive series of up to 20 cobble ridges, which in fact extend 900 m inland from the tip of the point. A half a dozen houses occupy the eastern side of the point.

Swimming: All three beaches are unsuitable owing to the cobbles, rock flat and very shallow water off the beaches.

Surfing: None.

Fishing: Generally too shallow to fish off the beaches.

Summary: Three very accessible beaches, which most people see as they drive by, but few stop.

36-40 BUNGALOO BAY, THE RAPIDS

		Unpatrolled		
No.	Beach	Rating	Type	Length
36	Bungaloo Bay	2	R+rock flats	800 m
37	Bungaloo Bay (W 1)	2	R+rock flats	80 m
38	Bungaloo Bay (W 2)	2	cobble+rock flats	800 m
39	The Rapids	2	cobble+rock flats	700 m
40	The Rapids (point)	2	R+rock flats	50 m

Pelican Point marks the beginning of a series of more crenulate bays and low points that extend for 5 km to Cape Banks. The crenulations are a result of both the dominating Tertiary limestone reefs, coupled with the Pleistocene calcarenite that increasingly dominates to the west as reef and then cliffs. The main road runs behind the beaches, with a caravan park, service station and store located behind Bungaloo Bay.

Bungaloo Bay (36) is a semicircular, 800 m long, southwest facing bay protected by reefs extending 1 to 2 km offshore. The reefs extend all the way to The Rapids. The low gradient sand beach is usually calm and covered in seagrass debris, with small boats moored just off the beach. About 40 shacks back most of the bayshore, including the low limestone bluffs to the west. In amongst the western bluffs is an 80 m long low energy beach (37) consisting of a cobble ridge, some sand and rock flats, and backed by shacks.

Immediately west of the bay the shore continues as a steep cobble beach (38) for 800 m, backed by about 20 shacks and fronted by shallow rock flats. At the western end it grades into a more sandy beach (39) at The Rapids, a semicircular, 700 m long bay with shacks at either end and a low foredune along the beach. A 10 m high calcarenite point protrudes 500 m to the south, with a pocket 50 m long, east facing beach (40) out on the eastern side of the point.

Swimming: More suitable for wading than swimming, owing to the shallow reefs.

Surfing: None.

Fishing: Most fishers use their boats to fish the outside reefs.

Summary: An older shack settlement gradually expanding with freehold development, and popular during the holiday periods.

41, 42 BUCKS BAY

		Unpatrolled		
No.	Beach	Rating	Type	Length
41	Bucks Bay	2	LTT	700 m
42	Bucks Bay (point)	2	R+rock flats	50 m

Bucks Bay is the bay for Carpenters Rocks, an older fishing settlement, with the fishing boats moored in the bay (Fig. 4.6). The settlement has been slowly growing since the 1970's and now includes a store, phone and basic facilities. Most visitors also launch their boat across the hard sand beach (**41**) from the ramp at the end of the main road, adjacent to which is a large car park.

Figure 4.6 *Bucks Bay has formed in a gap in the calcarenite reefs and headlands that dominate this part of the coast. The sheltered waters of the bay house the fishing boats of Carpenters Rocks.*

Bucks Bay is very protected from ocean waves and is usually calm at the shore, the waves breaking on reefs up to 1 km offshore. A 5 to 10 m high foredune and a few houses back the centre of the bay, with dune covered points extending about 500 m out either side. On the western point is a 50 m long pocket cobble beach (**42**) fronted by some sand and reef flats. A vehicle track winds through the scrub to the top of the beach.

Swimming: The sandy beach and bay floor make this the most suitable swimming beach in the area, however beware of the frequent boat traffic and swim well clear of the boat ramp.

Surfing: None.

Fishing: This is the fishing port for the area, with most shore fishing off the headlands.

Summary: Carpenters Rocks offers basic facilities for the visitor, with a caravan park in adjacent Bungaloo Bay. Its main use is however for boat launching.

43 CAPE BANKS

No.	Beach	Unpatrolled Rating	Type	Length
43	Cape Banks	4.5	LTT→TBR	2.4 km

Immediately west of Bucks Bay, the coast transforms as a deep nearshore zone permits the high southwest swell to reach the beaches, producing higher energy beaches backed by some substantial sand dune systems.

Cape Banks is a cuspate foreland protruding 500 m seaward in lee of calcarenite rocks and reefs. To the east is a 2.4 km long, curving, southwest facing beach (**43**) backed by nine active blowouts, and dunes extending up to 300 m inland and reaching heights of 20 m. The gravel road from Carpenters Rocks skirts the back of the dunes, providing access to the Cape Banks lighthouse (constructed 1882) and the beaches to either side of the foreland. The eastern beach receives waves averaging 1.5 m in the centre, which usually maintain about six rips along the beach and a surf zone up to 200 m wide. Waves decrease toward the eastern point as shallow reef dominates the nearshore.

Swimming: Be very careful on Cape Banks beach, where rips dominate. The safest swimming is in lee of the reefs along the eastern end of the beach.

Surfing: The best beach breaks are along the centre of the beach.

Fishing: A good spot for beach fishing into the rip holes.

Summary: Most visitors drive to the lighthouse and view the beaches to either side, with few but the locals actually using the beach.

Canunda National Park	
Established:	1959
Area:	9358 ha
Beaches:	23 (44-66)
Coast length:	45 km

Canunda National Park contains a 45 km long stretch of exposed, high energy beaches, including some of the most hazardous beaches in Australia. The beaches are backed by one of the most active and spectacular coastal dune systems in the state, totalling 95 km^2 in area (Fig. 4.7). The park is accessible to 4WD's with a well marked track through the deflation basins and dunes from Carpenters Rocks to Oil Rig Square, with additional access in the north from Bevilaqua Ford and Southend. In addition there are basic camping areas at Number Two Rocks, Oil Rig Square and Bevilaqua Ford (the latter two accessible by 2WD), picnic areas at Coola Outstation and McIntyre Beach, and a walking track around Cape Buffon.

Further information:
Park Ranger, Southend, (087) 35 6053

Figure 4.7   *A series of 20 m high transverse dunes (i.e. perpendicular to the westerly winds) in lee of Admella Beach, with Lake Bonney behind.*

44, 45 ADMELLA

No.	Beach	Unpatrolled Rating	Type Inner	Outer Bar	Length
44	Admella (1)	3.5	R→LTT		3 km
45	Admella (2)	8	LTT-TBR	RBB	6.6 km

Admella Beach runs for 9.6 km west of Cape Banks to the calcarenite reefs of Number Two Rocks. The first section of beach (**44**) extends for 3 km to a subdued cuspate foreland in lee of reefs located 500 m off the beach. Additional reefs parallel the beach and lower waves to about 1 m, which maintains a continuous bar usually free of rips east of the drain and small ephemeral rips developing between the drain and the foreland. The small overflow drain from Lake Bonney crosses the centre of the beach, marked by the bordering rock training walls. The drain is usually closed, only flowing following winter rains.

Admella Beach is named after the *SS Admella*, which was wrecked on the reefs in 1859 with a loss of 83 lives. Some of the survivors stayed on the wreck for a week awaiting rescue.

The main beach section (**45**) extends north of the foreland to the prominent reefs immediately south of Number Two Rocks. This 6.6 km long section faces southwest and apart from some waves breaking on scattered reefs, receives persistent high southwest swell. The waves maintain an energetic two bar system, with the inner bar cut by strong rips every 500 m, in addition to permanent rips adjacent to some of the reefs, with rip currents extending over 200 m out to sea. The entire surf zone is up to 400 m wide. It is backed by very unstable sand dunes that extend up to 1 km inland and consist of six large blowout/parabolics in the south, grading into a 2 km long section of transverse dunes (Fig. 4.7). A 4WD track winds along in the deflation basin between the foredune and the transgressive dunes.

Swimming: The main beach is very hazardous, with persistent strong rips and a deep trough and channels against the shore. The southern beach is the safer, though still watch for rips.

Surfing: Beach breaks over the inner and outer bars occur the length of the beach.

Fishing: A good fishing beach with near continuous gutters and holes associated with the rips.

Summary: A very accessible beach for 4WD visitors, offering excellent views of high energy surf and beach, together with the active dune system.

46-49 NUMBER TWO ROCKS

Unpatrolled			
No.	Beach	Rating	Type	Length
46	No. Two Rocks (1)	7	RBB	700 m
47	No. Two Rocks (2)	4	LTT	300 m
48	No. Two Rocks (3)	5	LTT/TBR	350 m
49	No. Two Rocks (4)	7	TBR	200 m

Number Two Rocks consists of three 20 m high calcarenite headlands and associated reefs extending up

to 500 m seaward and occupying a 2 km section of coast. The rocks break the long beach into four smaller beaches, each dominated to varying degrees by the headlands, rocks and reefs. The 4WD track runs along behind the beach, with access and parking in lee of each of the beaches, together with a camping area behind the second beach (47). The active dunes to either side of the rocks are moderately stable and largely vegetated in lee of the rocks.

The first beach (**46**) extends from a prominent calcarenite reef capped by sea stacks, to the first of the Number Two headlands. It curves around between the two, facing essentially southwest. It has a low gradient beach fronted by a 200 m wide surf zone, with permanent rips running out against the reefs and headland at either end. The surf zone continues out along the northern headland with the beach narrowing to form pockets of sand beneath the 10 m high bluffs, increasingly fronted by shallow reef. This area is the most protected and only area of the beach suitable for swimming.

The second beach (**47**) is the most protected of the four. It is 300 m long, curving between the two headlands and is fronted by a continuous calcarenite reef which forms a 50 m wide sandy lagoon in its lee (Fig. 4.8). Conditions are relatively calm at low tide, while lowered waves reach the shore at high tide. This is a relatively safe beach close inshore, however beware as reef-controlled rip currents drain out over the reef.

Figure 4.8   *The beach at Number Two Rocks (47) is well sheltered by the reefs that link the two headlands, providing calm conditions at the shore.*

The third beach (**48**) is 350 m long, also running between two calcarenite headlands, but fronted by deeper reefs which permit waves averaging about 1 m through to the shore. These maintain a continuous bar, with smaller rips permanently located at either end against the rocks.

The fourth beach (**49**) lies immediately north of the northern headland and extends north for 200 m to a calcarenite rock outcrop that runs across the beach and dips 100 m out into the surf. A 200 m wide surf zone fronts the beach, with a permanent rip running out along the headland.

Swimming: Be very careful swimming at these beaches, even on the apparently calmer sections, as all the water is draining back out to sea via rips of various sizes. Each of these beaches has permanent rips against the rocks, with the 'lagoon' beach draining out over the reef.

Surfing: Best chance is on the reef off the first beach (46), while beach 48 offers a lower shorebreak.

Fishing: There are many excellent fishing spots both around the reefs and headlands, as well as off the beaches into the permanent rip holes.

Summary: A relatively popular section of Canunda, owing to the variety of beaches and fishing opportunities.

50-52 NUMBER ONE ROCK, CANUNDA BEACH

	Unpatrolled			
No.	Beach	Rating	Type	Length
			Inner	Outer Bar
50	No. One Rock (S)	7	RBB	RBB 1.7 km
51	No. One Rock (N)	8	RBB	RBB 7.1 km
52	Canunda Beach	8	RBB	LBT 16.1 km

The 36 km of coast between Number One Rock and Geltwood Reef contains some of the highest energy beaches in South Australia and Australia. The beach faces directly into the prevailing high southwest swell and winds, which are expressed in the continuous 400 to 500 m wide high energy surf zone. The surf is dominated by large rips spaced approximately every 500 m and the beach is backed by active transverse dunes extending up to 1 km inland and reaching heights of 40 m (Fig. 4.9). The 4WD tracks follow the back of the beach from Number One Rocks to Oil Rig Square. There are beach access points at Number One Rock, in lee of Whale Rock, then every few kilometres until a major 2WD access and camping area at Oil Rig Square. The tracks from Southend extend down to Geltwood Reef, with the beach providing the only access between Oil Rig and Geltwood. This is however a very hazardous section to drive, owing to the soft sand, steep beach and high waves.

Number One Rock is a calcarenite reef extending 1 km seaward and capped by some low rocks. It forms a prominent cuspate foreland in its lee, which separates the two beaches. The south beach (**50**) is 1.7 km long,

terminating at Little Rock, a low calcarenite beachrock outcrop, with two prominent reefs lying off the beach. It receives waves averaging 1.6 m and usually has three to four reef-controlled rips flowing out through the 300 m wide surf zone.

Figure 4.9 *Typical high wave conditions along Canunda Beach. Waves break across the 400 m wide rip-dominated surf zone, with active dunes extending inland.*

The northern beach (**51**) is 7.1 km long, extending up to a beach calcarenite outcrop, with a major calcarenite reef called Whale Rock producing a sandy cuspate foreland toward the centre of the beach. This is an exposed high energy beach with rips spaced every 400 to 500 m.

Canunda Beach (52) begins on the north side of Little Rock and runs unbroken for 16.1 km to the first beachrocks of Geltwood Reef. This high energy beach receives waves averaging 2 m, which maintain two, to at times three, bars across a 500 m wide surf zone, with large rips draining the inner surf.

Swimming: Not recommended on any of these beaches owing to the high waves, deep inshore troughs and strong rips. Use extreme caution if swimming, stay inshore on the attached portions of bars and well clear of the rips.

Surfing: There is 36 km of wild beach breaks which are only for experienced surfers.

Fishing: This is a beach fisherman's paradise with rip after rip providing excellent holes and gutters.

Summary: An accessible and spectacular high energy beach and dune system, however beware of the strong rips if swimming.

53-58 GELTWOOD BEACH to McINTYRE BEACH

No.	Beach	Rating	Type Inner	Outer Bar	Length
		Unpatrolled			
53	Geltwood Beach	8	RBB	RBB	1.5 km
54	Pether Rocks	8	RBB	RBB	900 m
55	Cameron Rocks	8	rocks+RBB	RBB	600 m
56	Mounce & Battye Rocks	7	TBR		250 m
57	McIntyre Beach (E)	7	TBR		600 m
58	McIntyre Beach (W)	7	TBR		550 m

Geltwood Reef marks the beginning of 9 km of coast between the reef and Cape Buffon that is increasingly dominated by calcarenite reefs and cliffs. The first 4.5 km contains six beaches, all accessible by 4WD either from Southend or via Bevilaqua Ford. There is a camping area, accessible by car, one kilometre north of McIntyre Beach.

The largely bare sand dunes that extend all the way down Canunda Beach begin to stabilise as the calcarenite cliffs increase in dominance. The eroding cliffs and bluffs contain some interesting rock formations, including sea caves and stacks.

Geltwood Beach (**53**) extends for 1.5 km between the first rocks of Geltwood Reef and the more prominent 10 m high bluffs of Canunda Rock. It is a low gradient beach fronted by a 300 m wide surf zone, dominated by three to four large rips, including permanent rips against the rocks and reefs to either end. The reef is named after the barque *Geltwood* wrecked on the reef in 1876 with the loss of all 27 on board.

Pether Rocks beach (**54**) continues on the north side of Canunda Rock for 900 m to Pether Rocks, another 10 m high calcarenite bluff. Some rock also outcrops in the inner surf zone. The surf zone is 300 m wide, containing three permanent reef-controlled rips.

Cameron Rocks beach (**55**) is backed by a continuous, 10 m high scarped calcarenite bluff, with the low sand beach fronted by near continuous intertidal reef called Cameron Rocks, beyond which is a 300 m wide surf zone drained by two permanent rips.

Mounce and Battye Rocks beach (**56**) consists of two patches of sand contained in a 250 m long south facing break in the reefs. The high tide sand beaches are fronted by intertidal reefs, then a 200 m wide, rip-dominated surf zone. There are shell-rich middens in the exposed soils on the bluffs behind the northern end of the beach

McIntyre Beach (E) (**57**) is a curving, southwest facing, 600 m long beach bordered by calcarenite bluffs fronted by extensive intertidal rock platforms, in addition to a reef toward the southern end. Permanent rips at either

end drain the 200 m wide surf zone. The western McIntyre Beach (**58**) is very similar, 550 m in length, bounded by 20 m high calcarenite cliffs and with two permanent rips.

Swimming: Be careful on all these beaches owing to the prevailing high waves offshore, rocks, reefs and strong permanent rips.

Surfing: There are a variety of beach and reef breaks along this section.

Fishing: An excellent section of coast if you are looking for good access to beach and rock holes and gutters.

Summary: An interesting section of coast to walk and view the beaches and rock formations.

59-63 CULLEN BAY to CAPE BUFFON

No.	Beach	Rating	Type	Length
		Unpatrolled		
59	Cullen Bay	6	LTT/TBR	150 m
60	Eddy Bay (1)	5	LTT	120 m
61	Eddy Bay (2)	3	R	30 m
62	Double Island	4	R+rock flats	40 m
63	Back Beach	4	R+rock flats	50 m

Between Cullen Bay and Cape Buffon is a 1.5 km long section of high, irregular coast dominated by 10 to 20 m high calcarenite cliffs, in amongst which are five pockets of sand, all fronted by a maze of rocks and reefs. There are car access roads to Boozy Gully, next to Cullen Bay and to Back Beach, with 4WD tracks to 30 m high Stanway Point and Eddy Bay.

Cullen Bay (**59**) has a 150 m long, south facing beach which lies beneath 20 m high cliffs. While you can drive to the top of the overhanging cliffs, there is no safe foot access to the beach. Two reefs lie 200 m off the beach, lowering waves to about 1 m, which maintain a narrow attached bar, with one permanent rip draining the surf zone.

Eddy Bay contains two small beaches. The first (**60**) is a 120 m long, south facing sand beach, lying below 20 m high calcarenite cliffs, with no safe foot access. It has a narrow bar and receives waves averaging 1 m. The second beach (**61**) is a 30 m pocket of sand that faces east and is protected by its orientation and reefs, so that waves average only 0.5 m, with rocks and reefs dominating both on and off the beach.

In lee of **Double Island**, a small reef, is a 40 m patch of sand (**62**) below 10 to 20 m high bluffs, fronted by rock reefs and with no safe foot access. **Back Beach** (**63**) has 2WD access to the overhanging 10 m high bluffs. It is a

50 m long sand beach strewn with boulders and fronted by a rock reef. Both beaches receive only low waves at the shore.

Swimming: While these beaches receive lower waves, none are safe for swimming owing to the difficult access and/or presence of rocks and reefs.

Surfing: None at the beaches, though there is a bit of a left point break in Boozy Gully called Cullen Reef and a break off the point called Lighthouse Reef.

Fishing: A very popular fishing area with the many cliffs, bluffs and rock platforms providing a wide range of sites to fish the deeper rock holes and gutters.

Summary: The best way to see these beaches and this section of coast is to walk the Seaspray and Seaview walking trails, which take in both side of Cape Buffon.

64-66 CAPE BUFFON

		Unpatrolled		
No.	Beach	Rating	Type	Length
64	Cape Buffon (1)	3	R+rock flats	60 m
65	Cape Buffon (2)	3	R+rock flats	60 m
66	Cape Buffon (3)	2	R	50 m

On the north side of **Cape Buffon** a strip of sand runs along the base of the 10 to 15 m high calcarenite bluffs, broken by rocks into three small beaches, each backed and bordered by calcarenite bluffs, rocks and reef flats. The sealed road to the cape runs along the top of the bluffs with a car park on the point providing the best access to the beaches. However these three beaches are difficult to safely access and little used.

The first beach (**64**) is a 60 m long narrow strip of sand mainly awash at high tide, broken by protruding rocks and boulders and fronted by 100 m wide reef flats. The second beach (**65**) is about 10 m wide, 60 m long and more continuous with reef flats; while the third 50 m long beach (**66**) is fronted by sand and offers the best swimming.

67-70 SOUTHEND

		Unpatrolled			
No.	Beach	Rating	Type		Length
			Inner	Outer Bar	
67	Southend Jetty	2	R		50 m
68	Southend (west)	3	LTT		300 m
69	Southend (main)	3	LTT		300 m
70	Southend/Burdon	5	TBR	D	16.2 km

Southend is a small fishing village located at the southern end of Rivoli Bay (Figs.4.10 & 4.11). The fishing fleet is protected by Cape Buffon and the boats anchor in the bay, with a jetty providing access to the shore. The village is located immediately east of the cape, on low land behind the beach. The beach begins at the cape and runs for 16 km to Beachport at the northern end of the bay.

Figure 4.10   *View across Cape Buffon to Southend and the southern end of Rivoili Bay.*

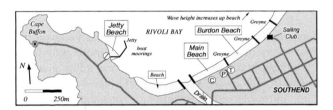

Figure 4.11   *Cape Buffon and the protected southern end of Rivoli Bay, where the small settlement of Southend is located.*

Southend has limited facilities consisting of a store, a beachfront caravan park and a beach reserve. The caravan park fills during the summer holiday period and the low energy protected beaches between the jetty and the village are popular for sunbathing and swimming.

There are four beaches in the vicinity of Southend. **Southend Jetty** beach (**67**) is a small pocket of sand, with usually calm conditions or low waves. **Southend (west)** beach (**68**) lies between the rocks of the cape and the Lake Frome drain entrance, and is a 300 m long, low, flat beach with usually low waves. It is used to launch small boats, in addition to the boat ramp at the jetty. The drain entrance is usually plugged with sand and only opens in winter following heavy rain.

The **Southend (main)** beach (**69**) is located in front of the caravan park, with a public car park right behind the beach. It runs from the eastern drain wall for 300 m to a second small rock groyne. Flags are placed on this beach during the holiday period, though there is no lifeguard on duty. It usually has waves averaging 0.5 m, which break over a low continuous bar.

North of the second groyne is the 16 km long **Burdon Beach (70)** that extends up past the Sailing Club, then all the way to the Lake George drain at Beachport. Vehicles, including cars, drive along this low, flat, hard beach, though a 4WD is recommended, as soft sand can be encountered. Wave height increases along this beach, causing the surf zone to widen and rips to prevail every few hundred metres.

The entire beach is backed by a series of 10 to 20 m high foredune ridges (Fig. 4.12), which outline the location of past shorelines. The up to 80 ridges extend 5 km inland, giving an indication of a massive accumulation of sand in the bay that has built out of the shoreline over the past 6,000 years.

Figure 4.12   *Burdon Beach has a moderate energy surf and is backed by a series of foredune ridges.*

Swimming: Southend (main) beach is the most suitable and most popular for swimming and catching the low waves. It has good access and is right next to Southend's facilities. If you are swimming up Burdon Beach, be careful of rips, stay inshore and on the bar and clear of any rip channels.

Surfing: The best surf is along Burdon Beach, where there are many kilometres of usually empty surf.

Fishing: The jetty and rocks around the cape are the most popular fishing areas, otherwise up the beach in the rip gutters and holes.

Summary: A nice, quiet, protected settlement and beach, more popular with the locals and regular visitors than the passing tourist.

70-72 **BEACHPORT**

No.	Beach	Rating	Unpatrolled Type		Length
			Inner	Outer Bar	
70	Surf (Burdon)	5	TBR	D	16.2 km
71	Town beaches	2	R		1.6 km
72	Back Beach	3	LTT		500 m

Beachport is an historic, well preserved and picturesque fishing and holiday town located just off the Princes Highway (alternate route) at the northwestern end of Rivoli Bay (Figs. 4.13 & 4.14). The town contains many old buildings constructed of dunerock. The main street leads to the 750 m long jetty, which is the focal point of the town and beach. The small town offers most facilities for travellers, including a beachfront caravan park. The port and town are protected from the high southwest swell by Glenns Point, Cape Martin and Penguin Island, together with a number of offshore reefs.

Figure 4.13   *Cape Martin and the more sheltered northern end of Rivoli Bay, including the town of Beachport with its long jetty and fishing fleet.*

The long Rivoli Bay beach, that once ran uninterrupted for 18 km from Southend to Glenns Point, is now crossed by two drains, the Lake Frome at Southend and the Lake George drain 1 km east of Beachport. In addition, between Glenns Point and the Lake George drain there are 15 low rocky groynes along the Beachport beachfront.

The Beachport end of the beach is called **Surf** or **Burdon** Beach (**70**). The main car park and beach vehicle access point is located 1 km east of the Lake George drain. There is also a memorial tower which provides a better view of the beach and surf. Waves averaging 1 m at the car park can increase slightly down the beach, with the surf zone widening to 200 m and rips tending to form every 400 to 500 m.

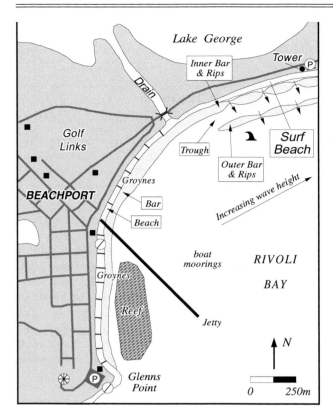

Figure 4.14 *Beachport's beaches are initially protected by Glenns Point, but become increasingly exposed to higher waves east of the Lake George drain.*

West of the drain, the groynes divide the 1.6 km long section of **Town Beach (71)** into 17 beach sections, each about 100 m long and bounded by the groynes. A road runs behind the entire beach, providing good access to all the beaches. The Town beaches usually have waves less than 0.5 m which break over a shallow attached bar. Rips only occur when waves exceed 0.5 m, with the rips running out against the groynes. The second beach west of the jetty has a concrete boat ramp for small boats.

Back Beach (72) is a 500 m long, southeast facing beach, protected by Cape Martin, Penguin Island and offshore reefs. It receives waves averaging 0.5 m and has a steep beach fronted by a narrow continuous bar, usually free of rips. The main fishing boat ramp crosses the eastern end of the beach, with a large car park behind the dune.

Swimming: The Back Beach and the small Town beaches are relatively safe, away from the boats and under normal low wave conditions. Higher waves produce a heavy shorebreak and rips against the groynes. The Surf Beach is moderately safe close inshore on the bar. Stay clear of the rips, which can extend a considerable distance out across the 100 to 200 m wide surf zone.

Surfing: Surf Beach is the most accessible and popular spot and offers a wide, low gradient surf zone, with usually 1 to 1.5 m high spilling breakers. There are also many kilometres of usually empty surf down the

beach toward Southend. During big swell there is a right hander in Beachport.

Fishing: There are many options at Beachport. The jetty is extremely popular, as is fishing from small boats in the bay and over the inner reefs. In addition there is rock fishing off the groynes and the cape, and beach fishing along Surf Beach.

Summary: A very popular summer destination offering sufficient facilities for the traveller, set in a picturesque town and physical setting.

The Robe Range

The Robe Range refers to the reefs, beaches and backing dunes that run for 45 km between Beachport and Robe. The range is in fact a partly submerged Pleistocene barrier system, which was formed about 40 000 years ago when sea level rose to a position about 20 m below present sea level. The beaches were laid down approximately where the outer reefs lie, that is up to 2 km seaward of the present shoreline, and the backing dune extended inland, up to and past the present shoreline. Sea level subsequently fell and the 'range' was lithified into dunerock. The most recent rise in sea level, ending about 6500 years ago, partly submerged the range and deposited the modern beaches and dunes on top. The older range is very prominent in the reefs, rocks, headlands and all lithified dunes (Fig. 4.15).

Figure 4.15 *The Robe Range is a Pleistocene barrier that is now a partly submerged shore-parallel reef system, with beaches forming between the headlands and in lee of the reefs.*

73-76 PORT WILLIAM, SNAPPER PT, SALMON HOLE

		Unpatrolled		
No.	Beach	Rating	Type	Length
73	Port William	3	R	450 m
74	Snapper Pt (1)	3	R	100 m
75	Snapper Pt (2)	3	R	200 m
76	Salmon Hole	3	R+rock flats	600 m

A sealed scenic road runs west of Beachport out along the backing, more exposed ocean beaches, providing good access along the first three kilometres of coast. Apart from car parking and beach access, there are no facilities.

Port William beach (**73**) is a 450 m long, southwest facing beach lying in lee of calcarenite reefs which lower waves at the shore to about 0.5 m and maintain a steep, barless beach. There is a car park at the western end, which provides a good view of the beach. The remainder of the beach is backed by active dunes, which extend 300 m inland and to heights of 20 m.

Snapper Point is a flat, 10 m high protruding calcarenite surface, which, combined with other rocks and reefs, forms two small beaches to the east, both backed by active 20 m high foredunes. The first beach (**74**) is a curving, east facing, 100 m long pocket of steep sand. The second (**75**) is 200 m long, but more irregular owing to rocks and reefs dominating the western half of the beach. Both beaches can be accessed from the road or a car park above Snapper Point.

Salmon Hole (**76**) is a 600 m long, curving beach, lying in lee of a near continuous, curving beachrock reef, which forms a relatively calm 30 to 50 m wide lagoon or hole between the reef and the beach. It is backed by a 10 to 20 m high, steep, scarped foredune. There is access to the beach from small car parks at either end.

Swimming: While waves are relatively low at all four beaches, be careful of the numerous rocks and reefs. Salmon Hole is the safest close inshore. Stay clear of the reef, as water drains out across and around the reef.

Surfing: None at the beaches.

Fishing: As the names 'Snapper Point' and 'Salmon Hole' imply, this is a popular stretch of coast to fish from the beach or rock platforms.

Summary: Four beaches located on the Beachport scenic drive and well worth viewing and spending more time to swim, walk or fish.

77-80 POST OFFICE ROCK, COWRIE IS, WOOLEY ROCK, PIGEON COVE

No.	Beach	Unpatrolled Rating	Type	Length
77	Post Office Rock	3	R	300 m
78	Cowrie Island	3	LTT	200 m
79	Wooley Rock	6	TBR+reefs	1.2 km
80	Pigeon Cove	5	R+reefs	400 m

The Beachport scenic drive continues on past Salmon Hole for another 2 km to a car park above Wooley Rock.

The sealed, then gravel, road provides excellent access to four beaches, with six car parks providing space for 200 cars.

Post Office Rock, which does have a makeshift postbox on its crest, is a prominent 15 m high calcarenite headland that separates Salmon Hole from the **Post Office** beach (**77**). This 300 m long, curving beach is well protected by Post Office Rock and the northern Cowrie Island head, which results in a low wave, steep barless beach and provides one of the safer swimming beaches on the drive.

Cowrie Island is a slightly detached headland, which forms the southern boundary of Cowrie Island beach (**79**), a 200 m long, curving, moderately protected beach, with a rocky shoreline forming the northern boundary and a rock-reef outcrop toward the centre. Waves average less than 1 m and a narrow bar runs the length of the beach. Watch for rips during higher waves.

Wooley Rock Beach (**80**), is the longest beach in this section at 1.2 km, with calcarenite reefs scattered along and just off the beach. During normal wave conditions, these maintain four strong permanent rips. The most protected section is the far northern end, below the car park, when the reef is attached to the rocks and provides a partly enclosed lagoon. Nonetheless, this is a hazardous beach.

Pigeon Cove beach (**80**) extends for 400 m north of Wooley Rock as a curving, southwest facing beach, which is partially protected by shoreline and offshore reefs, lowering waves to less than 1 m and maintaining a steep, barless beach. While waves are low at the beach, waves breaking over the reefs maintain a permanent rip which drains the cove.

Swimming: Of these four beaches, Post Office Rock is the calmest and safest, with the other three having higher waves, rocks, reefs and rips.

Surfing: There are a few close-in reef breaks along Wooley Rock beach.

Fishing: The rocks, rips and reef offer excellent fishing spots.

Summary: A very accessible and popular stretch of coast, heavily utilised during the holiday periods.

Beachport Conservation Park	
Established:	1968
Area:	710 ha
Beaches:	7 (81-87)
Coast length:	4 km

Beachport Conservation Park extends along the coast for 4 km between Pigeon Cove and Five Mile Rocks. It contains excellent examples of the reef and rock dominated beaches and headlands, backed by 600 ha of dunes, ranging from stable to active blowouts. There is 2WD access along the backing shore of Lake George, with 4WD required to access the open coastal section.

Further information:
Park Ranger, Southend, (087) 35 6053

81-87 BEACHPORT CONSERVATION PARK (THREE MILE ROCKS, FIVE MILE ROCKS)

No.	Beach	Rating	Type	Length
Unpatrolled				
81	Beach 81	6	TBR	700 m
82	Beach 82	6	TBR+reef	50 m
83	Beach 83	6	TBR+reef	80 m
84	Beach 84	6	TBR+reef	300 m
85	Three Mile Rocks (S)	6	TBR+reef	300 m
86	Three Mile Rocks (N)	6	TBR+reef	900 m
87	Five Mile Rocks (S)	7	RBB	1 km

The eastern boundary of **Beachport Conservation Park** is a 4 km section of sand and rocky shore, containing seven beaches. The scenic drive terminates south of the park boundary and a 4WD is required to access the park beaches and dune areas.

Beach **81** is a 700 m long, southwest facing beach, with calcarenite reefs and rocks forming the boundaries, together with some offshore reefs. Waves average 1 to 1.5 m and two permanent reef-controlled rips drain the 50 m wide surf. The rocks increase towards the northern end, where beaches **82** and **83** sit as two adjoining 50 and 80 m strips of high tide sand below 10 to 20 m high calcarenite bluffs, fronted by a mixture of sand and reef. Waves break more than 100 m off each beach, with each containing a reef-controlled rip.

Beach **84** is a convex sandy beach, building out in lee of a 100 m long reef lying 150 m offshore. The reef lowers waves in the centre of the beach, but still sufficient energy gets through to maintain rip currents flowing out on either side of the reef.

Three Mile Rocks form a low calcarenite headland, with rocks and reefs dominating the beaches to either side. The south beach (**85**) is 300 m long, with reefs extending up to 500 m off the beach and rocks occupying much of the inner surf zone. The northern beach (**86**) is 900 m long, with rocks and reefs increasing toward the northern end, forming a platform beach in places.

Five Mile Rock is the most prominent headland along this section, reaching a height of over 20 m and capped

by active clifftop dunes. The southern beach (**87**) is relatively high energy and sweeps for 1 km south of the rocks, with scattered reefs lying 300 to 400 m offshore. Sufficient wave energy reaches the wide, low gradient beach to maintain two beach rips, as well as permanent rips against the rocks-reef at either end. This is a hazardous beach, with a memorial on the southern headland to a youth who drowned in a rip at the beach.

Swimming: These are six potentially hazardous beaches dominated by rocks, reefs and permanent rips. Use extreme care if swimming here.

Surfing: There are numerous reef breaks amongst the beaches, with the best beach breaks on the Five Mile Rocks beach.

Fishing: There are excellent beach and rock holes the length of this section.

Summary: A popular section of coast explored by the 4WD fraternity, it provides easy access to some excellent beaches and dunes. All vehicles must stay on the marked tracks.

88-93 FIVE MILE ROCKS to STINKY BAY

No.	Beach	Rating	Type		Length
Unpatrolled					
			Inner	Outer Bar	
88	Five Mile Rocks (N)	6	TBR		700 m
89	Euro Point	7	TBR	RBB	2.1 km
90	Nine Mile Sandhill	7	TBR	RBB	5.5 km
91	Beach 91	6	TBR		1.7 km
92	Lurline Point	6	TBR		2.5 km
93	Stinky Bay	6	TBR-LTT		2 km

North of Five Mile Rocks, energetic sandy beaches and massive sand dunes dominate the coast for 12 km to Stinky Bay. The high energy beaches are backed by some of the most massive and active dune systems in the southeast, with the Nine Mile Sandhill extending up to 3 km inland (Fig. 4.16). A 4WD track runs through the dunes and along some of the beaches. However beware if driving on the beaches as some are soft and treacherous.

North of **Five Mile Rocks**, the beach (**88**) runs for 700 m to a 15 m high calcarenite bluff. This beach is partly protected by the rocks and reefs off the point and waves averaging 1.5 m maintain two beach rips, with permanent rips at either end. Beyond the bluffs the beach (**89**) continues for 2.1 km to the cuspate foreland called Euro Point, which has protruded in lee of reefs extending several hundred metres offshore. This section of beach receives higher waves and has an inner bar system usually cut by five to six rips, together with a permanent rip in lee of the reef. Active sand dunes have moved up to 1.5 km in from the beach.

Figure 4.16   *The massive Nine Mile Sandhills.*

On the north side of **Euro Point** is a 5.5 km long, high energy beach (**90**) backed by the massive Nine Mile Sandhill. The high energy waves maintain a 300 to 400 m wide double bar system, with rips cutting the inner bar every 400 to 500 m. The beach terminates at a protruding calcarenite bluff fronted by a reef. Extending north of the bluff is a largely submerged Pleistocene barrier, which parallels the coast about 1.5 km offshore. This reef increasingly lowers waves to the north, with just one bar occurring along the beach up to Stinky Bay. The steamer *Euro* struck the reef in 1881. As the ship began to sink, the captain steamed it toward the beach where it sank leaving only the masts and funnel showing above the waves.

Beach **91** continues for 1.7 km from the bluff to a second smaller bluff that juts out across the beach. The beach has a 100 m wide surf zone, cut by several more closely spaced beach rips between the bluffs.

North of the bluff, **Lurline Point** beach (**92**) continues on much the same, with a single bar cut by beach rips to its northern boundary at Lurline Point, a sandy cuspate foreland. The backing dune system is more stable, though the beach is backed by several large blowouts and parabolic dunes.

Stinky Bay (**93**) is a curving, 2 km long, low gradient beach lying between Lurline Point and the southern headland of Nora Creina. The name no doubt derives from the large rotting piles of seagrass that commonly cover the northern end of the beach. Wave height decreases from about 1.5 m down to less than 0.5 m at the western end, resulting in rips decreasing in size and a continuous low tide terrace in along the northern kilometre. This beach is backed by the Sunland resort, which has a 500 m long walking track through to the beach just north of Lurline Point.

Swimming: These are six exposed, remote beaches, all dominated by strong rips with a generally wide surf zone. Use extreme care if swimming along this stretch of coast, stay well clear of rips, close inshore and on attached portions of the inner bar. The safest swimming is at the protected northern end of Stinky Bay.

Surfing: There are beach and some reef breaks along the length of these beaches.

Fishing: The numerous rips and reefs also mean numerous inshore holes and gutters for beach fishing.

Summary: A more remote section of coast, requiring a good 4WD and knowledgeable driver to safely transit.

94   NORA CREINA

	Unpatrolled		
No. Beach	Rating	Type	Length
94 Nora Creina	2	LTT	1.3 km

Nora Creina is a small settlement of about 60 shacks and holiday houses, with the only facilities being a phone and an occasionally open store. The attractive bay contains a semicircular beach bounded by prominent 20 m high calcarenite headlands and reefs. The headlands are remnants of the Robe Range, that elsewhere has been eroded down to reefs (Fig. 4.17).

Figure 4.17   *Nora Creina Bay and its small shack settlement lie in lee of a breech in the Robe Range calcarenite reef system. Boats negotiate the reefs to the protective anchorage close to shore.*

The reefs and 500 m wide bay entrance lower waves at the shore to less than 0.5 m, which, with the fine white sand, maintain a firm low gradient beach and shallow bar (**94**). The beach is used as a car park and launching ramp, though vehicles are not permitted south of the access track. A few fishing boats are usually at anchor in the southern corner of the bay.

The bay is named after the brigantine *Nora Creina* that struck a reef off the bay in 1858.

Swimming: This is the best and safest swimming beach between Robe and Beachport, and a very popular spot during the holidays with locals and daytrippers.

Surfing: None at the beach.

Fishing: Most fishers head out from the bay to fish the reefs, however there is also good rock fishing off the adjacent platforms.

Summary: A picturesque little settlement and bay.

95-102 RABELAIS to ERRINGTON HOLE

No.	Beach	Rating	Type	Length
Unpatrolled				
95	Rabelais	5	LTT	750 m
96	Beach 96	6	TBR	1.6 km
97	German Pt (S)	6	TBR	1.2 km
98	German Pt	5	R+rock flats	150 m
99	German Pt (N)	3	R	150 m
100	Errington Hole (S)	5	LTT-TBR	950 m
101	Errington Hole	4	R+rock flats	80 m
102	Errington Hole (N)	4	R+rock flats	100 m

Cape Rabelais is the northern headland for Nora Creina Bay. It is 15 m high and protrudes 500 m seaward, forming the southern boundary of a more isolated 5 km section of coast between the cape and Errington Hole. This section is dominated by the calcarenite reefs lying about 500 m offshore, together with small headlands, bluffs and calcarenite outcrops, all of which influence the intervening six beaches to varying degrees. The beaches are backed by dunes extending up to 1.5 km inland and then private farmland, which make access difficult.

Rabelais beach (**95**) is a relatively straight, 750 m long, low to moderate energy beach, partially protected by the cape and several reefs. It extends from the base of the cape to a sandy foreland in lee of a close-in reef. Waves average just under 1 m and maintain a steep beach fronted by a narrow continuous bar.

Beach **96** is a 1.6 km long, double arcurate beach, with a sandy foreland in the centre. Reefs form the boundaries and some lie offshore, however sufficient wave energy reaches the shore to maintain three to four beach rips, with permanent rips adjacent to the reefs.

South of German Point is a 1.2 km long, double crenulate beach (**97**), also with a central reef-controlled sandy foreland, together with two beach rips and three reef and rock-controlled rips, particularly on the southern half of the beach.

German Point is a 10 m high calcarenite headland that protrudes 200 m seaward and has a high tide sand beach (**98**) along its southern face. The 150 m long beach is fronted by a 50 m wide intertidal rock platform. On the north side of the point is a second 150 m long pocket sandy beach (**99**) contained between the point and a smaller northern headland. This beach receives low waves and is steep and barless. It offers the safest swimming along this section of coast.

Errington Hole is a deep cut in a 20 m high calcarenite headland. Between German Point and the 'hole' is a 950 m long beach (**100**) which is afforded a moderate degree of protection by a series of reefs 500 m off the beach. Waves average 1 m at the beach and maintain a few small rips. Between the rocky end of this beach and the 'hole' is an 80 m long platform beach (**101**) below the headland. The beach is strewn with rock and boulders and fronted by a 50 m wide intertidal rock platform. On the north side of the hole is a second platform beach (**102**), a 100 m long high tide sand beach, lying below 20 m high bluffs and fronted by a 100 m wide reef flat.

Little Dip Conservation Park	
Established:	1975
Area:	1977 ha
Beaches:	24 (103 to 127)
Coast length:	14 km

Little Dip Conservation Park occupies 14 km of coast between Errington Hole and Cape Lannes. It has an area of 1977 ha which includes 24 beaches, with numerous calcarenite bluffs, rocks and reefs, together with the backing active dune system. A 4WD track runs the length of the park, with access from Nora Creina Drive at Errington Hole, Little Dip, Long Gully, Bishops Pate, Stony Rise and Cape Lannes, and 2WD access only at Long Gully. There are coastal camping areas at Little Dip, Long Gully and Stony Rise, and the camping and picnic area at the Little Dip Lake, right by the Nora Creina Road.

For further information contact:
Robe NPWS District Office (087) 682 543

Swimming: While waves tend to be a little lower along this section owing to the numerous reefs, only the northern German Point beach and to a lesser extent the beach south of Errington Hole, offer reasonably safe swimming. On all beaches beware of rips as well as the dominating rocks and reefs.

Surfing: Best beach breaks are on the 2 km of beaches south of German Point.

Fishing: Rip holes and deep water can be found along all the beaches and off the rock platforms and reefs.

Summary: A section of coast a little more difficult to access and consequently little used by the general public.

103-111   ERRINGTON HOLE to LONG GULLY

No.	Beach	Rating	Type	Length
		Unpatrolled		
103	Errington Hole	6	LTT/TBR	600 m
104	Little Dip (south)	4	LTT	500 m
105	Little Dip	4	R+rock flats	150 m
106	Perch Hole (1)	5	R+rock flats	200 m
107	Perch Hole (2)	4	LTT+rock flats	300 m
108	Perch Hole (3)	5	R+rock flats	100 m
109	Long Gully (1)	4	R/LTT	350 m
110	Long Gully (2)	6	R	150 m
111	Long Gully (3)	5	R+rock flats	150 m

The beaches of **Little Dip Conservation Park** are all characteristically short and dominated by the calcarenite reefs and rocks. The first 3 km section between Errington Hole and Long Gully contains nine beaches, half of them small sand beaches perched behind rocks/reef flats. A 4WD track runs along the coast, utilising the beaches and headlands. Be careful driving the beaches as they tend to be soft and steep. Additional tracks from the Nora Creina Road provide 4WD access to Errington Hole and Long Gully and 2WD access to Little Dip, with shacks, a camping area and beach boat launching at Little Dip.

Errington Hole beach (**103**) is a 600 m long sandy beach that curves north of Errington Hole, to a small reef-controlled cuspate foreland. Waves average 1 to 1.5 m and maintain a continuous bar, usually cut by two beach rips, with permanent rips against the boundary rocks. The 4WD track runs along the beach. On the north side of the sandy foreland, the southern Little Dip beach (**104**) continues for 500 m in two arcuate sweeps to the 20 m high bluffs of Little Dip. This beach is more protected by reefs so waves average about 0.5 m, which maintain a steep, barless beach, usually free of surf and rips. Active dunes extend up to 1.5 km in from the beaches.

Little Dip beach (**105**) lies at the end of the gravel road and is used to launch small boats. It is an irregular, 150 m long stretch of sand largely fronted by reefs, apart from a small sandy patch used as the boat ramp. The reef lowers waves to less than 0.5 m, providing the safe launching area.

North of Little Dip, the 4WD access reaches the coast between the first and second **Perch Hole** beaches (**106** and **107**). Beach 106 is a 200 m long curving strip of sand, fronted by near continuous reefs. A 100 m long protruding reef separates it from 107, which runs for

300 m to the side of a low protruding bluff. Waves average 1 m, some breaking over reefs, the remainder maintain a narrow bar, with one permanent rip draining the partly enclosed beach system.

Calcarenite bluffs control the next three beaches. Beach **108** is a 100 m long platform beach, consisting of a strip of high tide sand fronted by 50 to 100 m wide reef flats. Beach **109** is a 350 m long straight beach bordered by the bluffs, with a more open sandy shore and patchy reef offshore. Waves average about 1 m, with rips forming during higher waves. The third beach (**110**) is a curving, 150 m long, steep sand beach bordered and backed by calcarenite bluffs, with reefs forming the floor of the small bay. As a result, waves average less than 0.5 m at the beach.

Beach **111** lies at the end of **Long Gully**, with the track terminating on the low jagged bluffs overlooking the beach (Fig. 4.18). It consists of a 150 m long irregular strip of high tide sand, fronted by 30 m wide reef flats.

Figure 4.18 *Long Gully beach is a typical 'platform' sandy beach, formed at high tide in lee of an inter- to subtidal calcarenite reef or platform.*

Swimming: Be careful on all these beaches. The beaches fronted by rocks and reef flats, while usually sheltered, are unsuitable for swimming owing to the shallow, rocky water. The more open beaches usually have a lowered surf, with beach 104 being the safest and most free of inshore reef. Watch for rips, rocks and reef on all beaches.

Surfing: The best chances are the beach breaks on beaches 103 and 109.

Fishing: The coast abounds in beach, rock and reef holes, which can be easily accessed from the beaches and platforms.

Summary: This section is for the more experienced off-road drivers, and also someone who knows the local tracks, which can be hard to follow as well as potentially boggy.

112-120   **LONG GULLY to BEACH 120**

No.	Beach	Rating	Type	Length
Unpatrolled				
112	Long Gully (4)	6	LTT/TBR	1.2 km
113	Beach 113	6	TBR	350 m
114	Bishops Pate (S)	7	TBR	800 m
115	Bishops Pate	5	R-LTT-TBR	1.1 km
116	Beach 116	4	R+rock flats	70 m
117	Beach 117	2	R	350 m
118	Beach 118	3	R	100 m
119	Beach 119	3	R+rock flats	250 m
120	Beach 120	3	R+rock flats	30 m

Between Long Gully and Domaschenz beach is the 5 km long central section of the Little Dip Conservation Park. The coast continues to be dominated by calcarenite bluffs and reefs, with the shoreline consisting of a mixture of longer beaches receiving waves and smaller reef-bound beaches which are usually calm at the shore. Access to the nine beaches is via sandy 4WD tracks which for the most part utilise the beaches.

Looking north from the **Long Gully** bluffs toward Bishops Pate are three sandy beaches. The first is a 1.2 km long beach (**112**) fronted by several patches of reef extending 100 to 200 m offshore. Two of the reefs induce prominent sandy forelands (spits). While the reefs lower waves to about 1 m at the shore, there is sufficient energy, particularly along the southern half, to maintain one to two beach rips, as well as more permanent rips adjacent to the shore-attached reefs. The second beach (**113**) is a 350 m long strip of sand, cut into three small sections by four low protruding calcarenite bluffs. Permanent rips run out against each of the bluffs. The 4WD tracks skirt the back of the bluffs, running along the beaches to either side. The northern beach (**114**) is 800 m long and more continuous, with patchy reef off the beach, but sufficient wave energy to maintain three to four reef-controlled rips across the 100 m wide surf zone. This beach terminates at the more substantial bluffs of Bishops Pate. The vehicle track runs up off the beach and through some clifftop dunes to reach the track from the Nora Creina Road.

On the northern side of the bluffs is the near semicircular, 1.1 km long **Bishops Pate** beach (**115**), running round from the bluffs and reefs at the southern end to vegetated bluffs fronted by extensive reefs at the northern end (Fig. 4.19). The vehicle track runs along the central-northern section of beach, the southern 150 m cut off by a protruding bluff. Wave height increases to 1.5 m in the centre of the beach where two strong beach rips are

maintained, as well as permanent reef-controlled rips to either end. The safest swimming is in the protected lee of the northern reef, however even here, be wary as a rip drains the 'calm' area behind the reef.

Figure 4.19   *Bishops Pate (centre) is a calcarenite bluff, with the higher energy beach (114) to the south (top) and moderate energy beach (115) to the north.*

The next three beaches are contained in a 400 m wide embayment, bordered by 20 m high bluffs, with smaller bluffs cutting the shoreline into three small beaches, each also influenced by reefs in and across the bay mouth. The 4WD track winds through the dune just behind the beaches. Beach **116** is a curving, 70 m long pocket of sand bordered by low bluffs and fronted by reefs 100 m offshore, enclosing the beach in a protected 'lagoon', with only low waves at the shore and a steep barless beach. This is one of the safer beaches for swimming. Beach **117** is a 350 m long beach bordered by low protruding calcarenite bluffs and reefs, with a 100 m long beachrock reef paralleling the southern half of the beach, leaving a 20 to 30 m wide channel in between. Beach **118** continues on the other side of the northern rocks for another 100 m and becomes increasingly backed by vegetated bluffs rising to about 20 m high.

The 4WD track follows the top of the bluffs to the top of beach **119**, where there is a small parking area overlooking the centre of the 250 m long beach. The entire beach lies at the base of the 10 to 20 m high bluffs, with numerous large boulders on the beach and reefs dominating the floor of the 200 m wide embayment. Beach **120** is a 30 m wide pocket of sand lying below surrounding bluffs and fronted by intertidal reef flats.

Swimming: This is a reef and rip-dominated section of coast with the only relatively safe beach being beach 116, just north of Bishops Pate beach.

Surfing: There are beach and some reef breaks along the longer beaches between Long Gully and Bishops Pate.

Fishing: Reef and rip holes dominate this section, providing numerous good fishing spots.

Summary: A mixture of longer and more reef and bluff-bound beaches, all reasonably accessible by 4WD.

121-128   DOMASCHENZ to CAPE LANNES

No.	Beach	Rating	Type	Length
		Unpatrolled		
121	Domaschenz	5	LTT+reefs	400 m
122	Queen Head	6	TBR	1.3 km
123	Back Beach	6	RBB	1.6 km
124	Stony Rise (1)	7	R+rock flats	300 m
125	Stony Rise (2)	7	R+rock flats	200 m
126	Stony Rise (3)	3	R	100 m
127	Evans Cave	6	LTT/TBR	900 m
128	Cape Lannes	6	LTT/TBR	200 m

The northern section of Little Dip Conservation Park is a 5 km long section of generally longer and more open (reef-free) beaches, receiving higher waves, which produce wider surf and more beach rips, together with numerous reef-controlled rips. Vehicle access continues along the beaches and backing dunes, with a track leading to Stony Rise.

Domaschenz beach (**121**) is a curving, 400 m long beach backed by continuous eroding bluffs that rise to 30 m in the centre. The beach is also fronted by patchy reef which produces a hazardous rock and reef-strewn surf zone.

Queen Head beach (**122**) runs from the rocks at the northern side of Domaschenz beach for 1.3 km past one reef to the more prominent Queen Head reef, which forms a sandy foreland that protrudes 300 m seaward. The beach is backed by several active dunes and fronted by a 100 m wide bar broken by patchy reefs, resulting in two beach and four reef-controlled rips. On the northern side of the foreland is 1.6 km long **Back Beach** (**123**) which runs up to Stony Rise. This is the most energetic beach along this section, with waves averaging 1.6 m breaking across a 200 to 300 m wide surf zone, dominated by three to four large beach rips, together with boundary rips against the reefs. Active dunes back much of the beach.

Stony Rise is a 900 m long section of low calcarenite bluffs and reefs that contain three smaller beaches. The first (**124**) is an irregular, 300 m long strip of sand fronted by reef flats extending up to 100 m seaward of the beach. Beach **125** is a pocket 200 m long beach, bordered by protruding reefs and with reefs also scattered across the 100 m wide embayment. Beach **126** is a 100 m long pocket of sand, with a steep barless beach and small bluff and reef bordered embayment, largely free of reefs. It offers the safest swimming in this rip and reef-dominated section of coast. There is a 4WD track out from Robe to Stony Rise, with a camping area set in the scrub about

500 m behind the rise and tracks continuing through the dunes south to Queen Head and north to Cape Lannes.

Evans Cave beach (**127**) runs for 900 m between Stony Rise and Cape Lannes, a low sand and calcarenite bluff. The beach is relatively free of reefs and receives wave averaging 1.6 m which maintain a 100 to 200 m wide surf zone, dominated by two beach rips and boundary rips at either end.

Cape Lannes is fronted by a 200 m long sandy beach (**128**), bordered by the two sets of reefs, which extend up to 400 m off the cape. The beach receives waves averaging 1.5 m which maintain a 100 m wide bar, drained by a single reef-controlled rip. The cape marks the northern boundary of Little Dip Conservation Park.

Swimming: This is a hazardous section of rock, reef and rip-dominated coast. The safest beach for swimming is the northern Stony Rise beach (126). Be very careful swimming anywhere along this coast.

Surfing: There are numerous reef and beach breaks along the shore, with the best being along Queen Head, Back and Evans Cave beaches.

Fishing: A popular series of beaches and rocks to fish within easy reach of Robe, so long as you have a good 4WD and know the tracks.

Summary: An interesting section of coast offering a wide variety of beaches, bluffs and reefs. Most people drive through enjoying the views, while those who stay tend to fish.

129   WEST BEACH

No.	Beach	Rating	Type	Length
		Unpatrolled		
129	West Beach	7	LTT/TBR	1.6 km

West Beach (**129**) lies 1 km west of Robe and is the closest exposed surfing beach to the town. There is a car park off Adam Gordon Lindsay Drive, with a 150 m walk down to the northern end of the beach. Waves average 1.5 m at the northern end, decreasing slightly in lee of Cape Lannes toward the southern end. The waves produce a 100 m wide bar cut by several beach rips. The most hazardous section is just below the car park where reefs also occupy the surf. This beach is not recommended for swimming and should only be used by experienced surfers. It is however a popular spot for beach fishing.

130-135   FACTORY BAY

Unpatrolled			
No. Beach	Rating	Type	Length
130 Factory Bay (1)	4	R+rock flats	125 m
131 Factory Bay (2)	4	R+rock flats	30 m
132 Factory Bay (3)	4	R+rock flats	20 m
133 Factory Bay (4)	4	R+rock flats	30 m
134 Factory Bay (5)	4	R+rock flats	80 m
135 Factory Bay (6)	4	R+rock flats	30 m

Factory Bay is a 400 m wide rocky bay lying 1 km west of Robe township, with the scenic drive (Adam Gordon Lindsay Drive) running the length of the backing 10 to 20 m high bluffs. There is a 50 m wide bluff-top reserve between the drive and the bluffs, with houses on the other side of the drive. Below the bluffs are a series of six small pockets of sand (**130-135**), all bordered by rocks and bluffs and fronted by exposed and submerged reefs. None of the beaches are suitable for swimming, for while the waves are relatively low, the rocks and reef produce a hazardous seabed. They are however very popular for beach and rock fishing.

136-141   ROBE

Unpatrolled				
No. Beach	Rating	Type		Length
		Inner	Outer Bar	
136 Cape Dombey	2	R		50 m
137 Karatta Beach	2	R		130 m
138 Town Beach	2	R		400 m
139 Hoopers Beach	2	R		250 m
140 Fox Beach	3	LTT		150 m
141 Long Beach	5	TBR	D	10.5 km

Robe is South Australia's most popular coastal town. It has a well earned reputation based upon its historic, well preserved and attractive setting, its wide range of facilities and attractions and its relatively safe town beaches, together with the longer and more energetic Long Beach.

The town is nestled on low, undulating land in lee of Cape Dombey, a 30 m high calcarenite headland. The western side of the cape is wild and rugged, while the protected eastern side provided the first port and jetty site for Robe (Figs. 4.20 & 4.21). Today the fishing boats anchor in Lake Butler, which was connected to Guichen Bay by a short channel and breakwater. The town runs along the rocky southern shore of Guichen Bay. Robe has experienced a tourist and holiday boom in recent years and now contains numerous facilities for visitors and holidaymakers, while still retaining its original charm.

Figure 4.20   *Cape Dombey with the near landlocked Robe Harbour (a former lake), the town of Robe and the quieter southern shores of Guichen Bay.*

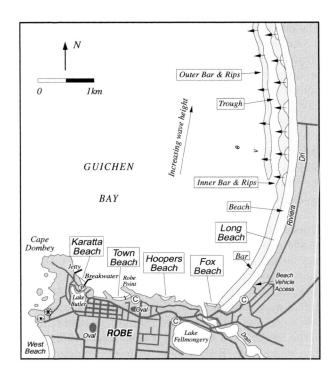

Figure 4.21   *Robe has four small protected north facing town beaches and the adjacent longer and more exposed Long Beach.*

The town area has six beaches, five located along the protected southern shore and then the more open Long Beach extending to the north. The first five, Cape Dombey, Karatta, Town, Hoopers and Fox beaches, all receive low waves averaging about 0.5 m and usually have a moderately steep beach and a narrow attached bar. Rips are usually absent, only forming when waves exceed 0.5 m. The Robe jetty is located between the first two beaches, with the harbour training wall forming the eastern boundary of Karatta Beach.

Cape Dombey beach (**136**) is a 30 m long pocket of sand lying immediately north of the Cape Dombey car park. It is a highly visible beach, though not a popular swimming location.

Karatta Beach (**137**) is a 150 m long, northeast facing beach that has built up against the channel training wall. A gravel road runs right to the wall, where there is limited parking. This beach is used more for launching sailboats, which are stored on the beach, than for swimming.

Town Beach (**138**) is the most popular of the town beaches. It extends from in front of the Robe Hotel for 300 m to 10 m high rocky Robe Point. A large beachfront caravan park backs the eastern end of the beach and the point. A small groyne crosses the last 50 m of the beach.

Hoopers Beach (**139**) is a smaller, 150 m long beach wedged in between two low rocky headlands. There is limited parking at the beach and it is used more by the beachfront property owners.

Fox Beach (**140**) is another small, 150 m long, north facing beach, located between the Drain L outlet and the low bluffs that form the southern boundary of Long Beach. Low reefs also lie off the beach.

Long Beach (**141**) is Robe's other popular beach and is used by people from the backing caravan park and by all those who want to drive to, and on, the beach. The southern 1 km is restricted to pedestrians, however north of the vehicle access point, cars and 4WDs have access to the rest of the 10 km of beach. During the summer, hundreds of cars can line the beach creating traffic jams, particularly at high tide, and adding an additional hazard to swimming at this beach.

Swimming: The five protected beaches are all relatively safe under normal wave conditions. However be careful of rips against the rocks when waves exceed 0.5 m and of the rocks themselves. Long Beach is safest at the southern pedestrian end, where waves are usually less than 1 m and rips absent. However, further up the beach wave height increases to more than 1 m, the surf zone widens and rips occur every few hundred metres.

Surfing: Long Beach is Robe's main and most popular surfing beach. It offers a wide surf zone, with usually moderate breakers spilling across the low gradient beach.

Fishing: Robe has a wide range of fishing locations, including the jetty, breakwater and low rocky headlands, beach fishing along Long Beach and boat fishing in the bay and over the inner reefs.

Summary: Robe has long been a popular holiday destination for people from Adelaide. It has now been discovered by the Victorians, causing it to be the most popular coastal town in South Australia.

Guichen Bay Conservation Park	
Established:	1967
Area:	82 ha
Beach:	part of beach 141

Toward the northern end of Long Beach is an 82 ha section of beach and backing foredune ridges, that preserves some of the natural vegetation which once covered the entire 38 000 ha foredune ridge plain. Access to the park is on foot from the beach.

142-147   **BOATSWAIN PT to CAPE JAFFA**

Unpatrolled			
No. Beach	Rating	Type	Length
142 Boatswain Pt	2	R	1.3 km
143 Cape Thomas	2	R	3.2 km
144 Aram Cove (1)	2	R	400 m
145 Aram Cove (2)	2	R	160 m
146 Wright Bay	3	R	12.7 km
147 Cape Jaffa	2	R	1.8 km

The northern end of Guichen Bay is just 10 km from Robe, however the landscape and degree of development are considerably different. The calcarenite bluffs that dominate the coast from Cape Banks to Robe lie submerged, reducing waves and maintaining usually calm conditions at the shore, but no longer dominating the coastal landscape. At Cape Thomas the only evidence of the calcarenite is the waves breaking on the shallow reefs that extend hundreds of metres offshore. The extensive shallow, seagrass covered reefs produce a low energy shoreline all the way to Cape Jaffa, then on to Kingston SE. There is a small residential development at Cape Thomas and a camping reserve at Aram Cove, but no other development or facilities.

Boatswain Point is a low 50 m long section of rocks that lies at the northern end of Long Beach and separates it from the crenulate, south facing, low wave beach (**142**) that runs for 1.3 km out to Cape Thomas. There is a gravel road to the point providing 4WD beach access to both beaches, but no facilities.

At Cape Thomas the coast turns and faces due west. The **Cape Thomas** beach (**143**) runs north in a crenulate fashion for 3.2 km. Waves average less than 0.3 m and a few small boats are usually moored in front of the small settlement, that extends for 1 km along behind the 15 m high foredune and foreshore reserve. Seagrass debris is commonly piled high on the beach.

At **Aram Cove**, the gravel road terminates at the junction of the two cove beaches. Both are bordered by low outcrops of dunerock. The first beach (**144**) is 400 m long, the second (**145**) just 160 m long and is directly in front of the camping reserve, which has toilets but no other facilities. Both beaches are usually calm, with seagrass piled along the shore.

Wright Bay beach (**146**) begins immediately north of the camping area. It is a relatively straight, but crenulate, 12 km long stretch of beach backed by a foreshore reserve, across which several blowouts have transported sand up to 500 m over the backing farmland. The beach receives waves up to 0.5 to 1 m high in the centre, which maintain a steep, cusped high tide beach. Past the blowouts a series of 300 m wide, vegetated foredune ridges back the beach all the way to Cape Jaffa.

The **Cape Jaffa light** was constructed on a reef off the cape in 1872. It was dismantled in 1973 and rebuilt at Kingston SE where it now stands as an historic tourist attraction.

At **Cape Jaffa** the shore trends north for 1.8 km and consists of a continuous low energy beach (**147**) contained in two sweeping arcs, with the central low sandy foreland being Cape Jaffa. The seagrass covered beach is backed by 300 to 500 m of densely vegetated foredune ridges. These ridges and those toward the northern end of Wright Bay contain the Bernouilli Conservation Park. Kings Camp is located 1 km west of the northern point and there is a vehicle track from the camp to the point.

Swimming: These are five low wave and relatively safe beaches, with usually calm conditions on the first four and low surging waves on the Wright Bay beach.

Surfing: None.

Fishing: Generally too shallow for beach fishing, except in Wright Bay. Most fishers head out in their boats to fish the reefs.

Summary: A quieter contrast to Robe, offering little used, low energy beaches backed by a more stable foredune.

Bernouilli Conservation Reserve

Established: 1993
Area: 242 ha
Beach: 148B

Bernouilli Conservation Reserve occupies a strip of naturally and densely vegetated foredune ridges extending from the northern end of Wright Bay to Cape Jaffa. The reserve is an important breeding area for the orange-bellied parrot. There are no facilities at the reserve and access is best along the tracks from the Cape Jaffa end.

---

148   THE COORONG

The Coorong (Cape Jaffa To Murray Mouth) (Figs. 4.22 & 4.23) Unpatrolled				
No.	Beach	Rating	Type Inner	Outer	Length Bar

No.	Beach	Rating	Type Inner	Outer	Length Bar
148A	Kings Camp	2	R	none	6 km
148B	Lacepede Bay	2	R	none	7.5 km
148C	Wyomi Beach	2	R	none	3.5 km
148D	Kingston SE	2	R/LTT	none	5 km
148E	Long Beach (S)	3	LTT	D	15 km
148F	The Granites	5	TBR	D	5 km
148G	Long Beach (N)	6	TBR	D	23 km
Coorong National Park (28 Mile Crossing to Murray Mouth)					
148H	28-32 Mile Crossing	6	RBB	D	12 km
148I	32-42 Mile Crossing	7	RBB	D	6 km
148J	42 Mile to Tea Tree	7	LBT	D	11 km
148K	Tea Tree-Murray Mouth	8	D	D	100 km
				Total length	194 km

The **Coorong** is the common name for Australia's longest beach, 194 km of continuous sand running from Cape Jaffa to the Murray Mouth, with only three rocks on the entire beach. The beach does not have a single official name, rather each section has local names. The common name Coorong comes from the long salty lagoon that backs about 140 km of the beach and is the name of the national park that also occupies approximately 140 km of the beach and dunes. The beach begins at low Cape Jaffa, a sandy promontory that has accumulated in lee of Margaret Brock Reef. The reef is named after the barque *Margaret Brock* which struck the reef in 1852. The first 20 km are devoid of ocean waves owing to extensive offshore reefs, with seagrass covered shallows lying off the steep narrow beach. Only beyond Kingston SE do ocean waves start to reach the beach. They gradually increase in height for the next 80 km until Tea Tree Crossing, beyond which the northern half of the beach is one of the highest energy in Australia. The beach finally terminates at Murray Mouth (the mouth of the Murray River).

There are only two settlements on the beach, the small fishing community at Kings Camp on Cape Jaffa and the sole town of Kingston SE. Most of the coast is given over to the magnificent surf, beach, dunes and backing lagoon.

Kings Camp is a small fishing community located 1 km east of Cape Jaffa. A small jetty services the fishing fleet which lies at anchor off the beach. There is also a store and caravan park. A reserve and road back the beach, providing beach parking and access. The first 6 km of beach (**148A**) is low, narrow, faces north and is usually covered in seagrass. Ocean waves are negligible, with only local wind waves washing against the shore. Breaking waves and currents are usually absent and only during large

Regional Map 2: The Coorong to Fleurieu Peninsula

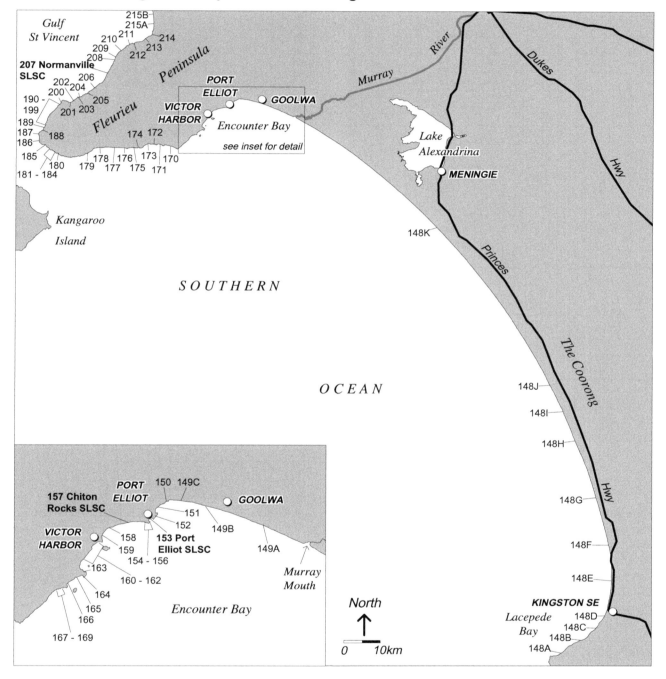

Figure 4.22   *Regional Map 2 – The Coorong and Fleurieu Peninsula*

seas do low ocean waves reach the beach. The beach slowly curves around to face northwest at the northern end and is backed by low spits and recurved spits, which grade seaward into beach-foredune ridges. At the northern end there are up to 40 Holocene beach ridges across a 1 km wide ridge plain, indicating the extent of shoreline progradation over the past 6 000 years.

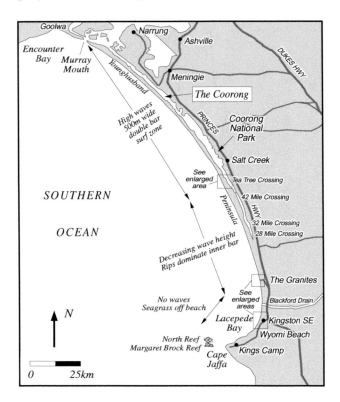

Figure 4.23   *Map of the coast between Cape Jaffa and Encounter Bay, which includes Australia's longest beach, the 194 km continuous stretch of sand from Cape Jaffa to Murray Mouth, which incorporates the Younghusband Peninsula and the backing Coorong lagoon.*

Lacepede Bay is the name of the low energy bay extending from Cape Jaffa to Kingston SE. While dominated by low waves, the bay contains large, seagrass covered sand waves, that apparently move very slowly eastward along the bay. Where the sand waves attach to the shore, the shoreline protrudes up to 300 m seaward. These protrusions give the bay shore a crenulate character (Fig. 4.24). The 7.5 km section of beach (**148B**) faces northwest, is low, narrow and usually covered with seagrass. The entire beach section is backed by 500 m of densely vegetated foredune ridges, including the salt Hog Lake, then cleared sheep paddocks. Butchers Gap Drain forms the northern boundary with Wyomi Beach and public access is limited to either end at Kings Camp or Wyomi Beach. The ridges and salt lakes associated with Butchers Gap Drain are part of the Butchers Gap Conservation Park.

Figure 4.24   *Lacepede Bay is dominated by shallow, shore-transverse sand waves with bare sandy crests and seagrass covered troughs. Where the crests attach to the low energy shoreline it protrudes up to a few hundred metres into the bay.*

Wyomi Beach (**148C**) is the name of a holiday-retirement settlement and beach located just east of Kingston SE. The 3 km long beach faces northwest and receives waves averaging less than 0.5 m, which build a low, flat beach. There is a beach boat ramp and vehicle access along the beach. Most of the houses are located behind a 200 to 300 m wide nature reserve, which runs the length of the beach and is crossed by several pedestrian and vehicle access tracks.

Kingston SE is the only town on the beach and offers most services. It is the site of Port Caroline, a stretch of open beach with a 300 m long jetty to service the fishing fleet (Figs. 4.25 & 4.26). The road from Wyomi Beach parallels most of the 7 km long beach (**148D**) with a 50 to 100 m wide, grassy foreshore reserve lying between the road and the beach. The beach faces west-northwest and usually has waves less than 0.5 m and low currents. However seagrass debris is a major problem and can be over a metre deep on the beach. Maria Creek (named after the tragic shipwreck) crosses the beach 400 m north of the jetty. Its mouth has been trained by two 150 m long rock walls, in an attempt to form a harbour entrance. This creek, together with Butchers Gap and Blackford drains, are the only streams crossing the entire Coorong beach. The fact that a jetty can be built on this west facing section of coast is owing to the extensive shallow reefs that, between Kingston and Cape Jaffa, extend up to 10 km offshore and lower ocean waves to essentially zero.

North of Kingston SE begins Long Beach, a 40 km long transition zone from zero to moderate ocean waves. The offshore reefs that have protected the Lacepede Bay beaches begin to deepen, allowing more wave energy to reach the shore. The backing barrier grades from a 1 km wide series of stable foredunes at Blackford Drain, to sporadic blowouts over the foredunes ridges at The Granites, to predominantly active dunes by the 28 Mile Crossing.

and maintain a surf zone up to 200 m wide, containing three low bars. There is a vehicle access track and car park behind the rocks, but no other facilities.

Figure 4.25 *Kingston SE is located on the low energy northern shores of Lacepede Bay, with seagrass growing to the shore, a jetty and the trained mouth of Maria Creek.*

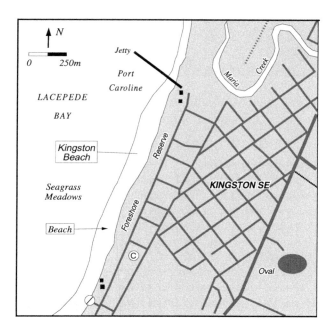

Figure 4.26 *Kingston SE and its beach and jetty.*

Figure 4.27 *The Granites are the only exposed bedrock between Cape Banks and Middleton. The handful of round rocks outcrop on the beach, providing a prominent landmark.*

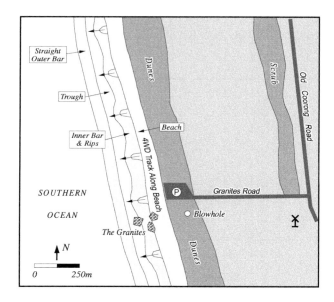

Figure 4.28 *The Granites beach usually has rips cutting the inner bar, with the outer bar lying 200 m off the beach.*

Blackford Drain is the main 4WD vehicle entrance to the southern Coorong. **Long Beach (S)** (**148E**) begins at the drain and is a 15 km long section of relatively straight, low, flat beach, extending north to The Granites. Waves average 0.5 m at the drain and are 1 m high by the northern end of this section. Because of the firm beach sand, 2WD as well as 4WD vehicles often use this beach. The beach is low and wide, particularly at low tide, and is fronted by a shallow surf zone which gradually widens to the north. As it does so a second bar forms off the beach, while rip currents and channels begin to cut the inner bar.

The Granites beach (**148F**) is the site of the only rocks on the entire beach, a few 2 m high, rounded granite knobs that lie in the intertidal swash zone (Figs. 4.27 & 4.28). Waves average over 1 m along this 5 km section of beach

By the northern end of The Granites beach section the shoreline begins to oscillate, owing to large beach rhythms, up to 3 km in length. Along the **Long Beach (N)** section (**148G**), these rhythms dominate the shoreline. They are produced by the increasingly higher waves and represent both points of bar attachment on the protruding horns and areas of large scale rip formation in the bays. The increasing wave height also widens the surf zone to over 400 m by 28 Mile Crossing.

The **28 Mile Crossing** is a 2 km long 4WD track that leaves the old highway and skirts most of the backing dunes to reach the beach at a point where coarse shell debris in the sand maintains a steep, narrow beach. The 12 km of beach between 28 Mile and 32 Mile Crossing (**148H**) is totally dominated by the coarse shell grit; as a result the beach is usually soft and steep and four-wheel driving is difficult. The surf zone however is a 400 m wide, low gradient, three bar system, with rips spaced every few hundred metres in the inner surf zone. The Coorong National Park begins 3 km south of the 28 Mile Crossing and runs along the beach for 132 km to Murray Mouth.

The **32 Mile Crossing** track begins about 1 km south of where the old highway rejoins the new highway. It crosses a causeway over the salt lake bed, then runs for 2 km through the dunes to reach the beach. The 6 km of beach (**148I**) between 32 Mile and 42 Mile Crossing is similar to that to the south, with the coarse shell continuing to dominate the beach face, fronted by a high wave, 400 to 500 m wide surf zone.

The **42 Mile Crossing** track leaves the main highway and skirts the southernmost end of the usually full Coorong Lakes for 2 km, before following a sandy track for another 2 km to the beach. The beach (**148J**) remains steep and coarse, fronted by a wide energetic surf zone between 42 Mile and Tea Tree Crossing, 11 km to the north.

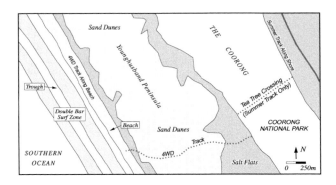

Figure 4.29   *Tea Tree Crossing crosses the usually dry Coorong salt lake, then winds through the 1.5 km wide sand dunes before reaching the high energy Coorong beach.*

Tea Tree Crossing is a summer-dry weather crossing. It begins off a section of the old highway 6 km south of Salt Creek and crosses a lake bed, which is usually flooded in winter. It then runs through an active dune sheet for 1 km before reaching a hard packed, 200 m wide deflation basin behind the beach (Fig. 4.29). This is the last crossing. North of here and up to Murray Mouth, the 100 km long Coorong Lagoon separates the beaches and dunes from the highway.

The beach, which begins with the coarse shell and wide, high energy surf zone, gradually changes over the first few kilometres north of the crossing into a finer sand, flatter, firmer beach, which continues on in this way to the **Murray Mouth**. The beach faces directly into the

persistent high southwest swell, with waves averaging over 2 m breaking over 500 m out to sea, across a wide double bar surf zone. Massive dune systems averaging 1.5 km wide back the beach (Fig. 4.30). During periods of lower waves, well developed rips, spaced approximately 500 m (though ranging in spacing from 300 to 900 m), dominate the inner surf zone. The beach usually narrows in width in lee of the rips (called the *rip embayment*) and may be scarped.

Figure 4.30   *View towards Murray Mouth across the massive active dune systems of the Coorong National Park.*

Driving the Coorong: It is possible to drive the Coorong from Blackford Drain to Murray Mouth, a distance of 165 km. It is a spectacular drive. Most of the beach is composed of fine sand and is low, flat and fast. However, there is the 32 km soft, shelly section beginning 30 km north of the Granites and extending a few kilometres past Tea Tree Crossing. Beach driving conditions can range from excellent to treacherous. Best conditions are during summer, at low tide and with low waves. Worst conditions are in winter, when the beach tends to be narrower, at high tide and any time during high waves. The high seas produce wave set-up and set-down. The periodic set-up occurs every 1 to 2 minutes and causes the swash to suddenly run high up the beach, a process which has trapped many cars. Finally, the beach rhythms or embayments cause the beach to narrow and steepen into the embayed sections. Be very careful when driving this section: check the tide and make sure you have sufficient time to drive the narrow section before the next swash or wave set-up occurs. Obtain a copy of the Coorong National Park guide which locates all access points and provides advice on driving conditions and rules. Always check with locals or other drivers before driving past The Granites.

Swimming: Swimming conditions range from relatively safe along the Lacepede Bay beaches, to moderately safe up to The Granites and then increasingly hazardous as the higher energy beaches north of The Granites are encountered. The biggest problems are large scale rips in the inner surf zone, spaced approximately every 500 m, strong set-up and set-down which produce strong seaward flowing currents and the big surf. Always be very careful if entering the surf anywhere north of Blackford Drain.

Surfing: There are 160 km of beach breaks along The Coorong, it's just a matter of driving along until you see a break that suits you. The better waves break on the outer bar, which can mean a 400 to 500 m paddle. So it's best to be fit and surf with friends.

Fishing: The Coorong is a very popular beach for fishing, with the majority of the 4WD vehicles belonging to fishers. During summer and weekends they set up many camps along the beach.

Summary: Australia's longest beach, which grades from no waves to one of the world's highest energy beaches. It is a spectacular beach backed by extensive dunes north of The Granites, very accessible by 4WD and with many kilometres of beach camping. However outside of Kings Camp and Kingston SE there are no facilities, apart from the camp sites.

Coorong National Park

Established: 1966
Area: 46 745 ha
Beaches: 148H-K
Coast length: 132 km (28 Mile Crossing to Murray Mouth)
Beach access: 28 Mile, 32 Mile, 42 Mile and Tea Tree Crossings.
Camp sites: 28 Mile, 32 Mile, 42 Mile and Tea Tree Crossings; camping is also permitted on the beach. Water and toilets are only available at the 42 Mile Crossing camp site.

Check *Coorong National Park* information brochure for full details.

Park Ranger
Private Bag 43
Meningie SA 5264
Phone: (08) 8575 1200

FLEURIEU PENINSULA DISTRICT
MURRAY MOUTH TO MYPONGA
(Fig. 4.22)

Length of Coast:	139 km
Beaches:	65 beaches (149-214)
Surf Lifesaving Clubs:	Port Elliot, Chiton Rocks, Normanville
Lifeguards:	Normanville
Major towns:	Goolwa, Port Elliot, Victor Harbor, Normanville, Second Valley

The **Fleurieu Peninsula** is an 80 km long extension of the South Mount Lofty Ranges that terminates at Cape Jervis. The peninsula has a maximum width of 25 km between Sellicks Beach and Victor Harbor, narrowing to the tip at Cape Jervis. It consists of a central hilly section which ranges up to 450 m in height, sloping down on the south and east to generally hilly, steeply sloping coastline, particularly west of Victor Harbor. The south coast is the major coastal area, between Goolwa and Victor Harbor, with largely undeveloped coast west to Cape Jervis, including the Newland Head and Deep Creek Conservation Parks. North of Cape Jervis the coast faces into the lower Gulf St Vincent, with the generally hilly, steep coastline dissected by several valleys, the larger ones containing the settlements of Rapid Bay, Second Valley and Normanville. The entire peninsula contains 139 km of bedrock-dominated coast and 65 generally small beaches. The south coast beaches are exposed to high waves, while the gulf beaches are increasingly protected.

149   GOOLWA (MURRAY MOUTH TO MIDDLETON)

No.	Beach	**Unpatrolled** Rating		Type		Length
		Inner	Outer	Inner	Outer	Bar
149A	Goolwa	6	8	D	D	12 km
149B	Surfers	6	8	D	D	3 km
149C	Middleton	7	8	D	D	2 km

The **Murray Mouth** produces a small but significant break in the long Coorong beach. On the north side of the mouth is an 11 km long beach and dune system backed by the Goolwa Channel and known as the Sir Richard Peninsula. The town of Goolwa is located at the western end of the channel. Goolwa Beach (149A) begins at the mouth and runs for 12 km, then for another 5 km as Surfers (149B) and Middleton beaches (149C), where it is backed by 10 m high bluffs. The beach finally ends against the rocks of Middleton Point. But for Murray Mouth (which occasionally closes), the entire beach system from Cape Jaffa to Middleton is 211 km in length.

The main beach access is at **Goolwa Beach (149A)**, 11 km northwest of the mouth (Fig. 4.31). The road runs straight out to the beach where there is a large car park, toilets and a kiosk. There is also 4WD access to the beach out to Murray Mouth. The other two access points are to the west through the **Surfers** and **Middleton** subdivisions, both spots providing car access to the low bluffs overlooking the long beach and wide surf zone (**149B & 149C**) (Fig. 4.32). There is a car park on Middleton Point, as well as toilets and a viewing platform.

Figure 4.31   *Goolwa Beach car park (left) lies toward the western end of the beach that terminates at Murray Mouth 11 km to the southeast. Note the typical high energy, 500 m wide dissipative surf zone.*

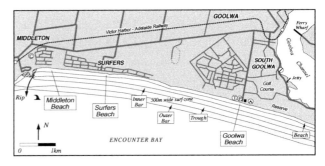

Figure 4.32   *The straight section of beach between Goolwa and Middleton represents the northern end of Australia's longest and one of its highest energy beaches.*

The entire beach is composed of fine sand and exposed to waves averaging over 2 m. These break across a 500 m wide double bar surf zone, characterised by numerous spilling breakers and substantial wave set-up and set-down at the shoreline and, during lower wave conditions, widely spaced rips. At Middleton a strong permanent rip runs out against the rocks.

Swimming: These are three of the most hazardous beaches in South Australia owing to the persistent high waves. However because of the very wide surf zone, it is moderately safe to swim in the inner surf zone on the bar. Do not, however, swim beyond the first line of breakers, as strong currents occupy the trough between the bars.

Surfing: Middleton is regarded as one of the state's main surfing beaches owing to the easier access across the wide surf zone, using the permanent rip which runs out past Middleton Point. There are numerous breaks both along and across this wide surf zone, with best conditions during moderate swell and northerly winds.

Fishing: Most beach fishers head for Murray Mouth and the beach east of Goolwa. There is also rock fishing at Middleton Point.

Summary: Goolwa is the most popular surfing beach south of Adelaide and attracts large crowds in summer. However take care if swimming here as it is potentially hazardous.

150-152   MIDDLETON ROCKS, BASHAM, CROCKERY BAY

Unpatrolled			
No. Beach	Rating	Type	Length
150 Middleton Rocks	5	TBR+rocks	400 m
151 Basham Beach	3	LTT	1.8 km
152 Crockery Bay	4	R+rocks	50 m

At Middleton are the first bedrock outcrops to occur on the coast since Cape Banks 320 km to the southeast. The Middleton residential area backs the 10 m high bluffs, which provide views of both the long Middleton beach as well as the rock-dominated Middleton Rocks beach. A sealed road runs along the back of Middleton Rocks beach and the eastern end of Basham Beach, while access to Crockery Bay is from a car park located 300 m west of the beach (Fig. 4.33).

Middleton Rocks beach (**150**) is dominated by dipping metasedimentary rocks which both border and front much of the 400 m long beach, with a central 30 m area clear

of rock providing access to the ocean. Waves break across the rock reefs and a permanent rip flows out in front of the sand pocket. There is a car park at the western end of the beach, as well as the car park and facilities on Middleton Point at the eastern end.

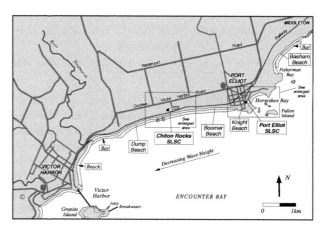

Figure 4.33   *Middleton to Victor Harbor, see Figures 4.35 and 4.36 for more detail of Port Elliot and Chiton Rocks.*

Basham Beach (**151**) is a curving, 1.8 km long, lower energy beach, which faces east, is sheltered by Commodore Point and receives waves decreasing from 1 m in the east to less than 0.5 m in the west. The protected western corner is called Fisherman's Bay. The low waves maintain a narrow continuous bar which decreases in width to the west. In addition, rocks outcrop along the beach and form shallow reefs off the beach. A bike track parallels the back of the beach, with vehicle access to car parks at either end. There is also a regional park backing the beach, with toilet and picnic facilities.

On Commodore Point is a 50 m pocket of southeast facing sand called **Crockery Bay** (**152**), that is largely fronted by rocks and boulders. The backing flat headland hosts a large caravan park, which extends down into adjoining Horseshoe Bay. The nearest public road and car park are 300 m west of the beach.

Swimming: Both Middleton Rocks and Crockery Bay beaches are dominated by rocks and are usually unsuitable for swimming. The eastern end of Basham, away from the rocks, offers the safest swimming area.

Surfing: There is no recognised surf at any of these three beaches, although surfers do paddle out from Middleton Rocks during big seas to surf the adjoining Middleton beach breaks.

Fishing: The rocks of Middleton Rocks and Commodore Point are the best locations. However, beware of large waves which wash over the rocks.

Summary: Three accessible beaches with no facilities.

153, 154   HORSESHOE BAY (PORT ELLIOT)

PORT ELLIOT SLSC

Patrols: November to March, weekends and
public holidays

No.	Beach	Rating	Type	Length
153	Port Elliot	3	R	700 m
154	Ladies Beach	3	R	60 m

Horseshoe Bay is a small, semicircular, southeast facing bay, wedged in between the prominent 30 m high granite headlands of Commodore Point and Freeman Nob. The bay is also afforded protection by Pullen Island and some granite reefs (Figs. 4.34 & 4.35). The moderately protected bay once served as the port for Port Elliot. While the small jetty remains, the bay is now a major recreation area and the site of Port Elliot Surf Life Saving Club, formed in 1935, a bowling club and a large beachfront caravan park. There is also a beachfront park with picnic and barbeque facilities.

Figure 4.34 *Port Elliot and Horseshoe Bay (foreground) with Chiton Rocks beach to rear.*

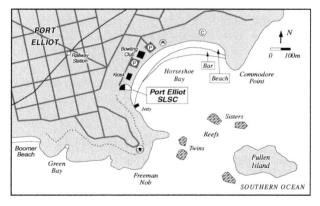

Figure 4.35 *Port Elliot and Horseshoe Bay are protected by Pullen Island and adjacent reefs.*

The **Port Elliot** beach (**153**) is 700 m long, faces east at the jetty and swings around to face south against Commodore Point. Because of the protection afforded the bay waves average 0.5 m, increasing slightly toward Freeman Nob. The low waves, together with the medium to coarse sand, produce a high, steep beach, with no bar following low waves and a narrow bar after high waves. These conditions cause the waves to surge strongly up the beach. Regular beach cusps are a feature of the shoreline.

Ladies Beach (**154**) is located in lee of the breakwater at the western corner of the bay. It is a 60 m long pocket of north facing sand, which was traditionally used by the ladies for bathing.

Swimming: This is a relatively safe beach under normal wave conditions. High swell will however produce a very heavy shorebreak. The safest swimming is always at the western surf club end, where waves are always smallest.

Surfing: Usually only a low shorebreak in the bay.

Fishing: The jetty and boundary rocks are the more popular spots to fish. The absence of a bar also allows deep water just off the beach.

Summary: This is a very accessible, well serviced little bay and beach. It is relatively safe, protected from westerly winds and very popular in summer.

To some extent, surf lifesaving in South Australia commenced at Port Elliot owing to several shipwrecks in the bay, all during strong winds and high wave conditions. The first was the *Solway* in 1837. There were many wrecks and groundings at this hazardous little port, with four vessels lost during 1856. As early as 1877 the bay was used as a site for sporting days, including aquatic events. The 1900 sports day at the bay included many swimming races.

155, 156   ROCKY BAY, GREEN BAY

	Unpatrolled		
No.	Beach	Rating	Type	Length
155	Rocky Bay	7	R+rocks	60 m
156	Green Bay	7	R+rocks	100 m

Rocky and **Green Bay** beaches (**155** and **156**) occupy two small rock bays eroded on the western side of Freeman Nob. A road runs along the top of the 20 m high bluffs backing the adjoining bays, with a well-made walking track around the back of Rocky Bay and a steeper track down to Green Bay. Both bays have high tide sand beaches and some sand in the surf, however rocks dominate the surf with a strong permanent rip running out of each bay. Neither is suitable for swimming or surfing. Their sloping granite rocks are used for rock fishing, however be very careful as they are prone to wave overwashing and can be extremely hazardous.

157-159   CHITON ROCKS to VICTOR HARBOR

CHITON ROCKS SLSC (Boomer, Watsons Gap and Chiton Rocks beaches)			
Patrols:	November to March, weekends and public holidays		
No. Beach	Rating	Type	Length
157 Chiton Rocks	5	TBR/LTT+rocks	3.6km
158 Hindmarsch River	4	LTT	900 m
159 Victor Harbour	3	LTT	1.6 km

To the west of Freeman Nob is a 3.6 km long, south facing beach (**157**) that goes under three different names. The first section is called Boomer Beach, then Watsons Gap around the small creek of the same name and finally Chiton Rocks after some rocks that outcrop in the surf. This beach is backed by a foreshore reserve, the old railway tracks and then residential development, much of it on low bluffs overlooking the beach. There is a large car park above Boomer Beach with stairs down to the beach and car parks either side of the Chiton Rocks Surf Life Saving Club (Figs. 4.36 & 4.37).

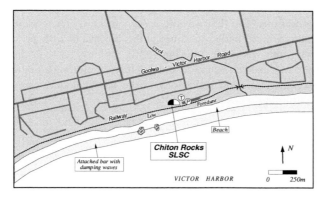

Figure 4.36   *Chiton Rocks Surf Life Saving Club and adjacent beach area, including the rocks and usually continuous attached bar.*

The **Boomer Beach** section (**157A**) begins amongst the rocks of Freeman Nob. Its name refers to the often heavily dumping waves on this reflective to low tide terrace beach. At **Watsons Gap** (**157B**), a usually closed small creek reaches the beach. Beyond this point, rock reefs run just off the beach, outcropping at Chiton Rocks 100 m east of the car park. The **Chiton Rocks** section (**157C**) usually has several small rips between the rocks and the

southern sandy foreland. The Chiton Rocks Surf Life Saving Club patrols this central 1 km long section of the beach. The main Goolwa-Victor Harbor Road runs 400 m inland, parallel to the beach. However the beach is not well sign posted and you might need a street map to find it. The road eventually reaches the beach at a long car park, with the surf club sitting below a 20 m high grassy bluff.

Figure 4.37   *Chiton Rocks Surf Life Saving Club and beach shown with waves breaking at high tide across the attached bar.*

The Victor Harbor Life Saving Club conducted the first patrols of Chiton Beach in 1933 and also erected a kiosk and changing sheds on the beach. An average of 34 people were rescued from the surf each year during the 1930s, with the first surf fatality occuring in 1936.

The reef-controlled sandy foreland at the western end of Chiton Rocks beach forms the eastern boundary of a 900 m long beach (**158**), that terminates at the usually closed mouth of the small Hindmarsh River (more a creek than a river). Waves average less than 1 m and maintain a narrow continuous bar, usually free of rips. The beach is backed by a 50 m wide scrubby reserve, then residential houses, with a reserve along the eastern side of the river mouth.

On the western side of the river, the beach continues for 1.6 km to the tip of the prominent sandy foreland in lee of **Granite Island** and the 500 m long causeway out to the island. This beach (**159**) becomes increasingly protected by the island. As it does so, it narrows and seagrass grows up to the beach. The entire beach is backed by a public reserve containing numerous recreation facilities, including a boat ramp near the point. The reserve is backed by the main shopping area of Victor Harbor.

The three beaches between Freeman Nob and Victor Harbor are partly protected by Granite Island and waves average 1 m, which produce a single continuous bar and moderately sloping beach. Weak rips are present under normal conditions and intensify when waves exceed 1 m.

Swimming: Boomer-Chiton Rocks is a moderately safe beach with a patrolled area, however watch for poorly defined rips, particularly against the rocks, and be careful of the often heavy shorebreak. Waves decrease toward the west, providing lower surf and usually calm conditions at Victor Harbor.

Surfing: There is usually a narrow beach break favoured more by body boarders than surfers.

Fishing: The rocks attract most fishers, while a good cast can reach the deeper water beyond the usually narrow bar.

Summary: Chiton Rocks is a patrolled beach missed by most tourists and frequented more by the locals. Most of the tourists head for the very accessible and low energy Victor Harbor beach.

160-163   **VICTOR HARBOR**

No.	Beach	Rating	Type	Length
Unpatrolled				
160	Victor Harbor (W)	2	R	1 km
161	Inman River	2	R	600 m
162	Yilki Beach	2	R+rock flats	3 km
163	Rosetta Head	3	R+rocks	300 m

Between Granite Island and Rosetta Head is an open, 3 km wide, southeast facing bay. The bay contains 5 km of predominantly sandy shore and four low energy beaches. The town of Victor Harbor (Fig. 4.38) backs the first beach and the road to Rosetta Head parallels the other three beaches, providing excellent access to all four beaches.

Figure 4.38   *Victor Harbor is located on a low, sandy, cuspate foreland formed in lee of Granite Island, with low energy beaches to either side, both backed by a wide foreshore reserve.*

Victor Harbor (W) beach (**160**) is a 1 km long, curving, south facing beach, which usually receives waves averaging about 0.5 m. The entire beach is backed by a foreshore reserve and car parking, then the town of Victor

Harbor. The beach usually has low to no surf, with the seagrass growing right to the beach.

Lifesaving on the south coast began at Victor Harbor, with the first display of lifesaving held in January 1910. The Victor Harbor Life Saving Club was formed in 1931.

The **Inman River** forms the western boundary of Victor Harbor township and the narrow and at times closed mouth forms the boundary between Victor Harbor and the river mouth beach, which extends for 600 m south to a low sandy foreland. This low energy beach (**161**) is fronted by seagrass beds and backed by a wide reserve in the west, with the eastern half consisting of a low sandy spit. Behind the spit is a large caravan park.

Yilki Beach (**162**) is the longest beach in the bay, running for 3 km in a southeast, then south, direction to the base of Rosetta Head. It is backed by a road and the settlements of Yilki and Encounter Bay. This is a very low energy, narrow beach, with shallow seagrass covered sand then rock flats fronting the entire beach. Calm conditions usually prevail. At the southern end is a small boat ramp and small boats often moor off the beach.

Rosetta Head is a 100 m high, conical granite headland. On its protected northern side is a narrow strip of sand largely fronted by rocks facing north toward West Island, 500 m offshore. A gravel road backs the winding beach (**163**) and leads out to a small wharf at its eastern end. The wharf is now mainly used for fishing.

Swimming: These are four low energy and relatively safe beaches. The most popular are Victor Harbor and the Inman River mouth beaches, which are both backed by large reserves.

Surfing: None in the bay, except during big seas when waves break at Victor Harbor beach.

Fishing: The Rosetta Head wharf is the most popular shore-based spot to fish from, along with the Inman River, particularly when the mouth is open.

Summary: Four beaches set in a low energy bay, with excellent access to all and the facilities of Victor Harbor right behind.

164-166   **PETREL COVE, RALGNAL, KINGS BEACH**

No.	Beach	Rating	Type	Length
Unpatrolled				
164	Petrel Cove	7	TBR+rocks	200 m
165	Ralgnal	7	TBR+rocks	500 m
166	Kings Beach	5	R/LTT	200 m

Rosetta Head marks a dramatic change in the nature of the coast. The predominantly sandy coast east of Victor Harbor is replaced by high rocky coast between Rosetta Head and Cape Jervis and up to Seacliff, just south of Adelaide, a total distance of 100 km. These are the often heavily folded metasedimentary rocks of the Fleurieu Peninsula. The rocks and the structural alignment of the coast dominate the beach systems.

The first section of coast between Rosetta Head and Newland Head, 10 km to the southwest, is essentially straight, bedrock-controlled and exposed to moderate to high southwest waves. There is a car park on the bluffs at Rosetta and a walking trail from there to the other beaches.

In the first 2.5 km are three small beaches, between Rosetta Head and King Head. The first is **Petrel Cove (164)**, a 200 m long, south facing beach, which is bordered and backed by the 20 m high steep bluffs of Rosetta Head. The steeply dipping metasedimentary rocks also form prominent jagged rock platforms to either side and along the eastern half of the beach. There is just a 50 m pocket of open sand and a bar, which is also the location of a permanent rip.

Just west of the cove is a 500 m long high tide sand beach **(165)**, also backed, bordered and fronted by rocks and jagged rock reefs, with a more open, sandy, 80 m long pocket toward the western end. Waves average about 1.5 m and two permanent rips drain the surf zone. This beach is backed by Ralgnal farm and a small rocky creek crosses the western end of the beach. Public access is via a foreshore walking track along the grassy bluffs from Petrel Cove.

King Head protrudes 500 m to the south and tucked on its eastern side is a 200 m long pocket of sand **(166)**, which receives lower waves and usually has a steep beach fronted by a narrow bar and is usually free of surf and rips. Grassy to rocky bluffs back and border the beach, and some rock reefs lie off the beach. A farm backs the beach with access via the coastal walking track.

Swimming: Kings Beach is the only relatively safe beach, as Petrel Cove and Ralgnal are dominated by rocks and rips.

Surfing: Chance of a short beach break at Petrel Cove, however watch the rocks.

Fishing: This is a popular spot with permanent rip holes and good access from both the beaches and rocks.

Summary: Three rock-dominated sand beaches, all lying along the coastal walking track to Newland Head.

167-169   KING HEAD

Unpatrolled			
No.	Beach	Rating	Type	Length
167	King Head (1)	10	boulder beach	300 m
168	King Head (2)	10	boulder beach	200 m
169	King Head (3)	10	boulder beach	100 m

King Head is a conical, 60 m high, grassy bluff surrounded by jagged, steeply dipping metasedimentary rocks. Between King Head and Newland Head is a straight, 7 km section of 100 m high, southeast facing cliffs. Tucked in beneath the first 1.5 km of cliffed coast are three small boulder beaches. The first **(167)** is 300 m long and is backed by a steep, V-shaped grassy valley which provides access to the beach. The second **(168)** is 200 m long and is backed by cliffs and steeper vegetated bluffs with no safe access. The third **(169)** is a 100 m pocket of sand beneath steep bluffs and cliffs, with no safe access.

The first of these three boulder beaches is accessible for fishing, however none are safe or at all suitable for swimming or surfing.

Newland Head Conservation Park
Established: 1985
Area: 1036 ha
Beaches: 2 (170 & 171)
Coast length: 7 km

Newland Head is a prominent, 90 m high, metasedimentary headland where the coast takes a 90^0 turn to face southwest into the higher southwest swell and winds. The conservation park is an area of dense natural vegetation covering the undulating surface behind the cliffs, as well as the beaches and backing dunes of Waitpinga and Parsons beaches. There is car access to Waitpinga and the cliffs above Parsons and a camping area at Waitpinga. A walking track following the cliffs, dunes and beaches runs the length of the park.

170, 171   WAITPINGA, PARSONS

Unpatrolled			
No.	Beach	Rating	Type	Length
170	Waitpinga	8	RBB	3.1 km
171	Parsons	8	RBB	1.2 km

Waitpinga Beach (170) is one of the more popular surfing beaches on the Fleurieu Peninsula. It is located 15 km west of Victor Harbor, with a 3 km long sealed road leading from the main road right to the beach. There is a car park behind the dune and an elevated walkway across the dune to the centre of the 3.1 km long beach. The beach is backed by largely vegetated dunes, with Waitpinga Creek and its elongated lagoon behind the centre of the beach. A second parking area is located on the western headland, reached via Parsons Beach Road.

The beach faces almost due south, exposing it to persistent high swell which averages about 2 m. Ninety metre high Newland Head forms the eastern boundary, with 40 m high Waitpinga Hill separating it from adjoining Parsons Beach. The combination of high swell and medium sand produces a steep beach and a surf zone dominated by eight to ten large rips, separated by bars which are usually detached from the beach (Fig. 4.39). The alternating bars and rips produce megacusps, large scallops in the beach in lee of the rips and protrusions in lee of the bars (Fig. 4.40). Strong permanent rips run out against each headland, in addition to rocks in the surf for 300 m south of Newland Head.

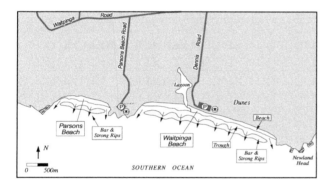

Figure 4.39 *Waitpinga and Parsons beaches have strong beach rips, including permanent rips against all four headlands.*

Figure 4.40 *Waitpinga Beach showing the typical wide surf zone, prominent strong rips and rhythmic shoreline.*

Parsons Beach (171), also known as **Pareena Beach**, lies immediately west of Waitpinga Hill headland and shares a clifftop car park with Waitpinga. There is a walking track from the car park down to the eastern end of the 1.2 km long beach. It receives moderate to high waves, which maintain a 150 m wide surf zone dominated by two permanent rips at either end and one to two central beach rips. Climbing vegetated dunes back the beach, with a small creek draining across the western end. Apart from the car park there are no facilities.

Swimming: These are two exposed, usually high wave and always rip-dominated beaches. There are deep rip channels and strong rip currents right off the beach. Only swim or surf here if you really know what you are doing and are a very experienced surfer.

Surfing: The alternating rips and bars on Waitpinga produce a series of beach breaks which can hold good waves up to 3 m, while Parsons can hold up to about 2 m. However only surf with friends as conditions here can be very hazardous.

Fishing: Waitpinga is a very popular spot for beach fishing owing to the deep, persistent rip gutters and holes, together with the rocks at each end.

Summary: Two relatively natural and wild beaches backed by rising dunes and a central small lagoon at Waitpinga, and both fronted by the rip-dominated surf. They are only suited for experienced surfers, otherwise it is worth the drive out for the view and a walk along the beach.

172-177   BEACH 172 to TUNKALILLA

| | **Unpatrolled** | | |
No.	Beach	Rating	Type	Length
172	Beach 172	7	TBR	220 m
173	Coolawang	7	TBR	210 m
174	Ballapanudda	7	TBR	190 m
175	Callawonga	7	TBR	250 m
176	Tunk Head	7	TBR	400 m
177	Tunkalilla	6	LTT	5 km

West of Parsons Beach is 15 km of bedrock coast that rises in steep, grassy slopes to elevations of over 200 m. Several streams have cut deep, steep valleys into these slopes, with small beaches forming in some of the drowned valleys. All the beaches face due south and receive waves averaging about 1.5 m, with increasing protection to the west from Kangaroo Island. The entire section is backed by private farmland, with the only public access via the Tunkalilla Road to the slopes above Tunkalilla Beach.

The first beach (**172**) lies 1 km west of Parsons. It is 220 m long, backed by gentle grassy slopes and a low bluff, with a small foredune and the beach. It has a 100 m wide surf zone, usually with two permanent rips against the jagged rock platforms to each end. During big seas the sand is stripped off, revealing a cobble-boulder beach and rock reefs in the surf.

Coolawang Beach (173) lies at the mouth of the creek of the same name. It is 200 m long, bordered by rock reefs and backed by a low foredune and small lagoon. A cobble-boulder beach also underlies its sand beach, with two strong permanent rips flowing out against each headland.

Ballapanudda Beach (174) is 190 m long and bordered on either side by cliffs rising over 60 m. Two small creeks reach the back of the beach, joining in a very small lagoon. The sandy beach has cobble-boulder beaches to either side, but a sandy surf zone dominated by a central bar and rips to either side.

Callawonga Beach (175) occupies the mouth of the grassy, V-shaped Callawonga Creek valley. The valley opens seaward, with the beach reaching 250 m in length. It has cobble-boulder beaches to each side, with a central sandy beach and bar, with usually two rips to the side and occasionally a central beach rip.

Tunk Head beach (**176**) lies immediately east of Tunkalilla Beach. It is a 400 m long sand beach, backed by steep, grassy slopes rising over 150 m. It is bordered by jagged rock platforms in the east and a jagged rock reef to the west. The 60 m wide surf zone usually has two permanent rips to either end and one central beach rip.

Tunkalilla Beach (177) is a straight, 5 km long beach, backed by a low foredune and wave overwash deposits, and continuous grassy slopes which rise in places to over 200 m. The surf zone averages 60 m in width and is cut by beach rips every 250 m in the west, with several reef-controlled rips to the east. This is the only publicly accessible beach, with a car park located on the bluffs above the western end of the beach.

Swimming: These are six potentially hazardous beaches, owing to the moderate to occasionally high waves, strong permanent rips and prevalence of rocks and reefs.

Surfing: There are beach breaks at all the beaches, the biggest problem is getting to them, with only Tunkalilla being reasonably accessible.

Fishing: Tunkalilla with its rip holes is popular with fishers who do not mind the steep walk back to the car park.

Summary: A scenic but largely inaccessible section of coast, offering several pocket beaches each bounded and backed by steep, grassy slopes.

Deep Creek Conservation Park

Established: 1971
Area: 4228 ha
Beaches: 3 (178 to 180)
Coast length: 14 km

Deep Creek Conservation Park covers 14 km of densely vegetated, steeply sloping coast and from 2 to 10 km of the backing hinterland. There are three beaches along the coast, with vehicle access limited to Boat Harbour and Blowhole (4WD only) and four camping areas on the top of the bluffs above the coast.

For park information contact:
Park Headquarters, C/- Delamere PO, Delamere SA 5204
Phone: 980 263

178-184   BOAT HARBOUR to BEACH 184

	Unpatrolled			
No.	Beach	Rating	Type	Length
178	Boat Harbour	6	boulder+LTT	100 m
179	Deep Creek Cove	6	cobble+LTT	50 m
180	Blowhole	6	LTT/TBR	120 m
181	Naiko Inlet (1)	5	LTT	40 m
182	Naiko Inlet (2)	5	LTT	40 m
183	Coalinga Creek	5	LTT	80 m
184	Beach 184	5	LTT	30 m

The southwestern section of the Fleurieu Peninsula consists of 20 km of steep, rocky and cliffed shoreline, extending from Boat Harbour beach to Lands End, the southwestern tip. The 200 to 300 m high slopes have been deeply incised by numerous small creeks, which in places have allowed small boulder, cobble and sand beaches to form at the creek mouths. The first 14 km of coast between Boat Harbour and Blowhole beach is part of the densely vegetated Deep Creek Conservation Park, with the coast to the west backed by usually cleared farmland. The only public access to the coast is at Boat Harbour and Blowhole beach, both of which require a 4WD to negotiate the steep descents. The more western beaches are backed by private farmland. These beaches all face south, but receive some protection from Kangaroo Island, resulting in waves usually below 1.5 m in height.

Boat Harbour beach (**178**) is a straight, 100 m long, steep, high tide cobble-boulder beach, fronted by a narrow low tide sand bar. It occupies a steep sided V-shaped valley, with the access track zig-zagging down the western spur (Fig. 4.41). Steep cliffs and jagged rocks and reefs border the beach, with Boat Harbour Creek draining across the western end.

Figure 4.41 *Boat Harbour beach and its backing valley and steep vehicle access track.*

Deep Creek Cove beach (**179**) is located at the mouth of a winding, deep, narrow valley, with the valley sides protruding 200 m seaward of the beach and offering a moderate degree of protection. The little cobble and sand beach is only 50 m wide at its mouth, but extends 100 m into the valley mouth as sand and cobble overwash flats and minor dunes, with the creek flowing against the western side of the beach. The only access to the beach is via the Heysen walking track, with the closest car park 2 km away at Tapanappa Lookout.

Blowhole beach (**180**) lies at the western boundary of the park. Access is via a steep 4WD descent to a car park on the eastern headland, otherwise there is car parking at Cobblers Hill 2 km inland. The beach lies at the base of a steep valley, with a creek draining across its eastern half and with headlands and rocks extending 100 m seaward. The beach is 120 m long and composed of sand, with a 70 m wide bar which usually has a permanent rip against the western rocks.

Naiko Inlet consists of two pockets of sand bordered by 50 m high cliffs, composed of steeply dipping metasedimentary rocks. A house lies behind the western beach. The two beaches (**181** and **182**) are both 40 m in length, separated by a 100 m wide headland. They both have high tide cobble-boulder beaches fronted by a sandy swash zone and narrow bar.

Coalinga Creek beach (**183**) lies 600 m west of Naiko Inlet. It is an 80 m long high tide cobble and sand beach, with a narrow bar. It occupies a small valley with a headland extending 200 m due south. Coalinga Creek emerges from a steep gully to cross the western end of the beach.

Beach **184** is a 30 m pocket of sand contained in a nick in the 30 m high backing cliffs. There is a farm track behind the bluffs above the beach, but otherwise no public access.

Swimming: While these beaches tend to receive moderate to low waves, their remoteness and prevalence of rocks, reefs and permanent rips make them potentially hazardous. Be very careful if there is any surf on any of the beaches.

Surfing: Only a few beach breaks on the beaches with sand bars.

Fishing: These beaches are more popular for beach and rock fishers, particularly those with 4WDs who tend to use Boat Harbour and Blowhole beaches.

Summary: These are seven difficult to very difficult to access and little used beaches. The first three lie in the Deep Creek Conservation Park and are accessible to the public, while the remainder are all backed by private farmland.

185-187   **FISHERY BAY, LANDS END, CAPE JERVIS**

Unpatrolled			
No.	Beach	Rating	Type	Length
185	Fishery Bay	5	rocks+LTT	500 m
186	Lands End	5	rocks+LTT	900 m
187	Cape Jervis (harbour)	2	R	40 m

The western tip of the Fleurieu Peninsula, terminating at Lands End, is a predominantly sloping, rocky shore. Unlike the coast to the east, this is a more accessible shoreline, with the main road terminating at the Cape Jervis ferry terminal and roads running out to Fishery Bay and to Lands End. These beaches are protected from high swell by Kangaroo Island, clearly visible just 14 km across Backstairs Passage. As a result, waves average about 1 m.

Fishery Bay beach (**185**) is an irregular, curving, south facing beach consisting of a strip of sand and cobbles, fronted by rocks in the east and a narrow sand bar in the west. A gravel road runs to the back of the beach.

At **Lands End**, the rocky shore turns 90⁰ and heads due north to Cape Jervis. In amongst the rocks is a 900 m long, disjointed, sandy beach (**186**), broken by rocks and rock reefs, together with high tide cobble beach sections. It is fronted in the sandy sections by a narrow bar and elsewhere by rock flats.

At **Cape Jervis** is the small harbour that houses the ferry terminal for the Kangaroo Island ferry, together with a boat ramp. Since the harbour was constructed, a small, steep, 40 m wide beach (**187**) has formed inside the harbour between the eastern harbour wall and the jetty rocks. The beach is now used for launching small boats.

Swimming: Be careful on these beaches. Fishery Bay and Lands End are sandy beaches, but dominated by rocks and reefs, while the little harbour beach is part of the harbour operations.

Surfing: Only low beach breaks in amongst the rocks at Fishery Bay and Lands End.

Fishing: This is a popular and accessible section of coast for beach fishing, with the addition of fishing off the rocks and the harbour jetty.

Summary: This section of coast is at the end of the road, Lands End. Most visitors are heading to or from Kangaroo Island, with mostly the locals visiting and using the beaches.

188-193   **MORGAN BEACH to BEACH 193**

Unpatrolled			
No.	Beach	Rating	Type	Length
188	Morgan Beach	3	R	800 m
189	Tea Tree Creek	3	boulder+LTT	250 m
190	Beach 190	3	LTT	500 m
191	Beach 191	3	boulder	150 m
192	Salt Creek	3	boulder+LTT	600 m
193	Beach 193	3	boulder+LTT	500 m

North of Cape Jervis is a relatively straight, northwest facing 11 km of rocky shoreline, backed by 150 to 200 m high steep, grassy bluffs and cliffs, all composed of steeply dipping metasedimentary rocks. In amongst the rocks are 12 small and often rocky beaches. Only the first beach, Morgan, is publicly accessible, the rest being backed by private farmland.

Morgan Beach (188) is an 800 m long, west facing white sand beach, nestled below 50 m high bluffs, up which a sand dune has almost succeeded in reaching the top. There is a gravel road from Cape Jervis to the bluff overlooking the southern end of the beach (Fig. 4.42). Rock reefs front the southern third of the beach, with a sandy seabed off the centre and northern end. Waves average about 0.5 m and the beach is usually steep and barless.

Tea Tree Creek flows onto the southern end of a 250 m long beach (**189**), which is predominantly composed of cobbles and boulders, with some sand along the northern shoreline, while elsewhere rock reefs dominate and 50 m high grassy bluffs border the beach. The beach is backed by farmland, with a farm track skirting the back of the beach.

Figure 4.42   *Morgan Beach usually has low waves, while steep sand dunes have climbed the backing slopes.*

Beach **190** is a 500 m long sandy beach with a scattering of rocks and rock reefs along the beach, and steep, 100 m high bluffs encircling the entire beach. There is a private vehicle track down the bluffs to a solitary house perched on a low bluff above the northern end of the beach. On the northern side of this bluff is a 150 m long, curving boulder beach (**191**), which lies below a 100 m high cliff.

The steep **Salt Creek** drains on to the northern end of a 600 m long boulder beach (**192**), backed by grassy slopes, with prominent 100 m high headlands to either end and a vehicle track to the top of the southern headland, but otherwise only foot access to the beach. A farm house is located 600 m east of the beach.

Beach **193** is a 500 m long mixture of sand, cobbles and rocks, with a 100 m long, open sandy section toward the northern end. There is a vehicle track down to a fishing shack on the slopes just above the centre of the beach.

Swimming: Only Morgan Beach is easily accessible to the public and also provides the safest swimming in the area, as it is relatively free of rocks and reefs. All the other beaches, while receiving relatively low waves, are dominated by rocks.

Surfing: Usually none owing to the low waves.

Fishing: This entire section offers many good spots to fish off the rocks. However access is a problem north of Morgan Beach.

Summary: You can glimpse this coast from the Morgan Beach car park. Access is however restricted by the backing private property and the rugged terrain.

194-200   BEACH 194 to RAPID HEAD

No.	Beach	Rating	Type	Length
Unpatrolled				
194	Beach 194	5	boulder	200 m
195	Beach 195	5	boulder	150 m
196	New Yohoe Ck	5	boulder	200 m
197	Stockyard Ck	5	boulder	1.6 km
198	Beach 198	5	boulder	170 m
199	Beach 199	5	boulder	600 m
200	Rapid Head (S)	4	R+rocks	150 m

Steep bluffs and cliffs averaging 150 m in height dominate the six kilometres of rocky coast south of Rapid Head. Along the base of the cliffs are six boulder beaches and one sand beach. Beaches **194** and **195** are virtually inaccessible, being backed and bordered by the high cliffs, with the best access by boat.

New Yohoe Creek flows down a more moderately sloped valley to the southern end of a 200 m long boulder beach (**196**). Cleared grazing land backs the beach, with 100 m high bluffs to either end.

Stockyard Creek (also known as Yohoe Creek) is the largest creek along this section of coast and flows out onto the middle of a 1.6 km long boulder beach (**197**), forming a small protruding boulder delta. The slopes rise steeply to over 100 m behind the beach, however there is a farm track down a smaller southern valley to the back of the beach.

Beach **198** is a 170 m long, curving boulder beach locked between two small protruding headlands, while beach **199** is an irregular, 600 m long boulder beach interrupted by rocks and rock reefs and backed by steep cliffs and bluffs. Finally, 500 m south of **Rapid Head** is a 150 m pocket of sand (**200**), backed by steep, 100 m high bluffs, bordered by rocks and interrupted by rocks and rock reefs (Fig. 4.43).

Figure 4.43 *One hundred metre high bluffs back sandy beach 200 (foreground) and cobble beach 199 (rear).*

Swimming: None of these beaches are suitable for safe swimming owing to the dominance of rocks on the beach and in the water. They are also difficult to access, all being backed by private farmland.

Surfing: None.

Fishing: Most fishing is from boats along this reef-dominated section of coast.

Summary: A spectacular section of steep rocky shore, with the 'beaches' basically the eroded remains of the cliffs piled into boulder beaches by the waves.

201, 202   RAPID BAY

No.	Beach	Rating	Type	Length
Unpatrolled				
201	Rapid Bay	3	R/cobble	1.5 km
202	Rapid Bay (N)	3	R/cobble	300 m

Rapid Bay occupies an open, north facing valley. It consists of the small settlement of Rapid Bay, fronted by a 400 m long jetty and a large rock quarry on the southern Rapid Head. There is a wide foreshore reserve given over to recreation and a beachfront camping and caravan park. The beach extends from the base of 130 m high Rapid Head for 1.5 km to the northern 100 m high headland. The entire beach and barrier system has been slowly building seaward owing to a supply of rock debris from the quarry site. The shoreline built out rapidly between 1940 and 1975 to maximum distance of 250 m, however erosion since then has seen it stabilise about 150 m seaward of the original shoreline. The beach (**201**) is coarse and steep, with usually low wave to calm conditions at the shore. It is also used to launch small boats. Yattagolinga Creek crosses the southern part of the beach.

Just around the northern headland, the supply of shingle has been building a 300 m long shingle beach (**202**) at the base of the 100 m high cliffs. This beach can only be accessed on foot from Rapid Bay.

Swimming: Rapid Bay is usually calm and relatively safe, just watch the deep water right off both beaches.

Surfing: None, other than a heavy shorebreak during bigger seas.

Fishing: You can fish deeper water right off the steep beach, or from the adjoining headland and jetty.

Summary: A popular holiday destination with beachfront camping and a relatively safe beach.

203, 204   SECOND VALLEY

Unpatrolled			
No.	Beach	Rating	Type	Length
203	Second Valley (1)	4	boulder	200 m
204	Second Valley (2)	3	LTT+rocks	100 m

Second Valley is an open valley that narrows at the coast to a 50 m gap through which flows the small Parananacooka Creek and the road to the beach. The small town and caravan park lie on the landward side of the gap (Fig. 4.44). Once through the gap, there is a small car park leading to a rock jetty with two small beaches to either side. To the west is a north facing, curving, 200 m long boulder beach (**203**) attached by rock reefs to a small headland, at the base of which are several boat sheds. Conditions are usually calm at the beach, however it is dominated by rocks and reef.

Figure 4.44 *Second Valley has a small beach and jetty, boat sheds (right) and backing township.*

The main beach (**204**) lies immediately west of the car park. It is a 100 m long sand beach fringed by boulders, together with a high tide cobble beach at the northern end. This is the main swimming beach for the town and caravan park.

Swimming: The main beach is popular in summer and is usually calm. The boat beach is unsuitable for swimming owing to the rocks and boat traffic.

Surfing: None.

Fishing: A very popular location with the jetty, rocks and beach all providing good spots to fish the surrounding reefs.

Summary: A picturesque valley, town and beaches.

205, 206   WIRRINA, YANKALILLA BAY

Unpatrolled			
No.	Beach	Rating	Type	Length
205	Wirrina	4	R+rocks	320 m
206	Yankalilla Bay	4	R+rock flats	4 km

The southern end of Yankalilla Bay is occupied by a long, narrow beach at the base of sloping, grassy bluffs that rise to over 150 m. Toward the southern end of the bay is **Wirrina Resort**, nestled in a valley behind the beach and a new marina. A road runs through a break in the bluffs to reach the southern end of the beach and the marina, where there is a large car park. The beach (**205**) now begins on the northern side of the marina and runs north for 320 m. It is a sand beach with increasing rock outcropping on the beach and as reefs toward the northern end. Waves are usually low and calm conditions common.

A prominent 10 m high bluff separates Wirrina Beach from the 4 km long **Yankalilla Bay** beach (**206**). This is a high tide sand beach fronted by continuous, 50 to 100 m wide intertidal rock flats and backed by steep, 100 m high grassy bluffs. The Cape Jervis Road runs along the northern 1 km of the beach, with a beach-side car park located just as the road turns inland. Otherwise there are no facilities and no other vehicle access along the beach.

Swimming: Two rock and reef-dominated beaches, mainly used for fishing.

Surfing: None.

Fishing: Both are popular and very accessible spots for fishing the rock reefs.

Summary: A long, narrow stretch of beach wedged between the bluffs and rock reefs, with a major marina newly built at Wirrina.

207-209   NORMANVILLE to CARRICKALINGA HEAD

NORMANVILLE SLSC

Patrols:	December to March
Surf Lifesaving Club:	Christmas school holidays then weekends to Easter
Lifeguards:	January

No.	Beach	Rating	Type	Length
207	Normanville	2	LTT	7.3 km
208	Carrickalinga	2	LTT	1.4 km
209	Carrickalinga Head	4	R+rocks	100 m

Normanville is the largest coastal settlement on the southern Fleurieu Peninsula. The town is located in the centre of the 7.3 km long beach (**207**), with Lady Bay and 100 m high Yankalilla Hill forming the southern boundary and low Haycock Point the northern (Fig. 4.45). Three small creeks, the Yankalilla River, Bungala River and Carrickalinga Creek, reach the beach, with their entrances usually blocked by sand. The main beach access is at Normanville where there is a large car park and toilets, together with a small jetty and a caravan park. There is also good access at Lady Bay and at Haycock Point.

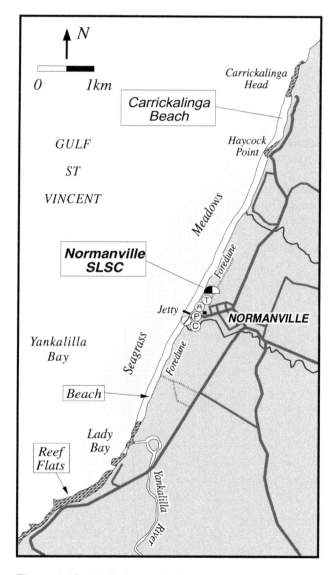

Figure 4.45 *Lady Bay and the continuous sand beach that runs up to Normanville, Haycock Point and Carrickalinga Beach.*

The beach faces northwest into St Vincent Gulf and usually receives low westerly wind waves and occasional southerly swell. Consequently waves average less than 0.5 m, the active beach is narrow and steep, and seagrass grows to within 50 m of the shoreline.

Carrickalinga Beach (**208**) lies immediately north of low Haycock Point and runs north for 1.4 km to 50 m high Carrickalinga Head (Fig. 4.45). A 50 m wide foreshore reserve, road and three rows of houses back the beach, the road terminating at a car park and reserve on the slopes of the headland. A small creek drains across the southern end of the beach. Between Haycock Point and the car park is a relatively safe sandy beach, while on the point and north of the car park, rock reefs front the beach.

On **Carrickalinga Head** is a 100 m long pocket of sand (**209**), largely fronted by rock reefs. This beach lies 500 m north of the car park and is only accessible on foot around the rocks.

Swimming: Normanville and Carrickalinga beaches are relatively safe under normal conditions, with best access and facilities at Normanville and Carrickalinga, with rock reefs at the southern Lady Bay and around Carrickalinga Head.

Surfing: Usually none, with sloppy wind waves only during strong westerlies.

Fishing: The best shore fishing is from the Normanville Jetty and Haycock Point. Most fishers head out into the gulf in small boats, which can be launched from the beach at Lady Bay and Normanville.

Summary: Two attractive, long, usually quiet beaches, with good access and facilities at Normanville.

210, 211   **BEACHES 210 & 211**

Unpatrolled			
No.	Beach	Rating Type	Length
210	Beach 210	6	boulder	180 m
211	Beach 211	6	boulder	80 m

Between Carrickalinga Head and Myponga are 6 km of northwest facing cliffs and slopes rising inland to over 200 m. Beach **210** lies at the mouth of a steep, meandering creek, and is a 180 m long boulder beach fronted by deep water. Beach **211** occupies a cliff-bound cove and is an 80 m long boulder beach, with cliffs and rock reefs to either end. While beach 210 is accessible on foot, neither is suitable for safe swimming.

212   **MYPONGA BEACH**

Unpatrolled			
No.	Beach	Rating Type	Length
212	Myponga	3	LTT	450 m

Myponga Beach (212) occupies a small valley that has been carved by Myponga Creek and is now partially filled with river and marine sediments, forming a flat plain occupied by a few houses and an elongate lagoon and wetland that flows across the southern end of the beach. Houses and shacks line the rear of the beach, so much so that a seawall has been built to protect some of the properties. In addition, there is a camping area on the backing flats.

Fifty-metre high grassy bluffs border each end of the beach, with rocks and reefs extending seaward. The beach is 450 m in length, composed of greyish sand, with a high tide shingle beach toward the rear. It is moderately steep and fronted by a narrow continuous bar, with waves averaging about 0.5 m. The beach was once used as a port and the ruins of the old jetty are still present.

Swimming: Relatively safe in the sandy central portion of the beach.

Surfing: Usually calm.

Fishing: A popular spot for both launching boats and for shore-based fishing off the adjacent rock reefs.

Summary: A quiet and picturesque little valley, settlement and beach, however there are no facilities for visitors.

and interrupted in places by several steep creek mouths and protruding bluffs and rocks. This beach can only be accessed at low tide and on foot. It is not suitable for swimming.

Cactus Canyon is a steep-sided canyon that reaches the coast 500 m south of Sellicks Beach. The canyon and neighbouring creeks have delivered rocks to the shore, that have been reworked by the waves to form a high tide shingle beach, overlain by a veneer of sand. The beach **(214)** extends south of the canyon for 1 km to where it merges with the boulders of beach 213. North of the canyon, houses line the top of the 60 m high bluffs, with the best access being on foot or by vehicle from Sellicks.

Swimming: The 1 km central sandy section of Cactus Canyon beach is relatively protected and usually calm, and fronted by a continuous shallow sand bar.

Surfing: Usually calm.

Fishing: The rock platforms along Beach 213 are used by the more adventurous and energetic rock fishers.

Summary: Two relatively isolated beaches, with Cactus Canyon offering seclusion as well as protection from southerly winds and waves.

213, 214   **BEACH 213, CACTUS CANYON**

	Unpatrolled		
No.	Beach	Rating	Type	Length
213	Beach 213	6	boulder	3 km
214	Cactus Canyon	4	LTT	1.4 km

East of Myponga, the high cliffs and steep bluffs continue for 5 km to Sellicks Beach. Running along the base of the cliffs is an irregular 3 km long boulder beach **(213)**, fronted for the most part by an intertidal rock platform

METROPOLITAN DISTRICT
ALDINGA TO NORTH HAVEN
(Fig. 4.46)

Length of Coast:	66 km
Beaches:	19 beaches (215-233)
Surf Lifesaving Clubs:	Aldinga Bay, Moana, South Port, Port Noarlunga, Christies Beach, Hallett Cove, Seacliff, Brighton, Somerton, Glenelg, West Beach, Henley, Grange, Semaphore, North Haven
Lifeguards:	Glenelg to Brighton
Major towns:	Adelaide and suburbs

The Adelaide Metropolitan District is the smallest of the state's seven coast protection districts, but by far the largest in terms of population and number of surf clubs. The nineteen beaches have fifteen surf lifesaving clubs between them, spread along the coast between Aldinga and North Haven, with eight of the clubs located along the near continuous 28 km stretch of sand between Seacliff and Largs Bay. This section of coast contains the state's most popular and heavily utilised beaches. Fortunately waves are usually low and conditions relatively safe for swimming, though higher waves do produce rips on a number of the more exposed beaches.

Regional Map 3: Adelaide Metropolitan Area

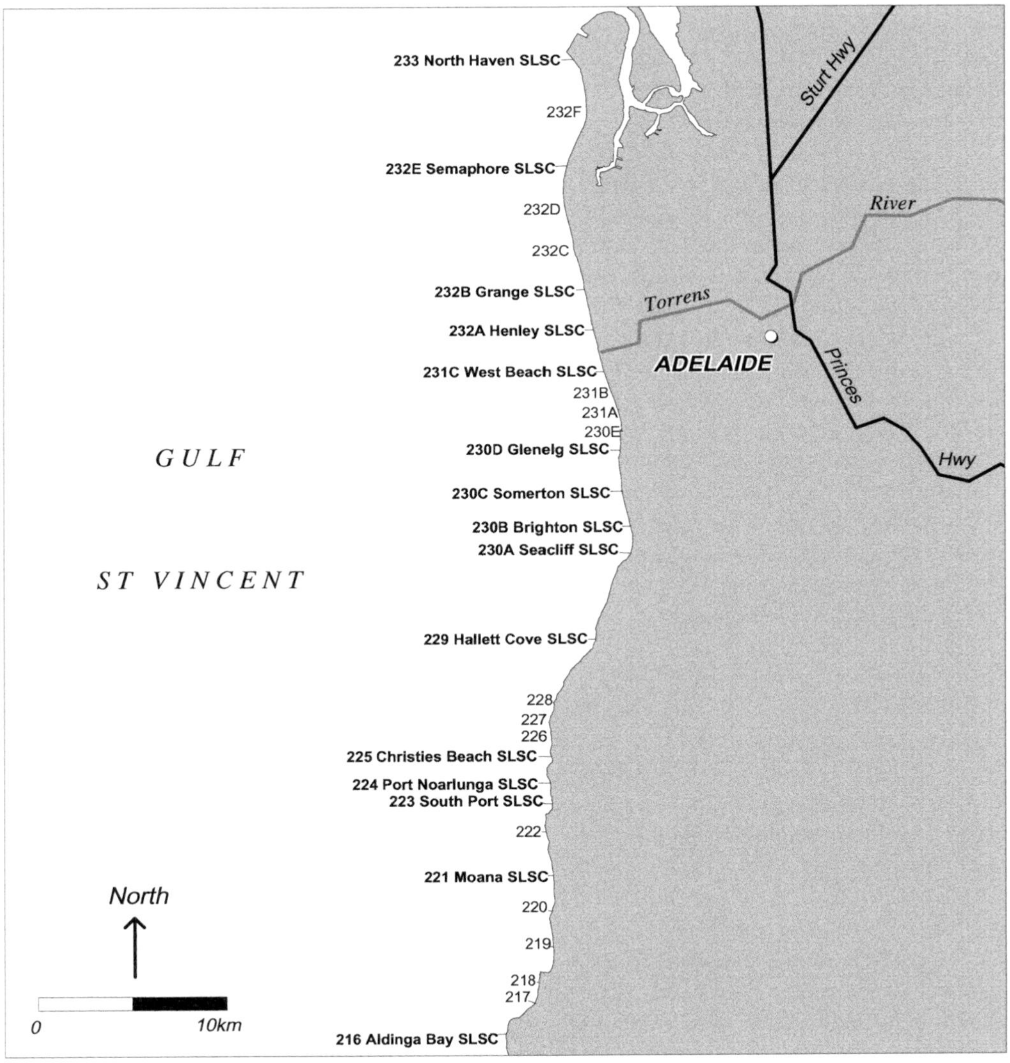

Figure 4.46 *Regional map 3 - The Adelaide Metropolitan District*

215, 216   ALDINGA BAY

ALDINGA BAY SLSC

Patrols:	November to March
Surf Lifesaving Club:	weekends and public holidays
Lifeguard on duty:	no lifeguard on duty or weekday patrols

No.	Beach	Rating	Type		Length
			Inner	Outer Bar	
215A	Sellicks Beach	4	LTT	LBT	3 km
215B	Silver Sands (Aldinga Bay SLSC)	4	LTT	LBT	3 km
216	Aldinga Beach	4	LTT/reef flats		1.5 km

Aldinga Bay is an open, 7.5 km long, west facing bay, bordered in the south by the 60 m high bluffs of Cactus Canyon and in the north by the lower bluffs backing Snapper Point. Two beaches occupy the shoreline, while the three settlements of Sellicks Beach, Silver Sands and Aldinga Beach back the beach (Fig. 4.47). Roads and houses back about half the beach area, with vehicles and parking also permitted on the main beach.

Aldinga Bay Surf Life Saving Club is located in the centre of a straight, 6 km long, west facing beach (**215**) that begins just north of Cactus Canyon, below the steep bluffs of southern **Sellicks Beach**. The beach runs almost due north for 3 km to **Silver Sands**, the site of the surf club, then on for another 3 km to the rock flats that form the border with Aldinga Beach. The beach is composed of fine sand, which, with the usually low gulf waves and swell (0.5 to 1 m), produce a low, flat beach, firm enough to drive a car on. Occasional bigger wind waves and higher swell maintain a shallow second bar that parallels the beach, a continuous trough separating it from the beach (Fig. 4.48). A shingle beach also backs the southern half of the beach. The shingle is eroded from the southern cliffs and reworked up the beach by higher seas.

Aldinga Beach (**216**) and adjoining Snapper Point is a 1.5 km long, narrow, crenulate high tide sand beach fronted by 300 to 400 m wide intertidal rock flats, that extend further offshore as reefs. The whole point and reef area is an aquatic reserve. Aldinga township backs the bluffs, with a bluff-top road paralleling the back of the beach and providing both views of and access to the beach and fronting platforms.

Swimming: The main beach is a moderately safe beach, particularly on the attached inner bar. Be careful of the deep trough between the inner and outer bars. When waves exceed 1 m, rips intensify in this trough. Also watch for cars when crossing the beach. Aldinga Beach is unsuitable for swimming owing to the rocky reef flats.

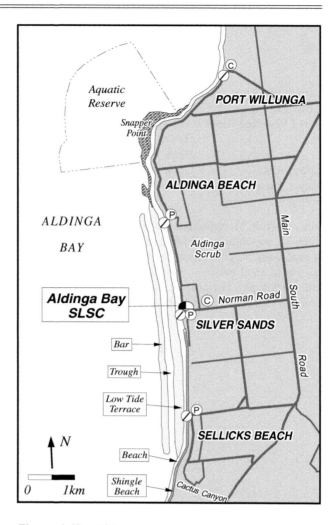

Figure 4.47 *Aldinga Bay contains Sellicks Beach, Silvers Sands (site of the Aldinga Bay Surf Life Saving Club) and Aldinga Beach. A low tide terrace is usually attached to the beach, with a shallow outer bar paralleling the beach.*

Figure 4.48 *Silver Sands beach, site of Aldinga Bay SLSC (left), showing the wide, low gradient beach, attached inner bar, trough, then outer bar, with usually low waves just breaking over the outer bar.*

Surfing: Aldinga Bay receives low swell and usually has low beach breaks over the inner and outer bars.

Fishing: Most beach fishers wade out on to the shallow inner bar to fish the longshore trough. There is also reef fishing at high tide at Aldinga Beach. However the Snapper Point area is an aquatic reserve.

Summary: A long, straight beach, very accessible for both people and vehicles, and popular during summer. It has some surf, but remains relatively safe under normal conditions.

217, 218   PORT WILUNGA

Unpatrolled			
No. Beach	Rating	Type	Length
217 Port Wilunga	4	R+rocks-LTT	1.5 km
218 Port Wilunga (N)	4	LTT-TBR	1.5 km

Port Wilunga is a growing coastal settlement that spreads 1.5 km along the low bluff that runs north of Snapper Point, to a small creek that drains across the beach. A road runs along the top of the bluff to a large car park and caravan park on the south side of the creek, which provides the best access to the beach (Fig. 4.49). The beach (**217**) extends for 1.5 km from Snapper Point past the creek to a 15 m high, reef fronted bluff that extends across the high tide beach. In the south it is fronted by a 100 m wide intertidal reef flat, then the ruins of the old port jetty and finally a natural sand beach which usually has an attached bar cut by one to two rips. North of the bluff, beach **218** continues on for 1.5 km to Blanche Point, a 30 m high headland cut in crumbling sedimentary rocks, with a 100 m wide intertidal rock platform and some sea caves cut into the base of the cliff. Bluffs and reefs fringe the beach, while a continuous bar usually cut by three rips parallels the beach.

Figure 4.49 *The northern section of Port Wilunga Beach showing the bluffs, large car park and bluff-top caravan park.*

Swimming: These are two moderately safe beaches under normal low wave conditions. Beware of the rips, which intensify when waves exceed 1 m.

Surfing: A chance of a low beach break during larger outside south swell or strong southerly winds.

Fishing: There is good fishing both off the platforms and into the rip holes.

Summary: Two bluff-bound, accessible beaches offering usually low surf.

219, 220   MASLIN BEACH, OCHRE POINT

Unpatrolled			
No. Beach	Rating	Type	Length
219 Maslin Beach	4	LTT/TBR	2.9 km
220 Ochre Point	5	R+platform	1 km

Maslin Beach is a small coastal settlement set on sloping bluffs behind the centre of the beach of the same name. There are two large car parks on the bluffs, which provide excellent access to the beach (**219**). The southern end of the beach, which is backed by bare bluffs rising to 50 m at protruding Blanche Point, is best known because of its optional dress code and is a popular location during much of the year. The northern end of the beach is less utilised and is backed by an abandoned bluff-top quarry. The 2.9 km beach faces west and terminates at the northern 50 m high Ochre Point (Fig. 4.50). It receives waves averaging about 1 m, with higher waves producing rips spaced about 100 m apart along much of the beach, the rip holes at times persisting for weeks afterwards.

At **Ochre Point** the beach narrows to a sandy high tide beach (**220**) that runs for 1 km along the base of the steep, 60 m high bluffs, fronted by 50 m wide intertidal rock rubble then reef flats. The only access is on foot from either end, while on the beach there is only one small sandy access gap through the reefs, located toward the southern end of the beach.

Swimming: Maslin is one of Adelaide's more popular beaches owing to the optional dress code. It is usually moderately safe, however beware of the chance of rip holes even under calm conditions and rip currents whenever there is surf.

Surfing: Only during higher outside swell.

Fishing: The rip holes, when present and the boundary rock platforms are the best locations.

Summary: A bluff-backed beach protected in the south by Blanche Point.

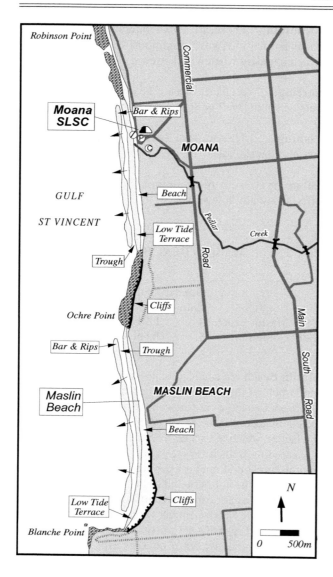

Figure 4.50 *Maslin and Moana beaches, the latter being site of the Moana Surf Club. Both beaches receive low to moderate gulf waves, which maintain a shallow attached inner bar, cut by rips during periods of higher waves.*

221, 222   MOANA, ROBINSON POINT

MOANA SLSC

Patrols: November to March, weekends and public holidays

No.	Beach	Rating	Type	Length
221	Moana	4	LTT/TBR	3.1 km
222	Robinson Pt	5	R+platform	2 km

Moana Beach (221) is a 3 km long, west facing beach, backed by the town of Moana and fringed in the north and south by the steep bluffs of Robinson Point and Ochre Point (Fig. 4.50). The Moana Surf Life Saving Club, founded in 1952, is located at the main access point to the beach. It is surrounded by large car parks, shops and facilities, a caravan park and Pedlar Creek (Fig. 4.51), which, combined with vehicle access to the beach, make this a popular beach.

Figure 4.51 *Moana Beach Surf Club (right) and large caravan park (far right). Note the cars on the beach and the continuous attached bar.*

The beach is relatively wide and flat, with an attached shallow bar, then a trough separating it from a shallow outer bar, the shape of which depends on wave conditions and at times has deep rip channels. Moana receives low swell and westerly wind waves, with waves averaging 0.5 to 1 m.

North of the main beach the bluffs gradually increase in height, reaching 20 m at **Robinson Point**, where reef flats also front the adjoining 2 km long beach (**222**) that is wedged between the bluffs and the flats. A road runs along the top of the bluffs, with several sloping access tracks down to the narrow beach.

Swimming: Moana is a moderately safe beach under normal conditions. However care should be taken if swimming in the trough or on the outer bar, as rip currents may be present. All currents and rips intensify when waves exceed 1 m. Robinson Point is unsuitable for swimming owing to the rock flats, though there are some tidal pools at low tide.

Surfing: There are usually low breaks over the inner and outer bars, with higher swell producing a better break.

Fishing: There are occasional rip holes, particularly following larger swell and rock fishing off Ochre and Robinson Points.

Summary: A popular and accessible beach, which receives some swell and usually has a low surf.

223 SOUTH PORT

SOUTH PORT SLSC

Patrols: November to March, weekends and
public holidays.

No. Beach	Rating	Type	Length
223 South Port	4	LTT/TBR	1.1 km

South Port Surf Life Saving Club patrols the southern
1 km of the 2 km long beach that includes Port Noarlunga
Beach (Fig. 4.52). The southern section consists of the
beach and sand dunes extending up to 200 m inland and
then the meandering Onkaparinga River. The river flows
out across the very southern end of the beach, against the
bluffs and rock flats of Robinson Point. A road parallels
the eastern side of the river. To reach the beach requires
parking on the main road and walking across a footbridge
and the dune, a total distance of 300 m. The surf club is
located in the sand dunes 400 m north of the river mouth.

South Port Beach (223) faces almost due west and
receives low ocean swell, as well as gulf wind waves.
Waves average 0.5 to 1 m and during higher wave
conditions produce a double bar system, with three to
four rips cutting across the inner and outer bars (Fig.
4.53). During normal low wave conditions, the inner rips
tend to infill and a continuous bar is attached to the beach.
In addition, the river flows across the southern end of the
beach building out a large tidal sand bar, cut by the deeper
tidal channel.

Swimming: South Port is one of the potentially more
energetic of the Adelaide beaches and care should be
taken here as rip channels and currents are common. Even
during low waves the deeper rip channels may persist.
Stay in the patrolled area and clear of the southern tidal
channels and any rip holes.

Surfing: The more exposed location makes South
Port one of the more persistent surfing beaches, with low
to moderate waves (0.5 to 1 m) breaking over both the
inner and outer bars.

Fishing: This is a popular spot with a choice of river,
inlet, rock and beach fishing, the latter often into rip
gutters and holes.

Summary: A relatively natural beach and river mouth,
with a more variable surf zone and a chance of bigger
waves.

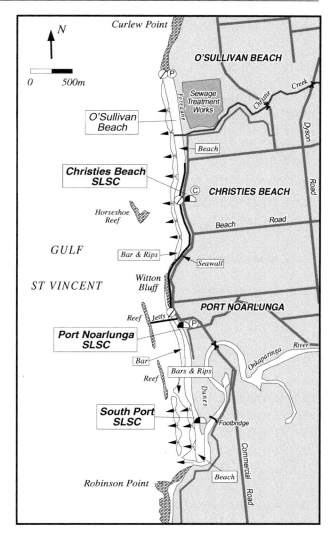

Figure 4.52 *South Port-Port Noarlunga and
Christies-O'Sullivan beaches. Surf clubs are located
at South Port, Port Noarlunga and Christies. The level
of wave energy and beach type range from more
exposed with rips at South Port, reef protected at Port
Noarlunga, to moderate protection at Christies.*

Figure 4.53 *South Port Surf Life Saving Club is
located on the sand spit that crosses the mouth of the
Onkaparinga River. Note the rhythmic bar and
(inactive) rip channels off the beach.*

Onkaparinga River Recreation Park

Established: 1985
Area: 284 ha
Beach: In lee of beach 223

The Onkaparinga River and recreation park begins immediately behind the dunes of South Port Beach (Fig. 4.52) and follows the river and its banks for about 10 km inland. The river system is the third largest estuarine system in South Australia.

224   PORT NOARLUNGA

PORT NOARLUNGA SLSC

Patrols: November to March, weekends and public holidays

No. Beach	Rating	Type	Length
224 Port Noarlunga	2	R	1.2 km

Port Noarlunga is an older settlement based around the port that used to operate off the beach. The port has long gone and now it is a growing residential area that spreads into adjoining Christies Beach. Residential and commercial development back the northern half of the beach, with the southern part a continuation of South Port Beach (Figs. 4.52 & 4.54). The Port Noarlunga Beach (**224**) faces west and extends for another 1 km to the red and white bluffs of Witton Bluff. Two straight calcarenite reefs lie 150 to 200 m off the beach and parallel the northern 700 m of the beach. A jetty, that used to service the port, runs out to the back of the northern reef. The main commercial area backs the jetty and the Port Noarlunga Surf Life Saving Club is located immediately south of the jetty, where it is backed by a large car park and fronted by a sea wall.

Figure 4.54 *Port Noarlunga was established last century in lee of the protective calcarenite reefs. Today the jetty is a popular fishing spot, while Port Noarlunga Surf Life Saving Club patrols the low energy reflective beach.*

The reefs lower waves at the beach to less than 0.5 m, which produces a single, continuous attached bar and no rips. Toward the southern end the beach protrudes seaward, forming a natural boundary with the more energetic South Port Beach.

Swimming: This is a relatively safe beach owing to the usually low waves and absence of rips. However you should stay inshore on the bar, as the reefs do pose a danger, owing to higher waves breaking over them and currents flowing around and between the reefs.

Surfing: The reefs lower the surf to a small beach break, most surfers head for adjoining South Port or Christies.

Fishing: The jetty is extremely popular, providing good access to deeper water and the reefs. The reefs however are an aquatic reserve. There is also deeper water off Witton Bluff, which can be fished from the seawall that protects the eroding rocks.

Summary: A very popular spot with good access, parking and facilities and a more protected, patrolled beach.

225, 226   CHRISTIES BEACH, O'SULLIVAN BEACH

CHRISTIES BEACH SLSC

Patrols: November to March, weekends and public holidays

No. Beach	Rating	Type	Length
225 Christies	4	LTT	550 m
226 O'Sullivan	4	LTT/TBR	1.6 km

Christies Beach (225) occupies the southern third of a 2 km long, west facing beach that extends from Witton Bluff in the south to Curlew Point in the north. The southern half is backed by the growing residential area called Christies Beach, while the northern half is known as **O'Sullivan Beach (226)** and is backed by a sewage treatment works (Fig. 4.52). There is excellent access at Christies Beach, where a road parallels the beach, together with a caravan park and the Christies Beach Surf Life Saving Club. A seawall protects the road, with a ramp to the beach in front of the surf club.

Christies Creek flows across the middle of the beach and separates Christies Beach from O'Sullivan Beach. The northern sewer works are fronted by a low sand dune, with the breakwater of a boat launching harbour forming the northern boundary.

Both beaches receive low ocean swell as well as gulf wind waves, with waves averaging 0.5 to 1 m. This is sufficient to produce a single bar, usually cut by rips every 200 m. The centre of Christies Beach, where the surf club is located, is partly protected by Horseshoe Reef, which lies 300 m offshore. This causes the beach to protrude seaward at this point and receive slightly lower waves (Fig. 4.55).

Figure 4.55 *Christies Beach and surf club (centre) and the southern half of O'Sullivan Beach. Note the attached bar grading to rip channels north of the surf club.*

Swimming: This is a moderately hazardous beach owing to the common occurrence of rips. The safest swimming is right in front of the surf club, where the waves are usually a little lower and the beach is patrolled in summer. Watch out for rips and stay inshore on the attached portion of the bar.

Surfing: This is a popular surfing beach owing to the low ocean swell that commonly reaches here. The swell, bars and rips can produce some reasonable beach breaks.

Fishing: The rips produce holes for beach fishing, while there is also rock fishing off the southern Witton Bluff or the northern breakwater. However most fishers use the boat ramp to fish the outer reefs and gulf waters.

Summary: A popular beach with some surf and good access and facilities.

227, 228   O'SULLIVAN BOAT HARBOUR, PORT STANVAC

No.	Beach	Rating	Type	Length
		Unpatrolled		
227	O'Sullivan Beach (harbour)	2	R+rocks	50 m
228	Port Stanvac	5	R+rocks/platform	2 km

The O'Sullivan Beach settlement is fronted by a rocky section of coast that extends up to Point Curlew and on to Seacliff. Immediately south of the rocks is a sand beach (**227**) backed by the sewerage farm, while to the north is the small **O'Sullivan Beach boat harbour**, built primarily to provide a safe, all weather boat ramp. Wedged in between the large boat ramp and the southern harbour wall is a 50 m long, north facing sand beach.

Immediately north of the harbour, the original beach continues up to **Port Stanvac** and on to the cliffs below the oil refinery. This is a narrow high tide sand and boulder beach (**228**), backed by 30 to 40 m high bluffs and fronted by 50 m wide rock platforms cut in steeply dipping metasedimentary rocks. The 300 m long Port Stanvac jetty is located in the centre of this beach and, while a road runs to the jetty, it is not accessible to the public.

Summary: Neither of these beaches are suitable for swimming, owing to the boat traffic and rocks.

229   HALLETT COVE

No.	Beach	Rating	Type	Length
229	Hallett Cove	5	LTT+rocks	1 km

Hallett Cove is a 1 km long break in the 40 to 60 m high cliffs that dominate the coast from Point Curlew to Seacliff. The break forms a natural amphitheatre along the shoreline, within which is a strip of sand called **Hallett Cove Beach (229)** (Figs. 4.56 & 4.57). The sloping land behind the beach has been partly developed for housing, with a park in the centre of the beach reserve for recreation and a boat ramp. Field River drains across the southern end of the beach.

The beach is 1 km long, faces west and is bordered by the cliffs of Curlew Point to the south and Black Cliff to the north. Black Cliff is a famous geological site where ancient glacial remains are clearly visible. It is now part of the Hallett Cove Conservation Park, which includes the northern half of the beach and bordering cliffs. The cliffs at either end are fronted by 50 to 100 m wide intertidal rock flats. These also extend along the length of the beach, where they are only partially covered by sand at high tide. Consequently the beach, while sandy at high tide, is predominantly rock flats and boulders at low tide, with a sand bar seaward of the rocks. The waves, which average 0.5 to 1 m, break over the bar and rock flats, usually maintaining two to three rips across the bar.

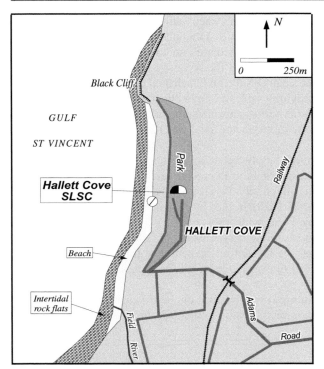

Figure 4.56   *Hallet Cove has a narrow high tide sand beach, fronted by an intertidal gravel-rock flat. The surf club has closed and no longer patrols the beach.*

Figure 4.57   *Black Cliff south to Hallett Cove.*

Swimming: One has to choose the time and place to swim at Hallett Cove. The best time is high tide when most of the rocks are covered by water and the best place is where most sand is available. The northern end of the beach near Black Cliff has less rocky beds and good sand coverage all year round is preferred by most beach goers.

Surfing: This is not a popular spot owing to the dominance of rocks.

Fishing: Most fishing is done from the rocks at either end. The boat ramp is really only useable at high tide, owing to the rocks.

Summary: An attractive cove, which unfortunately has lost much of its cover of sand, causing problems for swimmers and surfers.

Hallett Cove Conservation Park	
Established:	1976
Area:	51 ha
Beach:	229

This small conservation park includes the beach, cliffs and rock platforms north of the Hallett Park Surf Club, for a distance of about 1 km. The park was established to preserve the 270 million year old glaciated pavements, which are well preserved in the rocks.

THE ADELAIDE BEACHES

The last of the Adelaide seacliffs terminate at Marino Rocks on the southern boundary of Seacliff. Here the old seacliff turns inland and is fronted by an increasingly wide coastal plain. The plain is fringed by beaches all the way to Adelaide's Outer Harbour 29 km to the north. This stretch contains Adelaide's main beaches, patrolled by nine of South Australia's nineteen surf lifesaving clubs. The beach can be divided into three natural sections: the first 7 km from Seacliff to the mouth of the Patawalonga at Glenelg; the second, a 14 km stretch from the Patawalonga to Point Malcolm and the third, an 8 km section from Point Malcolm; to the Outer Harbour breakwater at North Haven. This entire coast is backed by Adelaide's seaside suburbs and together they offer a magnificent stretch of sand and coast. The public have responded with intense development along much of the coast, including six jetties, the nine surf lifesaving clubs, several sailing clubs and two marinas. Unfortunately much of the development was allowed too close to the shore and now threatens the beaches. Today the beaches are maintained with regular sand nourishment while the encroaching development is protected by seawalls.

230   SEACLIFF to GLENELG

SEACLIFF SLSC	BRIGHTON SLSC
SOMERTON SLSC	GLENELG SLSC

Patrols:	late October to early April, weekends and public holidays
Lifeguard on duty:	December to March (Glenelg – Brighton)

No.	Beach	Rating Inner	Type Outer Bar		Length
230A	Seacliff	3	LTT	LBT	1 km
230B	Brighton	3	LTT	LBT	2 km
230C	Somerton	3	LTT	LBT	2 km
230D	Glenelg	3	LTT	LBT	2 km

The southern section of the Adelaide coast extends for 7 km almost due north from Seacliff to Glenelg and incorporates the seaside suburbs of Kingston Park, Seacliff, South Brighton, Brighton, North Brighton, Somerton Park, Glenelg South and Glenelg. Surf life-saving clubs are located at Seacliff, Brighton, Somerton and Glenelg (Fig. 4.58).

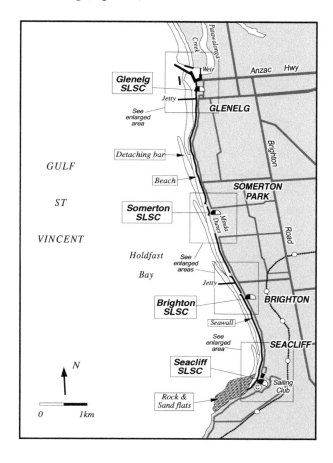

in the south, increasing to about 0.5 m past the clubs. The bar is usually continuous and rips are present only during bigger seas.

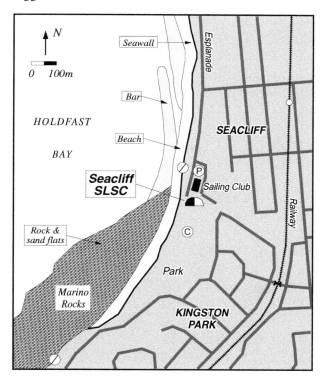

Figure 4.59　*Seacliff Beach is fronted by the extensive Marino rock flats, with the sand beach beginning at the Seacliff Surf Life Saving Club.*

Figure 4.60　*Seacliff Surf Life Saving Club (centre), sailing club and caravan park. Note the outer seagrass meadows, then rock flats and the inner bar and beach system.*

Figure 4.58　*The southern Adelaide beaches extend from Seacliff up to Glenelg. Southerly waves move sand north along the beaches, particularly in the shore parallel bars. See inserts for more details of Seacliff, Brighton, Somerton and Glenelg beaches and surf lifesaving clubs.*

Seacliff Beach (230A) begins amongst the broad rock flats of Marino Rocks. Several hundred metres north of the rocks, the rock flats thin as the sandy beach increases in extent (Fig. 4.59). By Seacliff Surf Life Saving Club and the adjoining Seacliff Sailing Club, the beach has almost replaced the rocks and a sandy beach continues to Glenelg (Fig. 4.60). In the south the beach is backed by a park below the old seacliff, then a caravan park and the two club houses, with a car park and boat ramp north of the sailing club. A road and housing then back the beach all the way to Brighton.

The beach itself consists of a high tide sand beach and fronting rock flats south of the surf club, with a shallow sand bar replacing the rocks to the north. Waves are low

Brighton Beach (230B) extends 1 km either side of the Brighton jetty (Figs. 4.61 & 4.62). The suburb of Brighton is centred on the 200 m long jetty, which is backed by the main street, while the beach to either side is backed by a seawall and road. The Brighton Surf Life Saving Club is located 200 m south of the jetty on the west side of the road. A ramp provides access down the seawall to the beach.

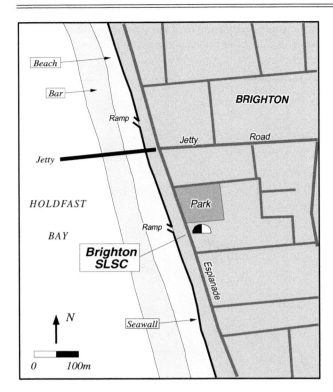

Figure 4.61 *Brighton Beach and jetty, with the Brighton Surf Life Saving Club located 200 m south of the jetty.*

Figure 4.62 *Brighton Beach showing the jetty, seawall fronting the central car park and the Brighton Surf Life Saving Club located south of the jetty. Low waves breaking across the continuous attached low tide bar are typical of beach conditions.*

Somerton Beach (230C) is located north of Brighton Beach, beginning at the Minda dunes and extending for 2 km to the north. The Somerton Surf Life Saving Club is located at the northern end of the Minda dunes (Fig. 4.63). At the surf lifesaving club there is a ramp to the beach, with a rock seawall and beachfront road beginning at the ramp and extending all the way to Glenelg. The beach is typically fronted by a continuous attached bar (Fig. 4.64).

Glenelg is Adelaide's main recreational beach (**230D**).

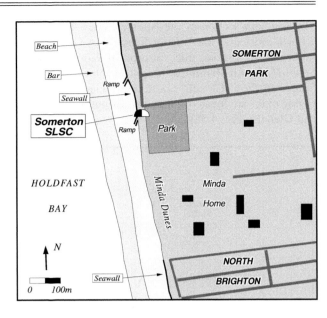

Figure 4.63 *Somerton Beach and surf lifesaving club.*

Figure 4.64 *Somerton Beach, with the Somerton Surf Life Saving Club located right centre, beyond which the road and seawall run north to Glenelg. Note the wide low tide bar and typically low waves.*

It is at the end of the tramline and has traditionally been Adelaide's favourite and most accessible beach. It is backed by parks and major hotels and various recreational facilities, including a marina in the river mouth (Figs. 4.65 & 4.66). The Glenelg Surf Life Saving Club is located in the foreshore park, 200 m north of the jetty. The beach terminates at the Patawalonga breakwater, where the breakwater and an artificial reef off the beach have trapped the sand, causing the beach to build over 100 m seaward.

Swimming: The Seacliff to Glenelg beaches provide relatively safe swimming, owing to the usually low waves and continuous shallow bar. However rips occasionally cross the bar, scouring deeper channels. Stay on the inner bar and clear of any deeper troughs. Care must also be taken near the rocks at Seacliff, around the two jetties, at the Patawalonga breakwater where there can be strong currents and at occasional breaks in the bar where there are deeper holes. The safest swimming is at the four areas patrolled by the Seacliff, Brighton, Somerton and Glenelg Surf Life Saving Clubs.

Surfing: Surf is usually low and sloppy along the Adelaide beaches. A high swell in the south or a strong westerly is required to produce waves over 1 m.

Fishing: The jetties attract most Brighton and Glenelg fishers, while Seacliff rock flats are also popular, as is the Glenelg breakwater. The water off the beaches tends to be shallow, with the best fishing at high tide.

Summary: This is Adelaide's and South Australia's most popular stretch of beach. It offers good access, a wide range of facilities and relatively safe swimming, with usually little surf.

Figure 4.66   *Glenelg jetty and large foreshore reserve which contains the Glenelg Surf Club, with the trained Patawalonga River and Marina. The jetty causes sand to build out against the southern breakwater.*

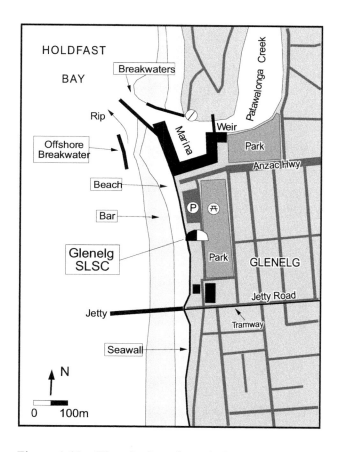

Figure 4.65   *Glenelg Beach and the mouth of the Patawalonga River.*

231   WEST BEACH

WEST BEACH SLSC				
Patrols:	November to March, weekends and public holidays			
No. Beach	Rating	Type Inner	Outer Bar	Length
231A West Beach (south)	3	LTT	LBT	1.8 km
231B West Beach (marina)	4	R		50 m
231C West Beach	3	LTT	LBT	2.5 km

The beach north of the Patawalonga Creek mouth runs relatively straight for 2 km to the new West Beach marina, then another 2.5 km to the Torrens River mouth (Fig. 4.67). The continuous beach is backed by a combination of dunes and seawalls and fronted by a surf zone containing two shallow bars and troughs. The beach faces west-southwest and receives low wind waves averaging 0.5 m and higher waves during strong westerly winds. The waves maintain a double bar system, with a usually attached inner bar, a shallow trough and shallow outer bar. The bars slowly migrate to the north, causing the configuration of the beach and surf zone to change over time. Rips occasionally cut the inner bar following higher waves and strong currents can flow in the trough, particularly on a falling tide.

West Beach (south) (231A) extends from the northern Patawalonga training wall for 1.8 km to the marina breakwater. This beach has been modified in the south by both the training wall, which extends 100 m seaward blocking southerly waves and the periodic dumping of sand from the build-up on the south side of the river. Residential development backs the first kilometre, followed by a sewage treatment works, then a caravan park and marina.

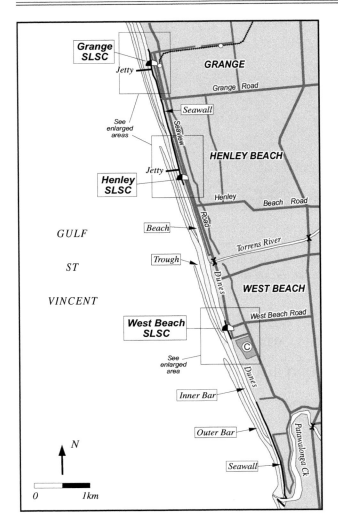

Figure 4.67   *The central Adelaide beaches, including West Beach, Henley Beach and Grange. See inserts for more details of these beaches.*

The **West Beach marina** consists of a southern breakwater that provides shelter for the boat ramp. Between the boat ramp and the smaller northern rock groyne is a 50 m long protected sand beach (**231B**). While this beach is usually calm, it is used for launching boats and jet skis and is often unsuitable and unsafe for swimming because of the boat traffic.

The main **West Beach** (**231C**) faces west and extends for another 2.5 km from the northern marina groyne to the usually closed mouth of the Torrens River. The West Beach Surf Life Saving Club is located at the southern end of the northern residential area, adjacent to the caravan park (Figs. 4.68 & 4.69). Just inside the Torrens breakwater is a large boat ramp which is very popular on weekends.

Swimming: The main beach at the surf club offers the best swimming conditions, well clear of the breakwaters, groynes and boat traffic to the south. There are occasional water pollution problems on southern West Beach and at the Torrens River mouth.

Surfing: Only during strong onshore winds.

Fishing: The training wall, marina breakwater and groyne are all very popular fishing spots.

Summary: A beach that has been modified by the river entrance walls and new marina, but still offers 2.5 km of natural beach and the added protection of the patrolled area.

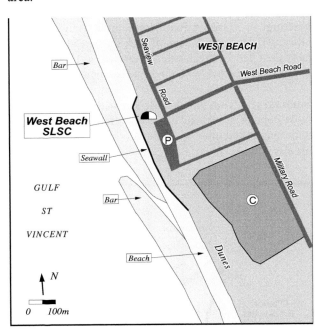

Figure 4.68   *West Beach and surf club before the last harbour development.*

Figure 4.69   *West Beach boat harbour.*

232A-D   **HENLEY to WEST LAKE**

HENLEY SLSC **GRANGE SLSC**	
Patrols:	November to March, weekends and public holidays

| No. | Beach | Rating | Type | | Length |
		Inner	Outer	Bar	
232A	Henley	3	LTT	LBT	2.6 km
232B	Grange	3	LTT	LBT	2 km
232C	Tennyson	3	LTT	LBT	2.6 km
232D	West Lakes	3	LTT	LBT	2 km

North of the Torrens River, the coast runs north-northwest for 10 km to the protruding sandy Semaphore foreland, then slowly curves into west facing Largs Bay. This entire 16 km of beach is backed by Adelaide's beach suburbs. It is bounded by the Torrens and North Haven breakwaters, with four jetties crossing the beach. Waves average less than 1 m at the southern Henley Beach and decrease even further north of Semaphore. At the same time, the beach has a continuous attached bar, fronted by a trough then a second bar. Past Semaphore and into Largs Bay a third bar is present, however waves only break on the outer bars during higher wind wave conditions. Under normal low wave conditions, these are all relatively safe beaches. Just be aware of the troughs between the bars if swimming out past the inner bar.

Henley Beach (232A) is an older suburb with a jetty backed by a plaza and the main shopping area (Fig. 4.70). The Henley Surf Life Saving Club is located 100 m south of the jetty. The entire beach is backed by residential development, including a seawall and road paralleling the back of the beach, with a continuous low tide bar attached to the beach (Fig. 4.71).

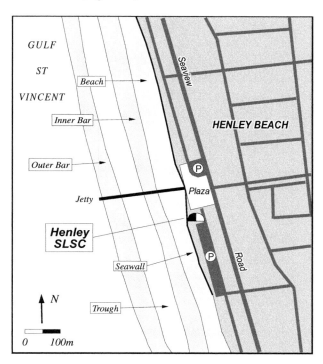

Figure 4.70 *Henley Beach usually has an inner (attached) and an outer bar.*

Figure 4.71 *View of Henley Beach plaza, jetty and surf club, located immediately south of the plaza. Note the attached inner bar and outer bar just off the jetty.*

Grange (232B) is another older seafront suburb, with a jetty flanked by old terrace houses and a large car park. The Grange Surf Life Saving Club is located just north of the jetty, adjacent to the terrace houses (Figs. 4.72 & 4.73).

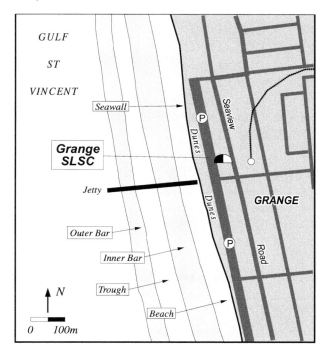

Figure 4.72 *Grange Beach, showing the jetty, beachfront car park, surf club and typically inner and outer bars fronting the beach.*

Tennyson (232C) and **West Lakes (232D)** beaches comprise a more newly developed stretch of shore, backed by beachfront houses and the West Lake development. The beach still retains much of its natural dunes, unlike the seawalls fronting the older suburbs.

Swimming: This section of coast is relatively safe inshore and on the attached portions of the inner bar. Care must be taken in the trough, particularly if occupied by currents and on the outer bar. Children sometimes get caught on the bars by the rising tide.

Surfing: The surf here depends on the winds, with strong westerlies required to produce a sloppy beach break. Occasionally high outside ocean swell reaches the beaches as a low swell.

Fishing: Many fishers use the Patawalonga and Torrens boat ramps to launch their boats for outside fishing, while the two jetties and the Patawalonga Creek mouth are the most popular shore locations. Along the beaches the migrating bars and higher waves produce a range of holes and gutters, which change over time.

Summary: A long section of beach with the older Henley and Grange in the centre and newer developments to the north and south.

Figure 4.73 *Grange Beach and jetty. Grange Surf Life Saving Club is located 100 m north of the jetty. Note the attached inner bar, trough and waves breaking over the outer bar.*

232E-F   **SEMAPHORE, LARGS BAY**

SEMAPHORE SLSC
LARGS BAY

Patrols: November to March, weekends and public holidays

No.	Beach	Rating	Type Inner	Outer Bar	Length
232E	Semaphore	3	LTT	LBT	3 km
232F	Largs Bay	2	LTT	LBT	3.6 km

At Point Malcolm the Adelaide coast protrudes more than 1 km seaward. The older suburb of Semaphore is located on the point, which, together with Largs Bay, make up 7 km of continuous residential development (Fig. 4.74). This entire area is situated on the Lefevre Peninsula, a

10 km long and up to 1 km wide accumulation of sand, that has been deposited over the past 6000 years by the northern movement of sand from as far south as Seacliff. This movement is continuing today, the results of which are clearly evident in the wide, low dune fronting the beach, the wide beach and multiple shallow bars. The sand is now accumulating under the jetties and against the North Haven breakwater, where the beach has built out 400 m seaward in the past 20 years.

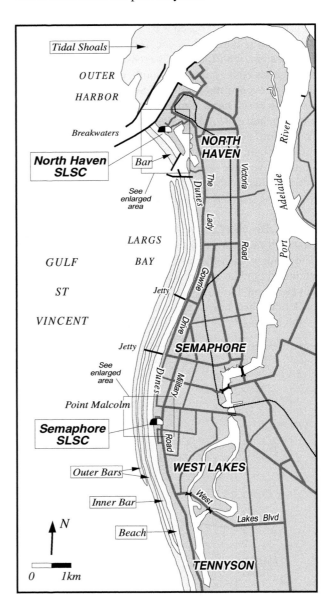

Figure 4.74   *The northern Adelaide coast, showing the continuous strip of sandy beaches and shore parallel bars between Tennyson and North Haven. See inserts for details of Semaphore and North Haven.*

This section of coast receives only low waves, usually less than 0.5 m. Occasional higher waves are sufficient to rework the fine sand into a wide, low, firm beach, fronted by two to three shore parallel shallow bars and troughs, extending up to 500 m seaward of the beach. Two jetties cross the beach at Semaphore and Largs Bay, the latter is now located within the shallow bar section.

The **Semaphore** Surf Life Saving Club is located on Point Malcolm (**232E**), south of Fort Granville and adjacent to the caravan park. It is fronted by a foreshore reserve, wide beach and wide, shallow surf zone with three shallow bars and a trough off the beach (Figs. 4.75 & 4.76).

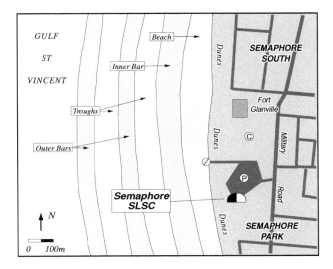

Figure 4.75 *Semaphore Surf Life Saving Club is fronted by three parallel bars, producing a wide, low gradient surf zone.*

Figure 4.76 *View north along Point Malcolm up to Largs Bay. Semaphore Surf Life Saving Club is located adjacent to the park and caravan park in the lower centre. Note the shore parallel bars that act as a conduit for northerly migrating sand into Largs Bay (top).*

Largs Bay (232F) is a 4 km long, west facing, curving shoreline, which receives low waves and is often calm, with the accumulating sand shoaling the bay floor as well as widening the beach. The end of the Largs Bay jetty now stands in shallow water.

Swimming: This is a relatively safe section of beach owing to the usually low waves and shallow water. Care need be taken with young children, as the water depth varies over the bars and troughs and there are some deeper holes.

Surfing: There is usually no surf up here. You need a very strong southwesterly or huge ocean swell to push rideable waves up into Largs Bay.

Fishing: The jetties are the most popular spots for fishing, as the beaches are very shallow.

Summary: A low energy but still dynamic section of beach that is continuing to grow seaward. There is excellent access and large areas of protected dune, plus the beach and bars to play on.

233   NORTH HAVEN

NORTH HAVEN SLSC			
Patrols:	November to March, weekends and public holidays		
No. Beach	Rating	Type	Length
233 North Haven	2	LTT	1.2 km

North Haven beach (**233**) is the product of a major redevelopment of the northern part of the Adelaide coast. In the 1980s, North Haven Marina development excavated a large marina behind the beach, with two breakwaters to protect the entrance. The North Haven beach is now located between the northern breakwater and the older breakwater for Adelaide's Outer Harbour (Figs. 4.77 & 4.78). The 1.2 km long beach is wedged between the two breakwaters and has been slowly building seaward. Residential development backs the beach, with the North Haven Surf Life Saving Club, a large car park and dunes occupying the beachfront.

The beach faces southwest and is composed of fine sand, which, together with the low waves, usually less than 0.5 m, has produced a wide, flat, firm beach and a wide, shallow bar attached to the beach.

Swimming: This is the safest beach on the Adelaide coast, with usually low wave to calm conditions and no rip currents.

Surfing: None, except during a huge ocean swell which can penetrate up the gulf as far as North Haven, where it arrives as a low, 0.5 m swell.

Fishing: The boundary breakwaters offer the best locations.

Summary: A relatively newly developed beach, with good access and parking, plus the surf club, in a low energy setting.

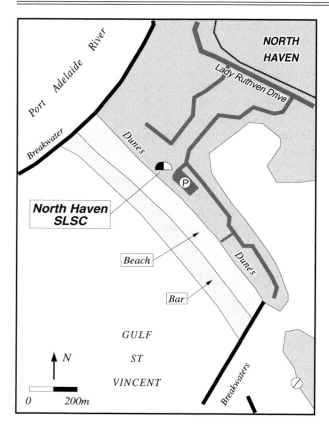

Figure 4.77   *North Haven has a beach that formed between the two training walls and is now patrolled by the North Haven Surf Life Saving Club.*

Figure 4.78   *The mouth of the Port Adelaide River and North Haven beach.*

YORKE PENINSULA DISTRICT
PORT GAWLER TO FISHERMAN BAY
(Figs. 4.79, 4.86, 4.99, 4.102)

Length of Coast:	635 km
Beaches:	315 beaches (234-548)
Surf Lifesaving Clubs:	none
Lifeguards:	none
Major towns:	Port Wakefield, Ardrossan, Port Vincent, Edithburgh, Point Turton, Port Victoria, Moonta, Wallaroo, Port Broughton

The Yorke Peninsula District commences north of Adelaide and includes both sides of upper Gulf St Vincent and all of Yorke Peninsula around to Fisherman Bay. Much of the 635 km of shoreline lie in the protected waters of St Vincent and Spencer gulfs, with the beaches dominated by low wind waves and increasingly, tides in the upper gulfs. The only section of the peninsula regularly exposed to ocean waves lies along the southwest section between about Waterloo Bay, Cape Spencer and Corny Point. Because of the generally low waves and remote location of many of the beaches, there are no surf clubs or patrolled beaches on the peninsula. However, care must be taken if surfing or swimming on any of the more exposed beaches, as most are remote and rips and rock reefs are common.

Regional Map 4: Upper Gulf St Vincent

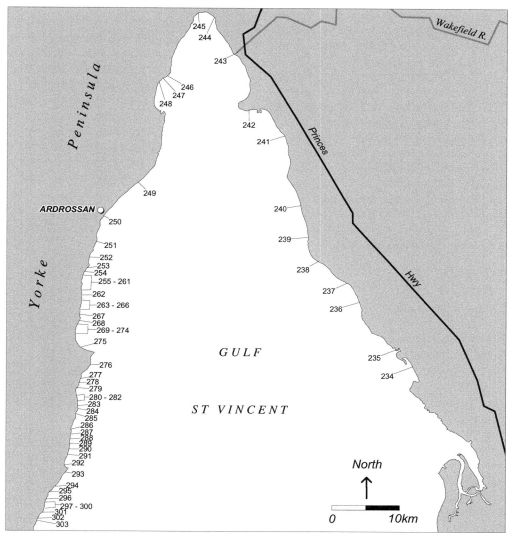

Figure 4.79   *Regional map 4   Upper Gulf St Vincent*

234-242   PORT GAWLER to PORT WAKEFIELD

No.	Beach	Rating	Type	Length
	Unpatrolled			
234	Port Gawler	1	R+tidal flats	4.5 km
235	Middle Beach	1	R+tidal flat	1.2 km
236	Light Beach	1	R+sand flats	2.2 km
237	Prime Beach	1	R+sand flats	4.8 km
238	Great Sandy Pt	1	R+sand flats	6.5 km
239	Webb Beach	1	R+sand flats	2 km
240	Parham Beach	1	R+sand flats	8.1 km
241	Lorne	1	R+sand flats	13.4 km
242	Bald Hill Beach	1	R+tidal flats	2.5 km

North of Adelaide there is a dramatic change in the nature of the coast and its beaches. Between Port Gawler, 35 km north of Adelaide, and Port Wakefield is a 50 km section of low gradient, tide dominated shoreline, containing a series of nine beaches. Six are accessible by car and two by 4WD, while Lorne Beach is part of the Military Firing Range and is off-limits to the public. Wave energy decreases to low wind waves and the coastal zone consists of a several kilometre wide zone of very low gradient tidal deposits. The beaches consist of low energy high tide shelly ridges, fronted by tidal and sand flats extending up to 3 km into the gulf and backed by equally wide salt flats. Mangroves fringe the lowest energy section of the coast and seagrass meadows grow seaward of the tidal flats, with their debris usually piled on the beaches. The main highway, the Port Wakefield Road, runs on the first high ground between 5 and 7 km inland. As a consequence, few travellers and tourists make the detour out to these low energy beaches.

Port Gawler Conservation Park

Established: 1971
Area: 418 ha
Beach: immediately south of beach 234

Port Gawler Conservation Park encompasses a 1 km wide strip of mangroves that extends for 4 km south of Port Gawler.

Port Gawler beach (**234**) is the first beach north of Adelaide and is located at the end of the Port Gawler Road. While a town was surveyed in 1839, little has changed in the area since. The 4.5 km long beach is backed by tidal flats, which have been developed as salt evaporation ponds. The beach consists of a high tide sand ridge, fronted by 3 km of tidal flats, with extensive mangroves on the flats south of the beach. The southern mangrove area is part of the Port Gawler Conservation Park. A 4WD track runs along the back of the beach and there is no development other than a camping area.

Middle Beach (235) is reached along the gravel Middle Beach Road. The 1.2 km long beach is almost encased in mangroves. Only the southern section at the mouth of the small Salt Creek is free of mangroves and is used for boat launching at high tide. There is a row of fishing shacks along the low, shelly beach, most facing into the mangroves, with a camping area at the northern end of the shacks.

Light Beach (236) is an undeveloped, 2.2 km long, low beach with 1 km wide sand flats, cut by three small tidal creeks, with only 4WD access across the backing salt flats.

Prime Beach (237) is a low, seagrass debris covered beach fronted by 1 km wide tidal flats. There is a gravel road to the beach, but no development. **Great Sandy Point (238)** extends for 6.5 km north of Prime Beach as a slowly curving, southwest facing foreland, backed by salt flats up to 4 km wide and fronted by 2 km wide tidal flats. There is only 4WD access across the salt flats to the undeveloped beach.

Webb Beach (239) and **Parham (240)** are both linked to the highway by the circuitous Parham Road. Webb Beach is 2 km long, bordered by Baker Creek in the south and a smaller creek in the north. It faces west and has a small subdivision at the northern end, set back from the beach. At the beach there is a boat ramp, toilets and a car park, fronted by 1 km wide tidal flats. At Parham, 1 km to the north, is a slightly larger subdivision boasting a telephone, toilets and car park (Fig. 4.80). The beach runs north for 8 km to a small sandy foreland.

Figure 4.80 *Parham Beach is backed by a small beachfront subdivision, then salt flats, with wide, low gradient intertidal sand flats fronting the low shelly beach, typical of many gulf beaches.*

Lorne is the name of an undeveloped settlement located on the centre of the 13.4 km long beach that extends from the north side of the foreland to Bald Hill. The entire beach (**241**) is located within the Port Wakefield firing range and is off-limits to the public, with a locked gate preventing entry.

Bald Hill Beach (242) is named after the backing 20 m high hill. It is a very low energy, south facing beach and tidal flat, which can be accessed via a gravel road to the western end. The road is closed to the public when the firing range is in use.

Summary: This section of coast is only used by fishers who use a variety of boats and elevated vehicles to fish the creeks, tidal flats and gulf. The entire shore is unsuitable for swimming except at high tide and there is no surf.

243   PORT WAKEFIELD

No.	Beach	Unpatrolled Rating	Type	Length
243	Port Wakefield	1	R+tidal creek	200 m

Port Wakefield 'beach' **(243)** offers the safest and most suitable swimming beach on Gulf St Vincent north of Adelaide. An arm of the Wakefield River has been dammed, creating a 200 m long and 50 m wide tidal pool and beach (Fig. 4.81). The eastern side of the beach is bordered by a shady picnic area and playground, a car park and the caravan park, with a footbridge across to the western side, which has some shade shelters.

Summary: The best place to cool off north of Adelaide and right next to all the necessary amenities, as well as 'downtown' Port Wakefield.

Figure 4.81 *Port Wakefield 'beach' is located either side of a dammed tidal creek. It does however offer good swimming on a coast dominated by shallow tidal flats.*

Clinton Conservation Park	
Established:	1970
Area:	1951 ha
Beaches:	2 (244, 245)
Coast length:	25 km

At the northern end of Gulf St Vincent lies the Clinton Conservation Park. It extends from either side of Port Wakefield to the top of the gulf and down to Port Clinton, with a shoreline distance of 25 km. Most of the very low energy shore is occupied by mangroves and fronted by tidal flats up to 3 km wide.

244, 245   CLINTON

No.	Beach	Unpatrolled Rating	Type	Length
244	Clinton (E)	1	R+tidal flats	3.2 km
245	Clinton (W)	1	R+tidal flats	1.6 km

At the very northern apex of the gulf are 5 km of open tidal flats backed by two low, high tide beach ridges. On the eastern side is 3.2 km long beach **244**, which is accessible via a 3 km long 4WD track across the backing salt flats. It is bordered on the south by mangroves and to the west by a tidal creek. To the west of the creek is the second beach **(245)**, a 1.6 km long, southeast facing beach ridge fronted by the wide tidal flats. Neither beach is suitable for swimming owing to the shallow flats.

246-248   PORT CLINTON

No.	Beach	Unpatrolled Rating	Type	Length
246	Port Clinton (N)	1	R+tidal flats	800 m
247	Port Clinton	1	R+tidal flats	700 m
248	Port Clinton (S)	1	R+tidal flats	800 m

Port Clinton is the first settlement on the eastern side of Yorke Peninsula. It is spread along 1.5 km of low energy sandy shoreline and 1 km wide tidal flats. The shore is divided into three beach areas by mangroves to either end and two central small forelands. The northern beach **(246)** is 800 m long and extends north from a small sandy foreland, formed in lee of a patch of mangroves, as a series of recurved spits that terminate at mangrove covered tidal flats. The main beach **(247)** runs south of the foreland for 700 m to a small protruding upland creek mouth. This beach is backed by the caravan park and foreshore reserve and is the main beach used for launching boats across the beach, and for wading and swimming when the tide is high enough. On the south side of the creek is the southern beach (**248**), which runs for 800 m, with about 20 shacks backing the southern half of the beach. The beach terminates at mangrove covered tidal flats, backed by 10 m high red bluffs which dominate the coast to the south.

Summary: Port Clinton offers good access to a low energy beach and tidal flats, with boating, wading and swimming only possible at high tide. The beach is mainly used by the locals to launch their small boats at high tide.

249, 250   MACS-TIDDY WIDDY-YOUNGS, ARDROSSAN

	Unpatrolled		
No.	Beach	Rating	Type	Length
249	Macs-Tiddy Widdy-Youngs	1	R+sand flats	11.2 km
250	Ardrossan	1	R+sand flats	150 m

Macs Beach (**249**) is the northern part of an 11.2 km long beach that begins as the mangroves of Mangrove Point are replaced by a low energy beach ridge, fronted by 300 m wide sand flats. The beach runs southwest for 8 km as Macs Beach, until Tiddy Widdy settlement, when the names change to **Tiddy Widdy Beach** and finally it is called **Youngs Beach** just before it terminates at the 10 m high red cliffs that continue on down to Ardrossan.

The beach receives occasionally low wind waves, which maintain the 300 m wide intertidal sand flats and the narrow shelly high tide beach. The waves also pile seagrass debris on the beach, the debris at times reaching 1 m deep and extending up to 50 m across the beach. There is access to the beach at Macs Beach where there is only a beach boat ramp, then at Tiddy Widdy where there is also a beach boat ramp but no other facilities.

South of Youngs Beach are 1.5 km long low bluffs which run to Ardrossan. **Ardrossan** is an interesting old town, with excellent facilities for visitors and those interested in fishing the jetty or gulf. The town straddles a small creek which has cut a gully though the red bluffs, at the end of which a 150 m long pocket of high tide sand (**250**) is located either side of the town jetty. Three hundred metre wide seagrass covered sand flats front the beach, which is not only bisected by the 400 m long jetty, but also crossed by three wooden groynes (Fig. 4.82). A park and amenities back the beach, which is mainly used for launching boats, with swimming only possible at high tide. The main deep water boat ramp is located at the new jetty 1.5 km south of the town.

Ardrossan was originally known as Clay Gully, with early settlers living in dugouts around the gully. It was proclaimed a town and named Ardrossan in 1873.

Swimming: Only possible at high tide owing to the shallow sand flats.

Surfing: None.

Fishing: Best at Ardrossan jetty or from a boat.

Summary: Two low energy beaches fronted by wide sand flats, with Ardrossan's beach surrounded by the jetty and the town facilities.

Figure 4.82   *A view from the southern bluffs along Ardrossan beach, showing the inner part of the jetty, wide, low beach and fronting sand flats. The main beach access is located at the jetty.*

251-253   PARARA to ROGUES POINT

	Unpatrolled		
No.	Beach	Rating	Type	Length
251	Parara	1	R+tidal flats	3.5 km
252	James Well	1	R+tidal flats	2.1 km
253	Rogues Pt	2	R+sand flats	1 km

South of Ardrossan the coast trends essentially due south, with the red bluffs forming the coast for 2 km south of the town, before they are fronted by beach deposits which run for 7 km down to just south of Rogues Point, where the bluffs again dominate the coast. There are three crenulate, low energy beaches along this section, with the southern two backed by numerous shacks.

Parara is the name of a low sandy foreland, accessible by a gravel road to the middle of the foreland where there is a tin shed but no facilities. It is named after the old Parara homestead located 3 km inland. The narrow high tide beach (**251**) is usually covered in seagrass debris and is fronted by 500 m wide tidal flats. The beach extends for 1.5 km north of the foreland and 1.5 km to the south, where it terminates at the protruding mouth of the small Pavy Creek. On the southern side of the mouth, **James**

Well beach (**252**) runs for 2 km to the tip of Rogues Point, a low sandy foreland that protrudes 300 m seaward in lee of some low intertidal rocks. About 20 shacks back the southern half of the beach and the point, with more shacks located west of the gravel road that parallels the back of the beach.

On the southern side of **Rogues Point** is a 1 km long, southeast facing beach (**253**) which extends down to the 10 m high bluffs at the base of Mennie Hill. There is a large sealed car park, a beach boat ramp and beach shelter at the northern end (Fig. 4.83), with 25 beachfront shacks backing the centre of the beach. Because of its orientation, this beach receives slightly higher waves, still averaging less than 0.5 m, which maintain three low ridges across the 100 m wide sand flats.

Figure 4.83   *The beach south of Rogues Point.*

Swimming: The southern side of Rogues Point offers the best swimming at mid to high tide.

Surfing: None.

Fishing: You can only fish from shore at high tide, with most fishing from boats out past the sand flats.

Summary: Three accessible, low energy beaches, two housing scores of shacks but no facilities other than the boat ramp.

254-266   **MENNIE HILL to THROOKA CREEK**

No.	Beach	Rating	Type	Length
Unpatrolled				
254	Mennie Hill (S 1)	2	R+sand flats	500 m
255	Mennie Hill (S 2)	2	R+sand flats	60 m
256	Mennie Hill (S 3)	2	R+sand flats	70 m
257	Muloowurtie Pt (N 2)	2	R+sand flats	400 m
258	Muloowurtie Pt (N 1)	2	R+sand flats	400 m
259	Muloowurtie Pt (S 1)	2	R+sand flats	400 m
260	Muloowurtie Pt (S 2)	2	R+sand flats	350 m
261	Muloowurtie Pt (S 3)	2	R+sand flats	400 m
262	Muloowurtie Pt (S 4)	2	R+sand flats	1 km
263	Throoka Creek	1	R+sand flats	600 m
264	Throoka Creek (S 1)	1	R+sand flats	300 m
265	Throoka Creek (S 2)	1	R+sand flats	450 m
266	Throoka Creek (S 3)	1	R+rock flats	300 m

Between Mennie Hill and just north of Pine Point is 7 km of east facing shore, backed by continuous 20 m high bluffs and cleared farmland. Thirteen narrow beaches lie along the base of the bluffs and all are fronted by partly seagrass covered sand flats ranging in width from 100 m at Mennie Hill to 500 m off Throoka Creek mouth. While the main road parallels the beaches approximately 500 m inland, the land between is all private farmland and there is no public access to any of the beaches, other than by foot along the shore. Nor are there any facilities at any of the beaches.

Beach **254** is a 500 m long, east facing strip of high tide sand located at the base of steep, 30 m high bluffs. It can be accessed on foot from the southern end of Rogues Point beach. Beach **255** is a 60 m long, northeast facing pocket of sand backed and bordered by bluffs, while beach **256** is similar, but 70 m long. Beach **257** is a straight, 400 m long, east facing strip of sand backed by steep, vegetated bluffs. Beach **258** is a curving, 400 m long, northeast facing beach that terminates at Muloowurtie Point.

On the southern side of **Muloowurtie Point**, beach **259** runs straight for 400 m to a small inflection in the bluffs. On the south side of the inflection is beach **260**, another straight, 350 m long beach. Beach **261** is a more crenulate and narrow beach, with a more irregular backing bluff. Beach **262** is the longest beach in this section at just over 1 km. It is relatively straight, faces east and has one unnamed creek backing the centre. Beach **263** is a narrow, crenulate beach backed by steeper bluffs, which terminates at the mouth of the usually dry Throoka Creek. The sand flats widen to 500 m in front of the creek, while creek-deposited boulders litter the creek mouth.

Beach **264** runs for 300 m south of the creek mouth to a small cove backed by 30 m high scarped bluffs. Beach **265** is a relatively straight, 450 m long, narrow beach backed by steep 30 m high bluffs. Beach **266** is a crenulate beach bordered by slightly protruding bluffs with rocks at their base.

Summary: These are 13 relatively safe beaches, fronted by shallow sand flats, however all are backed by private farms and can only be accessed on foot along the shore or by boat at high tide.

No.	Beach	Unpatrolled Rating	Type	Length
267	Pine Point (N)	1	R+sand/rock flats	1 km
268	Pine Point (S 1)	1	R+sand/rock flats	900 m
269	Pine Point (S 2)	1	R+sand/rock flats	350 m
270	Pine Point (S 3)	1	R+sand/rock flats	250 m
271	Pine Point (S 4)	1	R+sand/rock flats	350 m
272	Rocky Point (N 3)	1	R+sand/rock flats	450 m
273	Rocky Point (N 2)	1	R+sand/rock flats	200 m
274	Rocky Point (N 1)	1	R+sand/rock flats	500 m

Pine Point is a 2 km long sandy foreland that protrudes 300 m seaward of the backing 20 m high grassy bluffs. It has formed at the mouth of a small unnamed creek. Most of the low point area has been developed as a shack settlement, with approximately 50 small shacks. The main road runs along the top of the bluffs above the southern half of the settlement, with houses straddling the road for 2 km south of the point. Between Pine Point and Rocky Point, 3 km to the south, are eight low energy beaches generally located at the base of 20 m high grassy bluffs. There is vehicle access to the bluffs above all the beaches and tracks down to most of them.

The two **Pine Point** beaches (**267** and **268**) total 2 km in length, joined at the protruding sandy foreland. The backing sand flats are covered by the shacks, with a narrow foreshore reserve between the shacks and the beach. The beach is narrow, moderately steep and composed of shelly sand, with rock and sand flats extending 400 m into the gulf. The southern end of Pine Point beach narrows to the bluffs. Beach **269** lies below the 20 m high grassy bluffs and runs for 350 m to a landing that has been built across the beach and 50 m into the gulf. A gravel road connects the landing with the main road. On the south side of the landing is a sandy, 250 m long beach (**270**), bordered in the south by a small sandy foreland sitting on rock and sand flats.

Beach **271** is 350 m long and bordered by two small sandy forelands, formed in lee of rock flats and backed by 20 m high bluffs. There is no direct vehicle access to this beach or beach 272, though both are backed by bluff-top houses. Beach **272** is 450 m long and also bordered by rock flats and fronted by sandy forelands. Beach **273** can be accessed via a gravel road down the side of the backing bluffs. It is 200 m long and also bordered by small sandy forelands. Finally, beach **274** lies between the last sandy foreland and Rocky Point, a slight protrusion in the bluffs. A gravel track runs down the bluff to the beach, which, like all the others, is a 10 m wide high tide beach fronted by rocky/sand flats which extend 100 m offshore.

Swimming: You can only swim at high tide, however beware of the rocks on the foreland and along much of the tidal flats.

Surfing: None.

Fishing: Best at high tide off the landing and over the rock flats.

Summary: Pine Point is a quiet little shack-holiday settlement, with a store, caravan park and phone.

275, 276   **BLACK POINT**

No.	Beach	Unpatrolled Rating	Type	Length
275	Black Pt (N)	1	R+sand flats	4.3 km
276	Black Pt (S)	2	R+sand flats	4.2 km

Black Point is a large cuspate foreland that protrudes 3 km into the gulf. The northern side faces northeast into the more protected embayment called Port Alfred, whereas the eastern side faces straight into the gulf. The northern beach (**275**) curves around for 4.3 km from Rocky Point to Black Point. It is a continuous low energy sandy beach with seagrass growing almost to the shore. The beach is initially backed by sloping grassy bluffs as far as the landing. The old landing halfway along the beach is now the site of a concrete boat ramp, to the east of which is a continuous row of more than 100 beachfront shacks, some of which are using a variety of techniques to combat episodic erosion. Many of the shacks have boat ramps to the beach and numerous boats are moored offshore. A gravel road parallels the back of the shacks out to the point.

Beach **276** runs south of Black Point for 4.2 km to a low rocky bluff located 2 km north of Port Julia. This beach has no formal access and is backed by farmland. It receives slightly higher southerly wind waves and consists of a narrow, steep beach fronted by 200 to 300 m wide sand flats.

Swimming: The northern shack beach offers usually calm water and seagrass covered sand flats.

Surfing: None.

Fishing: Best along the northern beach with its deeper seagrass meadows close to shore.

Summary: The Black Point shack settlement faces a low energy, northeast facing beach, with a good boat ramp and usually calm anchorage.

277-283   PORT JULIA

	Unpatrolled			
No.	Beach	Rating	Type	Length
277	Port Julia (N 2)	2	R+sand/rock flats	600 m
278	Port Julia (N 1)	2	R+sand/rock flats	600 m
279	Port Julia	2	R+sand flats	1 km
280	Port Julia (S 1)	2	R+sand flats	400 m
281	Port Julia (S 2)	2	R+sand flats	200 m
282	Port Julia (S 3)	2	R+sand flats	500 m
283	Port Julia (S 4)	2	R+sand flats	180 m

Port Julia is an old wheat port, now the site of a few shacks, jetty and toilet, but no other facilities. Twenty metre high bluffs back the main beach and the coast to either side, with a total of seven beaches lying between the southern end of Black Point beach and Sheoak Flat.

Beach **277** is located 1 km north of the port. It is 600 m long, faces east, is backed by a grassy 10 m high bluff and has a narrow high tide sand beach partly covered in cobble and boulders. Boulder fronted bluffs form the boundaries. Beach **278** is also 600 m long, with more sand and less rocks, while both are fronted by 200 m wide, seagrass covered sand flats.

The **main beach** at Port Julia (**279**) is 1 km long, with the central 100 m long jetty backed by a wooden seawall and shacks to the south. A larger freehold settlement sits on the backing bluffs. The beach is 10 m wide at high tide, with 100 m wide sand flats, then deeper seagrass meadows (Fig. 4.84).

Figure 4.84   *Port Julia is located in an open embayment, with the beach backed by shacks and bluffs, and the jetty occupying the centre of the beach.*

South of the port, the bluffs dominate the coast. Beach **280** is a 400 m long strip of sand located at the base of steep, 30 m high eroding bluffs. Toward the southern end of the beach is a large pocket in the bluffs where about twelve beachfront shacks are located. Beach **281** lies

immediately to the south and is a 200 m long beach bordered and backed by 30 m high bluffs, with a solitary bluff-top house overlooking the otherwise undeveloped, narrow beach.

Beach **282** is a straight, 500 m long high tide sand beach, backed by steep, partly vegetated, 20 m high bluffs, with no vehicle or direct public access. Beach **283** is a 180 m long pocket of sand bordered and backed by steep, 20 m high bluffs, with two houses located to the lee of the bluffs, but no public access to the beach.

Swimming: The most popular spot is at Port Julia where there is good access, basic facilities, the jetty and some water at high tide.

Surfing: None.

Fishing: The Port Julia jetty provides the only shore-based access to the deeper seagrass meadows, otherwise it's a matter of waiting for high tide or using a boat.

Summary: A low energy, quiet section of the gulf, with a small settlement at Port Julia but no facilities for travellers.

284, 285   SHEOAK FLAT

	Unpatrolled			
No.	Beach	Rating	Type	Length
284	Sheoak Flat (N)	2	R+sand flats	750 m
285	Sheoak Flat	2	R+sand flats	1.8 km

Sheoak Flat is an irregular sandy foreland that extends for 2.5 km along the shore. It protrudes a maximum of 200 m and is backed by sloping, 20 m high grassy bluffs. The access road runs 200 m in from the beach and there are two small holiday settlements on the flat.

The northern beach (**284**) is a crenulate, 750 m long, east facing sandy beach, which has been building out into the gulf toward the southern end. The beach and foreland are in a natural state, apart from about twelve shacks set in the scrub toward the southern end. The main beach (**285**) runs for 1.8 km from the foreland through to a second curving foreland, that terminates at the base of the 20 m high Dowcer Bluff. The Sheoak Flat settlement occupies 200 m of the central flats, wedged in between the road and shore. As well as the houses, there is a beachfront car park. Both beaches receive low waves and are fronted by shallow sand flats and seagrass meadows.

Swimming: These beaches usually offer calm conditions at high tide.

Surfing: None.

Fishing: You can fish off the beaches at high tide, however most of the locals use their boats to fish the deeper water off the sand flats.

Summary: An accessible little settlement and quiet beach with no facilities.

286-291   **DOWCER BLUFF**

Unpatrolled			
No.	Beach	Rating	Type	Length
286	Dowcer Bluff (1)	2	R+rocks&flats	100 m
287	Dowcer Bluff (2)	2	R+rocks&flats	350 m
288	Dowcer Bluff (3)	2	R+rocks&flats	120 m
289	Dowcer Bluff (4)	2	R+rocks&flats	100 m
290	Dowcer Bluff (5)	2	R+rocks&flats	420 m
291	Dowcer Bluff (6)	2	R+rocks&flats	1.3 km

Dowcer Bluff is a 5 km section of 20 to 30 m high cliffed bluff that forms the northern boundary of Port Vincent Bay. A golf course occupies part of the northern bluffs, with farmland backing the rest down to Port Vincent. Tucked in along the base are six narrow, bluff-bound beaches, with varying degrees of access.

Beach **286** lies at the base of 20 m high bluffs immediately north of the golf course. It is a 100 m long, curving sliver of sand with no access other than by boat. Beach **287** lies below the golf club, with grassy slopes leading to the middle of the 350 m long beach. The crenulate beach is fronted by 100 m wide ridged sand flats, then seagrass meadows.

Beaches 288 and 289 are two inaccessible bluff-bound beaches, 120 m and 100 m long respectively. Beach **288** lies immediately south of the golf course and is surrounded by 20 m high bluffs, with no safe foot access. Likewise beach **289**, 500 m to the south, is backed and bordered by steep bluffs.

Beaches 290 and 291 are two longer, accessible beaches both housing a few shacks. Beach **290** is 420 m long and fronted by continuous, 50 m wide intertidal rock flats, then a 50 to 100 m wide strip of sand flats. There is a vehicle track to the southern bluff, with a walking track down to the two shacks. Beach **291** is 1.3 km long and runs from the southern side of the last prominent bluff, where four shacks are located, down to the end of the bluffs. The shack owners have built three rock groynes across the sand and cobble beach, which is fronted by a mixture of rock and sand flats, averaging about 150 m in width.

Swimming: While waves are usually low and calms are common, watch for the numerous rocks along these beaches.

Surfing: None.

Fishing: Best at high tide or out over the seagrass meadows from a boat.

Summary: Six small, narrow, bluff-bound beaches, located right below the bluff-top road, but mainly used by the few southern shack owners.

292, 293   **PORT VINCENT**

Unpatrolled			
No.	Beach	Rating	Type	Length
292	Port Vincent (N)	1	R+sand flats	1.9 km
293	Port Vincent (S)	1	R+sand flats	1.6 km

Port Vincent is a small town located on a prominent cuspate foreland called Surveyor Point (Fig. 4.85). The point protrudes 1.3 km into the gulf, with the town occupying the backing 1 km² of flat land between the shore and the 20 m high bluffs. Two beaches line either side of the sandy foreland.

Figure 4.85   *View from the northern bluffs across Port Vincent to Surveyor Point.*

Beach **292** lies on the north side and is backed by the main road. The 1.9 km long beach faces northeast and is sheltered by Surveyor Point. Consequently the jetty, boat ramp and boat moorings are located here and conditions are usually calm. The northern sand flats narrow and deepen, with seagrass meadows right off much of the beach. A channel has been dredged though the seagrass to the jetty, boat ramp and seawall, providing deepwater access at the shore. Beach **293** runs for 1.6 km south of Surveyor Point and, while facing southeast, is a low energy beach fronted by 200 m wide sand flats, with usually calm to low wave conditions along the beach. A gravel road and 50 m wide foreshore reserve backs the entire beach.

Swimming: These are two very accessible, low wave beaches, with the best swimming on the deeper northern beach and at high tide.

Surfing: None.

Fishing: Best off the jetty and seawall at high tide.

Summary: A small, attractive town nestled between the bluffs and the foreland, with most facilities for the visitor.

---

294 - 301   DEVIL GULLY to BEACH 301

No.	Beach	Rating	Type	Length
		Unpatrolled		
294	Devil Gully	2	R+sand/rock flats	200 m
295	Beach 295	2	R+sand flats	600 m
296	Beach 296	2	R+sand flats	50 m
297	Beach 297	2	Boulder+sand flats	150 m
298	Beach 298	2	R+sand flats	90 m
299	Beach 299	2	R+sand flats	60 m
300	Beach 300	2	R+sand flats	550 m
301	Beach 301	2	R+sand flats	750 m

South of Port Vincent, the coast trends relatively straight for 15 km to Stansbury. All of the intervening shoreline is dominated by 10 to 20 m high crumbling bluffs, fronted in places by narrow sand and boulder beaches, and then sand flats extending 50 to 100 m into the gulf. Waves average less than 0.5 m and calms are common. While the main road runs only 200 to 300 m in from the shore, farms back most of the bluffs and public access to the shore and beaches is very limited.

Devil Gully is a designated camping and recreation reserve with however no vehicle public access and no facilities. A 200 m long beach (**294**) lies across the mouth of the small gully. Beach **295** is a straight, 600 m long sand-boulder beach lying beneath 30 m high bluffs, with no direct access. Beach **296** is a 50 m pocket of sand and cobbles located at the mouth of a small gully, with 30 m high bluffs to either side. Beach **297** is a 150 m long beach accessible via a gravel track off the main road. It is classified as a recreation reserve, but there are no facilities. However the locals do use it to launch small boats off the beach.

Beaches **298** and **299** are two 90 and 60 m long pockets of sand and cobbles, backed by 20 m high bluffs. Beaches 300 and 301 front 15 m high bluffs that protrude slightly into the gulf. Beach **300** is 550 m long and faces east, while beach **301** is 750 m long and faces more southeast. Both beaches have built out a few tens of metres from

the bluffs, forming a narrow coastal flat. Farm tracks run to the southern end of each beach but there is no public access.

Swimming: More suitable at high tide when the sand and in places rock flats, are covered.

Surfing: None.

Fishing: Most fishers use Devil Gully for boat launching and occasional beach fishing at high tide.

Summary: Eight low energy, bluff-bound beaches, mainly backed by private farmland, with public access only to the recreation reserve at beach 297.

302-310   OYSTER BAY

No.	Beach	Rating	Type	Length
		Unpatrolled		
302	Beach 302	2	R+sand flats	100 m
303	Beach 303	2	R+sand flats	100 m
304	Beach Pt (N)	2	R+sand flats	350 m
305	Beach Pt (S)	2	R+sand flats	80 m
306	Oyster Bay (1)	2	R+sand flats	80 m
307	Oyster Bay (2)	2	R+sand flats	170 m
308	Oyster Bay (3)	2	R+sand flats	120 m
309	Oyster Bay (4)	2	R+sand flats	1.5 km
310	Oyster Bay (5)	2	R+sand flats	60 m

North of the Stansbury jetty is an essentially straight, 7 km long, east facing section of bluffed shoreline, fronted by continuous 50 to 100 m wide sand and rock flats, with nine narrow beaches wedged in at the base of the bluffs. All of the beaches face east into the open Oyster Bay, receive low waves and are often calm, and all are backed by farmland, with no public access to the shore.

Beaches **302** and **303** are two 100 m long pockets of sand backed and bordered by 20 m high bluffs. A small valley provides foot access to beach 303. Beach **304** is a straight, 350 m long beach at the base of 20 m high bluffs, with Beach Point protruding 100 m into the gulf at the southern end. Beaches **305**, **306** and **307** run south from Beach Point, occupying a total of 500 m of shoreline. They consist of small high tide beaches backed and bordered by bluffs, but linked at low tide by the sand flats. Beach **308** is a 120 m long beach occupying the mouth of a small valley, with a vehicle track reaching the rear of the beach. Beach **309** is a narrow, 1.5 km long beach lying at the base of the bluffs. Finally, beach **310** is a 60 m pocket of sand, backed by a bluff-top farm house and was the site of an old jetty.

Summary: Nine similar bluff-bound beaches, all backed by the bluffs and farmland with no public access other than by boat.

Regional Map 5: Yorke Peninsula

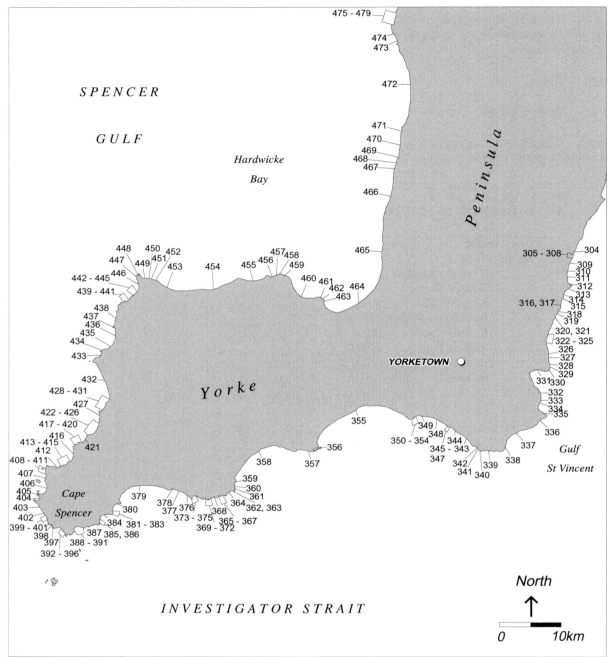

Figure 4.86   *Regional Map 5 – the lower Yorke Peninsula*

311-313   **STANSBURY**

Unpatrolled			
No.	Beach	Rating	Type	Length
311	Stansbury (jetty)	2	R+sand flats	900 m
312	Stansbury	1	R+sand flats	1.2 km
313	Stansbury (S)	1	R+sand flats	1.5 km

Stansbury is located on a long, sandy cuspate foreland called Oyster Point, that extends 1 km into the gulf. A foreshore reserve borders the entire foreland, with a road paralleling the beach and reserve. There are three beaches associated with the town, the northern jetty beach, the main northeast facing town beach and the southeast facing southern beach. The town has all facilities for visitors and travellers, and the bluffs behind the town offer views across the gulf to Adelaide.

Beach **311** runs from the **Stansbury jetty** for 900 m to the north, terminating amongst the 20 m high bluffs. For the most part, it is a narrow, sandy high tide beach, with width depending on the position of the bluffs. The southern 100 m of beach has built out into the gulf as sand has accumulated against the large jetty, car park and boat ramp facility. There is a reserve north of the jetty which provides the best spot for public recreation. The 300 m long jetty indicates the distance needed to reach deep water off the beach. The boat ramp also extends to water deep enough for small boats and combined with the large car park make this a popular location for boating, with many boats moored off the main beach.

The **main beach (312)** runs for 1.2 km from the boat ramp wall to the tip of the foreland at sandy Oyster Point. It is backed by a shady reserve and the main street, with a caravan park on the point. The beach is usually calm and seagrass grows in close to shore. Four low groynes cross the western end of the beach, with good access, parking and facilities along the beach.

On the southern side of **Oyster Point** is the 1.5 km southern beach **(313)**, which is backed by a 50 m wide reserve and a road, while 300 m wide, partly vegetated sand flats front the beach. As a consequence this beach is only suitable for swimming at high tide and the beach itself is usually covered in seagrass debris.

Swimming: The best locations are just north of the jetty and along the main beach.

Surfing: None.

Fishing: Best off the jetty and breakwater.

Summary: A very pleasant beachfront town with all the necessary facilities for visitors and travellers.

314-321   **BEACH 314 to KLEIN POINT**

No.	Beach	Rating	Type	Length
		Unpatrolled		
314	Beach 314	2	R+sand flats	700 m
315	Beach 315	2	R+sand flats	600 m
316	Beach 316	2	R+sand flats	150 m
317	Beach 317	2	boulder+sand flats	50 m
318	Beach 318	2	shingle+sand flats	200 m
319	Beach 319	2	R+sand flats	600 m
320	Klein Point (N)	2	R+sand flats	100 m
321	Klein Point (S)	2	R+sand flats	60 m

South of Stansbury the coast trends to the southwest, with 10 to 20 m high bluffs dominating the shoreline. Tucked in below and between some of the bluffs are eight generally narrow beaches, all fronted by sand flats and exposed to usually low wind waves out of the gulf. All are backed by the bluffs and while there is vehicle access to the top of all these beaches, some require a climb down the bluffs to reach them.

Beach **314** extends south of the southern Stansbury beach, backed only by 20 m high, sloping grassy bluffs and terminating at a small bluff headland. The low energy beach is fronted by 200 m wide sand and rock flats and is commonly covered in seagrass debris.

Beach **315** is a 600 m long, narrow high tide beach located at the base of steep, 20 m high bluffs and fronted by 100 m wide sand-rock flats. Beach **316** is a 150 m long, sandy high tide beach, bordered by protruding bluffs and fronted by narrow sand flats and then seagrass, with foot access down a steep backing bluff. A large boulder sits on the southern end of the beach. Beach **317** is a 50 m pocket boulder and rubble beach fronted by narrow sand flats.

Beach **318** is a 200 m long shingle beach, backed by steep, 20 m high bluffs. The beach has a high storm shingle ridge and a steep swash zone, then seagrass covered sand flats. Beach **319** is a 600 m long beach composed of boulders that grade to shingle in the south. It is backed by 20 m high, steep bluffs with seagrass meadows off the beach.

Beaches **320** and **321** lie on either side of Klein Point, whose backing bluffs are quarried for limestone, with the mine operations, stockpile, breakwater and jetty located between the two small beaches. The northern beach **(320)** is 100 m long and hemmed in between the bluffs and the jetty seawall and backed by the mine operations. Beach **321** is 150 m long and is likewise backed by mine operations.

Summary: None of these beaches are suitable for recreation owing to their usually difficult access down the bluffs, the dominance of rocks on some and the mining operations at Klein Point.

322-325   **WOOL BAY**

No.	Beach	Rating	Type	Length
		Unpatrolled		
322	Wool Bay (N 2)	2	R+sand flats	200 m
323	Wool Bay (N 1)	2	R+sand flats	100 m
324	Wool Bay	2	R+sand flats	700 m
325	Wool Bay (S)	2	R+sand/rock flats	700 m

Wool Bay is a small coastal settlement located on the top of 20 m high bluffs overlooking the old port of Wool Bay.

On the north side of the bay are two small beaches. Beach **322** is a 200 m long sand and cobble beach, backed by eroding 20 m high limestone bluffs and fronted by 100 m wide sand flats. Beach **323** is 100 m long, backed by steep, vegetated bluffs, with rocks off the southern bluff and shallow sand flats. The **main beach (324)** is a 700 m long sandy beach, with the 300 m long jetty located toward the southern end (Fig. 4.87). The jetty was built to service the lime kilns, the remains of which back the jetty. A large car park adjoins the jetty and there are toilets, a shelter and boat ramp at the beach, but no commercial facilities.

Figure 4.87 *Wool Bay beach and jetty are backed by the old lime kilns.*

The bluffs that border the main beach continue south to Giles Point. Along the base of the first 700 m of bluffs is a crenulate, sandy high tide beach (**325**) fronted by 50 to 100 m wide rock then sand flats.

Swimming: The main beach near the jetty is the most accessible and best place to swim.

Surfing: None.

Fishing: Best off the jetty.

Summary: Wool Bay is a very accessible sandy beach with the added interest of the historic jetty and lime kilns.

326-331   **PORT GILES to COOBOWIE**

No.	Beach	Rating	Type	Length
Unpatrolled				
326	Port Giles (N)	2	R+rock flats	60 m
327	Port Giles	2	R+rock flats	350 m
328	Port Giles (S)	2	R+rock flats	150 m
329	Giles Pt (boat ramp)	2	boulder+rock flats	200 m
330	Giles Pt	2	R+sand/rock flats	500 m
331	Coobowie	1	R+tidal flats	3.1 km

Port Giles consists of a 600 m long jetty, backed by tall, white wheat storage silos and loading facilities, but no commercial facilities. Either side of the port are 4 km of relatively straight, east facing, 20 m high bluffs,

terminating in the south at Giles Point. Along the base of the bluffs are a few low energy sand and boulder beaches, for the most part fronted by rock and some sand flats. The main road runs along the top of the bluffs providing good views and bluff-top access to all the beaches.

Beach **326** is a 60 m long pocket of sand 500 m north of the jetty, located below steep bluffs and the road and fronted by 50 m wide rock flats. Beach **327** begins on the south side of the jetty and runs for 350 m below the red bluffs. It consists of both sand and boulders, with a mixture of sand and rock flats extending up to 100 m off the beach. Beach **328** is a 150 m long, predominantly boulder beach, backed by steep eroding bluffs and fronted by 50 m wide rock flats.

Beach **329** is located on the northern side of Giles Point, as the bluffs decrease in height. A car park and boat ramp are located in the centre of the 200 m long sandy high tide beach, which is fronted by 100 m wide sand flats, providing access to the deeper gulf waters through the bordering rock flats. Beach **330** occupies 500 m of the point area and curves around to face southeast. It consists of a very low energy high tide beach with 200 to 300 m wide sand and rock flats.

Giles Point forms the northern boundary of Salt Creek Bay, a very low energy, 3 km wide, southeast facing bay, along the northern shore of which is the small settlement of Coobowie. The settlement backs the western half of a 3.1 km long, low energy high tide beach (**331**) which is fronted by 200 to 300 m wide rocky tidal flats (Fig. 4.88). A foreshore reserve backs much of the beach, then a 1 km long strip of beachfront houses, with a store, hotel and caravan park in the small town.

Figure 4.88 *Coobowie is fronted by a very low energy, seagrass covered high tide beach and a wide, low gradient rocky tidal flat.*

Swimming: None of these beaches are suitable for swimming except at high tide, owing to the shallow flats and rocks in places.

Surfing: None.

Fishing: Only toward high tide.

Summary: Six very low energy beaches, with the boat ramp and Coobowie being the only two really utilised by the public.

332-334   EDITHBURGH

Unpatrolled			
No.	Beach	Rating	Type	Length
332	Edithburgh	3	shingle+rock flats	70 m
333	Edithburgh (boat ramp)	2	R+sand flats	50 m
334	Edithburgh (jetty)	2	R+sand flats	100 m

Edithburgh is a small, well laid out town sitting on 10 m high bluffs that look east out over the southern gulf. Most of the foreshore consists of crenulate, crumbling rocky bluffs, backed by a continuous grassy foreshore reserve and the main road. Three small beaches are located below the bluffs.

On the north side of the town is a 70 m long, east facing, steep shingle beach (**332**) fronted by rock flats and backed by the reserve. In the town is firstly the small, 50 m long sand beach (**333**) on the south side of the boat ramp, with the caravan park on the backing bluffs and 200 m to the south, a second 100 m long, low beach (**334**) running south of the southern jetty. A slipway also crosses this beach, with boats both stored on the beach and moored over the sand flats.

Swimming: A tidal pool is located on the northern foreshore reserve and is the best place for swimming. The beaches are generally unsuitable, owing to the rocks, shallow flats and boat traffic at high tide.

Surfing: None.

Fishing: Best off the two jetties.

Summary: A well laid out little town offering most facilities for visitors and travellers.

335-338   HUNGRY POINT to GOLDSMITH

Unpatrolled			
No.	Beach	Rating	Type	Length
335	Hungry Pt	1	R+sand flat	1.5 km
336	Sultana Pt	1	R+sand flat	2 km
337	Sheoak	1	R+sand flats	7 km
338	Goldsmith	2	R+rock/sand flats	150 m

South of Edithburgh, wave energy begins to increase slightly as the more open waters of Investigator Strait are reached. As a consequence, substantial quantities of sand have been deposited along the shore in lee of the extensive Troubridge Shoals. The sand has formed a low sandy foreland called both Hungry Point and Sultana Point. West of the points is another 10 km of open, south facing coast out to Troubridge Point, which incorporates two beaches, including 7 km long Sheoak Beach.

Hungry Point is the northern point of the foreland and running northwest from the point is a 1.5 km long, low sandy beach (**335**), fronted by sand flats that merge toward the point with the 2 km wide Troubridge Shoals. A gravel road runs behind the beach out to the point, with over 50 shacks spread between the road and the beach. Fishers moor their boats over the sand flats, where they lie high and dry at low tide.

At the point the coast turns 90° and trends south for 100 m to **Sultana Point** (Fig. 4.89), then southwest toward Wattle Point and on towards Troubridge Point. Between Sultana Point and Wattle Point is a 2 km long, southeast facing beach (**336**), largely protected by the 5 km wide Troubridge Shoals. The beach is sandy, narrow and steep, with 50 to 100 m wide sand flats, then a mixture of seagrass meadows and the more tide dominated outer tidal shoals. A 3 m high foredune and gravel road run along the back of the beach, but there is no development.

Figure 4.89   *Low, sandy Sultana Point with Edithburgh in the distance.*

West of Wattle Point is crenulate, 7 km long, south to southeast facing **Sheoak Beach** (**337**). Despite its orientation, the narrow high tide beach is protected by sand flats, then extensive seagrass meadows extending kilometres out into the strait. Seagrass debris usually covers much of the beach, which is backed by a 3 m high foredune, then a gravel road. At the western end of the beach, low limestone bluffs extend for 2 km south to Troubridge Point. One kilometre out along a gravel track is 150 m long **Goldsmith Beach** (**338**). It faces east and consists of a high tide beach fronted by 300 m wide rocky sand flats.

Swimming: Only at high tide when the sand flats are covered.

Surfing: None.

Fishing: Most fishers heads out in their boats to fish the outer shoals and gulf.

Summary: A quiet corner of Yorke Peninsula, located at the heel of the peninsula.

339-342   TROUBRIDGE POINT to TROUBRIDGE HILL

No.	Beach	Unpatrolled		
		Rating	Type	Length
339	Troubridge Point (W)	4	R/LTT	100 m
340	Troubridge Hill (W 1)	4	R+platform	100 m
341	Troubridge Hill (W 2)	3	R/LTT	300 m
342	Troubridge Hill (W 3)	3	R/LTT	100 m

Troubridge Point is the southernmost point on the west coast of Yorke Peninsula. At the 10 m high cliffed limestone point, the coast turns and trends west, exposing the shore to low ocean waves that have travelled through Investigator Strait. As a consequence, the sand flats that dominate the entire Gulf St Vincent right round to Adelaide, are immediately replaced by low energy ocean beaches. Between Troubridge Point and the Clan Ranald memorial 5 km to the west are four small beaches, all lying below the bluffs and all accessible via a bluff-top gravel road.

Beach **339** lies 2.5 km west of the point and consists of a 100 m long pocket of sand lying below the bluffs, with low waves usually surging up the beach. Beach **340** is located immediately west of Troubridge Hill, a 30 m high clifftop dune. The 100 m long beach is fronted by a 50 m wide intertidal rock flat. To the west are two more beaches, **341** and **342**, 300 m and 100 m long respectively. They receive low swell and usually have a steep beach face with low surging waves. Access to all beaches requires a climb down the 10 to 15 m high bluffs.

Swimming: The two Troubridge Hill beaches with no platform offer the best swimming, with usually low waves and no rips.

Surfing: Usually too small.

Fishing: People fish both off the steep beaches and adjoining rock platforms.

Summary: Four very accessible beaches used occasionally by fishers and daytrippers.

343, 344   KEMP BAY

No.	Beach	Unpatrolled		
		Rating	Type	Length
343	Kemp Bay (1)	3	R/LTT	2 km
344	Kemp Bay (2)	3	R/LTT	1.9 km

Kemp Bay is an open, southwest facing bay located within the larger Waterloo Bay. It is bordered in the east by the Clan Ranald anchor and in the west by 500 m of 40 m high cliffs. Between the two are two similar straight, southwest facing beaches, both backed by 30 m high bluffs and a bluff-top gravel road.

Beach **343** is 2 km long and has a vehicle access track down the bluffs toward the eastern end, together with a bluff-top camping area, but no other facilities. The beach receives low swell and is usually moderately steep, with a narrow bar following periods of higher waves. Some of the bluffs and rocks protrude onto the beach. The beach is separated from beach **344** by a 200 m long bluff, with the second beach running in a more crenulate manner for 1.9 km to the higher cliffs. This beach is fronted by a scattering of calcarenite reefs which cause the crenulations in the shoreline and in places outcrop as rock flats.

Swimming: Both beaches are relatively safe under normal low waves, however watch for the deeper holes round the reefs on the second beach and rips if there is any surf.

Surfing: Usually too small.

Fishing: The beaches are used for beach and rock/reef fishing and occasional swimming.

Summary: Two relatively accessible and usually relatively calm beaches.

345-347   WATERLOO BAY CLIFFS

No.	Beach	Unpatrolled		
		Rating	Type	Length
345	Waterloo Bay cliffs (1)	5	R+platform	40 m
346	Waterloo Bay cliffs (2)	5	R+platform	150 m
347	Waterloo Bay cliffs (3)	5	R+platform	80 m

Midway along Waterloo Bay is a 700 m long section of 30 to 40 m high cliffs, fronted by a 50 m wide calcarenite platform. Tucked into crenulations in the cliffs are three high tide beaches, each bordered and fronted by calcarenite bluffs and platforms. There is a 300 m long vehicle track from the road to the top of beach 346, with a steep climb down the vegetated bluffs to reach the beaches. Beach **345** is just 40 m long and surrounded by the bluffs. Beach **346** has a narrow section that then widens into an amphitheatre, which has a deeper platform and is the only section providing access to deeper water. Beach **347** is an 80 m long section bordered and backed by the bluffs. These beaches and their platforms are only suitable for rock fishing, apart from swimming over the deeper reefs off beach 346 when the seas are calm.

348   WATERLOO BAY

No.	Beach	Rating	Type	Length
		Unpatrolled		
348	Waterloo Bay	3	R/LTT	4 km

The western end of **Waterloo Bay** is occupied by a relatively straight, southwest to south facing, 4 km long sand beach (**348**). The beach is backed by 40 m high bluffs in the east, covered by a thin veneer of clifftop dunes, with vegetated blowouts and parabolic dunes extending up to 300 m inland behind the centre and western end of the beach. The bluffs and dunes are backed by wheat fields, with no public access, other than via farm tracks. The beach receives low swell through the strait and usually has a steep beach with low surging waves and no rips. Occasional higher swell forms a narrow continuous bar and rips. Calcarenite reefs parallel part of the western beach, inducing additional rips.

Swimming: Usually relatively safe during normal low waves, however watch for rips if there is any surf, especially near the reefs.

Surfing: Usually only a low shorebreak.

Fishing: Best off the beach in the west, into the reef holes.

Summary: A longer, natural beach with no direct public access.

349-354   PORT MOOROWIE to POINT GILBERT

No.	Beach	Rating	Type	Length
		Unpatrolled		
349	Port Moorowie	2	R+sand flats	300 m
350	Port Moorowie (boat ramp)	2	R+sand flats	250 m
351	McLeod Hbr (1)	2	R+sand flats	80 m
352	McLeod Hbr (2)	2	R+sand flats	80 m
353	McLeod Hbr (3)	2	R+sand flats	180 m
354	Point Gilbert	2	R+sand flats	300 m

Port Moorowie is a small settlement with a scattering of houses spread along 2 km of low bluffs in the western corner of Waterloo Bay, called McLeod Harbour. The protected section of the bay has a shallow, seagrass covered floor with shallow sand flats bordering the bluffs and beaches. Apart from a boat ramp there are no facilities at the port. Six small beaches are located along the base of the 5 to 10 m high bluffs, with vehicle access to the top of each of the beaches.

Beach **349** lies below the bluffs at the end of the Minlaton Road. The 300 m long high tide sand beach wraps around a low protruding bluff, capped by low clifftop dunes. Beach **350** is the location of the port boat ramp, with the access road running the length of the 250 m long beach to the ramp at the eastern end.

Beaches **351, 352** and **353** are three pockets of sand, 80 m, 80m and 180 m long respectively. They are located below low bluffs along the western southeast facing section of the settlement. A gravel road runs along the bluffs providing good access to each of the beaches. The beaches are usually calm, composed of sand, with rubble from the eroding bluffs slipping onto the ends and shallow sand and seagrass flats right off the shore.

Beach **354** is located on **Point Gilbert** and consists of a 300 m long, south facing high tide sand beach, backed by some stable dune covered bluffs and fronted by deeper seagrass meadows in the centre, with a 200 m wide reef flat in the west. Waves are usually low to calm at the shore. Access is from the bluff-top car park at the western end of the settlement, above beach 353.

Swimming: Six relatively safe and usually calm beaches, just beware of the rocks and reefs off some of them.

Surfing: None.

Fishing: Best off the point beach.

Summary: A quiet settlement with good access and a boat ramp but no other amenities.

355, 356   STURT BAY, POINT DAVENPORT

No.	Beach	Unpatrolled Rating	Type	Length
355	Sturt Bay	2→4	R→LTT/TBR	19.3 km
356	Pt Davenport	2	R+sand flats	600 m

Sturt Bay is a 15 km wide, south to southwest facing bay containing a gently curving, 20 km long, generally low energy beach system. It extends from 20 m high dune-capped Point Gilbert in the east to low, sandy Point Davenport in the west. The entire beach is backed by a combination of up to 30 foredune ridges grading to more stable, vegetated parabolic dunes in the east, which extend up to 1 km inland. While the gravel south coast road runs along the rear of these dune deposits, it is usually a few hundred metres from the beach, with only one direct public access point in the centre, at the end of the Sturt Bay Road. However apart from a boat shed, there are no facilities along the entire beach. For the most part it is backed by a mixture of cleared grazing land and scrubby, stable dunes reaching a maximum height of 20 m in the east.

Sturt Bay beach (**355**) usually receives waves averaging 0.5 m, which maintain a low energy, steep beach fronted by a narrow bar and low surf. There are three lower energy sections, in the east in lee of the reefs off Point Gilbert, in the centre on either side of the Sturt Road access and in the west in lee of Sandy Point. In these locations the lower waves permit seagrass to grow almost to the beach, while in lee of Sandy Point the sand flats widen to 200 m. Occasional periods of higher waves along the higher energy sections produce a one to two bar system, with rips spaced every 100 to 200 m along the inner bar. While these are usually inactive, they do however result in a variable bar and trough topography close to shore.

Point Davenport beach (**356**) is a low, sandy recurved spit that is slowly growing toward the east. A small tidal creek separates the end of the spit from the main bay beach. The east facing end of the spit has a 600 m long, low energy beach, fronted by 100 m wide sand flats then seagrass meadows. Waves are usually calm. Five kilometres of coast on either side of the point and the backing spit and lagoon are contained in the Point Davenport Conservation Park.

Swimming: Relatively safe under normal low wave conditions. However if there is any surf, beware, as rips may be present along the eastern section.

Surfing: Usually none, only chance is during occasional high outside swell.

Fishing: The beach has a mixture of close-in seagrass meadows and more variable bars and troughs, while offshore are the deeper seagrass meadows and some reefs.

Summary: A long, relatively natural beach, with no development, but limited access.

Point Davenport Conservation Park	
Established:	1987
Area:	239 ha
Beaches:	356 & 357
Coast length:	5 km

Point Davenport Conservation Park was established to protect the lagoon impounded by the sandy spit that forms the point (Fig. 4.90).

357, 358   FOUL BAY

No.	Beach	Unpatrolled Rating	Type	Length
357	Foul Bay (E)	2	R	3.6 km
358	Foul Bay	2	R+sand flats	12.8 km

Foul Bay is probably named after the decaying seagrass that usually litters the long beach. The open, southeast facing 16 km of bay shore is protected from the southwesterly swell by offshore shoals that lower waves to near calm at the shore and permit seagrass meadows to grow close to shore for the length of the bay. Along the more protected western shore of the bay, sand flats extend 100 to 200 m off the beaches. The south coast road provides good access to the western half of the bay, which also contains a 2 km long section with approximately 50 beachfront shacks.

Figure 4.90 *Point Davenport consists of multiple recurved spits that have grown east to impound the backing lagoon. Most of this area is part of the Point Davenport Conservation Park.*

There are two beaches in the bay, with the eastern 3.6 km long section (357) lying between Point Davenport and a substantial sandy foreland that protrudes 400 m into the bay. Davenport is the terminus of a growing 3 km long recurved spit. The spit is slowly extending to the east, in the process leaving behind the remains of more than 20 former spits. Much of the spit and the backing lagoon it impounds, is part of the Point Davenport Conservation Park (Fig. 4.90). The beach is a low, narrow sandy beach, often covered in seagrass debris, with extensive seagrass meadows lying just 50 m off the shore.

The main **Foul Bay** beach (358) extends from the foreland to the shack settlement of Foul Bay. The entire beach receives low waves and is usually covered by seagrass debris, owing to the meadows right off the beach, except in the west where shallow sand flats front the beach. Apart from the shacks and a concrete boat ramp, there are no facilities in the bay.

Swimming: Usually calm along the shore.

Surfing: None.

Fishing: Best in a boat out over the deeper seagrass meadows and reefs.

Summary: An exposed but still relatively calm bay, dominated by the reefs and seagrass meadows, with the conservation park in the east.

359-366   POINT YORKE to FOUL HILL

No.	Beach	Rating	Type	Length
Unpatrolled				
359	Point Yorke (N 3)	3	R	700 m
360	Point Yorke (N 2)	3	R	200 m
361	Point Yorke (N 1)	3	R	300 m
362	Point Yorke (S 1)	5	R+rocks	150 m
363	Point Yorke (S 2)	5	R+rocks	50 m
364	Point Yorke (S 3)	5	R+rocks	40 m
365	Foul Hill (1)	6	R+platform	300 m
366	Foul Hill (2)	6	R+platform	50 m

The Foul Bay boat ramp marks the end of the long, sandy Foul Bay beach and the beginning of 15 km of rocky shoreline composed of bedrock metasediments and for the most part capped by calcarenite bluffs and cliffs up to 40 m high. The first 10 km round to Foul Hill contains eight rock-controlled beaches backed by bluffs and farmland with no public access.

The first three beaches lie between the boat ramp and Point Yorke. They face east and receive low waves. Each is backed by vegetated, 20 m high bedrock bluffs and then

clifftop farmland. The first (359) is a crenulate, 700 m long, narrow high tide beach at the base of the bluffs and is fronted by seagrass meadows. The second (360) is a crenulate, 200 m long sand beach with extensive patches of rocks and seagrass meadows. The third (361) is 300 m long and bordered by prominent rock platforms.

On the southern side of **Point Yorke** the coast trends toward the west, with the bedrock continuing to dominate the shoreline. The beaches face south and receive more persistent low swell, as well as occasional higher waves. The first beach (362) is a 150 m long sand beach, backed and bordered by 30 m high bluffs. Its neighbour (363) is similar, only 50 m in length. Minor eroded clifftop dunes lie on the backing bluffs. Beach 364 is a solitary 40 m long pocket of sand, also backed by 30 m high bluffs and disturbed clifftop dunes.

Foul Hill is a 55 m high vegetated dune atop of the bedrock cliffs. Immediately west of the hill are two platform beaches. The first (365) is a crenulate, 300 m long high tide sand beach, backed by 40 m high cliffs and fronted by a 50 m wide intertidal rock platform and rock reefs. Beach 366 is its 50 m long neighbour, similar in characteristics and locked in by the bluffs and protruding headlands.

Swimming: These are eight isolated and difficult to access beaches, all lying at the base of bluffs and cliffs, and mostly dominated by rocks. While moderately safe under low waves, they are very hazardous when any surf is breaking.

Surfing: Usually none.

Fishing: The rock platforms provide a good base for rock fishing the deeper water and reefs.

Summary: Eight beaches close to the main road, but difficult to access and little utilised.

367-375   METEOR BAY to HILLOCK POINT

No.	Beach	Rating	Type	Length
Unpatrolled				
367	Meteor Bay	5	R	200 m
368	Bangalee	5	R	600 m
369	T Cove	4	R	150 m
370	boulder 370	4	R boulder	80 m
371	boulder 371	4	R boulder	60 m
372	Coffin	4	R	350 m
373	Butlers	4	R	600 m
374	Butlers (W)	4	R	230 m
375	Hillock Pt	4	R	100 m

The Hundred Road terminates at the cliffs 1 km west of Foul Hill. To the west of the road, winding along the clifftop, is a 2WD track called Hillocks Drive that runs for 5 km to Butlers Beach, providing access to the cliffs and bluffs above these eight essentially south facing beaches. The drive is on private farmland and is open from 18 August to 18 June. Apart from the track there are no facilities, other than a small camping area above Butlers Beach and a sign warning of the dangerous surf.

Meteor Bay (367) is a curving, 200 m long beach, backed by 30 m high sloping bluffs, partly covered in sand, together with rocks and reefs along and off the beach. Access is via the sandy bluffs. **Bangalee** beach (**368**) is a crenulate, 600 m long sand beach, partly fronted by platforms and reefs, with only a small clear opening to the sea toward the western end. It is backed by a foredune, then steep, vegetated bluffs, with access via a steep track from a clifftop car park toward the western end. **T Cove (369)** is a 150 m long pocket beach surrounded by 30 m high bluffs, with headlands and rock platforms extending 100 m seaward at either end. Access is via a steep descent down the bluffs.

Beaches **370** and **371** are two boulder beaches 80 m and 60 m long respectively. Both lie in a bedrock controlled cove, bordered by prominent sloping rock platforms and backed by 30 m high bluffs.

Coffin beach (**372**) is a 350 m long, curving sandy beach, bordered by sloping rock platforms and backed by 20 m high, sand covered bluffs, with a bluff-top car park overlooking the beach. The beach receives slightly higher waves and usually has a low surging shorebreak, with a single strong rip draining the small cove during higher seas.

Butlers Beach is the main beach in the area. It consists of a main 600 m long beach (**373**) and a 230 m long western section (**374**), located past some rocks that protrude across the beach. The main car park is above the central section. This is a steep reflective beach, usually containing well developed beach cusps and a low to moderate shorebreak. The western beach has slightly lower waves.

In lee of **Hillock Point** is a 100 m long, east facing narrow beach (**375**) that widens slightly toward the point. A 10 m high bluff backs the beach with rocks scattered the length of the beach. Waves are usually low and calms often prevail.

Swimming: Butlers and Coffin are the two more accessible and popular beaches, but usually have a low surging shorebreak and are free of rips. However be very careful if there is any surf, as all the beaches are dominated by rocks and reefs and strong rips can quickly develop.

Surfing: Usually too small.

Fishing: Butlers Beach is a popular spot for both beach and rock fishing.

Summary: An accessible and scenic section of coast offering several small beaches, together with the small caravan park at Butlers Beach.

376-378   **SALMON BEACH, BUTTERFISH BAY**

No.	Beach	Rating	Type	Length
Unpatrolled				
376	Salmon Beach	3	R	1.4 km
377	Butterfish Bay (1)	4	R	250 m
378	Butterfish Bay (2)	5	R+reefs	350 m

Between Hillock Point and Meehan Hill is a 4 km section of southwest facing shore dominated by 20 to 30 m high cliffs, but with sufficient sand to form Salmon Beach and its backing clifftop dunes, and two smaller beaches at the base of the cliffs. Hillock Drive terminates at a car park above Salmon Beach, with access via a walk down the dunes, while the two Butterfish Bay beaches are backed by the cliffs and private farmland and have no public access, as well as no safe access down the cliffs to the beaches.

Salmon Beach (**376**) is 1.4 km long, faces southwest and receives waves averaging just under 1 m. These combine with the medium sand to form a steep, cusped reflective beach, with a surging shorebreak. During periods of higher waves the beach has a heavy shorebreak. It is bordered in the east by the sloping rocks of Hillock Point and steep calcarenite cliffs in the west. The southwest winds have blown the sand up onto the backing bluffs and up to 1 km inland as a series of clifftop parabolic dunes, with the road skirting the rear of the dunes.

Butterfish Bay is an open, cliffed section of coast containing two sections of sand at the base of the dune-capped cliffs. Beach **377** is 250 m long and consists of a narrow beach fronted by patchy deeper reefs. Beach **378** is 350 m long, awash at high tide and fronted by shallower reefs and surf.

Swimming: Salmon Beach is a popular spot offering usually low surging waves. Beware of the shorebreak when waves exceed 1 m.

Surfing: Usually too low.

Fishing: Salmon Beach is a favoured spot for both beach and rock fishing.

Summary: A slightly more exposed section of coast producing the surging shorebreaks and dunes.

379, 380   MARION BAY, PENGUIN POINT

Unpatrolled				
No.	Beach	Rating	Type	Length
379	Marion Bay	3	R→sand flats	9.5 km
380	Penguin Pt	3	R	1.25 km

Marion Bay is an open, 8 km wide, south to southeast facing bay, containing 9.5 km of slightly crenulate shoreline, that grades from a cusped reflective beach in the east at Meehan Hill, to a protected low energy beach with sand flats in the west, in lee of Penguin Point. The south coast road parallels the back of the dunes and links with the Stenhouse Bay road which runs into the small Marion Bay settlement, located at the east facing, low energy section of the beach. Beach access is provided in the east at Meehan Hill, where there are two car parks (one overlooking the beach) and in the west at the settlement, which includes a 200 m long jetty, beachfront shacks and a newer subdivision on Penguin Point.

The bay beach (**379**) receives waves averaging less than a metre in the east, which maintain a steep cusped beach. The waves gradually decrease to the west, permitting the seagrass to grow to the shore, finally widening into 150 m wide sand flats south of the jetty in lee of Penguin Point.

On the south side of **Penguin Point** is a 1.25 km long, south facing beach (**380**), bordered by the point in the east and 20 m high cliffs in the west. Deeper reefs fronting the beach lower waves to less than 1 m and maintain a steep reflective beach face, with stable dunes climbing the backing bluffs, backed by a road and a Marion Bay subdivision. There are three car parks on the bluffs and point and steep tracks down to the beach.

Swimming: The two beaches are relatively safe, with deeper water off the bay beach in the east, while at Marion Bay, shallow flats front the shore.

Surfing: Both usually have only low shorebreaks, however there are good waves over the reefs west of Penguin Point during higher swell.

Fishing: The Meehan Hill and Penguin Point car parks provide access for beach fishing, while at the settlement, the jetty is the most popular shore-based location.

Summary: Marion Bay is the largest settlement on the south coast and offers a limited range of facilities and accommodation, including a caravan park.

Innes National Park

Established:	1970
Area:	9232 ha
Beaches:	381 to 420
Coast length:	40 km

Innes National Park covers over 9000 ha of the 'toe' of Yorke Peninsula. The park includes 40 km of often spectacular coastline, including 39 beaches, as well as numerous headlands, together with an extensive dune and backing salt lake system. Entry to the park is at Stenhouse Bay and camping areas are available at Stenhouse Bay, Cable Bay, West Cape, Pondalowie Bay and Shell Beach.

Further information:
Innes National Park Information Office
Stenhouse Bay (08) 8854 4040

381-386   JOLLYS BEACH, RHINO POINT, STENHOUSE BAY

Unpatrolled				
No.	Beach	Rating	Type	Length
381	Jollys Beach	2	R+sand flats	250 m
382	Rhino Point (N)	5	R+rocks	50 m
383	Rhino Point (S)	5	R+reef	100 m
384	Stenhouse Bay (N)	4	R+reef/rocks	1.2 km
385	Stenhouse Bay (centre)	4	R+platform	200 m
386	Stenhouse Bay (jetty)	4	R+boulder	80 m

The eastern end of Penguin Point beach marks the boundary of Innes National Park, which encompasses the next 40 km of coast. Much of the coast is dominated by rocky shoreline, with usually small headland-bound beaches. The first 6 km of shore between Penguin Beach and Stenhouse Bay is dominated by calcarenite bluffs, with six small rock and bluff-bound beaches. A gypsum mine operated at Stenhouse Bay from 1912 until the closure of the town in 1974. It was then incorporated into the national park.

Jollys Beach (**381**) is a discontinuous, crenulate, low energy, 250 m long beach, fronted by sand flats that widen to 100 m at its south end in lee of Rhino Head. There is a gravel track to the rear of the beach, which usually has calm conditions. Bush camping is permitted on the bluffs overlooking the centre of the beach.

Rhino Head is bordered by 40 m high calcarenite cliffs, at the base of which are two rock-bound beaches, both of which require a steep climb to reach. Beach **382** faces east and is a 50 m pocket of sand, partially covered in rocky cliff debris and fronted by an intertidal platform-reef. Beach **383** faces south and is a 100 m long sand beach lying at the base of the bluffs, with a mixture of sand and reefs fronting the beach.

Stenhouse Bay is an open, southeast facing bay that extends for 2.5 km west of Rhino Head. It contains three small beaches. The main beach (**384**) is 1.2 km long, faces south and occupies the northwestern corner of the bay. The narrow beach is backed by low, dune covered bluffs in the west, narrowing to the east where it is fronted by an intertidal rock platform and backed by increasingly irregular bluffs, eventually giving way to the bluffs and platform. The Stenhouse Bay road clips the western end of the beach and there are fuel storage tanks between the road and the beach, but otherwise no facilities.

Stenhouse Bay is the port for a gypsum works located in the backing Marion Lake. Much of the port facilities back the bay, with the port jetty located on beach 386. Beach **385** is a 200 m long, east facing sand beach, partially fronted by an intertidal platform-reef. A vehicle track runs along the 10 m high bluff backing the beach. Beach **386** is an 80 m long, east facing, irregular boulder beach fronted by a sandy bed and located at the base of 20 m high bluffs, with the 150 m long port jetty crossing its southern end.

Swimming: Jollys Beach with its camping area is the most popular. All except the Rhino Head beaches are relatively safe under normal low wave to calm conditions, however the Stenhouse Bay beaches are more difficult to access and surrounded by the mine facilities.

Surfing: The reefs west of Rhino Head provide some good breaks during higher swell.

Fishing: Rock platforms border all the beaches, with reefs and rocks lying off most of the beaches.

Summary: Six accessible but generally little used beaches.

387-391   **CHINAMANS HAT, CABLE BAY**

No.	Beach	Unpatrolled		
		Rating	Type	Length
387	Chinamans Hat	5	R+reef	550 m
388	Cable Bay (E 2)	5	R+platform/reef	300 m
389	Cable Bay (E 1)	5	R+platform/reef	300 m
390	Cable Bay (main)	2	R+reef	350 m
391	Cable Bay (W)	3	R+reef	120 m

West of Stenhouse Bay the Pondalowie Road runs along behind the cliffs and bluffs for 4 km, providing views and access to five bluff-backed and reef-bound beaches that occupy Cable Bay. The first beach is in lee of **Chinamans Hat**, a small, distinctive island 300 m offshore. There is

a bluff-top car park and a steep walk down the 20 m high bluffs to the 550 m long, south facing beach (**387**), which protrudes seaward in lee of the 'hat'. The beach is also protected by extensive reefs and usually has low waves to calm conditions. However, 200 m offshore is the famous Chinamans surfing break, a hard right that breaks over the shallow reef.

Immediately west of the break is the open, south facing, 1.5 km wide Cable Bay. Bluffs back the entire bay and reefs dominate the bay floor. The first two bay beaches lie immediately west of the Chinamans break. The first (**388**) is a straight, 300 m long sand beach, backed by the bluffs and fronted by a 50 m wide platform reef, with a usually small but heavy surf break. Its neighbour (**389**) is similar, also 300 m long, but with a protruding central section and also fronted by shallow reef and a heavy surf. While the road runs along the top of the bluffs and small car parks overlook both beaches, they can only be accessed down the steep bluff.

The main **Cable Bay** beach (**390**) is a 350 m long, low energy sand beach, protected by extensive reefs, with usually calm conditions at the shore. The main road runs to the back of the beach where there is bluff-top parking, a camping area and beach boat launching.

Two hundred metres to the west is beach **391**, a 120 m long, narrow sand beach, lying at the base of 20 m high bluffs. There is a small bluff-top car park, with steep access down to the beach.

Swimming: While all four beaches have relatively low waves at the shore, the presence of the shallow reefs makes them generally unsuitable for swimming.

Surfing: Chinamans has an international reputation and is one of the best waves on the peninsula.

Fishing: You can fish the shallow reefs from the shore, with most fishers heading further out in boats.

Summary: Four beaches very accessible by vehicle, with only Chinamans and the main Cable Bay beach being utilised on a regular basis.

392-399   **CAPE SPENCER to REEF HEAD**

No.	Beach	Unpatrolled		
		Rating	Type	Length
392	Cape Spencer (N 1)	7	TBR	450 m
393	Cape Spencer (N 2)	7	TBR	150 m
394	Cape Spencer (N 3)	7	TBR	160 m
395	Cape Spencer (N 4)	7	TBR	100 m
396	Cape Spencer (N 5)	7	TBR	450 m
397	The Gap	6	LTT	100 m
398	Howling Cove	6	LTT	100 m
399	Reef Head (N)	10	TBR+rocks	200 m

At **Cape Spencer** the coastal orientation turns to face squarely towards the southwest, exposing the coast to the full force of the prevailing swell and winds. The result has been the accumulation of massive cliffs composed of past (now lithified) dune systems, with small remnants of once larger beach systems tucked in along their base. Cape Spencer, at 84 m high, typifies the massive calcarenite cliffs that dominate most of the exposed sections of the South Australian coast. The cape does however sit, like most similar headlands, on a foundation of ancient granite bedrock. A gravel road runs out to a small car park, with a walking track out to the lighthouse.

Immediately north of the cape are five beaches, all linked by a common 1.7 km long surf zone, but separated by small, but high, protruding calcarenite bluffs (Fig. 4.91). Only the first beach (**392**) is accessible via a steep descent from the cape car park. The other four beaches (**393 to 396**) are essentially inaccessible on foot. All receive waves averaging 1.5 m which maintain a rip-dominated, 100 m wide surf zone, together with strong topographic rips against the rocks and bluffs.

Figure 4.91 *Cape Spencer lighthouse, 50 m high calcarenite cliffs and bluffs and five small bluff-bound beaches.*

The Gap beach (**397**) lies 1 km west of the end of beach 396 and is located in a prominent gap in the bedrock and calcarenite, with the 50 m high cliffs overhanging the inaccessible pocket beach. Likewise, its neighbour on Reef Head is another inaccessible, 100 m long pocket beach (**398**) lodged deep inside a cliff-bound cove and appropriately named **Howling Cave** beach.

Seven hundred metres north of the head is an inaccessible, 200 m long high tide beach (**399**), wedged in at the base of 50 m high cliffs and fronted by a rock strewn surf zone, resulting in one of the most dangerous beaches on the coast.

Swimming: None of these beaches are suitable for safe swimming, they are dominated by rips and rocks and are generally inaccessible.

Surfing: There are beach breaks on the first of the Cape Spencer beaches, with a steep climb down required to reach the beach.

Fishing: All the beaches are generally too inaccessible for shore-based fishing.

Summary: A spectacular coast to view, particularly from the Cape Spencer lookout, but unsuitable for recreational pursuits.

400-403   ETHEL WRECK to WEST CAPE

No.	Beach	Rating	Type	Length
Unpatrolled				
400	Ethel Wreck	7	TBR	450 m
401	Ethel (N 1)	8	TBR+rocks	60 m
402	Ethel (N 2)	8	TBR+rocks	300 m
403	West Cape	8	TBR	1.95 km

The 'Ethel' is one of two ships that have come to grief on a 450 m long sandy beach, now known as **Ethel Wreck** beach (**400**). The ship 'Ethel' was wrecked in 1906, followed by the 'Ferret' in 1920. A gravel road runs out to a car park above the beach, with a steep track down the 40 m high backing bluffs to the beach. The remains of the wreck lie at the base of the northern end of the beach. The beach receives waves averaging 1.5 m which usually maintain two beach rips, together with rips against the rocks at either end. A walking track runs along the top of the bluffs north of the car park, providing a view of the beach. Immediately north of the northern rocks is a second bluff-bound, 60 m long beach (**401**), which can only be reached around the rocks on calm days.

Two hundred metres further on is a 300 m long beach (**402**), also bluff-bound, with rocks in the surf, but connected by a common surf zone to the longer beach 403. Beach **403** runs for almost 2 km straight to West Cape and is the most exposed and highest energy beach on the Yorke Peninsula. It receives waves averaging over 1.5 m, which usually maintain eight beach rips, together with two strong rips against the boundary rocks. The beach can be accessed on foot from the Ethel car park, or the northern car park on West Cape, both requiring a steep climb down the dunes to the beach. Massive dunes and blowouts extend up to 1 km behind the beach.

Swimming: Be very careful if swimming on Ethel or West beaches, as both are rip-dominated, while the two pocket beaches are extremely hazardous.

Surfing: Best chance is along the northern end of West Cape beach, where the rips can carve some good banks and excellent beach breaks.

Fishing: Both Ethel and West Cape usually have good rip holes right against the beaches.

Summary: Two exposed, energetic beaches suitable only for experienced fishers and surfers.

404, 405   **GROPER BAY**

Unpatrolled			
No.	Beach	Rating	Type	Length
404	Groper Bay	4	R→LTT	1.5 km
405	Groper Bay (N)	3	R+reef	300 m

Groper Bay is a semicircular, west facing bay, containing a 1.5 km long, curving beach (**404**). It is bounded by 50 m high West Cape in the south and an island-tied tombolo in the north. A gravel road runs out to the West Cape car park, providing access to both the West Cape and Groper Bay beaches. West Cape and reefs within the bay reduce waves to about 0.5 to 1 m in the bay, resulting in a reflective to low tide terrace beach, with usually a low and narrow surf zone. Rips do occur however during higher waves. Dunes extending up to 1 km inland back the beach.

On the north side of the tombolo is the smaller northern beach (**405**) which is protected by both the prominent island and northern headland, as well as shallow reefs linking the two. Consequently calm conditions often prevail at the shore and seagrass debris usually covers the beach. A car park is located at the northern end of the beach.

Swimming: Both beaches usually have low waves, with a chance of rips on the main beach and the seagrass on the northern beach.

Surfing: The main beach has a beach break during moderate to higher swell.

Fishing: There are usually a couple of rip holes on the main beach, with the rocks bordering both beaches used to fish the adjoining reefs.

Summary: Two moderately protected beaches mainly used for fishing.

406, 407   **PONDALOWIE BAY**

Unpatrolled			
No.	Beach	Rating	Type	Length
406	Pondalowie Bay	3	R+sandflat→LTT	4 km
407	Pondalowie Bay (N)	5	LTT to TBR	900 m

Pondalowie Bay is semicircular in shape and opens to the west and north. While it is exposed to westerly winds, the southern point, together with South Islet and Middle Island, afford it substantial protection from ocean waves, particularly toward the southern end (Fig. 4.92).

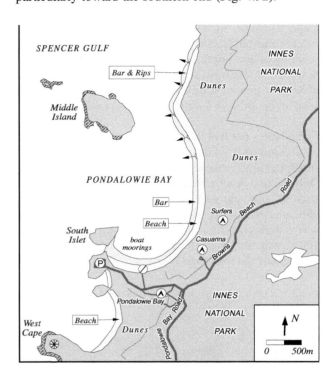

Figure 4.92 *Pondalowie Bay is the focus of camping and water-based activities in Innes National Park, offering several camp sites, boat launching and increasing surf up the beach.*

The bay lies at the end of the Pondalowie Bay Road. The popular bay area has a large parking area on the southern headland, beach access and beach boat launching in the southern corner and three camping areas. The southern camping area is called Pondalowie Bay and has facilities for caravans; the second is Casuarina; and further up the beach off the Browns Beach Road is the third, called Surfers, adjacent to the higher energy part of the beach. In addition to the national park facilities, there are a few fishing shacks near the boat area and fishing boats moored in the southern bay.

The **main beach** (**406**) is 4 km long. It initially faces north and receives waves less than 0.5 m, making the southern half ideal for both boat launching and moorings. The southern beach is composed of fine sand which, with the low waves, produces a flat, firm beach, usually partly covered in seagrass. Wave height slowly increases up the bay, reaching about 1 m at the sandy foreland in lee of Middle Island.

North of the foreland the beach (**407**) continues for 900 m and becomes increasingly exposed to the southwesterly waves, resulting in a wider surf zone and the formation of two to three beach rips, together with a permanent rip against the northern boundary rocks. Dunes extending up to 1.5 km inland back both beaches, with a large sand blow immediately north of the foreland.

Swimming: The best place to swim is in front of the Casuarina camping area, where you are away from the boats and much of the seagrass, but the waves are usually relatively low and rips absent. Be careful up the beach, particularly if waves exceed 0.5 m, as rips are usually present.

Surfing: There are two recognised surfing areas at Pondalowie, accessible on foot from the Casuarina and Surfers campgrounds, with Richards break a few hundred metres further up the beach. Waves are usually less than 1 m at Pondalowie, increasing up the beach and during bigger swell.

Fishing: Most fishers go outside in boats, or fish from the rocks around the headland. The best beach fishing is up the beach, particularly if rip holes are present.

Summary: It's a long drive down and around to Pondalowie, but worth it if you are after a picturesque bay, with good park facilities and a few waves up the beach.

408-411   ROYSTON HEAD

Unpatrolled			
No.	Beach	Rating	Type	Length
408	Royston Hd (S 2)	7	TBR+rocks	30 m
409	Royston Hd (S 1)	7	R+rocks	60 m
410	Royston Hd	7	R+rocks	200 m
411	Royston Hd (E)	3	R+reef	180 m

Royston Head is an irregular, 50 m high calcarenite headland, sitting atop a metasedimentary base, the latter manifest as prominent sloping rock platforms, capped by clifftop and parabolic dunes. The only access to the headland and its adjoining beaches is on foot from Pondalowie Bay 3 km to the south, or from the Dolphin Beach car park 2 km to the east. A coastal walking track links both access points. Consequently the beaches are little visited by anyone other than walkers.

South of the head is a predominantly calcarenite coast with rocks and reefs, in amongst which are three beaches. The first (**408**) is a 30 m pocket of sand, fronted by rocks and reefs, with a permanent rip draining the beach. The second (**409**) is 60 m long, bordered by small headlands and with a reef-strewn surf zone and a permanent rip. In lee of Royston Island are two beaches. The southern one (**410**) is a 200 m long, west facing sand beach, together with beachrock along the northern half and reefs off the

beach. Immediately to the east is a 180 m long, more protected reflective beach (**411**), backed by dunes climbing onto the 20 m high bluffs.

Swimming: The beach east of the head is the most protected, least rocky and safest for swimming. The beaches south of the head are all dangerous owing to the rocks, reefs and rips.

Surfing: There are waves, but the beaches tend to be too rocky.

Fishing: Excellent rip and rock holes along the entire section, just beware of the higher waves.

Summary: This section is well worth the 6 km walk, but be cautious if going for a swim and even then only use the protected northern beach.

412-416   DOLPHIN BEACH to BROWNS BEACH

Unpatrolled			
No.	Beach	Rating	Type	Length
412	Dolphin	2	R	800 m
413	Shell Beach	2	R	300 m
414	Beach 414	2	R	1 km
415	Beach 415	2	R	480 m
416	Browns	2	R to LTT	2.4 km

For 4 km east of Royston Head the coast is dominated by north facing, calcarenite-capped, low granite headlands, before turning to face west at Browns Beach. In amongst the headlands are four rock-bound, lower energy sandy beaches. Browns Beach Road runs 500 m in from the beaches, providing vehicle access to the first two and to Browns Beach, where the road terminates.

Dolphin Beach (**412**) is a curving, 800 m long, north facing beach, which receives usually low waves and has a moderately steep reflective beach face and no surf. The beach is bounded by sloping granite platforms, with vegetated bluffs in the west fronted by a stable to increasingly unstable foredune to the east. There are two old shacks at the western end, together with vehicle access to a small car park.

Shell Beach (**413**) is a 300 m long, curving, low energy reflective beach, bordered by protruding granite rock platforms, with beachrock outcropping along the centre of the beach. It is backed by a climbing 10 m high foredune. There is vehicle access to a small car park and solitary shack at the western end and a camping area in behind the beach.

Beach **414** is a more irregular, 1 km long sand beach, with calcarenite bluffs and bedrock outcrops dominating the western half and small climbing dunes on the eastern half. Beach **415** is a straight, northwest facing, 480 m long beach, the eastern half of which is fronted by a narrow beachrock reef. An unstable foredune backs the beach, with bare dunes extending 300 m in from the eastern end. There is foot access to the beach from the Browns Beach car park.

Browns Beach (416) faces due west and consists of two parts. The southern 500 m has two beachrock reefs lying 50 and 100 m off the beach, impounding a continuous lagoon, backed by a sheltered beach. The northern 600 m is free of the beachrock and has a more energetic reflective to low tide terrace beach, with waves often reaching 1 m in height. The entire beach is backed by active blowouts, which increase in size to the north, where they extend up to 400 m inland. A car park and camping area are located above the southern end of the beach.

Swimming: These are five relatively safe beaches during normal low wave conditions, with the Browns Beach 'lagoon' offering the quietest water, while small waves usually surge up most of the steep beach faces. Beware of bedrock and calcarenite reefs and rocks on some of the beaches.

Surfing: Usually too low.

Fishing: These are all good sites for both beach and rock fishing.

Summary: Five natural beaches with good access to three and a basic camping area at Browns Beach.

417-420   **BEACHES 417-420**

Unpatrolled			
No.	Beach	Rating	Type	Length
417	Beach 417	3	R/LTT	300 m
418	Beach 418	2	R	550 m
419	Beach 419	2	R	250 m
420	Beach 420	2	R	350 m

Between the northern end of Browns Beach and Gym Beach is 2 km of bedrock dominated, northwest facing coast containing four small beaches, with only the easternmost beach accessible by vehicle. They are also the northernmost in the national park, with the park boundary bisecting the northernmost beach.

Beach **417** is a 300 m long, northwest facing beach receiving sufficient waves to maintain a narrow attached bar. It is bordered by low rocky to boulder covered bedrock points and backed by active dunes extending

300 m inland. Beach **418** is 550 m long and also faces northwest, but is sheltered by a 300 m long western headland, with usually low waves lapping against the steep beach. It is backed by moderately active dunes extending 300 m inland. Beach **419** is a curving, sheltered, 250 m long, north facing beach, fronted by a sandy bay containing small bedrock reefs and islets. The Gym Beach car park is located 150 m east of the beach. Beach **420** is a crenulate, 350 m long, sheltered, north facing beach. The western half is fronted by bedrock outcrops and reefs and a 20 m high foredune backs the entire beach, with three car parks and a camping area backing the foredune.

Swimming: These are four usually low wave and relatively safe beaches, apart from the rocks and reefs along some of the shores.

Surfing: Chance of a reef break off beaches 417 and 418 otherwise usually calm beaches.

Fishing: There is good rock fishing off all four.

Summary: Beaches 419 and 420 are the most accessible and more popular, particularly with the camping area and are mainly used by fishers and hikers.

421-423   **GYM BEACH to BEACH 423**

Unpatrolled			
No.	Beach	Rating	Type	Length
421	Gym Beach	3	R/LTT	1.2 km
422	Beach 422	2	R+rocks/reef	200 m
423	Beach 423	2	R+rocks/reef	150 m

Gym Beach (421) lies immediately north of the park boundary and is accessible along a 3 km road off the Marion Bay road, with a car park at the southern end. The 1.2 km long beach faces northwest and receives waves averaging 0.5 m, increasing slightly up the beach. There are deeper reefs off the beach, some of which provide surf during larger swell.

The northern end of the beach terminates at a low bedrock platform and reef, backed by 10 m high dunes. On the north side of the rocks are two small rock-bound beaches. Beach **422** is a 200 m long sand beach bordered by the rocks and fronted by deeper reefs, with surf breaking over the western reefs. Its neighbouring beach **423** is similar, 150 m long with rocks and reefs increasing to the north. Both beaches are backed by 10 to 20 m high dunes then cleared farmland, with no public access, other than along the beach from Gym Beach.

Swimming: Gym Beach is a relatively popular and

accessible spot with usually low waves and a shorebreak.

Surfing: The reefs at the northern end of Gym Beach are worth checking out in a low to moderate swell.

Fishing: Gym Beach is used for both beach and rock fishing.

Summary: Gym Beach is one of the few publicly accessible beaches north of the national park and, combined with the camping area, make it a popular spot.

424-427   POINT MARGARET

		Unpatrolled		
No.	Beach	Rating	Type	Length
424	Beach 424	4	R+rocks	1 km
425	Pt Margaret (S 3)	5	R/LTT	1 km
426	Pt Margaret (S 2)	6	LTT/TBR	1.1 km
427	Pt Margaret (S 1)	6	TBR	1.1 km

South of Point Margaret is a gently curving 4 km stretch of west facing shore, dominated by a mixture of calcarenite bluffs tied to low bedrock headlands at either end, with four 1 km long beaches spread in between the more prominent rocks and bluffs. All four are backed by 10 to 20 m high dunes extending 100 to 200 m inland and then generally cleared farmland. As a result there is no public access.

Beach **424** is partly protected by the rocky point and reefs to the west and its northwest orientation. It receives usually low waves, increasing slightly up the beach to where it terminates at a low calcarenite bluff and bedrock reef. Rocks and reefs also outcrop along much of the southern half of the beach, with deeper reefs offshore.

Beach **425** lies 2 km south of Point Margaret and is a 1 km long, west facing, lower energy beach backed by a mixture of dunes and dune covered bluffs, with some rocks along the shore and a narrow attached bar with a low surf. Beach **426** extends for 1.1 km from a protruding calcarenite bluff to a series of crenulate bluffs which eventually replace the beach with rocks. In addition, there are extensive shallow reefs off the beach. Waves average about 1 m and up to four strong rips form against some of the rocks and reefs. Beach **427** runs from the northern side of the bluffs for 1.1 km to Point Margaret, a 30 m high, dune covered bedrock point and reef. The bluffs in the south give way to a 20 m high foredune in the north and in lee of the point. Waves average just over 1 m and usually four strong beach rips dominate the surf (Fig. 4.93).

Swimming: The southern two beaches receive lower waves but still have much rock and reef to contend with,

while the northern two are dominated by strong beach and topographic rips, so use caution on all four.

Surfing: The best beach breaks are immediately south of Point Margaret, while there are also right hand reef breaks over the reefs off the point.

Fishing: The reefs, rocks and rips provide a range of good locations for shore-based fishing.

Summary: A generally inaccessible section of coast offering four natural, rock and reef-dominated beaches.

Figure 4.93   *Immediately south of Point Margaret the beach is backed by dune draped bluffs and contains a surf zone dominated by strong rips, as shown here.*

428-432   CONSTANCE BAY, BABY LIZARD, FORMBY BAY

		Unpatrolled		
No.	Beach	Rating	Type	Length
428	Constance Bay	4	R	600 m
429	Constance Bay (N)	4	R/LTT	650 m
430	Baby Lizard	7	R+platform	100 m
431	Baby Lizard (N)	6	TBR+reefs	250 m
432	Formby Bay	6	TBR/RBB	6.5 km

Between Point Margaret and Daly Head is the open, 8.5 km wide, west facing Formby Bay, including the smaller southern Constance Bay. The bays contains five exposed and energetic beaches, including the longer rip-dominated bay beach. Access is via the Ilfracombe Road in the south and Daly Head Road in the north.

Constance Bay lies immediately north of Point Margaret and is an open, northwest facing bay, protected in the south by the point, in the centre by small Ella Rock 800 m offshore and by shallow reefs on much of the bay floor including the right hand reef break called *Trespassers*. As a consequence waves average less than 1 m. The 600 m long southern beach (**428**) has a steep reflective beach broken by some rocks and reefs and is backed by a narrow dune-draped bluff, while the 650 m long northern beach (**429**) is a continuous strip of sand fronted by a narrow bar and backed by active dunes extending up to 600 m inland. Cleared farmland backs both the dune systems.

Baby Lizard is the name of the area at the end of the Ilfracombe Road where there is a car park and basic camping area on the 20 m high calcarenite cliffs. Immediately below the car park is a 100 m long high tide sand beach (**430**) fronted by an intertidal calcarenite platform, then deeper reefs. Three hundred metres north of the car park is a second 250 m long beach (**431**) which is backed and bordered by the bluffs and fronted by a mixture of rocks, reefs and rips which includes the right hand *Baby Lizard* surf break.

Formby Bay beach (**432**) begins immediately north of this beach and runs for 6.5 km to 60 m high Daly Head, a dune and calcarenite-capped bedrock headland and point. The west-southwest facing beach is one of the highest energy on the peninsula and the 1.5 m high waves combine with the medium sand to produce a 200 m wide, rip-dominated surf zone, with up to 20 strong rips along the beach. The beach is backed by a foredune, then up to 3.5 km of active and stable sand dunes. There is good access to the Daly Head car park, but a walk or 4WD is required to access most of the beach.

Swimming: These are five hazardous beaches because of the higher waves, prevalence of rips and in places the rocks and reefs. Use extreme care if swimming here.

Surfing: There are good right reef breaks in Constance Bay at *Trespassers* and at *Baby Lizard*.

Fishing: These are all good spots for both rock, reef and beach fishing, with usually good rip holes along Formby Bay.

Summary: Five reasonably accessible beaches, with a camping area at Baby Lizard, mainly used by fishers and more experienced surfers.

433-438   **DALY HEAD to POINT ANNIE**

No.	Beach	Rating	Type	Length
		Unpatrolled		
433	Daly Head (N 1)	4	LTT	800 m
434	Daly Head (N 2)	3	R+reefs	350 m
435	Gleesons Landing	2	R+platform	1.8 km
436	Swincers Rocks (S)	4	LTT→TBR	2.2 km
437	Swincers Rocks	5	TBR	200 m
438	Point Annie	5	LTT-TBR	2.1 km

On the north side of **Daly Head** the coast trends northeast and bedrock dominates the shore for 3 km to Gleesons Landing, where the coast turns and trends due north for 6 km up to Point Annie. This section can be accessed via Gleesons Road to the landing and the Swincers Rocks and Point Annie roads, with 4WD tracks leading to the other beaches.

Beach **433** is bordered by prominent calcarenite-capped bedrock points, together with some rock outcrops along the 800 m long beach and deeper reefs offshore. Waves average about 1 m and maintain a sandy beach fronted by a 50 m wide bar, which is cut by three to four rips following periods of higher waves, most of the rips are located against the headlands and reefs. A 4WD track leads from the landing to the northern headland, while largely stable dunes back the beach.

Beach **434** is a 350 m long pocket of sand lying bounded by 300 m long bedrock points, together with rocks and reefs dominating much of the area between. It faces northwest and receives only low waves at the beach, with some surf on the outer reefs.

Gleesons Landing is a popular fishing, boating and camping area. The landing is a low energy high tide beach (**435**) fronted by 200 to 300 m wide intertidal calcarenite flats, which produce calm conditions at the shore with the left hand *The Spit* surf break out on the southern reef. The beach extends for 1.8 km to the north, where the flats give way to the open, higher energy Swincers Rocks beaches.

Beach **436** runs for 2.2 km to the north, terminating at the first of the Swincers Rocks. Wave height increases up the beach, averaging just over 1 m at the northern end. As the waves pick up, the bar widens and is cut by several rips along the northern half. A 4WD track leads to the northern rocks and stable dunes back the beach.

Beach **437** lies on the southern side of Swincers Rocks. It is a 200 m long, west facing beach, bordered by 10 m high calcarenite bluffs and backed by both stable and active dunes extending up to 400 m inland. The beach receives waves averaging over 1 m and has permanent rips against the rocks at either end and occasionally a central beach rip flowing out through the low gradient, 100 m wide surf zone.

On the north side of **Swincers Rocks** is a 2.1 km beach (**438**) that runs due north to Point Annie, a 20 m high granite point and reef. There is a rock reef 400 m off the southern end of the beach and patches of reef along the base of the beach. Wave height increases up the beach, transforming a narrow bar into a wider bar with usually several rips along the northern kilometre of beach. The point lies at the end of the Point Annie Road and provides good access to the northern end.

Swimming: Swimming ranges from calm at Gleesons Landing and beach 434, to rip-dominated on the other beaches. Use caution on the exposed beaches, particularly if waves exceed 1 m, as rips will be present.

Surfing: Best chance is on the first Daly Head beach, along the Swincers Rocks beaches and at Point Annie, all of which usually have low to moderate beach breaks.

Fishing: Gleesons Landing is a major base for both shore and boat based fishing, together with fishing off the rocks and beaches.

Summary: A relatively popular series of beaches centered on the Gleesons Landing beachfront camping and boat launching area.

439-445   POINT DEBURG, BERRY BAY

No.	Beach	Unpatrolled Rating	Type	Length
439	Pt Deburg (N 1)	4	LTT	230 m
440	Pt Deburg (N 2)	4	R/LTT	250 m
441	Pt Deburg (N 3)	3	R+rocks/reef	400 m
442	Berry Bay (1)	3	R+rocks/reef	150 m
443	Berry Bay (2)	3	R+rocks/reef	300 m
444	Berry Bay (3)	3	R+rocks/reef	150 m
445	Berry Bay (4)	3	R+rocks/reef	60 m

Between Point Annie and West Beach is 5 km of rocky coast dominated by the 10 to 20 m high, rounded granite points, boulders and reefs, all backed by dune-capped calcarenite bluffs. The West Coast Road follows the coast north from Point Annie and there are numerous tracks and parking bays off the road, providing access to the bluffs above the seven small beaches along this section.

At **Point Deburg** the coast turns and trends toward the northeast. The point itself consists of granite points and coves, some of which are relatively calm and suitable for diving or snorkelling. The beaches begin 1 km east of the point. Beach **439** is a 230 m long sand beach, bordered by sloping granite rocks and backed by vegetated calcarenite bluffs. The beach usually has a narrow attached bar and low waves, with a few rocks and reefs off the beach. Its neighbouring beach (**440**) is 250 m long, with a narrow attached bar, while it is backed by steep, 15 m high bluffs. A car park on the headland that separates the two provides access down the bluffs to both. Beach **441** lies immediately to the north and is an irregular, protected, 400 m long strip of high tide sand lying beneath the bluffs, with a foreshore completely dominated by granite boulders and reefs and a small open access to the sea at the northern end.

Berry Bay is the name of the open, 6 km wide section of curving coast between Point Deburg and Corny Point. Along the central section, the rocks of Point Deburg continue to West Beach. Amongst the rocks extending for 1 km south of the beach are four small rock and reef-bound beaches. Beaches **442** and **443** are neighbouring curving, 150 m and 300 m long pocket beaches with granite boulders extending off either end and usually calm conditions at the shore. Beach **444** is a small cuspate beach, protruding seaward in lee of a shallow granite reef, while beach **445** is a 60 m long pocket of sand bordered by granite platforms.

Swimming: These seven generally sheltered beaches are usually calm and relatively safe away from the rocks and reefs.

Surfing: None.

Fishing: Excellent rock and beach fishing the length of this coastal section.

Summary: An accessible, scenic section of coast consisting of the granite rocks and pocket sandy beaches.

446-448   NORTH BERRY BAY, CORNY POINT

No.	Beach	Unpatrolled Rating	Type	Length
446	North Berry Bay	5	TBR	2.4 km
447	Corny Point (W)	7	R+platform	500 m
448	Corny Point	2	R+reefs	320 m

NORTH BERRY BAY (446) lies immediately south of Corny Point, a major boundary between the more energetic beaches to the south and the low to very low energy beaches of the upper gulf. The beach faces west and receives waves averaging about 1 m, which maintain a rip-dominated sandy beach, with usually ten rips spread along the length of the beach, together with two reef patches in the south and centre (Fig. 4.94). The West Coast Road provides good access in the south, toward the northern end and at Corny Point. In between, the beach is backed by 20 to 30 m high calcarenite bluffs.

Running along the western side of **Corny Point** is a crenulate, 500 m long high tide sand beach (**447**), backed by the 20 to 30 m high bluffs and fronted by an irregular, 50 to 100 m wide bedrock platform and reefs.

On the tip of Corny Point is a semicircular, north facing bay, with a 100 m wide, reef-strewn mouth, containing a 320 m long, curving sand beach (**448**). This beach is well protected by the headlands, rocks and reefs, receives low swell and usually has calm conditions at the shore.

Ketches used to use this beach to load grain until 1942, while the backing Corny Point lighthouse was built in 1882.

Figure 4.94   *North Berry Bay has a moderate energy rip-dominated surf zone.*

Swimming: North Berry Bay (446) has a surfing beach with persistent beach rips, so be careful, while the rocky point beach (447) is unsuitable for safe swimming. The protected point beach (448) offers the safest swimming.

Surfing: Usually good beach breaks, with some reef breaks off Corny Point.

Fishing: Excellent rock fishing off the points and beach fishing in the rip holes.

Summary: Depending on which way you are travelling, these are either the first or last relatively high energy beaches on the western peninsula, with good views from the backing bluffs and Corny Point.

449-452   **CORNY POINT to DANDY BEACH**

No.	Beach	Unpatrolled Rating	Type	Length
449	Corny Point (E 1)	1	R+sand flats	1.1 km
450	Corny Point (E 2)	1	R+sand flats	350 m
451	Greig Lookout	1	R+sand flats	100 m
452	Dandy Beach	1	R+sand flats	500 m

To the east of Corny Point is the large, open *Hardwicke Bay*, the shore of which is totally protected from the southwest swell and is only lapped by onshore west to northerly wind waves, with calm conditions prevailing during all offshore wind conditions. The coast responds by permitting the accumulation of wide intertidal sand flats, backed by very low energy high tide beaches and fronted by extensive seagrass meadows. These conditions dominate most of central to upper Spencer Gulf.

Immediately east of Corny Point are three low energy beach-sand flats, all paralleled by the Lighthouse Road, with access tracks off the road to each of the beaches. Beach **449** extends for 1.1 km to the east of Corny Point. It consists of a high tide sand beach, backed by stable, 10 to 15 m high dunes and fronted by sand flats that are 300 m wide next to the point, narrowing to 100 m in the east. Calcarenite rocks separate it from beach **450**, which is 350 m long and bordered at either end by calcarenite. It has a high tide beach, fronted by 100 m wide sand flats. Beach **451** is a 100 m long patch of sand bordered by calcarenite bluffs and rocks, with 'Greig Lookout' at the eastern end atop the backing 20 m high stable dunes. **Dandy Beach (452)** is a 500 m long crenulate stretch of high tide sand, extending from the rocks below the lookout to a low calcarenite bluff, which separates it from the longer beaches to the east.

Swimming: These are four low wave and usually calm beaches, which can only be used for swimming at high tide.

Surfing: None.

Fishing: Best at high tide or in a boat over the seagrass meadows.

Summary: Four readily accessible, low energy beaches, with the most activity at Corny Point where there is a camping area and beachfront parking.

453-455   **DUNN POINT, COUCH'S, LEVEN/BURNERS**

No.	Beach	Unpatrolled Rating	Type	Length
453	Dunn Point	1	R+sand flats	4.2 km
454	Couch's Beach	1	R+sand flats	9.5 km
455	Leven/Burners	2	R+sand flats-LTT	4.7 km

The southwestern shore of Hardwicke Bay, between Dunn Point and Burners Beach, consists of 18 km of low energy, north facing sand flats backed by three narrow high tide beaches. The Corny Point Road, then Brutus Road and finally the North Coast Road provide access to the three beaches respectively.

Dunn Point beach (**453**) is a 4.2 km long, north facing strip of high tide sand fronted by 300 to 400 m wide intertidal sand flats, then seagrass meadows. Three small linear settlements of Corny Point, The Dairy and Collins Beach back most of the beach, with good access and numerous tracks to the usually calm beach. Collins Beach settlement sits atop low calcarenite bluffs, which form the eastern boundary of the beach.

Couch's Beach (454) extends east of the calcarenite bluffs for 9.5 km to a large sandy foreland. The Pines settlement occupies 1 km of the beachfront, while much of the shore on either side of the foreland is part of Leven Beach Conservation Park. The foreland has built out into the gulf up to 3.5 km, leaving behind a series of up to 60 low foredune ridges, largely covered in natural vegetation. The beach itself has a low energy to calm high tide beach (Fig. 4.95), fronted by flat to ridged sand flats up to 400 m wide.

Figure 4.95 *Couch's Beach has usually calm conditions, as shown here, with tractors used to launch fishing boats from the beach.*

Leven Beach Conservation Park

Established:	1988
Area:	493 ha
Beaches:	454 & 455
Coast length:	4 km

Leven Beach Conservation Park covers the central part of a 10 km long, cuspate foreland which is fronted by the low energy beach and sand flats and backed by a 2 km wide series of densely vegetated low beach to foredune ridges.

On the eastern side of the foreland, the beach continues as **Leven Beach (455)** for 4.7 km to Burners Beach, located below grassy, 30 m high bluffs and finally a steeply dipping metasedimentary rock platform. There is good access at **Burners Beach** where there is also a basic beachfront camping area, but no facilities. The sand flats that are 300 m wide at the foreland narrow to 100 m at Burners Beach as it becomes more exposed to westerly wind waves, with even low surf and some rips occurring during strong wind and wave conditions.

Swimming: Three usually calm to low wave beaches, with only Burners Beach having some low surf during strong onshore westerly winds.

Surfing: Usually none, except at Burners during strong winds.

Fishing: Only at high tide from the beach and the Burners Beach rock platforms.

Summary: Three accessible and natural low energy beaches mainly used by the local residents and shack owners.

456-460   GALWAY BAY, PT SOUTTAR, BRUTUS

Unpatrolled			
No. Beach	Rating	Type	Length
456 Galway Bay	3	LTT+bar	650 m
457 Galway Bay (E)	3	R+sand flats	550 m
458 Pt Souttar (E 1)	3	R+sand flats	180 m
459 Pt Souttar (E 2)	3	R+sand flats/rocks	500 m
460 Brutus	2	R+platform	7.3 km

Galway Bay is a 650 m long, north facing beach bordered by protruding, low metasedimentary rock platforms. A boat ramp is located on the western platform, while about 20 shacks and the North Coast Road back the usually low energy beach **(456)**. The beach is however exposed to west through northerly wind waves, which are sufficient to maintain an attached bar, with two to three rips cutting through a shallow second bar, in addition to rock outcrops in the surf.

Immediately to the east is a second 550 m long beach **(457)**, also bordered by low rock platforms, with rock reefs along half the beach. The remainder has an attached bar, with one rip operating during higher wind waves. Several shacks and the road back the beach.

Point Souttar is a 25 m high calcarenite bluff, along the base of which are two narrow sandy beaches. Beach **458** lies below 10 m high bluffs to the west of the point and is 180 m long, with seagrass covered sand flats 20 m offshore. Beach **459** faces northeast and is a 500 m long high sand beach fronted by 50 to 100 m wide sand flats, dissected by linear bedrock ridges, which extend off the beach. It is backed by the road and several shacks.

Beach **460** extends from the Point Souttar beaches south and then west to Fish Point, a distance of 7.3 km. The entire shoreline is dominated by the low bedrock bluffs and 100 m wide rock platform, with the continuous sandy high tide beach wedged in between the two. The Brutus road and a few shack settlements back the beach. The only water access is at the very eastern end just before Fish Point, where there is a beachfront car park and a small cove located in a gap in the rocks that has two pontoons for swimming.

Swimming: Galway Bay offers the best swimming, as long as the waves are low, otherwise take care of rips. All the beaches are best at high tide, while beach 460 is unsuitable owing to the continuous rock platform and reefs, apart from the cove at Fish Point.

Surfing: The only chance is at Galway Bay during strong onshore winds.

Fishing: There is excellent fishing the length of the shore into the Galway Bay rip holes and off the rocks and platforms that dominate this section of coast.

Summary: A very accessible stretch of coast occupied by a few score of shacks.

461-464 **FISH PT, PT TURTON, FLATTERY BEACH**

Unpatrolled				
No.	Beach	Rating	Type	Length
461	Fish Point	3	R-cobble	60 m
462	Pt Turton	3	LTT	30 m
463	Pt Turton (jetty)	2	R+sand flats	350 m
464	Flattery Beach	2	R+sand flats	9.6 km

Point Turton is the first medium sized settlement on the west coast and consequently a focus for much fishing and tourist activity. The small settlement spreads south of the point for 2 km as a series of older beachfront shacks and newer freehold subdivisions, including one on the slopes behind the point. The point is an old wheat port and still boasts a jetty (built in 1877) and a small high tide boat harbour.

Fish Point lies 500 m northwest of Point Turton and in between are two small beaches. Beach **461** is a 60 m pocket of sand and cobbles at the base of 20 m high calcarenite bluffs, with seagrass growing to the shore. Beach **462** is a smaller 30 m pocket immediately west of the point, also backed by sloping bluffs and fronted by a 50 m wide bar, then the seagrass. During higher wind waves a rip runs across the bar.

The main settlement beach (**463**) is a crenulate, low energy, usually calm high tide beach, fronted by ridged sand flats that widen to 400 m in the south (Fig. 4.96). The 100 m long jetty runs off the point at the northern end of the beach, with a small hooked groyne providing some shelter for boat launching. A beachfront caravan park is located next to the jetty.

Flattery Beach (**464**) begins just south of the settlement and runs for 9.6 km almost to the settlement of Hardwicke Bay. The northwest facing beach receives all westerly wind

waves, however they break across 700 m wide sand flats covered by up to 16 low ridges and runnels, with usually calm conditions resulting at the shore. There is access to the beach at either end and a central car park (also known as **Town Beach**), but no facilities. The beach is backed by a 2 km wide coastal plain containing up to 20 low foredune ridges in the west and centre, which grade into generally stable parabolic dunes in the east.

Figure 4.96 *Point Turton jetty, boat ramp, low energy beach and ridged sand flats.*

Swimming: Best along Flattery Beach at high tide.

Surfing: None.

Fishing: The jetty and groyne, as well as the rocks around the two points are the most popular locations.

Summary: Point Turton is an older, but still relatively small and quiet settlement, offering basic facilities for the traveller and holiday-maker. Popular with fishers.

465-468 **HARDWICKE BAY to RUDIS BAY**

Unpatrolled				
No.	Beach	Rating	Type	Length
465	Hardwicke Bay	2	R+tidal flats	10.4 km
466	Parsons Beach	2	R+sand/rock flats	8.3 km
467	Bluff Beach	2	R+sand flats	240 m
468	Rudis Bay	2	R+sand/rock flats	1.9 km

At **Hardwicke Bay** settlement the coast turns 90⁰ and trends north for the next 50 km to Port Victoria. While this section of coast faces due west into the prevailing winds and gulf wind waves, the beaches remain very low energy owing to the continuous 200 to 500 m wide rock and sand flats that front the shoreline. Consequently the beaches remain narrow strips of high tide sand, usually covered in piles of seagrass debris. A grid of gravel roads backs the shoreline, with access provided to most of the shore along formed gravel roads or vehicle tracks following the back of the beach or dunes.

Hardwicke Bay beach (**465**) begins 1 km south of the Hardwicke Bay settlement, which occupies 1.5 km of the crenulate, 10.4 km long beach. The settlement has a store and petrol station and consists of about 80 generally beachfront shacks and more recent freehold subdivisions, that sit atop a 10 m high foredune (Fig. 4.97). The beach is very low energy and much of the 400 m wide sand/rock flats are covered with sarcocornia. A gravel road parallels the back of the beach, known locally as Sheriffs Beach, up to Port Minlacowie, site of a wheat jetty from 1877 to 1971 and now only offering a large car park and boat ramp, located at a gap in the rock flats. North of the old port the beach is known as **Cockle Beach**.

Figure 4.97 *Hardwicke Bay settlement and the low energy bay beach and ridged sand flats.*

Parsons (or Watsons) Beach (**466**) extends for 8.3 km from 2 km north of the old port to the southern rocks of Bluff Beach. The dunes backing the beach increase to 10 to 20 m in height and extend up to 500 m inland. They restrict vehicle access to the southern end, at the Chenoweth settlement, site of boat ramp and via the northern Bluff Beach settlement. For the most part the beach is fronted by 150 to 300 m wide sand and rock flats and backed by the dunes, which increase in size and instability to the north, with the largest called Mount Pisca.

Bluff Beach (**467**) is a 240 m long pocket of sand, bordered by 10 m high calcarenite bluffs and fronted by a gap in the rock flats. This gap has been used for landing boats since the still small settlement was established in 1836. Today it has a score of shacks on the southern bluffs, some beachfront shacks along the beach and a camping area.

Rudis Bay (**468**) is a 1.9 km near straight section of shore between Bluff Beach and Brown Point, which are also the only two access points, as 500 m wide dunes back the low energy beach. It is fronted by continuous 200 to 300m wide sand/rock flats and has no facilities.

Swimming: All these beaches experience generally low wind waves or calm conditions, with swimming only possible at high tide when the rock and sand flats are covered, however be careful of the extensive rock flats.

Surfing: None.

Fishing: Most fishers head out into the gulf, with shore fishing only possible at high tide.

Summary: A quiet section of coast with relatively good access and a few facilities at some of the small settlements.

469-474   **BARKER ROCKS to RIFLE BUTTS BEACH**

No.	Beach	Unpatrolled Rating	Type	Length
469	Barker Rocks	2	R+sand flats	160 m
470	Barker Rocks (N)	2	R+rock/sand flats	3.9 km
471	Port Rickaby	2	R+sand flats	1.1 km
472	The Bushes Beach	2	R+rock flats	15.2 km
473	Second Beach	2	R+sand/rock flats	1.7 km
474	Rifle Butts Beach	2	R+sand flats	200 m

North of Barkers Rocks is a 22 km long, west facing section of coast dominated by the 15 km long Bushes Beach, with several smaller beaches to either side, including the beach and small settlement of Port Rickaby. All the beaches receive low wind waves and are dominated by a combination of shallow intertidal rocks and sand flats, with continuous seagrass meadows paralleling the shore.

Barkers Rocks beach (**469**) is a 160 m long sand beach located in a gap in the rock flats, with elevated beachrock bordering the back and either end of the beach. There is a gravel road right to the beach and a large car park to service the boat launching off the beach.

On the north side of the beach, the 300 m wide rock flats recommence and extend for another 2 km, before the beach is fronted by seagrass covered sand flats up to Port Rickaby. This 3.9 km long beach (**470**) is backed by continuous, 300 to 400 m wide dunes, with access only at either end, or via several 4WD tracks across the largely stable dunes.

Port Rickaby is a small settlement located in lee of a 1.1 km long sand beach (**471**), fronted by relatively narrow, 60 m wide sand flats (Fig.4.98). A 100 m long jetty crosses the flats, providing access to deeper water and, in past days, a berth for the wheat ships. The port is dominated by a large beachfront caravan park and a large car park to service the boat ramp and beach users. There is also a general store and fuel available.

Figure 4.98 *Port Rickaby jetty crosses a wide, low energy high tide beach, which is also used for launching boats.*

North of the port is a 15 km long beach, known from south to north as **The Bushes, The Bamboos, Kemps Beach** and **Wauraltee Beach** (**472**), which terminates at the prominent Renowdens Rocks, a tombolo-backed reef. This beach is dominated in the south by 5 km of 400 to 600 m wide rock flats. These give way to 200 m wide sand flats, all fronted by continuous seagrass meadows. Ten to twenty metre high dunes back the entire beach. They widen to 2 km in the north and are dominated by largely stable parabolic dunes. A rough vehicle track runs along the back of the dunes, with a mixture of vehicle and 4WD tracks both along and through the dunes to the beach. There is no development or facilities on the beach.

On the north side of Renowdens Rocks is 1.7 km long **Second Beach** (**473**), with access via a northern car park. The beach has a mixture of rocks and sand flats, with a predominantly sandy section 200 m south of the car park. **Rifle Butts beach** (**474**) is a smaller, 200 m long, relatively steep sand beach with flats, bordered by elevated beachrock. It is one of the more popular swimming beaches for the adjacent town of Port Victoria.

Swimming: Port Rickaby offers one of the nicer swimming beaches on the western peninsula, as well as a well developed caravan park, jetty and other facilities. Most of the beaches can only be used at high tide. Be careful of the prominent rock flats and beachrock along much of the shore.

Surfing: None.

Fishing: The jetty is the most popular location for shore-based fishing.

Summary: A quiet section of coast, with good access and basic facilities at Port Rickaby.

475-479   PORT VICTORIA

No.	Beach	Rating	Type	Length
		Unpatrolled		
475	Point Gawler	1	R+sand/rock flats	300 m
476	Port Victoria (S 2)	1	R+sand flats	250 m
477	Port Victoria (S 1)	1	R+sand/rock flats	150 m
478	Port Victoria (jetty)	1	R+sand/rock flats	350 m
479	Port Victoria (N)	2	R+rock flats	60 m

Port Victoria is one of the larger settlements on the western shore of the peninsula and provides a jetty, boat ramp, caravan park and hotel, as well as limited facilities in the small beachfront town. To either side of the main jetty beach are several small, very low energy beaches and rock-sand flats, with the northernmost beach grading into the tidal flats in lee of Point Pearce.

Point Gawler is a 15 m high calcarenite bluff that forms the southern boundary of Port Victoria. There is a boat ramp out on the point providing access to deeper water, while inside the point 200 m wide sand and rock flats front the low energy, 300 m long, curving beach (**475**). The Port Victoria golf course runs south of the point.

Beaches **476** and **477** are 250 m and 150 m long respectively, lying between the point and port, and are backed by a sealed road. They are both bounded by raised beachrock bluffs and consist of a low energy high tide beach fronted by 50 m wide sand flats and seagrass meadows.

Port Victoria is a former wheat port which still maintains its 200 m long jetty, now mainly used for fishing. The small town lies on 10 to 20 m high bluffs and overlooks the jetty and small, 350 m long beach running to the north (**478**). There is a beachfront caravan park and foreshore reserve backing the shoreline and beach. There is relatively deep water off the port and seagrass extending almost to the base of the beach.

Eight hundred metres north of the jetty is a moderately steep, 60 m long pocket of sand (**479**) bordered by bedrock boulders and fronted by deeper seagrass meadows.

Swimming: The jetty beach in front of the caravan park is the most accessible and most popular town beach. Watch for the rocks and rock flats on most of the other beaches.

Surfing: None.

Fishing: The jetty and off the rocks at high tide are the best locations.

Summary: Port Victoria is a popular holiday destination and provides good facilities for visitors as well as the added recreational attractions.

Regional Map 6: Upper Yorke Peninsula

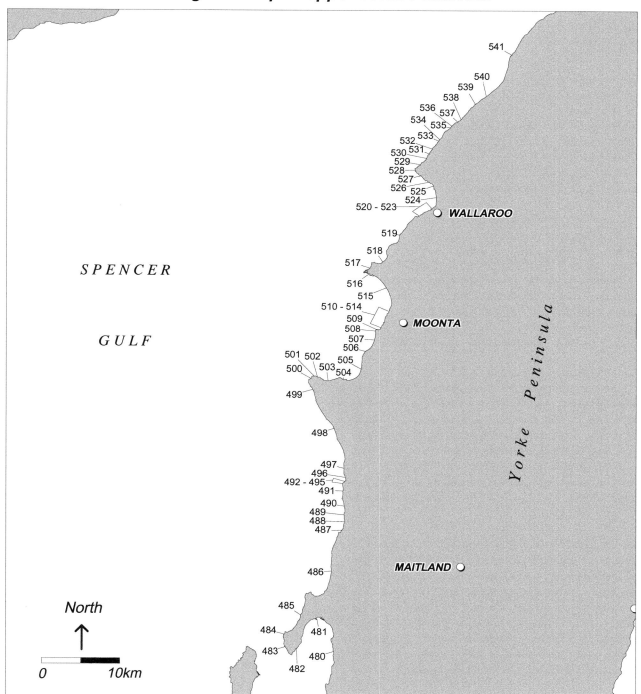

Figure 4.99   *Regional Map 6 – Upper Yorke Peninsula*

480-486   POINT PEARCE ABORIGINAL LAND

Unpatrolled			
No.	Beach	Rating	Type	Length
480	Pt Pearce (E)	1	R+rock/sand flats	3.4 km
481	Pt Pearce (bay)	1	R+tidal flats	8 km
482	Pt Pearce (landing)	1	R+rock flats	2 km
483	Pt Pearce	1	R+rock&sand flats	3 km
484	Island Pt	1	R+rock flats	300 m
485	Island-Reef Pts	1	R+rock/sand flats	6.8 km
486	Chinaman Wells	1	R+rock/sand flats	10.8 km

Point Peace and Wardang Island provide protection for the curving 14 km of shoreline from north of Port Victoria round to Point Pearce. This area is all part of the **Point Pearce Aboriginal Land**, that occupies the coast from immediately north of Port Victoria, round Point Pearce and up to Chinaman Wells, a shoreline distance of 28 km. The coast is dominated by low energy high tide beaches fronted by a mixture of tidal, sand and rock flats.

Beach **480** lies immediately north of Port Victoria where the sand flats widen progressively to 800 m. The backing 3.4 km long high tide beach receives occasional low waves and at its northern end grades into the vegetated tidal flats of beach **481**. This beach curves around to face south then east in lee of Point Pearce. It consists of a narrow high tide strip of sand fronted by mangrove and seagrass covered flats up to 1.5 km wide. Vehicle tracks parallel the back of the beaches, with no facilities.

Point Pearce lies at the end of a 5 km long, 1 to 2 km wide, generally low bedrock peninsula, which is tied to the mainland by 3 km wide, low beach ridges and high tide flats extending from the southern shore to Chinaman Wells. The southeastern shore of the point consists of a crenulate, bedrock dominated, low shoreline, with a narrow high tide beach (**482**) fronted by 200 to 300 m wide rock and tidal flats. An old landing is located at the northern end of the beach, while vehicle tracks parallel the rear.

The southeast side of the point faces southwest, but remains protected by Wardang Island (4 km offshore) from the prevailing wind and waves. The first 3 km of shoreline consists of a crenulate high tide beach (**483**) fronted by a mixture of rock and tidal flats, that widen to 1 km in lee of Green Island and link it to the mainland at low tide. A 300 m wide beach-foredune ridge plain has formed a cuspate foreland in lee of the island. A vehicle track parallels the back of the shore and plain.

Island Point is connected to the shore at low tide. On the southern side of the point is a 300 m long high tide sand beach (**484**), fronted by 100 m wide rock flats. A landing crosses the southern end of the beach. On the north side of the point bedrock continues to dominate

the coast, with a 6.8 km long, crenulate, west facing beach (**485**), fronted by a mixture of bedrock flats up to 400 m wide and narrower sand flats. Low, partially active dunes back the beach, with an access track at the southern end and access through the dunes via 4WD tracks.

At **Reef Point** begins the 10.8 km long sandy beach that includes **Chinaman Wells** and terminates at Point Warrene, the site of Balgowan settlement. The Aboriginal land extends as far as Chinaman Wells. This long sandy beach (**486**) contains a wide, low gradient high tide beach, fronted by sand flats that reach 2.5 km in width in the south, narrowing gradually to 500 m off Point Warrene. The Chinaman Wells Road parallels the northern boundary of the Aboriginal land, then turns at the coast to run for 2 km up to the Chinaman Wells shacks, with the fishing boats moored on the sand flats. Four-wheel drive tracks extend north of the shacks all the way to Balgowan.

Swimming: Chinaman Wells is the only beach suitable for swimming and then only at high tide. The remainder are more tidal and rock flats than beaches.

Surfing: None.

Fishing: Only possible at high tide owing to the wide tidal flats, with the two landings offering the best access to deeper water.

Summary: This entire area (south of Chinaman Wells) is Aboriginal land and permission may be required to enter.

487-490   BALGOWAN

Unpatrolled			
No.	Beach	Rating	Type	Length
487	Balgowan	2	R+sand flats	1.15 km
488	Balgowan (N 1)	2	R+sand flats	750 m
489	Balgowan (N 2)	2	R+sand flats	500 m
490	Balgowan (N 3)	2	R+sand flats	1.8 km

Balgowan is a small town located at Point Warrene. The remains of the old jetty are located at the tip of the small point, with a large beachfront car park. The town sits on 20 m high bluffs overlooking both the southern beach (**486**) (Fig. 4.100) that runs down to Chinaman Wells and the first of four west facing beaches that extend north of the town for 5 km to Tiparra Rocks. All receive low wind waves and are fronted by continuous seagrass meadows. The **town beach** (**487**) is 1.1 km long and extends north of the point. It is accessible from the car park and is used to launch boats, with tractors and cars often parked on the beach. Seagrass meadows lie 20 m off the shore. Steep red alluvial bluffs back the beach, with a second bluff-top car park at the northern end.

Figure 4.100 *View south from Point Warrene, showing some of the Balgowan shacks and the wide beach fronted by shallow sand and rock flats.*

Beach **488** extends north of the second bluff for 750 m to a more continuous series of bluffs. This seagrass-fronted sandy beach is also backed by bluffs and is accessible from car parks at either end. Beach **489** runs for 500 m along the base of the bluffs and consists of a narrow high tide beach, broken in places by the protruding bluffs and rocks, with seagrass just off the shore.

Beach **490** runs for 1.8 km from the northern side of the bluffs to Tiparra Rocks, a 600 m long reef. This beach is only accessible by a rough vehicle track that runs behind the 100 to 200 m wide bluff-top dunes. It consists of a wide sand beach, with seagrass meadows less than 50 m offshore.

Swimming: These are usually low wave and relatively safe beaches, apart from the deeper water over the seagrass meadows.

Surfing:　　None.

Fishing:　　Point Warrene and the jetty ruins provide the best access to deeper water.

Summary:　　Balgowan is a small town offering good beach access and limited facilities for travellers, but no accommodation.

491-496　　TIPARRA ROCKS, THE BAMBOOS

Unpatrolled			
No.	Beach	Rating	Type	Length
491	Tiparra Rocks	2	R+sand flats	1.8 km
492	The Bamboos (pt 1)	2	R+sand flats	50 m
493	The Bamboos (pt 2)	2	R+sand flats	50 m
494	The Bamboos (pt 3)	2	R+sand flats	50 m
495	The Bamboos (pt 4)	2	R+sand flats	50 m
496	The Bamboos	2	R+sand flats	1.1 km

Tiparra Rocks is the first of a series of more prominent bedrock reefs, that extend seaward of the 10 to 20 m high, dune-capped bluffs. Between the rocks and The Gap, the dunes are backed by a rough vehicle track, which provides access to The Bamboos beaches.

Tiparra Rocks beach (**491**) faces due west, is 1.8 km in length and fronted by a mixture of sand and rock flats, while 300 to 400 m wide dunes back the bluffs. A car park and camping area are located behind the northern end of the beach. The beach terminates at The Bamboos reefs. On the north side of the reef, northwest facing, 15 m high bluffs dominate the shore for 500 m. Along the base of the bluffs are four pockets of sand (beaches **492, 493, 494** and **495**), each 50 m in length and separated and backed by eroding bluffs and rock debris. There is a solitary shack above the first beach.

The main **Bamboos** beach (**496**) is 1.1 km long, extending from the southern bluffs to a 300 m wide northern reef. Red, 15 m high bluffs back the entire beach. Bluff-top dunes widen from 100 to 300 m in lee of the reef, with a vehicle track behind the dunes. The Bamboos car park and camping area are located in the dunes behind the northern end of the beach.

Swimming: These beaches usually receive low waves and are relatively safe, apart from the deeper water over the seagrass meadows and the presence of rocks and reefs off some beaches.

Surfing:　　None.

Fishing:　　The two reefs are the most popular areas to fish and attract the campers.

Summary:　　A relatively natural section of coast, apart from the access tracks and two basic camping areas.

497-500　　THE GAP, CAPE ELIZABETH

Unpatrolled			
No.	Beach	Rating	Type	Length
497	The Gap (S)	2	R+sand flats	1.5 km
498	The Gap	2	R+sand flats	9.8 km
499	Cape Elizabeth (S)	2	R+sand flats	3 km
500	Cape Elizabeth	2	R+sand flats	700 m

The Gap is a break in the near continuous dunes that extend from Balgowan to Cape Elizabeth. A gravel road runs out to the beach where there is a car park that provides access to both Gap beaches and a camping area. The southern beach (**497**) extends for 1.5 km south of the car park to the northern Bamboos reef. It is a slightly crenulate sandy beach, fronted by 200 to 300 m wide sand and rock flats, with dunes extending up to 500 m in lee of the beach.

The main Gap beach (**498**) begins where the rock reef diminishes, basically at the car park, and trends north-northwest toward Cape Elizabeth for 9.8 km. It includes two sandy forelands in lee of shoaler reefs, with rock reef generally increasing in occurrence toward the north. The beach faces into the strong southwesterly winds and several active blowouts and parabolic dunes extend up to 2 km inland. Apart from the car park and camping area, there is no other vehicle access or facilities. Several 4WD tracks do however cross the dunes.

Beach **499** begins at a small protruding bluff that separates it from The Gap beach. It then runs on to Cape Elizabeth, a dune-capped foreland in lee of reefs extending 300 m offshore. This beach is fronted by a mixture of rock and sand flats, backed by 600 m wide dunes and is only accessible by 4WD. Just around the corner is beach **500**, located right on the cape. It is a curving, 700 m long, northwest facing sand beach, fronted by shallow sand flats, with a private boat landing at the northern tip and dunes then farmland to the south.

Swimming: These are relatively safe beaches during normal low wave conditions, just beware of the many rock flats and reefs along the beaches.

Surfing: None.

Fishing: The Gap beach is popular for fishing over both the reefs and seagrass meadows.

Summary: The Gap is the most popular spot and only area accessible by car, as well as having the camping area.

501-504   **CAPE ELIZABETH (E)**

Unpatrolled			
No.	Beach	Rating	Type	Length
501	Cape Elizabeth (E 1)	1	R+sand flats	400 m
502	Cape Elizabeth (E 2)	1	R+sand flats	500 m
503	Cape Elizabeth (E 3)	1	R+sand flats	2.5 km
504	Cape Elizabeth (spit)	1	R+sand flats	1.1 km

At **Cape Elizabeth** the coast turns 90° into Tiparra Bay and the shoreline, for the first several kilometres, faces north. The whole cape area is backed by dunes in the west, beach ridges and spits in the north and then farmland, resulting in restricted public access. The shoreline is a continuous, low energy sandy beach fronted by sand flats, with prominent sandy forelands separating the individual beaches.

Beach **501** lies on the very northern tip of the cape and is backed by a beachfront farm house at its western end. The 400 m long beach faces north, with sand flats widening from 50 m in the west to 200 m in the east. It terminates

at a 100 m wide, protruding sandy foreland, on the east side of which begins 500 m long beach **502**, a similar curving, north facing beach. It is located between two low sandy forelands, with remnants of beach ridges and spits backing the beach, while the sand flats widen to 600 m off the beach.

Beach **503** is a 2.5 km long, curving beach with a slightly crenulate sandy shoreline and sand flats reaching 1.1 km in width off the centre. It is backed by low beach-foredune ridges and farmland. Beach **504** is a 1.1 km long sand spit that extends east from the foreland that separates it from beach 503. Remnants of earlier spits back the beach, while the sand flats average 1 km in width.

Swimming: These are four low energy, north facing beaches dominated by wide, shallow sand flats and are only suitable for swimming at high tide.

Surfing: None.

Fishing: meadows.   Best off the beaches over the seagrass

Summary: farmland.   Four isolated beaches all backed by

505-507   **TIPARRA BAY, SOUTH BEACH**

Unpatrolled			
No.	Beach	Rating	Type	Length
505	Tiparra Bay (S)	1	R+sand flats	3.4 km
506	Tiparra Bay (foreland)	1	R+sand flats	1.6 km
507	South Beach	1	R+sand flats	2.5 km

South of Port Hughes the southeast shore of **Tiparra Bay** consists of a 7 km long, continuous sandy shoreline, broken into three beaches by a prominent central foreland formed in lee of reefs a few hundred metres offshore. The shoreline and backing land from here up to Moonta Bay is dominated by stabilised and now farmed longitudinal dunes.

The southern beach (**505**) is accessible via a 4 km long gravel road which runs out to the coast between two longitudinal dunes. The 3.4 km long north to west facing beach has a low gradient and is fronted by tidal flats up to 1.5 km wide, the inner portions of which are vegetated, while the sandy section contains up to 28 low ridges. Beach **506** is bordered by small sand protrusions which are in line with old dune ridges. A narrow strip of coastal sand dunes caps the ridges, while the sand flats narrow to 400 m.

South Beach (**507**) is the southern beach of Port Hughes. It faces west and extends for 2.5 km from the northern protrusion up to the rock flats of Port Hughes. It is accessible via a sealed road and large car park at the northern end, and consists of a sandy beach fronted by sand flats, which narrow to 100 m at the port.

Swimming: South Beach is the more popular, with good access to a relatively safe, sandy beach, while the gravel road provides public access to the southern beach.

Surfing: None.

Fishing: Only possible at high tide owing to the shallow sand flats.

Summary: Two accessible but generally little used beaches.

508-513   **PORT HUGHES to ROSSITERS PT**

No.	Beach	Unpatrolled Rating	Type	Length
508	Port Hughes (jetty)	2	R+platform	300 m
509	Port Hughes (hbr)	1	R+sand flats	50 m
510	Port Hughes	1	R+sand flats	950 m
511	Harry Point	1	R+sand flats	350 m
512	Sim Cove	1	R+sand flats	850 m
513	Rossiters Point	1	R+sand flats	650 m

Port Hughes was developed as the port for the once flourishing Moonta copper mines, located 5 km to the east. While the 400 m long jetty no longer services ships, the port area remains a popular recreational centre and focus of boating in the adjacent gulf. The port settlement sits partly on a low, 400 m wide promontory bordered by South Beach to the east, the jetty and beach to the west and the harbour and main beach to the north. All beaches are accessible by sealed roads, with beachfront parking and access.

Port Hughes jetty extends west across a 300 m long high tide sand beach (**508**) fronted by 50 m wide intertidal rock flats, then seagrass meadows. It is backed by a caravan park and a large car park, which also services the adjoining boat ramp in the small, 50 m long **harbour beach (509)**. This beach was created when the harbour was excavated in the rocks and tidal flats, with two 200 m long harbour walls almost encircling the usually calm beach. The boat ramp is located at the western end of the beach.

Immediately north of the harbour is the 950 m long main **Port Hughes** beach (**510**), which faces west and runs up to the low Harry Point, the first in a series of truncated, red longitudinal dunes, which form a series of five points up

to Moonta Bay. The beach is backed by a foreshore reserve, road and houses, and fronted by low, ridged, 200 m wide sand flats. On the north side of **Harry Point** is a 350 m long beach (**511**) backed by a low foredune and also fronted by the continuous sand flats, with car parking to either end. The same sand flats continue on along **Sim Cove** beach (**512**), an 850 m long sand beach, bordered by the truncated dune ridges. This beach is backed by a mixture of beachfront houses of Moonta Bay and low dunes, with public access at either end. **Rossiters Point** is another truncated dune ridge and forms the southern boundary of beach **513**, a 650 m long sand beach fronted by a mixture of rock and sand flats, the latter widening to 250 m at the northern boundary of Moonta jetty.

Swimming: All these usually low wave to calm beaches can only be used for swimming toward high tide, with the sand flat beaches offering the safest waters.

Surfing: None.

Fishing: The jetty, rocks and harbour walls are the most popular locations.

Summary: Six very accessible and relatively safe beaches fronting both Port Hughes and neighboring Moonta Bay.

514, 515   **MOONTA BAY**

No.	Beach	Unpatrolled Rating	Type	Length
514	Moonta (jetty)	1	R+sand flats	700 m
515	Moonta Bay	1	R+sand flats	4.9 km

Moonta Bay is the second 'port' for Moonta, located just 2.5 km to the southeast. The bay settlement spreads along the gulf shore from Sim Cove for 2 km to north of the jetty. The 500 m long, hook-shaped jetty lies at the end of the main road and has a boat anchorage on its northern side, as well as supporting a swimming enclosure toward the outer 250 m wide sand flats. The jetty beach (**514**) extends for 700 m to the north and is backed initially by a caravan park, then a foreshore reserve and houses. The sand flats widen to 500 m at its northern end.

Moonta Bay is a curving, west to southwest facing, open bay, containing a 4.9 km long sand beach (**515**), fronted by ridged sand flats up to 800 m wide, which in the north are vegetated by mangroves in lee of Warburto Point. Some minor dunes that have transgressed over earlier foredune ridges back the beach. There are a few houses and good access in the south, and a gravel road, then rough vehicle track, to the northern end, with 4WD tracks through the dunes.

Swimming: The jetty swimming enclosure is a popular spot, with water depth increasing into the gulf.

Surfing: None.

Fishing: The outer jetty is the main location for shore-based fishing.

Summary: Moonta Bay offers good access to usually quiet beaches, as well as most facilities for holiday-makers and travellers.

516-519   MOONTA BAY to POINT HUGHES

No.	Beach	Rating	Type	Length
Unpatrolled				
516	Moonta Bay (N)	1	R+tidal flats	300 m
517	Warburto Pt (N)	1	R+tidal flats	1.6 km
518	Bird Is	1	R+tidal flats	3.6 km
519	Pt Hughes (S)	1	R+tidal flats	6.1 km

Between the north end of Moonta Bay and Point Hughes, 9 km to the north, is 12 km of crenulate, very low energy shoreline, containing a narrow high tide beach. It is entirely fronted by wide, usually vegetated tidal flats and backed by a mixture of low recurved spits and beach ridges. The entire section is only accessible by 4WD, apart from a gravel road out to the southern end of beach 519 and a vehicle track toward the northern end.

Beach **516** is a 300 m long, south facing beach bordered by extensive mangroves extending out across the 1 km wide tidal flats. Beach **517** runs along the northern side of low Warburto Point for 1.6 km and consists of low recurved spits fronted by 1 km wide tidal flats. Beach **518** lies in lee of Bird Island and is a curving, 3.6 km long, northwest facing, low beach with 1.5 km wide tidal flats. Finally beach **519** runs for 6.1 km from near the gravel road up to Point Hughes. The southern half is fronted by 1 km wide tidal flats covered with sarcocornia and scattered mangroves, which narrow to low ridged, 50 m wide flats at the point.

Summary: Four usually calm beaches fronted by extensive tidal flats and backed by low beach ridges then farmland.

520-525   POINT HUGHES, WALLAROO BAY

No.	Beach	Rating	Type	Length
Unpatrolled				
520	Point Hughes (N 1)	1	R+sand flats	600 m
521	Point Hughes (N 2)	1	R+sand flats	500 m
522	Office Beach	1	R+sand flats	150 m
523	Wallaroo (boat ramp)	2	R+sand flats	100 m
524	Wallaroo Bay	1	R+sand flats	2.3 km
525	North Beach	1	R+sand flats	700 m

Wallaroo is the 'port' for the former copper mining town of Kadina, located 9 km to the southwest. The port was established in lee of Point Hughes, with the copper exports now replaced by a large grain terminal, with a jetty extending 800 m out to deeper water. The crenulate shoreline between Point Hughes and the jetty, and the more open Wallaroo Bay, provide six low energy beaches for the town.

At **Point Hughes** the extensive tidal flats to the south give way to narrower sand flats, with seagrass meadows just offshore. Beach **520** extends for 600 m north of the point and consists of a low energy sand beach fronted by a mixture of sand and cobbles across the 200 m wide sand flats, together with intertidal rock reefs to either end. There is a gravel road backing the beach and a car park toward the southern end. Beach **521** is backed by low red bluffs with a scattering of buildings on the bluffs and a northern car park. It is 500 m long, with 50 m wide sand flats and some seawalls and groynes protecting buildings along the northern 200 m of beach.

On the north side of the jetty is 150 m long **Office Beach** (**522**), wedged in between the jetty seawall and a large protruding car park. This small beach is backed by a stepped seawall and the caravan park, and fronted by 200 m wide sand flats. The car park provides access to the boat ramp, which is located next to the adjoining beach (**523**). This small, 100 m long beach contains shade shelters and is popular with campers from the caravan park. It is bordered on its northern side by the old mine slag heaps, which used to be dumped into the bay.

On the north side of the slag heaps is the main 2.3 km long **Wallaroo Bay** beach (**524**), a sand beach fronted by ridged sand flats up to 400 m wide. A caravan park and beachfront houses back most of the beach. Most beach users park their cars right on the beach and at low tide out on the sand flats (Fig. 4.101). The beach terminates at protruding, 200 m wide rocks and reef, on the northern side of which is 700 m long **North Beach** (**525**), essentially a continuation of the beach and sand flats, which is also backed by the beachfront settlement of North Beach.

Figure 4.101 *Beachgoers follow the falling tide at Wallaroo Bay and park their cars out on the sand flats.*

Swimming: The caravan park beach, main Wallaroo Bay and North beaches are the most accessible and most popular, with all requiring high tide for water over the sands flats. To provide all-tide swimming, there is a swimming enclosure out on the jetty.

Surfing: None.

Fishing: Good fishing off the jetty and the Point Hughes rocks at high tide.

Summary: Wallaroo offers excellent beachfront camping/caravan facilities and a range of low energy, relatively safe beaches.

526-528   RILEY POINT

No.	Beach	Unpatrolled Rating	Type	Length
526	Riley Pt (S 2)	3	R+rock flats	850 m
527	Riley Pt (S 1)	3	R+rock flats	1.8 km
528	Riley Pt	3	R+rock flats	370 m

Riley Point is a gently sloping bedrock promontory that forms the northern boundary of Wallaroo Bay. Between the point and the sand of North Beach is 3 km of bedrock-controlled coast, containing three high tide sand beaches fronted by 50 to 100 m wide intertidal rock flats and boulders. Access is via North Beach in the south and the Riley Point gravel road in the north.

The first beach (**526**) is backed by the growing subdivisions of North Beach. It is 850 m long, sweeping in two arcs, with a reserve then houses backing the length of the beach. Beach **527** is similar, 1.8 km in length, becoming more crenulate toward the point, with a high tide beach and 50 m wide rock flats and farmland behind. Out on the tip of the point is 370 m long, west facing beach **528**, which is fronted by 100 m wide rock flats. Two fishing shacks are located at the northern end.

Summary: Three kilometres of rocky coast and three beaches only suitable for rock fishing.

529-538   RILEY PT to BLACK ROCK

No.	Beach	Unpatrolled Rating	Type	Length
529	Riley Pt (N 1)	3	R+rock flats	200 m
530	Riley Pt (N 2)	3	R+rock flats	500 m
531	Riley Pt (N 3)	3	R+rock flats	300 m
532	Riley Pt (N 4)	3	R+rock flats	500 m
533	Riley Pt (N 5)	3	R+rock flats	100 m
534	Riley Pt (N 6)	3	R+rock flats	100 m
535	Myponie Pt (N 1)	3	R+rock flats	150 m
536	Myponie Pt (N 2)	3	R+rock flats	300 m
537	Black Rock (N 1)	3	R+rock/sand flats	250 m
538	Black Rock (N 2)	3	R+rock/sand flats	200 m

At **Riley Point** the coast turns to face northwest and runs relatively straight for 6.5 km to Myponie Point, then another 2 km to Black Rock and finally another 4.5 km to Tickera Bay. The entire section is dominated by sloping bedrock, with the shoreline consisting of 10 to 20 m high bluffs and some steep gullies, most capped by stable longitudinal dunes, which have been cleared for farmland. In amongst the many crenulations along the shore are 10 small sand-cobble-boulder beaches (**529-538**), all fronted by 50 to 100 m wide rock flats, with some sand flats beyond the rocks north of Black Rock. A rough bluff-top vehicle track follows the first 9 km of shore, until it is washed out in a gully, with a second track reaching the northern shore from Tickera.

Several shacks are located in lee of some of the beaches, particularly those that provide clearer access through the rock flats for boat launching. There are approximately 15 shacks between Riley Point and Myponie Point, 6 shacks between Myponie and Black Rock, and 4 shacks at Black Rock.

Summary: While this section of coast is accessible along a rough vehicle track, it is primarily used by the shack owners for rock and boat fishing, and is generally unsuitable for safe swimming owing to the predominance of rocks.

539-541   LOCHMORE, TICKERA BAY

No.	Beach	Unpatrolled Rating	Type	Length
539	Lochmore	1	R+sand flats	2.5 km
540	Tickera Bay	1	R+tidal flats	1.2 km
541	Tickera Bay (N)	1	R+tidal flats	12.2 km

Tickera is the site of an old, now demolished wheat jetty, which used to extend several hundred metres across the tidal flats to reach deep water. Today the small settlement sits atop the 20 to 30 m high bluffs looking out across the gulf and only a few small boats lie moored off the beach.

Three low energy beaches front the town. On the southern side, accessible by a gravel road off the main Tickera Road, is a 2.5 km long, crenulate high tide sand beach (**539**), fronted by a mixture of rock and increasingly sand flats, which widen to 300 m in the north. The beach is backed by 20 m high bluffs, cut by a series of eight steep gullies, together with approximately ten shacks spread along the back of the beach. The main **Tickera Bay** beach (**540**) fronts the golf course and settlement, and consists of a 1.2 km long, relatively straight section of bluff-backed high tide beach, fronted by tidal flats up to 600 m wide. A car park has been built across the northern end to provide access for parking and boat launching from the beach. North of the car park is the northern beach (**541**), a 12 km long, very low energy beach fronted in places by 2.5 km wide tidal flats. The beach is initially backed by the bluffs and a gravel road, which after 5 km give way to a series of low beach-foredune ridges accessible by 4WD tracks and finally multiple recurved spits, backed by high tide salt flats.

Summary: This is the beginning of the very low energy northern section of Spencer Gulf, with low gradient tidal flats dominating the shore. There is no surf and calm conditions usually prevail along the shore, with swimming only possible toward high tide. Most fishers use boats to fish the deeper gulf, while crabbing from the beach is popular on beach 541.

542   WEBLING POINT

	Unpatrolled		
No.	Beach	Rating	Type	Length
542	Webling Pt	1	R+tidal flats	15 km

Webling Point is located 4 km west of Port Broughton and is a low, protruding inflection along a very low energy shore dominated by extensive tidal flats, ranging from 1 to 2.5 km in width. The shore itself consists of a crenulate high tide sand beach, backed for most of its length by a mixture of low beach-foredune ridges and recurved spits, then the east-west trending longitudinal dunes, now cleared for farmland. There is a gravel road to the coast 3 km south of the point and a 4WD track along much of the 15 km long 'beach', (**542**) but no development or facilities other than the farmland. This is another popular beach for crabbing.

543-547   PORT BROUGHTON

	Unpatrolled		
No.	Beach	Rating	Type	Length
543	Port Broughton (jetty)	1	R+tidal flats	1.2 km
544	Port Broughton	1	R+tidal flats	400 m
545	Port Broughton (N)	1	R+tidal flats	1 km
546	Munderoo Bay	1	R+tidal flats	3.5 km
547	Fisherman Bay	2	R+tidal flats/creek	800 m

Port Broughton is a protected small town and port in lee of Webling Point. The entire port and its five beaches are protected by up to 5 km of tidal flats and shoals. The town is a popular holiday destination and much of the 3 km of town foreshore is given over to holiday houses and caravan parks, together with a 400 m long jetty and 300 m long dredged channel for a boat ramp.

There are three west facing beaches fronting the town proper. The **jetty beach** (**543**) is 1.2 km long and extends south of the boat ramp, down past the jetty to the southern caravan park. It is fronted by 300 to 400 m wide tidal flats, then the deeper channel for Mundoora Arm. It is backed by the caravan park, a foreshore reserve and the town centre. On the north side of the **boat ramp** is beach **544**, a 400 m long, curving beach, entirely backed by beachfront houses. Four hundred metre wide tidal flats and then the channel front the beach. Immediately to the north is the 1 km long northern beach, a more crenulate high tide beach (**545**), also backed by beachfront houses, with 200 m wide tidal flats then the channel. It terminates at a low sandy foreland, beyond which is the 3.5 km long, relatively straight beach (**546**) that runs up to the entrance to Fisherman Bay. It is developed at either end, with most of the beach backed by low foredunes, while 200 to 400 m wide tidal flats front the beach.

Figure 4.102 *View of north facing Fisherman Bay beach, with the backing shacks and mangroves of the bay.*

At **Fisherman Bay**, the waves and tidal currents have formed an 800 m long, north facing spit, which is entirely occupied by houses and shacks. The beach (**547**) is fronted by a narrow sand flat, then the deeper bay tidal channel (Fig. 4.102). Fishers use this channel for launching boats and reaching the deeper gulf waters.

Swimming: All swimming is dependent on the high tides covering the tidal flats, except at Fisherman Bay where the deeper channel is close inshore, however beware of strong tidal currents and deep water in the channel.

Surfing: None.

Fishing: The jetty in Port Broughton is the most popular location, otherwise most fishers head out to the deeper channel and gulf waters.

Summary: A popular destination for fishers, many of whom have holiday homes and shacks at the port or bay.

UPPER SPENCER GULF DISTRICT
FISHERMAN BAY TO PLANK POINT
(Fig. 4.103)

Length of Coast:	273 km
Beaches:	54 beaches (549-602)
Surf Lifesaving Clubs:	Whyalla
Major towns:	Port Pirie, Augusta, Whyalla

The Upper Spencer Gulf District contains 273 km of very low energy shoreline that extends north of Fisherman Bay on the eastern Yorke Peninsula side and north of Plank Point on the western Eyre Peninsula side, up to the top of the gulf at Port Augusta. This is a region of decreasing wave height and increasing tide range, with the state's highest tide of 3 m at Port Augusta.

Regional Map 7: Upper Spencer Gulf

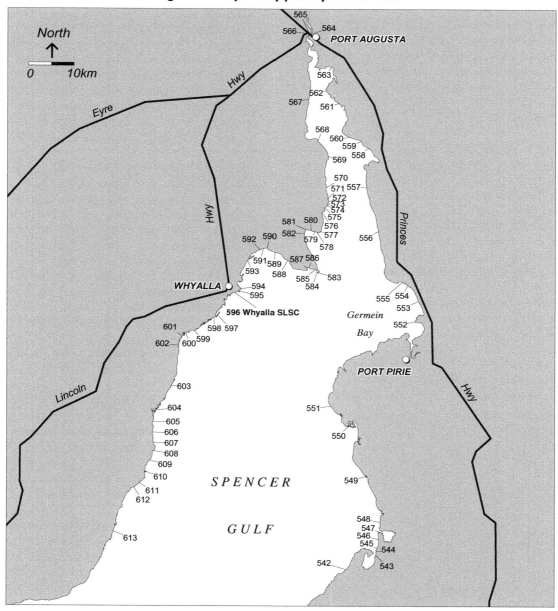

Figure 4.103 *Regional Map 7 - Upper Spencer Gulf*

Australian Beach Safety and Management Program

548-551   **FISHERMAN BAY to POINT JARROD**

No.	Beach	Rating	Type	Length
	Unpatrolled			
548	Fisherman Bay (N)	1	R+sand flats	3.5 km
549	Wood Point	1	R+sand flats	18 km
550	Beach 550	1	R+sand flats	5.8 km
551	Point Jarrod	1	R+sand flats	7 km

The coast north of Fisherman Bay becomes increasingly dominated by a very low gradient coastal tidal plain, backed by several kilometres of salt flats and low country, with very low energy high tide beaches fronted by tidal sand flats up to 4 km wide. Between Fisherman Bay and Deep Creek is 37 km of low sandy coast, followed by another 30 km of mangrove-lined coast round to Port Pirie. This section is only accessible via farm tracks in the south leading to the northern side of Fisherman Bay and by a gravel road, then 4WD track, out to Wood Point. The northern two beaches are essentially inaccessible by vehicle, except across the tidal flats. Apart from a few shacks on Fisherman Bay inlet and backing farmland, there is no development.

On the north side of Fisherman Bay inlet is a 3.5 km long, relatively straight, west facing sand beach (**548**), fronted by sand flats up to 2 km wide. A low foredune and farmland back the beach, which can only be accessed via the farm. At the southern end of the beach, facing into the 300 m wide inlet, are seven shacks.

The protruding, 18 km long **Wood Point** beach system (**549**) is a very low energy sand shoreline composed of up to 25 beach ridges, fronted by 3 km wide sand flats and backed by 2 to 5 km wide salt flats, with vehicle access to the point (Fig. 4.104). Beach **550** is a crenulate, 5.8 km long, low energy series of spits and beach ridges, backed by 5 km of salt flats. Isolated **Point Jarrod** (**551**) is another protruding, 7 km long, 200 to 300 m wide beach ridge plain, backed by up to 8 km of salt flats, including Deep Creek and fronted by 3km wide sand flats.

Figure 4.104 *Wood Point is a low, sandy foreland backed by up to 25 low beach ridges and fronted by two (or more) kilometre wide sand flats.*

Summary: This is an isolated and difficult to access section of coast, dominated by the very low gradient salt flats and sand flats, with the narrow beaches only active at high tide.

552-555   **GERMEIN BAY**

No.	Beach	Rating	Type	Length
	Unpatrolled			
552	Weeroona Island	1	R+tidal flats	1.6 km
553	Telowie Beach	1	R+sand flats	7.5 km
554	Port Germein	1	R+sand flats	1.7 km
555	Port Germein (N)	1	R+sand flats	1.4 km

Germein Bay is an open, west facing, 14 km wide bay that extends north of Port Pirie. Port Pirie is located on the banks of the Port Pirie River, a large tidal creek that drains into the southern end of the bay. Extensive mangrove woodlands dominate the southern Port Pirie shores, which extend north for 4 km to Weeroona Island. The western side of the island and the central-northern section of the bay are free of mangroves and dominated by low energy beaches, fronted by wide sand flats.

Weeroona Island beach (**552**) wraps around the western half of the 1 km wide island for 1.6 km, the eastern half being surrounded by mangroves. The island, while detached from the mainland at high tide, is linked by the tidal flats at low tide, with a 300 m long causeway to the mainland. The beach borders the southern, western and northern shore and generally consists of a narrow high tide beach, backed by 20 shacks along the southern shore and larger freehold buildings along the northern shore, all backed by grassy slopes rising to the 30 m high island crest. The beaches are fronted by a narrow strip of sand, then vegetated tidal flats which extend for hundreds of metres north and south, except on the western side where they are truncated by the tidal channel.

On the north side of the island, the mainland **Telowie Beach** (**553**) extends in a relatively straight direction for 7.5 km up to the small tidal creek on the south side of Port Germein. The highway runs 300 to 400 m east of the beach and there are several access tracks across the low backing beach ridges to the undeveloped beach. It is fronted by 1 km wide ridged sand flats.

On the north side of the creek is the main **Port Germein** beach (**554**), a 1.7 km long, straight, southwest facing beach, which terminates at the base of a 1 km long jetty. The small beachfront town spreads either side of the jetty. On the north side of the jetty, the northern beach (**555**) continues on for another 1.4 km, terminating in amongst a mangrove woodland. Both beaches are fronted by 900 m wide sand flats, covered by up to 40 low sand ridges.

Swimming: Because of the wide, shallow sand flats you can only swim toward high tide.

Surfing: None.

Fishing: It's a long walk, but the end of the Port Germein jetty is still the best location.

Summary: Four accessible, very low energy beaches with facilities at Port Germein.

556, 557   MOUNT MAMBRAY, MOUNT GULLETT

		Unpatrolled		
No.	Beach	Rating	Type	Length
556	Mount Mambray	1	R+sand flats	13.3 km
557	Mount Gullett	1	R+sand flats	8.4 km

Mount Mambray and Mount Gullett are two small, isolated, conical hills, 20 and 40 m high respectively, located on an otherwise flat, 2 km wide coastal plain, dominated by salt and sand flats. Mount Mambray lies on the southern side of Mambray Creek, an upland creek that also drains the salt flats. **Mount Mambray** beach (**556**) extends 13.3 km south of the mount and creek mouth, terminating at a mangrove woodland. The very low energy beach consists of a series of low beach ridges, fronted by 1500 m wide sand flats. The more crenulate 8.4 km long **Mount Gullett** beach (**557**) begins on the north side of the creek mouth and terminates in the mangroves of Yatala Harbour. The sand flats widen from 2 km to 4 km off the northern end. Both beaches can be accessed by vehicle tracks off the highway, which runs between 3 and 6 km to the east.

Summary: Two beaches occupying a 22 km long section of low energy coast dominated by sand flats, beach ridges and salt flats.

558-560   MOUNT GRAINGER, RED CLIFF POINT

		Unpatrolled		
No.	Beach	Rating	Type	Length
558	Mount Grainger	1	R+tidal flats	3.7 km
559	Mount Grainger (N)	1	R+tidal flats	550 m
560	Red Cliff Point	1	R+tidal flats	5.2 km

Between the mangrove-dominated Yatala Harbour and Red Cliff Point is a 9 km long section of southwest facing shore, dominated by the central 60 m high Mount Grainger, which contains three continuous low energy beaches fronted by wide tidal flats. The highway runs a few kilometres inland, with access to the two main beaches via a gravel road and farm tracks. There is also a track to the fishing landing in Chinamans Creek, in lee of Red Cliff Point.

The main **Mount Grainger** beach (**558**) emerges from the Yatala Harbour mangroves 3 km south of the mount and continues on around the base of the mount to a small tidal creek on the north side. A vehicle track backs the beach south of the mount. On the northern side of the mount, between two small tidal creeks, is a 550 m long beach ridge (**559**), which at high tide is surrounded by the creek and flooding salt flats. The **Red Cliff Point** beach (**560**) begins on the northern side of the second creek and runs for 5.2 km up to the low Red Cliff Point, where it merges with a disjointed series of mangroves and detached sand ridges. A vehicle track follows the northern half of this beach. Tidal flats up to 1 km wide front all three beaches.

Summary: A very low energy, little used section of coast, backed by grazing land, with a small fishing landing in lee of the northern point.

561-563   POINT PATERSON, PORT PATERSON

		Unpatrolled		
No.	Beach	Rating	Type	Length
561	Pt Paterson (S)	1	R+tidal flats	1.6 km
562	Pt Paterson	1	R+tidal flats	3.4 km
563	Port Paterson	1	R+tidal flats	1.4 km

Point Paterson is a very low, beach rimmed tidal flat that protrudes into the gulf, narrowing the width of the gulf to just 4 km. The high tide beach is crenulate, discontinuous and in places replaced with mangroves. It consists of a southern 1.6 km long section (**561**) and the 3.4 km section (**562**) that curves around the west facing point. Both are backed by high tide flats that flood on each tide and there is no vehicle access. They are fronted by several hundred metre wide intertidal flats.

In lee of the northern side of the point is **Port Paterson**, a largely mangrove-filled, shallow embayment. Along the northern, southwest facing shore is a 1.4 km long, low energy beach (**563**) fronted by 500 m wide tidal flats. It is backed by a mixture of beach-foredune ridges and an older dune transgression.

Summary: These are three little used beach-tidal flats, with vehicle access only possible to the 'port' beach.

564-566   PORT AUGUSTA

Unpatrolled			
No.	Beach	Rating	Type	Length
564	Port Augusta (E)	2	R+sand flats	250 m
565	Port Augusta (W 1)	2	R+sand flats	160 m
566	Port Augusta (W 2)	2	R+sand flats	150 m

On the long drive across southern Australia the highway rarely touches the coast. If you leave from the east coast, the first time you actually see the 'sea' is as you cross the bridge across the upper reaches of Spencer Gulf. The gulf narrows to 300 m at the **Port Augusta bridge** and what one sees is more a river than the sea. However if you do wish to cool off, there are three small sandy beaches either side of the bridge, with the main swimming area located immediately south of the eastern bridge approaches (Fig. 4.105). The eastern beach (**564**) lies at the foot of the main shopping area and is backed by a car park, park and dressing sheds. A fishing jetty runs out across the beach, with a floodlight for night fishing and swimming. The old dock area forms the southern boundary of the 250 m long beach and the old and new bridges the northern boundary.

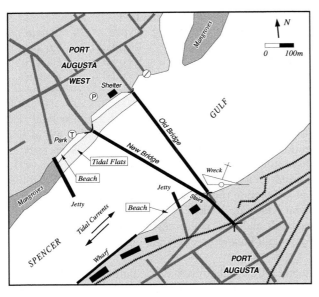

Figure 4.105 *Port Augusta is located at the tip of Spencer Gulf and on either side of the port are three small high tide beaches, each fronted by sand flats, then the deeper tidal channel of the gulf.*

On the western side of the gulf are two beaches, one on either side of the bridge. The northern (**565**) is bordered by the old and new bridges. It is 160 m long, backed by a large floodlit car park, with a shelter on the coarse shelly beach and a boat ramp on the other side of the old bridge. A few mangroves line the new bridge causeway. The southern beach (**566**) runs for 150 m from the new bridge

down to a jetty, with a grassy park behind the beach and a hotel next to the jetty.

These three sandy beaches and low tide sand flats are usually calm, with only the 2 to 3 m tide range causing changes in the shoreline. While there are no waves or rips, the water drops off steeply at high tide and off the sand flats at low tide, and there are strong tidal currents in the main 400 m wide channel.

Swimming: These are popular spots to cool off in summer. However stay close inshore, owing to the tidal currents off the beaches.

Surfing: None, ever.

Fishing: You can reach deep tidal water from the jetties, boundary docks or even with a good cast off the beach.

Summary: Three quiescent beaches. More a tidally driven swimming pool than an 'ocean' beach.

567   COMMISSARIAT POINT

Unpatrolled			
No.	Beach	Rating	Type	Length
567	Commissariat Pt	1	R+rock&sand flats	13 km

Commissariat Point is the easternmost point in a 13 km long section of bedrock-dominated coast, that extends from the mangrove-dominated shoreline of the upper gulf south to mangrove-filled Blanche Bay. The shoreline is backed by slopes that rise in places to 300 m within 3 km of the coast, with over 30 small, usually dry, creeks running down the slopes to the shoreline, resulting in a creek approximately every 400 m, each fronted by a small rocky delta. In amongst the bedrock and rocky deltas is a near continuous high tide, highly crenulate sand and gravel beach (**567**), also containing a scattering of mangroves. The shore is fronted by the rocky deltas, rock flats and, further out, sand flats, averaging 300 m in width.

The entire 13 km shoreline therefore consists of small, crenulate sandy bays, usually bordered by the small deltas and the mangroves. Immediately behind much of the shore are approximately 300 beachfront shacks, usually occurring in clusters, with a number fronted by small jetties (Fig. 4.106). The coast and shacks are accessible along a gravel military road that runs for 8 km and a network of vehicle tracks off the road to the shore and south to Blanche Bay. To the west of the military road is an army training reserve.

Figure 4.106 *The Commissariat Point shoreline contains irregular, crenulate, low energy beaches backed by shacks and gentle bedrock slopes, while fronted by sand and rock flats, interspersed with mangroves.*

Swimming: This is a relatively safe beach close inshore at high tide. Beware of the strong tidal current seaward of the sand flats.

Surfing: None.

Fishing: Most of the shack owners come to fish the gulf, usually from boats, but also from shore at high tide.

Summary: One of the more extensive shack settlements on the coast, located within 30 minutes of Port Augusta.

568-570   MANGROVE POINT to MONUMENT HILL

No.	Beach	Rating	Type	Length
		Unpatrolled		
568	Mangrove Pt	1	R+sand flats	900 m
569	Two Hummock Pt	1	R+sand flats	1.2 km
570	Monument Hill	2	R+rock-sand flats	3.7 km

Between Blanche Harbour and Fitzgerald Bay, 20 km to the south, is a protruding, north-south coastal range that peaks at 200 m high Monument Hill. This low energy, bedrock dominated section of coast faces into the upper gulf, which is only 5 km wide at Mangrove Point, widening to 9 km at Fitzgerald Bay. Access to the coast is restricted to a gravel military road and tracks paralleling the back of the shore and leading down to some of the beaches. Along the first 13 km of shore are three low energy beaches and sand flats.

Mangrove Point, as the name suggests, is dominated by mangroves to either end, with a 900 m long, more open beach (**568**) bordered by mangrove-fringed recurved spits to either end and fronted by sand flats. Four-wheel drive

tracks reach the point from the north and south. Three kilometres to the south is **Two Hummock Point**, backed by two 30 m high conical hills. A 1.2 km long beach (**569**) fronts the southern hummock and itself is fronted by 500 m wide sand flats. Mangroves extend for another 5 km south of the hummock, to beach **570**. This 3.7 km long section of coast consists of a highly crenulate high tide beach, broken up by nine small dry creeks and their rocky deltas, fringed by rock flats, then sand flats, extending on average 300 m into the gulf. A 4WD track parallels most of the beach, with the military road a few hundred metres inland.

Summary: These are three very low energy beaches backed by the military training area. They are accessible by 4WD and have no development.

571-576   DOUGLAS POINT, BEACH 576

No.	Beach	Rating	Type	Length
		Unpatrolled		
571	Douglas Pt (N 3)	2	R+rock-sand flats	150 m
572	Douglas Pt (N 2)	2	R+rock-sand flats	70 m
573	Douglas Pt (N 1)	2	R+rock-sand flats	300 m
574	Douglas Pt (S 1)	2	R+rock-sand flats	450 m
575	Douglas Pt (S 2)	2	R+rock-sand flats	1.05 km
576	Beach 576	1	R+sand flats	600 m

Douglas Point is the easternmost tip of the small coastal range. The 3 km of bedrock dominated shoreline to either side is highly crenulate and contains five beaches, separated by bedrock shores and rock flats, four of which contain a total of 50 shacks. The point is at the end of a scenic drive that originates 10 km north of Whyalla and runs as a sealed road to Lowly Point, then as a gravel road to Fitzgerald Bay and Douglas Point.

Beach **571** is a curving, 150 m long, east facing beach fronted by rock, then sand flats. Beach **572** is a 70 m long pocket of southeast facing sand, backed by four shacks. Beach **573** lies on the north side of Douglas Point and is a crenulate, 300 m long beach with rock and sand flats and two shacks near the point. On the south side of the point is 450 m long beach **574**, a more open sandy beach with 24 shacks occupying most of the back beach. Beach **575** curves around for just over 1 km and contains two clusters of shacks at either end. The sand flats widen to 500 m off the centre of the beach. Three hundred metres to the south is a second southeast facing beach (**576**), backed by low foredunes and fronted by 500 m wide sand flats.

Swimming: These are relatively safe beaches apart from the presence of the rock flats, with swimming best toward high tide.

Surfing: None.

Fishing: You can fish the rocky points at high tide, with most shack owners heading out into the gulf in boats.

Summary: The Whyalla water pipeline comes ashore immediately south of Douglas Point and the pipeline road provides good access from Whyalla for the shack owners.

577-582   **BACKY POINT, FITZGERALD BAY**

No.	Beach	Rating	Type	Length
	Unpatrolled			
577	Backy Point	2	rock+sand flats	150 m
578	Fitzgerald Bay (1)	1	R+rock-sand flats	300 m
579	Fitzgerald Bay (2)	1	R+rock-sand flats	400 m
580	Fitzgerald Bay (3)	1	R+rock-sand flats	800 m
581	Fitzgerald Bay (4)	1	R+rock-sand flats	1.2 km
582	Fitzgerald Bay (5)	1	R+rock-sand flats	1 km

At 50 m high Backy Point the bedrock coast turns 90⁰ and heads west into Fitzgerald Bay. **Backy Point** beach (**577**) is located in a small bay on the north side of the point. The bay has a rocky shoreline, fronted by 150 m wide sand flats, with ten shacks spread around the shore and a gravel road clipping the back of the bay. Once round the point, the south facing bay receives slightly higher waves which have pushed sediment into the northern shore of the bay, forming four beaches, with a fifth running down the western shore. A gravel road runs the length of the shore providing access to all beaches.

Beach **578** is a 300 m long, south facing high tide beach, with 200 m wide rock then sand flats. Its neighbour, beach **579**, is identical except 400 m in length. Beach **580** is contained within an 800 m long embayment and is fronted by 600 m wide ridged sand flats. Beach **581** is the largest beach system and is located in a deeper embayment, which it has partially filled with 20 beach ridges and is now fronted by 800 m wide sand flats. At the western end of the beach, the bedrock shore trends south and beach **582** runs along the bedrock shore for 1 km, before giving way to bedrock. It consists of a high tide beach, fronted by rock flats, then 300 m wide sand flats. Several shacks back the beach.

Summary: A slightly more exposed, south facing bay, with some low waves at high tide, but wide sand flats at low tide.

583-588   **LOWLY POINT to BLACK POINT**

No.	Beach	Rating	Type	Length
	Unpatrolled			
583	Lowly Pt (E)	2	R+rock-sand flats	80 m
584	Lowly (1)	1	R+sand flats	300 m
585	Lowly (2)	1	R+sand flats	450 m
586	Weeroome Bay (1)	1	R+sand flats	50 m
587	Weeroome Bay (2)	1	R+sand flats	300 m
588	Black Pt	1	R+sand flats	900 m

Lowly Point is a prominent projection on the west side of the gulf, north of which the gulf narrows to less than 15 km and south of which it widens initially into False Bay. Bedrock shoreline dominates from Fitzgerald Bay for 9 km down to the point, with just one small beach (**583**) located on the eastern tip of the 10 m high point. The beach is 80 m long and consists of a high tide sand beach, fronted by 50 m wide rock then sand flats. The Lowly Point lighthouse, built in 1883, is located right behind the beach. A sealed road runs out to the jetty at Port Bonython, located immediately west of the point.

On the western side of the point, more sand has accumulated and four beaches are located within the first 2.5 km. **Lowly Beach** (**584**) is 300 m long and faces southwest into False Bay. It is fronted by 100 m wide sand flats and backed by nine shacks. Its neighbouring beach **585** is 450 m long, with 150 m wide sand flats and is backed by an older dune transgression that has crossed the point, with 30 shacks located on the points to either side.

To the east of Stony Point is 500 m deep **Weeroome Bay**, which contains two adjoining pocket beaches. Beach **586** is a 50 m long pocket of sand backed by a 200 m long blowout, while beach **587** is 300 m long and also backed by dunes extending 400 m inland (Fig. 4.107). The two beaches share 200 m wide ridged sand flats. The Port Bonython jetty extends several hundred metres off the point, out into False Bay. The jetty delivers liquid gas from the Moomba gas field 660 km to the north. It is not open to the public.

Between Stony Point and Black Point is 3 km of 20 m high bluffs, fronted by rock flats. In the lee of **Black Point** is a 900 m long, south facing sandy beach (**588**), backed by several shacks and a camping reserve, and fronted by 300 m wide ridged sand flats, with rock flats to either end. This beach is accessible off the Lowly Point road.

Summary: Six relatively safe and accessible beaches, receiving some low waves and low surf during southerly conditions, and backed by more than 50 shacks.

Figure 4.107 *Weeroome Bay contains two beaches fronted by wide ridged sand flats and backed by two parabolic dune systems, aligned to the prevailing southerly winds.*

589-593   FALSE BAY

	Unpatrolled			
No.	Beach	Rating	Type	Length
589	False Bay (1)	1	R+sand flats	3.5 km
590	False Bay (2)	1	R+sand flats	300 m
591	False Bay (3)	1	R+sand flats	600 m
592	False Bay (4)	1	R+sand flats	600 m
593	False Bay (5)	1	R+sand/tidal flats	3.9 km

False Bay, as the name implies, is an 11 km wide, open, south facing bay, which is occupied entirely by very wide, low gradient tidal, sand and salt flats, in places up to 7 km wide. While much of the curving bay shore faces into the south through southeast, the tidal flats afford sufficient protection to maintain five very low energy beaches along the shore. Only the two boundary beaches are accessible by vehicles, the central three are surrounded by salt and tidal flats and tidal creeks.

Beach **589** is the most substantial beach system in the bay. It is 3.5 km long, faces south and is accessible off the Lowly and Black Point roads. The beach fronts up to 20 earlier beach ridges which widen to 900 m at its western end and is in turn fronted by 2 km wide ridged sand flats (Fig. 4.108).

Beaches **590, 591** and **592** are three small barrier islands occupying 1.5 km of shore. Each consists of a series of beach ridge-spits, backed by 3 to 4 km of salt flats, fronted by the wide sand flats and bordered by meandering tidal creeks. They are inaccessible by vehicle.

Beach **593** is a 3.9 km long sand spit that is slowly growing northward into the bay. The spit begins amongst the mangroves in lee of the Whyalla harbour training walls and runs north as a series of at times discontinuous spits,

to terminate at the first of a series of five small tidal creeks, which separate it from beach 592.

Summary: Five very low energy, little-used beaches dominated by the low energy, low gradient bay environment.

Figure 4.108 *False Bay is a very low gradient embayment containing multiple beach ridges that have partially filled the bay, fronted by 2 km wide sand flats, as shown here.*

594-596   WHYALLA

WHYALLA SLSC				
Patrols:	November to March, weekends and public holidays			
No.	Beach	Rating	Type	Length
594	Whyalla (jetty)	1	R+sand flats	70 m
595	Whyalla (boat ramp)	1	R+sand flats	80 m
596	Whyalla	1	R+sand flats	1.8 km

While **Whyalla** is renowned for its iron ore processing, it does have three beaches located between the main jetties and the southern mangroves (Fig. 4.109). Right under the main jetty is a 70 m long pocket of sand (**594**) that has accumulated between the jetty seawall and the backing low bluffs of Hummock Hill. On the southern side of the small 40 m high hill is a second 80 m long pocket of sand wedged between the hill and the 300 m long rock jetty that is part of the boat launching area, which includes the small beach (**595**). Both beaches are fronted by 100 to 200 m wide ridged sand flats, while there is a large car park behind the boat ramp beach.

On the southern side of the rock jetty is the main 1.8 km long, southeast facing Whyalla beach (**596**), site of the Whyalla Surf Life Saving Club and the focal point for water-based recreation at Whyalla. The beach is backed by a continuous seawall and road, with a large foreshore park, also site of a Foreshore Centre offering a range of facilities, with picnic and barbeque facilities backing the northern half, and a caravan park behind the southern half. The jetty forms the northern boundary, while the southern end grades into mangroves. The beach consists of a narrow high tide beach, which contains several shade shelters and 500 m to 1000 m wide sand flats, that are exposed at low tide.

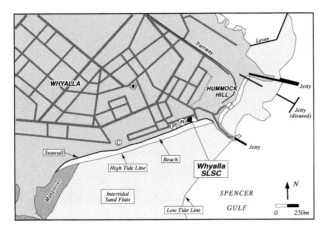

Figure 4.109 *The Whyalla shoreline has three beaches, with the Whyalla Surf Life Saving Club patrolling the main beach.*

Swimming: You should only swim at the main beach at high tide, as at mid to low tide it is a long walk to waist-deep water. So check the tide times before you go to the beach.

Surfing: None.

Fishing: It is possible to fish the beach at high tide, however the northern jetty is the most popular spot.

Summary: This is a wide, low gradient, usually calm beach, backed by a wide range of facilities.

597-602   **BEACH 597 to COWLEDS LANDING**

No.	Beach	Rating	Type	Length
Unpatrolled				
597	Beach 597	1	R+sand/tidal flats	300 m
598	Murrippi Beach	1	R+sand/tidal flats	1.2 km
599	Mount Young (1)	1	R+sand/tidal flats	1.3 km
600	Mount Young (2)	1	R+sand/tidal flats	1.5 km
601	Mount Young (3)	1	R+sand/tidal flats	500 m
602	Cowleds Landing	1	R+sand/tidal flats	100 m

The coast south of Whyalla, like the coast to the north, is a low energy, difficult to access and little used section of shore, extending from Whyalla for 100 km to Cowell at Franklin Harbour.

Immediately south of Whyalla's main beach is 5 km of mangrove shore, before the first of the low energy

beaches is reached. Beach **597** is a 300 m long section of open sand, located on a 500 m long beach ridge-spit, bordered and backed at either end by small tidal creeks and mangroves, and fronted by 1 km wide sand flats. Its neighbour, **Murrippi Beach** (**598**), is 1.2 km long and is accessible off the Eight Mile Road at its southern end, which is used for launching boats. It is also an official nude bathing beach. The beach is part of a series of beach ridge-recurved spits, backed by wide salt flats and fronted by sand flats, with fringing mangroves and tidal creeks.

After 3 km of mangroves are three more isolated beaches, only accessible by vehicles across the salt flats. The first (**599**) lies 3 km due south of 130 m high Mount Young. This 1.3 km long beach is backed by 2 km of salt flats and fronted by 1 km wide sand flats. Beach **600** is a curving, 1.5 km long beach of a similar nature, while beach **601** is 500 m long and faces east.

Two kilometres to the south is the shack settlement of Cowleds Landing, essentially a mangrove-dominated beach ridge (**602**) fronted by 500 m wide sand flats. To the south the mangroves extend unbroken, apart from small tidal creeks, for another 8 km.

Summary: Only beach 598 and Cowleds Landing are used on a regular basis, for launching boats at high tide to fish the deeper gulf waters. The other beaches are essentially inaccessible by vehicle.

EYRE PENINSULA DISTRICT
PLANK POINT TO WILSON BLUFF (WA BORDER)
(Figs. 4.103, 4.110, 4.121, 4.150, 4.164, 4.181)

Length of Coast:	1726 km
Beaches:	849 beaches (606-1454)
Surf Lifesaving Clubs:	none
Lifeguards:	none
Major towns:	Cowell, Arno Bay, Port Neill, Tumby Bay, Port Lincoln, Coffin Bay,
	Elliston, Venus Bay, Streaky Bay, Smoky Bay, Ceduna, Fowlers Bay

The Eyre Peninsula District is the largest of the state's coast protection districts, extending for 1726 km from Plank Point in Spencer Gulf to the SA-WA border at Wilson Bluff. It contains 849 beaches and a wide variety of coastal and beach systems, ranging from the protected gulf waters to some of the most exposed, high energy beaches in Australia, together with the largest coastal dune systems in southern Australia. While there is the large Port Lincoln and a number of small towns scattered along the eastern and southern sections of the peninsula, most of the coast and beaches are relatively undeveloped and little visited. Extreme care should be taken if surfing or swimming on any of the more exposed beaches, as strong, often permanent rips dominate and there is often no one around to provide assistance. There have been a number of unfortunate drownings on Eyre Peninsula beaches in recent years, most involving unwary visitors caught in rips on remote beaches.

Regional Map 8: Eastern Eyre Peninsula

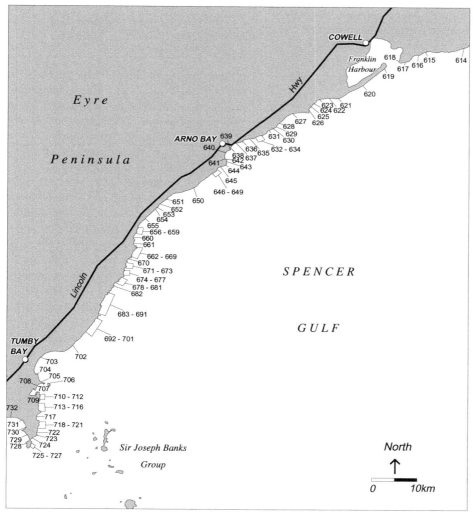

Figure 4.110 *Regional Map 8 – Eastern Eyre Peninsula*

603-612   BEACH 603, MURNINNIE, PLANK POINT

No.	Beach	Rating	Type	Length
Unpatrolled				
603	Beach 603	1	R+sand/tidal flats	3.5 km
604	Murninnie (N)	1	R+sand flats	1 km
605	Murninnie Beach	1	R+sand flats	2.5 km
606	Murninnie (S 1)	1	R+sand flats	2.8 km
607	Murninnie (S 2)	1	R+sand flats	1.7 km
608	Plank Point (N 3)	1	R+sand flats	2.5 km
609	Plank Point (N 2)	1	R+sand flats	3 km
610	Plank Point (N 1)	1	R+sand flats	4 km
611	Plank Point	1	R+sand flats	3.5 km
612	Plank Point (S)	1	R+sand flats	1 km

A discontinuous series of low beaches backed by 200 to 1000 m wide beach ridge plains, often terminating with recurved spits, extends for 20 km north of Plank Point. The ridges are backed in turn by a mixture of salt flats and truncated longitudinal dunes, and are fronted by sand flats between 800 and 1800 m wide, then the deeper seagrass meadows of the gulf. The beaches are generally low and narrow with moderate slopes, with seagrass debris along the swash line and a scattering of mangroves. The only formed access is a 15 km track from the highway to the shacks at Murninnie Beach, the remainder usually requiring a 4WD.

Beach **603** is a 3.5 km long beach ridge spit, backed by 500 m wide beach ridges and salt flats. It is separated from beach **604** by 4 km of mangroves, with mangroves dominating either end of the 1 km long beach, which also has small tidal creeks at either end.

Beach **605** marks the beginning of a more continuous sandy shoreline, with the individual beaches defined by more prominent sandy forelands. The **Murninnie** shacks are located on this beach, which, while extending for 2.5 km, has a scattering of mangroves along much of the beach. Beach **606** runs for 2.8 km between two sandy forelands, while beach **607** is 1.7 km long, terminating at a prominent sand foreland, with a vehicle track reaching the centre of the beach.

Beach **608** is a 2.5 km long, curving, east facing beach bordered by low sand forelands. Beach **609** is 3 km long, becoming more protected toward its southern end. Beach **610** lies in lee of Plank Point and is a protected, very low energy beach.

At Plank Point the coast turns and faces southeast. The Plank Point beach (**611**) runs for 3.5 km down to a sandy foreland, on the south side of which is 1 km long, south facing beach **612**. The point beach is accessed by a vehicle track to the northern end and contains the beach, backing beach ridges up to 400 m wide and ridged tidal flats up to 1 km wide.

Summary: This 20 km long section of low energy coast can only be accessed via a single gravel road to the shacks, by vehicle tracks along the beaches and in some cases across the salt flats to some of the other beaches. Apart from the shacks there is no development along this section.

Munyaroo Conservation Park
Established: 1977
Area: 12 334 ha
Beaches: 606 to 609
Coast length: 10 km

Munyaroo Conservation Park covers 10 km of coast between Murninnie Beach and Plank Point, including the backing coastal samphire flats and vegetated longitudinal dunes.

613   PLANK POINT – SHOALWATER POINT

No.	Beach	Rating	Type	Length
Unpatrolled				
613	Plank Pt-Shoalwater Pt	1	R+tidal flats	22.5 km

Four kilometres south of **Plank Point** the sand flats widen to between 1 and 2 km, further decreasing wave energy at the shoreline. The result is a 22.5 km section of very low energy shore dominated by the wide sand flats and in places tidal flats, with the shoreline consisting of an irregular, discontinuous series of approximately 20 multiple, northward migrating recurved spits, between one and a few kilometres in length. The spits are usually separated by a combination of mangroves and tidal creeks. In addition to very slowly moving sediment northward, they are also building the shoreline out across the sand flats, leaving behind a series of beach ridge-spit deposits on the backing samphire flats, with usually between six and ten relic beach ridge-spits deposited over a few hundred metres of backing flats. These ridges interfinger with southeast trending, stable longitudinal dunes, which were last active during the drier, windier, colder 'glacial' period 20 000 years ago and which were partially drowned by the sea level rise 6500 years ago.

The 'beaches' (**613**) along this section usually consist of a narrow (10 m wide) high tide beach face, backed by low, vegetated beach ridges, then usually a mixture of samphire flats and older ridges and are fronted by the wide, often vegetated sand and tidal flats. Small wind waves are only present at high tide during onshore wind conditions.

This section of coast is backed by some formed roads running between Midgee and Mitchellville, however only the end of the Mitchellville Road (6 km north of Shoalwater Point) provides direct public access, while there are two 4WD access tracks approximately 13 and 11 km north of the point.

Summary: A very low energy, difficult to access and little used section of coast.

| 614-618 | **SHOALWATER POINT to VICTORIA POINT** |

Unpatrolled				
No.	Beach	Rating	Type	Length
614	Shoalwater Pt	1	R+sand flats	16.5 km
615	Beach 615	1	R+sand flats	700 m
616	Lucky Bay	1	R+sand flats	3.1 km
617	Red Cliff	1	R+sand flats	3.8 km
618	Victoria Point	2	R+tidal flats	1.6 km

Shoalwater Point is a major landmark on the western gulf shore, as the trend of the coast swings toward the southwest, resulting in the gulf doubling in width from 45 km at the point, to 100 km south of Port Franklin. The wider gulf and more exposed shoreline results in higher wind waves and consequently higher energy gulf beaches, particularly south of the port.

The first 20 km of coast between Shoalwater Point and Victoria Point, the northern entrance to the port, is protected by wide sand flats, backed by generally higher beach-foredune ridges which have prograded the shoreline usually a few hundred metres. A mixture of longitudinal dunes and intervening swales, occupied by samphire flats, backs the ridges. Away from the shore the dunes and swales have been cleared for grazing and crops. Access to this section of coast is via the Shoalwater Point Road and the Lucky Bay Road, with a number of 4WD tracks also reaching the shore, particularly in the Shoalwater Point region. The only development along this coast is the large shack settlement at Lucky Bay.

Shoalwater Point beach (**614**) begins at the end of a 4 km long spit that runs north of the point. It continues on for 16.5 km down around the point and then for another 13 km to the west until it is cut by a small tidal creek, that drains a 500 m long lagoon. The beach is crenulate alongshore, fronted by sand flats up to 2 km wide and backed by the beach-foredune ridges and swales (Fig. 4.111), then the farmland.

Lucky Bay beach (**616**) begins on the western side of the tidal creek. The beach is part of a 3.1 km long protruding sand spit, that is slowly extending to the east. The spit averages about 100 m in width, with a causeway across the salt flats providing access to the 135 beachfront shacks, that occupy 1.2 km of the beach. Many of the shacks are at risk from undercutting by the dynamic spit. A worse site could not have been chosen to build 'permanent' shacks so close to the shore. However the Lucky Bay spit has also built out into the gulf across the

former sand flats, thereby providing good access to deep water, as the sand flats narrow to 50 m in places.

Figure 4.111 *View west of Shoalwater Point, showing the 2 km wide sand flats and backing scrub-covered low dunes and beach ridges, typical of this section of coast.*

Beach **615** is a 700 m long sand island, located 2 km east of Lucky Bay. It is bordered by small tidal creeks to either side and backed by salt flats. There is no dry access to the beach.

Red Cliff beach (**617**) is another 3.8 km long sand spit, that begins in lee of the low Red Cliff dunes, the northern entrance to Franklin Harbour and extends to the east as a crenulate, 100 to 200 m wide, low sand spit. It is backed by salt flats and fronted by the 3 km wide ebb tidal delta of Franklin Harbour. Access is via 4WD tracks along the backing 'mainland' shore and across the drier samphire flats to the spit.

Just past Red Cliff the shoreline turns north at **Point Victoria** and trends into Franklin Harbour as a crenulate, 1.6 km long sand shore against the point, then as a 600 m wide series of recurved spits that terminate in amongst the northern mangroves of the port. The beach (**618**) is fronted by 50 to 100 m wide sand flats, then the deeper seagrass covered waters of the bay. A 4WD track runs around the point.

Swimming: Lucky Bay is the best location, with relatively calm but deep water close to shore. Beware of the strong tidal currents off Red Cliff and Victoria Point.

Surfing: None.

Fishing: Again, Lucky Bay with its numerous shacks is the place most locals head for to fish the beach and gulf waters.

Summary: A transition section of coast, from the very low energy upper gulf, to the start of the more exposed lower gulf western shores.

619-623 OBSERVATION POINT to PORT GIBBON

No.	Beach	Rating	Type	Length
		Unpatrolled		
619	Observation Pt	2	R+tidal channel	1.8 km
620	Windmill	2	R	12.8 km
621	The Knob	2	R/LTT	800 m
622	Port Gibbon (E)	3	R+ outer TBR	3.25 km
623	Port Gibbon	2	R	1.4 km

Observation Point (619) is located at the end of a 1.8 km long spit, that has extended from Point Germein into the bay. The spit faces north into the deep tidal channel and can only be reached via the 13 km long Windmill Beach.

Windmill Beach (620) is the first higher energy beach, not fronted by sand flats, encountered on the western gulf shore. It faces southeast into the windier gulf and receives higher wind waves during onshore and southerly conditions. These maintain a relatively straight, 13 km long beach, usually reflective, with deeper water off the beach, then the seagrass meadows usually within 100 m of the shore. The beach is part of a Holocene barrier that built northward as a series of 15 recurved spits, that narrowed the entrance of Franklin Harbour from 7 to 2 km, then built seaward as a series of ten beach-foredune ridges, a few hundred metres in width. A vehicle track extends for 9 km along the back of the beach.

The Knob (621) lies at the southern end of Windmill Beach and is an 800 m long, curving cobble and sand beach, bordered in the north by rock flats, locally known as Flat Rocks and the low bedrock Knob in the south. There are three car parks behind the beach, which is usually a reflective beach with low waves and a narrow surf zone, developing an attached bar and wider surf during higher wave conditions. The northern end of the beach is backed by the first active dunes on the western gulf shore, the retreating dunes exposing older cobble beach ridges.

On the south side of The Knob is a relatively straight, 3 km long beach (**622**) that runs down toward a protruding bluff. The Port Gibbon Road runs along the back of the beach with several car parks at the beach and on top of the bluffs. The south facing beach receives periodic higher waves, which toward the Knob end can produce a series of small rips and an outer bar (Fig.4.112). Further down the beach, waves decrease and the bar is attached to the shore. On the south side of the bluff, the **Port Gibbon** beach (**623**) continues on for another 1.4 km to the port and small jetty. This beach is backed by 10 m high red bluffs composed of Pleistocene alluvium, with the small port settlement sitting on the bluffs overlooking the relatively sheltered beach, which is often covered with seagrass debris (Fig. 4.113).

Figure 4.112 *The beach south of The Knob (622) is one of the first on the western gulf shore to receive low ocean waves, which maintain a low to occasionally moderate energy surf zone and rip system, here highlighted by the concentration of seagrass debris.*

Figure 4.113 *The small settlement of Port Gibbon sits on the bluffs overlooking the remains of the old jetty and low energy beach.*

Swimming: These are relatively safe beaches under normal low wave conditions. However be aware of the strong tidal currents off Point Germein and Observation Point, and during higher waves, rips along the eastern Port Gibbon beach.

Surfing: This is the first spot for occasional surf on the western gulf shore.

Fishing: The deeper waters of Port Franklin and off some of the beaches, together with the Port Gibbon jetty, provide a range of shore-based fishing locations.

Summary: This is the beginning of the more exposed gulf shore and beaches, and is also the first section normally seen by travellers reaching the coast at Cowell and taking the detour out to Port Gibbon.

Franklin Harbour Conservation Park

Established:	1976
Area:	1333 ha
Beaches:	619-622
Coast length:	16 km

Franklin Harbour Conservation Park covers the mangrove-dominated southern shore of the port and includes Observation Point, Windmill Beach and The Knob. This takes in the Windmill recurved spit and beach-foredune ridge system, as well as the extensive backing mangrove systems, together with the interesting Knob cobble ridges and active dune system.

624-627   POINT GIBBON, MILLS BEACH

Unpatrolled				
No.	Beach	Rating	Type	Length
624	Point Gibbon (1)	2	R	1.1 km
625	Point Gibbon (2)	2	R	500 m
626	Point Gibbon (3)	2	R	1.4 km
627	Mills Beach	4	R+RBB	7.7 km

At **Port Gibbon** the red bluffs curve around to face southeast and run for 4 km out to the sand-draped Point Gibbon. Between the port and the point are three narrow, lower energy, bluff-dominated beaches. The gravel road running out along the top of the bluff provides good access to the second two beaches.

Beach **624** is a 1.1 km long beach lying at the base of the 15 m high bluffs, with access only down the bluffs. Beach **625** is the site of the Port Gibbon boat ramp, actually a vehicle access down the bluff to the beach, where the boats are launched off the beach into the usually calm waters. The 500 m long beach is backed by the bluffs and bordered by two slightly protruding bluffs. Beach **626** runs out to the point. The 1.4 km long beach is backed by bluffs for 600 m, which then give way to active sand dunes that move across the point from Mills Beach. The car park for both beaches is located 600 m from the point, with a 4WD required to reach the point and Mills Beach. The point itself and eastern end of Mills Beach are fronted by rock flats and rock reefs.

Mills Beach (**627**) is a south facing, more exposed sandy beach, which receives sufficient high waves along its eastern shore to have a double bar system, with up to 20 rips spaced every 150 m commonly cutting the outer bar. Even when waves are low the rip channels often persist. Toward the west the beach becomes more protected and reflective, eventually being fronted by rock flats along the westernmost 1 km of beach. The entire beach is backed by foredunes, grading into active transgressive

dunes along the eastern half, known as the Point Price sandhills. This is in fact the most extensive sand dune system on the western gulf shore, a function of its exposed, south facing orientation and higher gulf waves.

Swimming: The Point Gibbon beaches are usually calm and relatively safe, as is Mills Beach during low waves. However beware of the rips and rock reefs if any surf is running.

Surfing: Chance of a beach break at Mills during higher southerly waves.

Fishing: The Point Gibbon ramp is the focus for boat fishing, with the best beach fishing in the persistent rip holes along Mills Beach.

Summary: The boat ramp and Mills Beach are both relatively popular locations.

628-636   POVERTY BAY to RED BANKS

Unpatrolled				
No.	Beach	Rating	Type	Length
628	Poverty Bay (N 3)	3	R+reefs	850 m
629	Poverty Bay (N 2)	3	R+reefs	800 m
630	Poverty Bay (N 1)	4	LTT+reefs	750 m
631	Poverty Bay	3	R	1 km
632	Poverty Bay (S)	3	R+platform	900 m
633	Beach 633	3	R+platform	500 m
634	Beach 634	3	R	750 m
635	Red Banks (N)	3	R	900 m
636	Red Banks	3	R+reefs	100 m

South of Mills Beach, southeast facing, low bedrock bluffs dominate the coast for 10 km down to Red Banks. The bluffs not only form much of the shoreline, but also result in sections of intertidal rock platforms, reefs and a scattering of rocks along some of the beaches. There are nine sandy beaches along this section, all dominated or controlled by the bluffs and reefs, which generally lower waves to less than 0.5 m, resulting in steep reflective beaches. The Lincoln Highway runs between 2 and 3 km inland, and there is vehicle access to the coast at Poverty Bay, beach 634 and Red Banks, with 4WD tracks leading along the bluffs to the other beaches. Apart from the backing farmland there is no development or facilities.

There are three beaches lying 2 km north of Poverty Bay. Beach **628** curves round following the backing low bluffs, with reefs off the beach at either end, the southern reefs producing a right hand surf during higher swell. Beach **629** is 800 m long, with two protruding sections in lee of more prominent rock reefs, with a right hand break over the southern reef. It is backed by low bluffs, with the southern bluffs covered by small dunes. Beach **630** is 750 m long and fronted by slightly deeper reefs, which permits slightly higher waves, resulting in a small bar fronting the beach and a chance of beach rips during higher waves. The bluffs parallel the back of the beach, with low dunes spreading up and over the northern bluffs.

Poverty Bay (**631**) is a 1 km long, southeast facing sand beach, that has filled a small valley. Winditite Creek flows into the valley and breaks out across the middle of the beach following heavy rain. Offshore rock reefs lower waves to about 0.5 m, which surge against the usually steep beach. There is a car park on the southern bluffs.

South of Poverty Bay, the bluffs, rocks and reefs again dominate. Beach **632** is a slightly pointed, 900 m long strip of sand wedged in between the low bluffs and a rock platform and reef-strewn nearshore, which results in low waves and calm conditions at the shore. Beach **633** is similar in form and runs for 500 m down to more prominent protruding rock bluffs and reef. On the south side of the reef is beach **634**, a 750 m long sand beach, with a small central foreland attached to reefs. Low dunes extend up to 200 m in from the beach, into a small central blocked valley and up and over the backing bluffs.

Beach **635** is a 900 m strip of sand lying below the 10 m high bluffs, with usually low waves at the shore. It terminates at the start of an exposed rocky bluff section of shore which runs for 1.5 km down to Red Banks. At Red Banks there is a 100 m long sand beach (**636**) located at the base of the bluffs and fronted by reefs. The road terminates above the beach at a large car park. The car park provides good access to the southern end of beach 635, while a steep climb down the rocks is required to reach Red Banks beach.

Swimming: Poverty Bay and beach 634 offer the best beaches, being freer of reefs and rocks. Red Banks is also popular because of the good access and tidal pools amongst the red rocks. The rocks are part of a geological monument.

Surfing: During higher swell there are some right hand breaks along the northern reefs.

Fishing: These are all relatively popular with the locals and occasional holiday-makers, who come to fish the rocks and reefs.

Summary: An accessible, but still out of the way and little used section of coast.

Unpatrolled				
No.	Beach	Rating	Type	Length
637	Arno Bay (N 3)	2	R	2.6 km
638	Arno Bay (N 2)	2	R	800 m
639	Arno Bay (N 1)	2	R	200 m
640	Arno Bay	2	R	2.2 km
641	Arno Bay (S)	2	R+sand flats	700 m

Arno Bay is a small town located just off the highway, on the northern shore of the largely infilled bay of the same name. There is a causeway from the town across the usually dry samphire flats to the main beach (640), where the settlement consists of about 90 shacks lined up in two rows behind the northern half of the beach (Fig. 4.114). There is also a boat ramp, jetty, caravan park and foreshore reserve.

Figure 4.114 *Arno Bay jetty, caravan park and usually low energy beach.*

There are three other near continuous beaches running north of the main beach. The first (**637**) can be accessed in the north via the Red Banks road. This beach sweeps for 2.6 km from south of the bluff-top car park down to small protruding bluffs, with three beachfront shacks, lying just a few hundred metres east of the highway. Apart from the shacks there are no facilities. Beach **638** curves round on the south side of the bluffs for 800 m to a small sandy foreland, which is occupied by 200 m long beach **639**. A substantial and protected boat ramp has been built at the northern end of the beach and is backed by a large bitumen car park and access road. A protected boat ramp also exists at Arno Bay Creek at the southern end of the beach, but has limited use due to the shallow depth of the creek making it unnavigable by boats at low tide.

The main boat ramp also forms the northern boundary of the main **Arno Bay** beach (**640**), a 2.2 km long, curving beach that runs down past the shacks, caravan park and jetty, to a sandy 900 m long spit that terminates at the small inlet for the bay. This is the most accessible and by far most popular beach, particularly during the summer holiday periods.

On the south side of the inlet, beach **641** continues on for 700 m to a low sandy point. It is a protected, northeast facing beach fronted by 300 m wide sand flats, grading to intertidal rock flats off the point. The beach is backed by semi-stable dunes extending up to 200 m inland, with access only by 4WD.

Swimming: Arno Bay is a very popular summer and weekend destination and provides relatively safe swimming right along the beach, apart from the inlet.

Surfing: Generally too protected.

Fishing: The jetty and boat ramp breakwater are most popular, while the two boat ramps are used by boat fishers.

Summary: A small town offering a wide range of beachfront facilities for both holiday-makers and travellers.

642-649   **CAPE DRIVER, MOKAMI (N)**

No.	Beach	Rating	Type	Length
		Unpatrolled		
642	Cape Driver (N)	2	R+rock flats	180 m
643	Cape Driver	2	R+sand/rock flats	950 m
644	Cape Driver (S 1)	2	R+reefs	550 m
645	Cape Driver (S 2)	3	R-LTT	1.9 km
646	Cape Driver (S 3)	3	R+rocks/reefs	200 m
647	Mokami (N 3)	3	R+reefs	300 m
648	Mokami (N 2)	3	R+reefs	80 m
649	Mokami (N 1)	3	R/LTT+reefs	200 m

South of Arno Bay the coast trends to the southwest from Cape Driver toward Port Neill. The 30 km of shore consists of a mixture of usually low, Quaternary, fluviatile bluffs, capped by southeast trending longitudinal dunes, now cleared for farmland, with rocks, reefs and rock flats dominating much of the intertidal zone. The Lincoln Highway runs parallel to the coast about 3 km inland, with continuous farmland between the highway and coast, apart from public access at the Dutton River mouth. As a consequence, most of this section, while backed by 4WD farm tracks, is difficult to access and little used by the public. The first 5 km is occupied by eight bluff and reef-dominated, generally low energy beaches.

Cape Driver is a low, sand-capped sedimentary headland, fronted by extensive intertidal rock flats and shallow reefs. Two beaches lie between the cape and Arno Bay.

Beach **642** is a 180 m long, northeast facing high tide sand beach, fronted by 200 m wide rock flats, containing patches of sand. It is backed by unstable dunes that have blown over from beach 643. The main **Cape Driver beach (643)** is a curving, 950 m long sand beach that begins in lee of the cape and spirals up to the northern point. It is fronted by a mixture of rock and sand flats, with waves breaking out over the reefs and off the cape. It is backed by a low bluff, with a large shed located on the bluffs toward the southern end.

South of the cape the shore trends to the southwest and is dominated by the low sedimentary bluffs and fronting rock flats and reefs, with the beaches scattered between the bluffs and rocks. Beach **644** lies immediately south of the cape and is a 550 m long, protruding high tide beach, lying below the bluffs, and broken in places by projecting bluffs and rocks. Fifty metre wide rock flats, then reefs, front the beach, with the usually low waves breaking over the reefs 50 to 100 m offshore. Beach **645** is a 1.9 km long beach, bordered by low bluffs and reefs, but is predominantly a sand beach and surf zone, backed by a low foredune. Waves average just over 0.5 m and break across an attached bar, with a chance of beach rips during higher waves. There is 4WD access at either end of the beach.

Beaches 646 to 649 are four bluff and rock-dominated beaches occupying a 1.5 km section of the shore. Beach **646** lies immediately past the southern bluff of beach 645. It is a 200 m long, east facing strip of sand, backed by grassy bluffs and fronted by an irregular mixture of rock flats and reefs. Beach **647** is a 300 m long, narrow high tide beach, wedged in between steep, low bluffs and reefs and rock flats. Beach **648** is an 80 m long pocket of sand lying below the bluffs and is fronted by a more sandy seabed. Beach **649** is a 200 m long, more exposed sand beach, with low waves usually breaking across a narrow bar and rips forming against the boundary rocks and reefs during higher wave conditions.

Swimming: The most suitable is beach 645, which is generally free of reefs and rocks and has a relatively safe low surf during normal wave conditions.

Surfing: Only low beach breaks on the open beaches, with reef breaks off Cape Driver.

Fishing: There are numerous rock and reef holes right along this section.

Summary: A little used section of coast dominated by the low bluffs and fronting rocks, with a scattering of sand beaches.

650 MOKAMI

Unpatrolled			
No.	Beach	Rating	Type	Length
650	Mokami	3	R	10.8 km

The central section of coast between Arno Bay and Port Neill is occupied by a continuous, 10 km long, low energy sand beach (**650**). The beach is initially backed by low grassy and dune-covered bluffs, then a mixture of salt flats and longitudinal dunes, through the southern portion of which winds the usually dry Driver River, fronted by a small sand delta (Fig. 4.115). For 3 km south of the river mouth, a 500 to 600 m wide foredune ridge plain backs the beach, which terminates below 10 to 20 m high grassy bluffs. The beach is backed by farmland in the north, then the salt lakes, with the only public access along a rough road reserve which reaches the shore on the southern bluffs. Four wheel drive tracks run north through the foredunes to the river mouth.

Figure 4.115 *The usually dry Driver River reaches the coast at Mokami Beach.*

Mokami beach (650) faces southeast to south in places, but is protected from higher waves by near continuous offshore reefs, which lower waves at the shore to 0.5 m and less in places, with seagrass meadows usually lying just off the beach. Apart from the southern access there is no development and no facilities at the beach.

Summary: A partly accessible, though little used, lower energy sand beach.

651-655 BRATTEN CAIRN to DUTTON RIVER

Unpatrolled			
No.	Beach	Rating	Type	Length
651	Bratten Cairn (N)	4	R+platform	2.4 km
652	Bratten Cairn (S)	4	R+rocks/reef	500 m
653	Dutton Bay (1)	3	R	70 m
654	Dutton Bay (2)	3	R+rocks	90 m
655	Dutton River	2	R to R+sand flats	5.1 km

At the southern end of the longer Mokami beach, the coast is again dominated by low bluffs, which extend 9 km to the southwest in an essentially straight line to the Dutton River mouth. This generally open section of coast is called Dutton Bay. The coast is backed by cleared farmland, with the highway passing 2 km inland. Access is provided along the road reserve from Bratten Cairn and at the Dutton River mouth, with no development or facilities at the coast. Waves average less than 1 m and are lowered further at the shore by the extensive nearshore reefs.

Bratten Cairn is located at an inflection in the otherwise straight highway, leading off from which is a 1.5 km vegetated road reserve and vehicle track that runs straight out to the top of the low sedimentary bluffs. Immediately below the end of the track is the southern end of beach **651**, a crenulate, 2.4 km long high tide beach fronted by a continuous rock platform up to 50 m wide and backed by 10 to 15 m high, steep bluffs (Fig. 4.116). The beach terminates at the southern end of the road reserve, where the bluffs protrude across the platform. On the southern side is 500 m long beach **652**, a slightly more open sandy beach, backed by a low foredune and steep grassy bluff.

Figure 4.116 *The road reserve at Bratten Cairn is highlighted by the remnant native vegetation. It reaches the coast above the low energy high tide sand beach fronted by a rocky intertidal flat.*

Five hundred metres to the south is the pocket beach **653**, a 70 m long strip of sand located at the mouth of a very small, usually dry creek and bordered by bluffs, rocks and reefs. Another 500 m to the south is the similar beach **654**, 90 m long, with rocks dotting the sand beach and narrow surf zone.

The **Dutton River** is one of the larger creeks (or smaller rivers) on the Eyre Peninsula. It reaches the coast 6 km north of Port Neill, where, besides the 600 m wide, usually blocked, river mouth, it has built a small foredune ridge barrier on the south side of the mouth. The barrier is backed by an earlier seacliff, while a narrow, bluff-backed sand beach extends for 3 km to the north. In all, the beach (**655**) is 5.1 km long, grading from a low wave reflective beach north of the river, to a high tide beach fronted by 300 m wide sand flats from the river mouth to the south, the flats representing the Dutton River 'delta'.

Swimming: The southern Bratten Cairn beach and the Dutton River beach at high tide offer the safest and most rock-free swimming. Beware of the rocks and reefs on all the other beaches.

Surfing: No recognised surf, though there are many reefs which may break during periods of higher swell.

Fishing: Numerous spots for rock fishing along the northern beaches.

Summary: A partly accessible, though little used section of coast.

656-660   DUTTON RIVER (S), PORT NEILL (N)

No.	Beach	Rating	Type	Length
Unpatrolled				
656	Dutton R (S 1)	2	R+sand flats	200 m
657	Dutton R (S 2)	2	R	300 m
658	Dutton R (S 3)	2	R	150 m
659	Port Neill (N 2)	3	R/LTT	500 m
660	Port Neill (N 1)	3	R/LTT	600 m

South of Dutton River, ancient Proterozoic granites begin to dominate the coast for 120 km down to the tip of the peninsula at West Point, as well as the gulf islands. The granites produce an overall straight coast, but more resilient and with occasional higher headlands. In detail however, it also contains irregularities, resulting in numerous small headland-bound beaches. Waves average less than 1 m on the more exposed beaches, while calms often prevail on the more protected beaches. The first five beaches are all backed by bluffs and then farmland, with public access only to the southernmost beach.

Beach **656** lies 1 km south of the river mouth and is a curving, 200 m long, narrow sand beach, backed by 20 m high bluffs and protected by the southern extension of the river mouth sand flats, which extend 200 m off the beach. Beaches **657** and **658** are neighbouring isolated beaches, 300 m and 150 m long respectively. They are located below steep, 20 m high bluffs, face east and receive waves averaging about 0.5 m which maintain reflective beaches, grading into sand flats at the protected southern end of beach 658.

Beach **659** is a 500 m long, east facing beach backed by a low foredune and farmland, with its neighbouring beach **660** similar at 600 m in length. It has a small, usually dry creek crossing its southern end, with sand flats wedged between the creek mouth and the southern headland. A

gravel road from Port Neill runs to the southern end of the beach and up to a lookout on the 25 m high southern headland.

Summary: Five isolated, natural beaches, close to Port Neill, but all backed by farmland and dominated by usually low waves.

661-663   PORT NEILL

No.	Beach	Rating	Type	Length
Unpatrolled				
661	Port Neill	3	R/LTT	1 km
662	Cape Burr	3	R/LTT	300 m
663	Back Beach	4	LTT/TBR	350 m

Port Neill is a small beachfront town, located 2 km off the Lincoln Highway (Fig. 4.117). It lies in lee of Cape Burr, whose protection led to its establishment as a wheat port last century. The 200 m long jetty remains, while today the town caters to the residents, shack owners and the seasonal influx of holiday-makers.

Figure 4.117 *Back Beach (foreground) with the town of Port Neill, its jetty and lower energy main beach.*

Three beaches abut the town and cape. The 1 km long **Port Neill** beach (**661**) begins as a low energy, north facing beach in lee of the cape, fronted by 50 m wide sand flats. It curves round to face east at the northern end, where waves average less than 1 m and maintain a narrow attached bar. Fifty beachfront shacks line the beach either side of the jetty, with a caravan park on the cape and the town extending in behind the shacks. A deep water, all-weather boat ramp is located in lee of the southern point.

On the north side of the cape is 300 m long, east facing **Cape Burr** beach (**662**), which is bordered by granite rocks. The curving beach usually has a narrow attached bar and low waves. It can be accessed via the boat ramp car park. On the south side of the cape is the 350 m long, southeast facing **Back Beach** (**663**), which receives occasionally higher waves and usually has rip holes against the boundary rocks, as well as a central reef. It is backed by dunes up to 200 m wide, with the best access at the southern end.

Swimming: Best north of the jetty on the main beach, while the cape and Back beaches are relatively safe under low waves, but will have rips when waves exceed 1 m.

Surfing: Only during higher swell, with the Back Beach offering a chance of beach breaks.

Fishing: The jetty, cape rocks and Back Beach rip holes are all popular locations.

Summary: A quiet little town just off the highway, providing a range of sandy beaches, the jetty and a good boat ramp.

664-668   **GOLF COURSE**

Unpatrolled			
No.	Beach	Rating	Type	Length
664	Golf Course (1)	3	R/LTT	100 m
665	Golf Course (2)	3	R/LTT	160 m
666	Golf Course (3)	4	LTT	130 m
667	Golf Course (4)	3	R/LTT	300 m
668	Golf Course (5)	3	R/LTT	70 m

On the south side of Port Neill is a series of five small rock-bound beaches. They occupy 1.3 km of shore and are backed for the most part by the Port Neill golf course, then by a grassy foreshore reserve accessed via a gravel road that terminates 500 m south of the golf course. All the beaches face east into the gulf and are afforded varying degrees of protection from the boundary rocks and headlands, as well as two small islets lying 100 to 200 m offshore.

Beach **664** is a 100 m long pocket of sand backed by a large car park. This is a relatively safe beach and popular for both swimming and rock fishing. One hundred metres to the south is beach **665**, a 160 m long sand beach that protrudes seaward in lee of a rocky islet 100 m offshore. The islet and other reefs provide considerable protection and seagrass grows to the shore. Immediately to the south is beach **666**, a 130 m long pocket sand beach, bounded by protruding granite rocks, with a narrow bar and sandy seafloor fronting the beach. Just past the end of the golf course is beach **667**, a 300 m long beach, also protected by a rock islet lying 200 m offshore, but connected to the centre of the beach by a shallow rocky reef. Rocks and reefs dominate much of the beach with some surf over the reefs during higher swell. Beach **668** lies at the end of the golf course track and is a 70 m long pocket of northeast facing sand bordered by prominent rocks and reefs.

Swimming: These are five usually low wave beaches which are relatively safe under normal conditions. However watch for the rocks and reefs and the presence of rips when waves exceed 1 m.

Surfing: Chance of a reef break during periods of higher south swell.

Fishing: Excellent rock and beach fishing along the entire section of shore.

Summary: Five accessible and relatively popular beaches during the holiday periods.

669-672   **CARROW WELL**

Unpatrolled			
No.	Beach	Rating	Type	Length
669	Carrow Well (N)	3	R/LTT	300 m
670	Carrow Well	4	LTT	1.2 km
671	Carrow Well (S 1)	3	LTT	600 m
672	Carrow Well (S 2)	3	R/LTT	300 m

South of the Port Neill golf course is a low, protruding rocky point, with the Carrow Well beaches to either side. The golf course vehicle track terminates at the point and provides good access, while 1 km to the south is another access road, via the actual Carrow Well, which terminates in the dunes just behind the main beach.

The northern Carrow Well beach (**669**) is a 300 m long, slightly curving, east facing beach, afforded some protection by the southern point. It usually has a continuous attached bar and receives waves less than 1 m, however higher waves will generate up to three rips along the beach. On the south side of the point is the main **Carrow Well** beach (**670**), a 1.2 km long, southeast facing, more exposed beach. It not only receives slightly higher waves, but is also more exposed to the southerly winds which have blown active sand dunes up to 400 m inland, to heights of 25 m. It is bordered by low rocky points and when waves exceed 1 m, several rips form along the beach and against the rocks.

On the south side of the southern rocks are two more rock-bound beaches, with access only possible by foot or 4WD. Beach **671** is 600 m long, faces southeast and has a continuous sand bar, which during higher waves is cut by two to three rips. The southernmost beach (**672**) is a curving, 300 m long, east facing beach, protected by a more prominent southern headland. Waves are usually low and break across a narrow continuous bar.

Swimming: These beaches receive slightly higher waves and usually have a low surf breaking across a continuous bar, with rips forming when waves exceed 1 m.

Surfing: Usually only low beach breaks on the more exposed sections.

Fishing: Best off the rocks and into the occasional rip holes.

Summary: Four natural beaches with limited access, backed by dunes and then farmland.

673-679   **CAPE HARDY**

No.	Beach	Rating	Type	Length
		Unpatrolled		
673	Cape Hardy (N 5)	4	LTT	400 m
674	Cape Hardy (N 4)	4	LTT	100 m
675	Cape Hardy (N 3)	3	LTT	200 m
676	Cape Hardy (N 2)	3	LTT	120 m
677	Cape Hardy (N 1)	3	LTT	160 m
678	Cape Hardy (S 1)	4	LTT	600 m
679	Cape Hardy (S 2)	3	LTT	500 m

Cape Hardy is a 20 m high, dune-capped granite headland, located 7 km south of Port Neill and accessible via a gravel road that runs from the highway due east for 8 km to the cape. On either side of the cape are a series of low, headland-bound, white sandy beaches, all backed by low dunes and then farmland, with public access only via the Cape Hardy track and with no development or facilities. All the beaches tend to face east to southeast and usually receive low swell and wind waves less than 1 m high.

Beach **673** is a 400 m long, southeast facing sand beach bordered by low, granite boulder headlands. A vehicle track winds through the backing farmland to a car park, adjacent to the small creek that occasionally flows across the southern end of the beach. A continuous bar fronts the beach, with rips forming during higher waves.

Four hundred metres to the south is the 100 m long pocket beach **674**, which is bordered and backed by granite boulders, with the sand only forming the lower beach and narrow bar. This is a more hazardous beach when waves are breaking, owing to the boulders. Just to the south is beach **675**, a 200 m long sand beach, backed by a small foredune and bordered by rocky granite headlands. The track from Cape Hardy reaches the back of this beach.

On the north side of Cape Hardy are two small, more protected beaches. Beach **676** is a 120 m long pocket of sand, with some large rocks on the beach and seagrass close inshore, while its neighbouring beach **677** is 160 m long, with granite rocks and reefs extending off either end.

On the south side of the cape are two slightly longer beaches, both backed by narrow foredunes and farmland almost to the beach. Beach **678** is 600 m long, with some rocks toward each end and a continuous sand beach and bar in between. During higher waves a few rips form along the beach. A vehicle track follows the usually dry creek to a small car park at the southern end. Beach **679**, immediately to the south, is a curving, 500 m long beach, the northern half consisting of a sand beach and narrow bar, while the southern half is dominated by rocks and reefs. A small, usually dry creek, paralleled by a farm track, backs the centre of the beach.

Swimming: These are seven relatively safe beaches during normal low wave conditions, however watch the many rocks and reefs, and rips when waves exceed 1 m.

Surfing: Only a chance of a beach break on the more exposed beaches during higher swell.

Fishing: Mainly frequented by fishers who fish the rocks and occasional rip holes.

Summary: Seven accessible, though little used, natural beaches.

680-688   **KIANDRA to SHEEP HILL**

No.	Beach	Rating	Type	Length
		Unpatrolled		
680	Kiandra (1)	4	LTT	100 m
681	Kiandra (2)	4	LTT	1.1 km
682	Kiandra (3)	5	R+reef	200 m
683	Beach 683	3	LTT	300 m
684	Ponto Creek	3	LTT	400 m
685	Beach 685	3	LTT	120 m
686	Sheep Hill (1)	3	LTT	220 m
687	Sheep Hill (2)	3	LTT	850 m
688	Sheep Hill (3)	4	R+rocks	30 m

South of Cape Hardy the backing grassy slopes gradually steepen toward Lipson Cove, peaking at 120 m high Sheep Hill, 9 km south of the cape. The coast becomes increasingly dominated by the granite bedrock, with several small creeks carving valleys that have been partially filled with sand, to produce nine beaches between Kiandra and Sheep Hill. A gravel road runs approximately one kilometre inland of the coast, with a few access tracks, usually through farmland, out to some of the beaches.

Beaches 680 and 681 share an open, southeast facing embayment, with 100 m long beach **680** located toward the northern end. It is bordered by granite rocks and rock reefs. On the south side of the rocks, beach **681** runs to the south for 1.1 km to a more prominent 20 m high granite headland. The beach has a scattering of rock toward the northern end, but otherwise a continuous attached bar and usually low waves. A small creek flows through a steep gully to reach the southern end of the beach, to the north of which is a gravel road which runs out from the highway. On the southern side of the headland is a curving, 200 m long pocket beach (**682**), bordered by 300 m long granite headlands and near continuous rock reefs fronting the sandy high tide beach. While waves are low at the beach, the rocks and reefs are a hazard.

Sloping granite bedrock dominates the next 2 km of coast, with beach **683** also located in a 300 m deep and 200 m wide rocky bay. The 300 m long sandy beach is free of rock and usually has an attached bar, with a rip forming against the northern rocks during higher waves. A small creek drains across the centre of the beach, with access from the highway via a farm track.

Ponto Creek flows into the centre of a curving, 400 m long sand beach (**684**), bordered by sloping 20 m high granite headlands, with a continuous bar fronting the beach. A vehicle track reaches the beach from the main Sheep Hill beach. Five hundred metres to the south is beach **685**, a 120 m long pocket of sand bordered by granite rocks extending 50 m offshore. The vehicle track runs round the slopes behind the beach.

Sheep Hill is fronted by moderate slopes, which run down to three beaches. Beach **686** is a straight, 220 m long, exposed beach, with bordering granite rocks, a continuous bar and rips against the rocks when waves exceed 1 m. The vehicle track leads to a car park at the southern end. The main Sheep Hill beach (**687**) is a curving, 850 m long sand beach, backed by 100 m wide active sand dunes and a southern, usually dry 'lagoon'. It is bordered by granite rocks, which protrude eastward in the south to afford some additional protection. A continuous bar usually fronts the beach, with a chance of rips toward the northern end during higher waves. Beach **688** is a 30 m pocket of sand located in a small V-shaped bay on the southern headland. It is bordered and fronted by rocks and reefs and is unsuitable for swimming.

Swimming: This 7 km section of coast offers several relatively safe beaches during normal low wave conditions. Stay clear of the boundary rocks and the two rock-dominated beaches and watch for rips during higher wave conditions.

Surfing: Chance of beach breaks only when higher swell is running up the gulf.

Fishing: There are many excellent locations along both the rocks and beaches, with reasonable access if you know which tracks to follow.

Summary: A relatively accessible section of

undeveloped coast backed by farmland, but with local knowledge required to reach some of the beaches.

689-693   **ROGERS BEACH, LIPSON COVE**

No.	Beach	Rating	Type	Length
Unpatrolled				
689	Rogers Beach	3	LTT	300 m
690	Lipson Cove (N)	3	LTT	90 m
691	Lipson Cove	3	LTT	1.1 km
692	Lipson Cove (S 1)	3	LTT	100 m
693	Lipson Cove (S 2)	3	LTT	400 m

Lipson Cove is the site of an abandoned mine and the remains of its jetty lie submerged off the southern end of the beach. Today a gravel road runs out to the cove, which is a reasonably popular spot for swimming, beach and rock fishing, launching fishing boats and occasionally surfing.

Rogers Beach (**689**) is a 300 m long, headland-bound beach, located 500 m north of the cove, with access only via the backing farmland. It usually has a continuous attached bar, with a few rocks in the surf. Beach **690** is a curving, embayed, 90 m long pocket of sand immediately north of the cove, with an access track from the cove road to the southern point overlooking the beach. The point is used for rock fishing.

Lipson Cove beach (**691**) is 1.1 km long and initially faces east, curving around to face north in lee of Lipson Island. The small island is tied to the beach at low tide and is a conservation park. The beach usually has low waves breaking across a 50 m wide continuous bar (Fig. 4.118). Bare sand flats back the southern half of the beach and are used as a car park, with some low dunes to the north.

Figure 4.118 *Lipson Cove beach, with Lipson Island located off the southern end.*

Lipson Island Conservation Park	
Established:	1967
Area:	1 ha

Five hundred metres south of the cove are the two southern beaches, both of which can only be accessed via private farmland. Beach **692** is a 100 m long pocket of southeast facing sand, bordered by sloping granite rocks. On the southern side of the rocks is 400 m long beach **693**, which has a continuous attached bar and terminates in lee of a granite reef that protrudes 100 m seaward. A farm track leads to the point above the reef.

Swimming: Lipson Cove is by far the most accessible, safest and popular of these beaches and is used by locals and some travellers during the summer.

Surfing: Chance of usually low beach breaks at Lipson Cove.

Fishing: The rocks at the northern end of the cove are a popular location, with some fishing also off the beach.

Summary: Lipson Cove is one of the few publicly accessible beaches between Port Neill and Tumby Bay. It also has a beach access point suitable for boat launching and as a result is relatively popular.

694-701   **OSWALD**

No.	Beach	**Unpatrolled** Rating	Type	Length
694	Oswald (1)	4	R+reef	50 m
695	Oswald (2)	3	LTT	150 m
696	Oswald (3)	4	R+rocks	30 m
697	Oswald (4)	3	R/LTT	80 m
698	Oswald (5)	3	R/LTT	150 m
699	Oswald (6)	3	R+rocks	50 m
700	Oswald (7)	3	R/LTT	100 m
701	Oswald (8)	3	R/LTT	150 m

South of Lipson Cove is a 5 km long section of moderately steep coastal slopes, which peak at 70 m high Oswald trig station. Along the shore a series of small creeks have cut indentations into the slopes, which are occupied by eight small beaches. All are backed by the cleared sloping farmland, with no public access to the beaches.

Beach **694** is a narrow, 50 m long sliver of sand bordered by granite slopes, together with rocks and reefs off the beach. Beach **695** lies across the mouth of a small creek and is a 150 m long sand beach, with rocky coast to either end. Beach **696** lies 400 m to the south and is another sliver of sand, just 30 m long, with more sand off the beach than along the shore. Beach **697** is 100 m further to the south and is an 80 m long pocket of sand, backed by a very small creek and bordered by rocky coast. One hundred metres to the south is beach **698**, a 150 m long,

more open sand beach, with a narrow bar and sloping granite rocks to either end.

Four hundred metres to the south are the three southern beaches, each located in slight indentations in a relatively straight stretch of rocky shore. Beach **699** is a 50 m pocket of sand, with rocks and reef dominating the narrow surf zone. Beach **700** is a 100 m long strip of sand, while beach **701** is 150 m long, with a small creek crossing the centre of the beach and depositing rocks at the shore.

Swimming: These are eight usually low wave and relatively safe beaches, apart from those with rocks off the beach. If waves are breaking, rips will form in each of the small bays.

Surfing: Usually none, apart from a possible beach break on the longer beaches.

Fishing: There are good locations for rock and beach fishing the length of this section.

Summary: A sloping, rock-dominated section of coast, with essentially no formal public access.

702-708   **TUMBY BAY**

No.	Beach	**Unpatrolled** Rating	Type	Length
702	Salt Creek	2	R/LTT	7.5 km
703	Tumby Bay	2	R to sand flats	10 km
704	Tumby Inlet	2	R+sand flats	300 m
705	Tumby Pt (N)	2	R+rock/sand flats	2 km
706	Tumby Is	2	R+sand flats	350 m
707	Tumby Pt (S 1)	3	R+rock flats	350 m
708	Tumby Pt (S 2)	2	R+sand flats	2.4 km

Tumby Bay is one of the larger and more popular towns on the western shore of the gulf. The town spreads along the southern 2 km of a 10 km long beach, the location chosen to afford protection for shipping last century. Today it is a major service centre and holiday destination.

There are seven beaches in the vicinity of the town, two long beaches to the north and five low energy beaches to the south. Beach **702** extends for 7.5 km north of Salt Creek, one of the large creeks to reach the coast 10 km north of the town. The sand deposited by the creek today forms a 400 m wide sand delta and over time has contributed to the building out of a low, 2 km wide coastal plain. The beach curves around the coastal plain and consists of a low gradient white sand beach fronted by 50 m wide sand flats, then continuous seagrass meadows down to the delta. It is backed by a mixture of stable and active low sand dunes and flats, then salt flats and farmland, with vehicle access along road reserves in the north and centre.

Tumby Bay beach (**703**) begins on the west side of Salt Creek, where it faces south and curves around for 10 km to face east at Tumby Bay town. Low dunes, then beach ridges, back the north and central part of the beach, with the town built on stable beach-foredune ridges. The entire town is fronted by a foreshore reserve, with public access and facilities the length of the town shore. The beach is protected in the south by Tumby Island and sand flats gradually widen to the south, with the Tumby Bay jetty extending 300 m into the bay to reach deep water. Fishing boats are moored between the jetty and Tumby Inlet at the southern end of the beach. The inlet has been dredged to provide a deep water channel for the marina and boat ramp (Fig. 4.119).

Figure 4.119 *Tumby Inlet, at the southern end of Tumby Bay, is fronted by wide, shallow sand flats that have been dredged to provide access for the marina and boat ramp.*

Beach **704** lies on the south side of the inlet. It is an irregular, 300 m long, north facing, very low energy high tide beach, fronted by 10 to 100 m wide rock and sand flats, then the dredged channel. There is a car park toward the eastern end. Beach **705** begins amongst the low rocky bluffs that extend south of the inlet, the sand gradually increasing along this curving, 2 km long beach, which faces north in lee of Tumby Island, where it is backed by a 300 m wide series of foredune ridges. A newer subdivision and foreshore reserve back the northern bluff-dominated 1 km of the beach, with a road running the length of the beach.

Tumby Island is a 25 ha, low bedrock island lying 500 m off the tip of Tumby Point and connected to the point at low tide by curving sand and rock flats. The island was named by Matthew Flinders in 1802, after his native parish in England. On the northern side of the island is a 350 m long sand beach (**706**), fronted by 100 to 200 m wide rock and sand flats. The island is a conservation park and can only be reached by boat or on foot at low tide. The remainder of the island's shore has low rock bluffs.

Tumby Island Conservation Park	
Established:	1969
Area:	35 ha
Beach:	705
Coast length:	2 km

On the southern side of the tip of **Tumby Point** is a 350 m long high tide sand beach (**707**), fronted by irregular granite rock flats. The gravel road from Tumby Bay reaches the back of the beach, with 4WD tracks through the dune ridges to the eastern end of beach 705.

Most of the southern shore of the protruding Tumby Point consists of 2.4 km long beach **708**, a curving, southeast facing sand beach lying between the northern low rocks and the southern 400 m wide Second Creek inlet. Waves remain low, despite its orientation, with the low, wide beach fronted by 50 m wide sand flats, then seagrass meadows, with the sand flats widening to 300 m off the inlet. The southerly winds have, however, blown low sand dunes up to 500 m in from the beach. There is vehicle access to the northern end of the beach, with a 4WD required to reach the inlet.

Swimming: The Tumby Bay foreshore is the most accessible and popular location and usually has low waves. All the other beaches also have usually low waves, with the sand flats restricting swimming to higher tides.

Surfing:		Usually none, owing to the low waves.

Fishing:		There are a number of popular locations, beginning with the Tumby Bay jetties, Tumby Inlet and rock fishing on the point.

Summary: Tumby Bay is a major regional tourist centre and provides a wide range of attractions, accommodation and beaches for visitors.

709-716		**SECOND CREEK to RED CLIFF**

Unpatrolled				
No.	Beach	Rating	Type	Length
709	Second Ck	2	R+tidal flats	950 m
710	Cape Euler (N)	2	R+rock/LTT	900 m
711	Cape Euler	2	R+LTT	700 m
712	Trinity Haven	2	R+LTT	300 m
713	Trinity Haven (S)	2	R+LTT	1.2 km
714	Red Cliff (N 2)	2	R+LTT	100 m
715	Red Cliff (N 1)	2	R+LTT	1.3 km
716	Red Cliff	2	R+LTT	250 m

Between the Second Creek inlet and Cape Bolingbroke is 20 km of east facing, low energy shore, consisting of low bedrock points and generally curving beaches in between. The islands of the Sir Joseph Banks Group provide considerable protection from much of the ocean swell, with only low swell and wind waves reaching the shore. As a consequence, seagrass grows to within 50 m of most of the shore. A gravel road from Tumby Bay runs 2 to 3 km inland. This links with the Trinity Haven scenic drive, which begins at the inlet and follows the coast down to Red Cliff and further south to Massena Bay and Thuruna youth camp. Apart from two youth camps, a basic camping area at Red Cliff and a few farms, there are no facilities or development along the shore.

Beach **709** faces north into 400 m wide Second Creek inlet, which includes a southern 50 m wide tidal channel, then tidal flats. The irregular, 950 m long beach is fronted by 100 m wide sand flats and some rock flats, then the deeper channel. There is a large car park at the western end, with access to the beach suitable for 4WD boat launching.

Beach **710** extends for 900 m north of **Cape Euler**. The high tide sand beach is dominated by low tide rock flats which widen to either end and extend as rock reefs for 500 m off the cape. On the south side of the cape, beach **711** curves round for 700 m to more rock reefs at the southern end (Fig. 4.120). For the most part the beach consists of a reflective high tide beach, fronted by a 30 m wide sand terrace. A small car park is located toward the southern end. Beach **712** lies between two rock terraces and is a curving, 300 m long sand beach with a narrow sand terrace. The small Trinity Haven youth camp is located in lee of the southern rocks.

Figure 4.120 *View south of Cape Euler along beach 711.*

On the south side of the youth camp, beach **713** runs essentially due south for 1.2 km, as a clean, white sand beach, backed by 100 m wide active dunes and fronted by a 50 m wide, shallow sand terrace. Rocks again form the southern boundary. Beach **714** is a 100 m long pocket of sand located in amongst the rocks. On the south side of the rocks, beach **715** curves round for 1.3 km as a continuous sand beach, backed by low, 100 m wide, vegetated dunes and fronted by a continuous 50 m wide sand terrace. Toward the southern end is a beachfront

camping area with basic facilities. Rock reefs extending 150 m offshore from Red Cliff form the southern boundary, with beach **716** located in lee of these rocks, as a curving, 250 m long high tide sand beach fronted by the irregular rock flats.

Swimming: The sandy beaches all usually receive low wave to calm conditions and are relatively safe. Just beware of the rock flats that border most of the beaches.

Surfing: Usually none.

Fishing: Best at high tide off the points and in the inlet.

Summary: An accessible section of shore with boat launching access in the north and several usually quiet sandy beaches.

717-727   MASSENA BAY to POINT BOLINGBROKE

No.	Beach	Rating	Type	Length
Unpatrolled				
717	Massena Bay	3	R to R+LTT	2.2 km
718	Massena Bay (S 1)	3	R+rock flats	200 m
719	Massena Bay (S 2)	2	R+LTT	400 m
720	Red Point (N 3)	3	R+LTT+rocks	100 m
721	Red Point (N 2)	3	R	1.1 km
722	Red Point (N 1)	3	R+rock flats	900 m
723	Red Point (S 1)	3	R+rocks	90 m
724	Red Point (S 2)	3	R	1.5 km
725	Pt Bolingbroke (N 2)	2	R+LTT	300 m
726	Pt Bolingbroke (N 1)	2	R+LTT	80 m
727	Pt Bolingbroke (E)	3	R+rock flats	300 m

The coast south of Red Cliff consists of 9 km of low, bedrock-controlled coast terminating at 20 m high Cape Bolingbroke. It contains 11 sand beaches, all bordered by low red granite rocks and reefs. A gravel road extends as far as the Thuruna youth camp at the southern end of Massena Bay, with only farmland and scrub backing the beaches to the south.

Massena Bay is a curving, east facing, 2.2 km long sand beach (**717**), receiving sufficient low swell along the central-northern section to maintain a relatively steep, often cusped, reflective beach, with seagrass about 50 m offshore and backed by up to 200 m of active dunes. In the south it curves round in lee of the Thuruna headland and is more protected. The youth camp backs the southern corner and marks the end of the gravel road, which parallels the southern 1 km of beach.

The southern low Thuruna granite rocks extend 500 m into the gulf, with a small, 200 m long beach (**718**) located 300 m east of the youth camp in amongst the rocks. It is bordered and partly fronted by the rocks, with 4WD access to the rear of the beach.

On the south side of the rocks is a 400 m long, slightly curving, reflective sand beach (**719**), bordered by low granite rocks and reefs and backed by a low foredune then farmland. Just south of the southern rocks is 100 m long beach **720**, set in amongst low rocks to either end, with some rocks also outcropping along and off the beach. Beach **721** extends for 1.1 km to a more substantial rock reef, containing a small islet 300 m offshore. On the south side of this reef is 900 m long **Red Point beach** (**722**) which terminates at the point, a low bluff fronted by scattered rocks and reefs up to 300 m offshore. In addition, rocks and reefs lie off the beach resulting in a low energy shoreline with seagrass growing to the shore and farmland right behind the beach. On the southern side of Red Point is a 90 m pocket of south facing high tide sand (**723**) fronted by continuous rocks and reef.

Beach **724** lies 400 m south of Red Point and consists of a 1.5 km long, curving, more exposed sand beach, with some rocks and reef outcropping at either end. Occasional swell waves maintain a reflective beach, with narrow sand flats in the protected southern corner. Farmland backs the northern half of the beach, with scrub and 4WD tracks behind the southern half.

The southernmost tip of **Point Bolingbroke** consists of a 2 km long, 20 m high bedrock point, capped by low, vegetated, longitudinal dunes and joined to the mainland by a 300 m wide isthmus consisting of beaches 724 and 729. Along the eastern side of the point are three small beaches, all accessible by 4WD tracks around the point. Beach **725** is a 300 m long, southeast facing sand beach, bordered by rocks and fronted by a 50 m wide low tide terrace. Beach **726** is just 80 m long and bordered by rocks, including rock reefs extending 100 m off the northern end, with seagrass growing 50 m off the shore. Beach **727** is a cuspate, southeast facing, 300 m long sand beach, with rocks dominating each end and extending 300 m seaward as reefs, resulting in a low energy beach with seagrass growing to the shore.

Swimming: Massena Bay is the most accessible and popular beach, not that many people travel to this relatively isolated location. All the other beaches are relatively low energy and safe, except for near the numerous rocks and reefs.

Surfing: None, apart from a low shorebreak on the more exposed beaches.

Fishing: There are numerous locations along the rocky sections of shore, with much of the sandy shore being relatively shallow off the beaches.

Summary: Past Massena Bay this is a relatively isolated and difficult to access section of coast, used mainly by the farmers and locals.

728-734   PEAKE BAY

Unpatrolled			
No.	Beach	Rating	Type	Length
728	Pt Bolingbroke (W)	3	R+reef	400 m
729	Pt Bolingbroke (NW)	3	R+reef	3.5 km
730	Picnic Rocks (S)	2	R	2 km
731	Picnic Rocks (N)	2	R	1.6 km
732	Peake Bay	2	R	4 km
733	Peake Point (N 2)	2	R	1.2 km
734	Peake Point (N 1)	3	R+reef	800 m

Peake Bay is a south facing, 6 km wide bay, bounded by Point Bolingbroke and Peake Point. Between the two points is 17 km of predominantly sandy shoreline, with low granite headlands and reefs forming the boundaries of the seven beaches (728-734) within the bay. All the beaches receive waves averaging less than 0.5 m and calms often prevail. There is limited access to the beaches, with a gravel road off the Massena Bay road reaching the Picnic Rocks beaches and the central Peake Bay beach. The other beaches are backed by private farmland, except for those on Point Bolingbroke, which require 4WD access. There are no facilities at any of the beaches.

Beach **728** is located on the western side of Point Bolingbroke, approximately 500 m north of the point. It faces southwest into Peake Bay and consists of a 400 m long, curving sand beach, bordered by granite rocks and reefs. Stable, vegetated, longitudinal dunes back the beach and cap the point. Beach **729** is a 3.5 km long, curving, west to southwest facing beach, which consists of stretches of continuous sand bordered by sand and rock or reef patches, particularly in the protruding central section. The southern half forms the western side of the 300 m wide isthmus which links the point with the mainland, with stable transgressive dunes extending up to 400 m in from the beach. The northern half is backed by bluffs, a foredune and farmland.

Picnic Rocks form a low, subdued point in lee of granite reefs, accessible by a gravel road which leads to about ten shacks located in a cluster on the north side of the point. On either side of the point are two low energy beaches. The southern beach (**730**) is 2 km long, faces west and curves round in lee of bordering rock reefs. It is backed by a foredune up to 100 m wide, then farmland. The northern beach (**731**) is a southwest to south facing, 1.6 km long sand beach, bordered by rock reefs, with some stable transgressive dunes extending 300 m in from the beach, then backing farmland.

Regional Map 9: Lower Eyre Peninsula

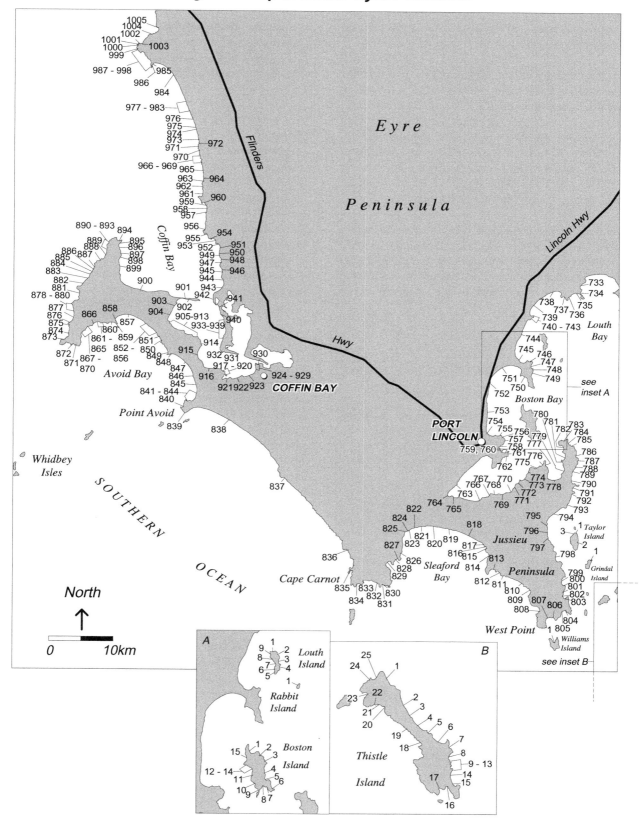

Figure 4.121 *Regional Map 9 – lower Eyre Peninsula*

Australian Beach Safety and Management Program

The main **Peake Bay** beach (**732**) is a 4 km long beach that faces south toward the wide bay mouth. It can be accessed via a gravel road at the western end, where a farm house is also located. The low energy beach is backed by a stable foredune, then continuous 300 to 500 m wide samphire and salt flats.

The western shore of the bay contains two smaller beaches north of Peake Point. The first (**733**) is a curving, east facing, 1.2 km long sand beach fronted by rock reefs, except for the southern end, which has 300 m of open sand beach backed by a low foredune and small dry lagoon. Beach **734** is an 800 m long, curving, east facing sand beach, also fronted by rock flats and reefs to either end. Together with beach **735** it forms a 1 km long sandy isthmus joining Peake Point to the mainland.

Swimming: These are all relatively safe beaches under normal low wave conditions, with the sandy beaches preferable to those fronted by rocks and reefs. Only the Picnic Point and main Peake Bay beaches are used by the general public. A fatal shark attack occurred in the shallows of the bay in the 1980s.

Surfing: None.

Fishing: Excellent fishing off the many rocks and reefs along the shore.

Summary: Seven rarely visited beaches offering usually calm conditions but no facilities.

735-749   LOUTH BAY

No.	Beach	Rating	Type	Length
		Unpatrolled		
735	Peake Pt (W 1)	2	R	1.8 km
736	Peake Pt (W 2)	3	R+reef	450 m
737	Beach 737	3	R+platform	600 m
738	Louth Bay (N)	2	R+LTT	3 km
739	Louth Bay	2	R+LTT/SF	4.5 km
740	Louth Bay (jetty)	2	R+reef	20 m
741	Point Warna (1)	3	R+reef	30 m
742	Point Warna (2)	2	R+LTT	100 m
743	Point Warna (3)	2	R+LTT	150 m
744	Poonindie	2	R+LTT	5.3 km
745	Tod River	1	R+sand flats	4 km
746	Pt Boston (N 4)	2	R+LTT	400 m
747	Pt Boston (N 3)	2	R+LTT	800 m
748	Pt Boston (N 2)	2	R+LTT	100 m
749	Pt Boston (N 1)	3	R+LTT/reefs	600 m

Louth Bay is an open, southeast facing bay bordered by the low protruding granite headlands of Peake Point in the north and Boston Point 16 km to the south. The bay contains 30 km of shoreline, including 15 beaches. The small settlement of Louth Bay is located on Point Warna in the centre of the bay and provides the only formal access to the shore. It consists of about sixty houses, a golf course and tennis courts. Louth and Rabbit Islands protect the southern end of the bay and, together with the headlands and the bay's location, contribute to a low energy bay, with waves usually less than 0.3 m high and the beaches usually fronted by shallow sand bars and sand flats.

The northern end of the bay is dominated by the low **Peake Point** and adjacent low headlands, which contain three beaches. Beach **735** extends for 1.8 km west of Peake Point and consists of a curving, southwest to south facing sandy beach backed by some unstable dunes that have blown up to 400 m inland, some reaching the back of beach 734. Cleared farmland backs the dunes. Immediately west is beach **736**, a 450 m long, southeast facing sand beach, fronted by near continuous rock reefs and backed by a narrow dune strip then farmland. Beach **737** is a narrow, curving, 600 m long high tide sand beach fronted by continuous 50 m wide rock flats and backed by a strip of vegetated slope, then farmland.

The northern Louth Bay beach (**738**) begins in lee of the westernmost granite headland and sweeps around for 3 km, facing initially west, then south and finally southeast. It is backed by a foredune and salt flats and terminates at some low bluffs and fronting rock reefs. A creek flows out of the salt flats and across the southern end of the beach following heavy rain. The main **Louth Bay** beach (**739**) begins on the south side of the bluffs and runs for 4.5 km down to Warna Point. It faces southeast, swinging round to face north in lee of the point. The settlement of Louth Bay overlooks the southern end of the beach and provides access for boat launching. The beach is composed of fine sand, with a firm, low gradient beach slope, fronted by a 50 m wide bar, with seagrass meadows growing 100 m off the shore.

Warna Point consists of low, 5 m high bluffs fronted by 50 m wide rock flats. On the north side of the point next to the jetty is a 20 m pocket of sand (**740**) fronted by rock flats and reef. On the eastern side are three beaches, the first (**741**) is a 30 m pocket of sand fronted by patchy rock reef, while on the southeast side of the point are two adjoining sand beaches. Beach **742** is a 100 m long sand beach bordered by low bluffs, rocks and a car park, while beach **743** is similar, at 150 m in length.

South of Warna Point is a 9 km stretch of predominantly east facing sand beach, divided in two by the Tod River mouth. Between the point and river is beach **744**, a low energy, 5.3 km long beach, backed by twelve low beach-foredune ridges, with seagrass growing to the shoreline. The beach bulges seaward in lee of Louth Island and toward the river mouth the sand flats widen to 400 m. There is vehicle access to the northern end and centre of the beach, but no facilities. On the south side of the 100 m wide river mouth is a 4 km long high tide beach (**745**), fronted by sand flats which widen to 700 m in the south, where the beach swings to face north in lee of the Point Boston headland.

Point Boston is the southern tip of a 4.5 km long, up to 50 m high, 1 to 2 km wide granite ridge that forms the boundary between Louth Bay and Boston Bay. Beaches 745 and 750 form an isthmus to connect the ridge to the mainland. Three narrow beaches are located along the base of the bedrock slopes. On the north side is north facing beach **746**, a low energy, curving beach with seagrass growing to the shore. It is backed by a low foredune in the west, with low bluffs backing and encroaching on the eastern half. Seagrass grows to within 30 m of the shore. On the more exposed eastern side of the point is a curving, 800 m long sand beach (**747**), which is bordered by 10 to 20 m high, sloping headlands, with a solitary house at the southern end. Halfway down the eastern side is beach **748**, a 100 m long pocket of sand bordered by grassy headlands. Two hundred metres to the south is beach **749**, a 600 m long, east facing, narrow high tide sand beach fronted by sand then a near continuous rock flat.

Swimming: All beaches usually have low waves and are relatively safe, apart from those fronted by rocks and reefs. The main Louth Bay beach is the most popular swimming location.

Surfing: None.

Fishing: The Louth Bay jetty is the most popular spot, with much of the shore fronted by shallow sand flats and seagrass meadows.

Summary: A 30 km section of low energy coast centred on the quiet settlement of Louth Bay.

750-753   **BOSTON BAY (N)**

No.	Beach	Rating	Type	Length
	**Unpatrolled**			
750	Pt Boston (W)	2	R+sand flats	1.2 km
751	North Shields	2	R+sand flats	6.5 km
752	Beach 752	2	R+cobbles	200 m
753	Port Lincoln (N)	2	R+cobbles	4 km

Boston Bay is an open, although protected, southeast facing bay, bordered by Point Boston in the north and Kirton Point in the south. Seven kilometre long Boston Island and Cape Donington afford protection to most of the southern half of the bay, with only the northern North Shields area exposed to some low swell, as well as wind

waves. The bay is 9 km wide at the entrance, with 23 km of shoreline. Port Lincoln is located on the southern shores of the bay, where the protection noted above, plus its northerly aspect, have permitted the development of the town's port facilities.

The northern section of the bay between Point Boston and Boston Heights contains the solitary settlement of North Shields, as well as a string of houses and farms following the highway north from Port Lincoln. The highway hugs the shore to North Shields, beyond which the shore curves round to the east and become more difficult to access.

Beach **750** is located 2 km northwest of Point Boston and faces south into the protected section of the bay. It does receive some swell between Boston Island and the point, however low wind waves tend to dominate the shore. The beach is 1.2 km long and bordered by sloping granite bluffs, with some rocks outcropping in the centre, to which there is 4WD access. It has a low gradient beach and 100 m wide sand flats (Fig. 4.122).

Figure 4.122 *North Boston Bay contains low energy beaches fronted by wide, ridged sand flats. The backing beach helps tie Point Boston to the mainland.*

North Shields beach (**751**) is a curving, 6.5 km long sand beach, that begins in the east against bedrock slopes that gradually rise to 50 m. Here the beach is fronted by 400 m wide sand flats containing up to twelve low ridges. The beach gradually trends west, then southwest toward the small North Shields settlement and jetty. The sand flats narrow to 100 m at the jetty. Most of the beach is backed by a foredune, with some dune activity in the north, then 1 km wide salt flats. At the settlement the beach is eroding into the dunes and low bluffs south of the jetty. As a consequence, a rock wall has been constructed to protect the road. A small creek crosses the beach just south of the jetty, south of which is a beachfront caravan park.

One kilometre south of North Shields, 100 m high dissected slopes back the shore down to Port Lincoln and the beach not only narrows but becomes dominated by rock debris from the eroding slopes.

Beach **752** is a 200 m long, east facing sand and cobble beach lying 2 km south of North Shields. It is backed by a gravel access road, then the highway and fronted by a mixture of rocks, rock debris and narrow sand flats. To the south is an irregular, 4 km long, narrow beach (**753**) that extends down to the northern Port Lincoln boat ramp. This beach is also located below the highway and consists of a mixture of pockets of sand, largely fronted by 50 to 100 m wide rock flats and patches of sand flats.

Swimming: North Shields is the best location, with high tide offering more water over the low gradient sand flats.

Surfing: None.

Fishing: Best off the North Shields jetty and some of the rock outcrops below the highway.

Summary: A relatively protected, southerly facing section of shore dominated by the highway in the west and the curving North Shields beach in the north.

754   PORT LINCOLN (TOWN BEACH)

Unpatrolled			
No.	Beach	Rating	Type	Length
754	Town (Port Lincoln)	1	R+sand flats	2.5 km

Port Lincoln is the largest town on the Eyre Peninsula and a major commercial and service centre, as well as the base for a large fishing fleet that roams the Southern Ocean. The town spreads over Kirton Point and the rising slopes to the west. At the base of the main shopping area is the curving, 2.5 km long Town Beach. The Lincoln Highway parallels the northern half of the beach, with streets and the town centre backing the middle and the port facilities for the bulk grain terminal forming the southern boundary. A boat jetty, with a tidal swimming pool, lies off the centre of the beach.

Port Lincoln beach (**754**) initially faces east and swings round in the south to face north into Boston Bay, an arm of Spencer Gulf. The bay is usually calm, with low wind waves only accompanying strong north through easterly winds. The northern part of the beach is narrow, gravelly, backed by 20 m high bluffs and fronted by 100 m wide gravel and sand flats. The beach becomes sandy and widens in the town section, with a seawall, pine trees and a park backing the beach. Shallow sand flats front the beach, increasing to 100 m wide at the eastern end (Fig. 4.123).

Swimming: This is a relatively safe beach, with usually calm conditions and a shallow bar attached to the beach and exposed at low tide. The best swimming off the beach

is at mid to high tide, otherwise at any time in the tidal pool off the jetty.

Surfing: None.

Fishing: The jetty is the best spot to fish the seagrass meadows.

Summary: This beach presents a pleasant backdrop to Port Lincoln. It is a popular and relatively safe aquatic recreational area, backed by all the facilities of Port Lincoln.

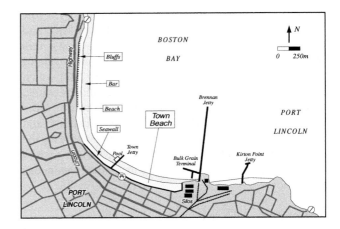

Figure 4.123 *Port Lincoln's Town Beach parallels the centre of town and terminates at the port.*

755-762   KIRTON POINT to BILLY LIGHTS POINT

Unpatrolled			
No.	Beach	Rating	Type	Length
755	Kirton Pt	2	R+sand flats	50 m
756	Shelly Beach	2	R+sand flats	100 m
757	Kirton Pt (caravan park)	2	R+sand flats	100 m
758	Porter Bay (N)	1	R+tidal flats	120 m
759	Porter Bay (S)	1	R+tidal flats	300 m
760	Billy Lights Pt (W)	1	R+cobble/tidal flats	300 m
761	Billy Lights Pt (S)	2	R+sand flats	600 m
762	Billy Lights Pt jetty	2	R+sand flats	300 m

The eastern half of Port Lincoln spreads over the slopes between Kirton Point and Porter Bay. The shoreline is dominated by low bluffs at the base of the slopes, with a few narrow beaches and sand flats at the base of the bluffs. All the beaches are protected from ocean swell and receive only low wind waves. The Parnkalla walking track follows the shore, with a variety of reserves, roads and commercial activities on the backing slopes.

The **Kirton Point** beach (**755**) was separated from the main Port Lincoln beach by the construction of the grain loading facilities. It now remains as an impounded 50 m long pocket of high tide sand, bordered by the large grain facility and jetty to the west and the low, rocky point to the east. The walking track commences at a car park above the beach. Immediately east of the point is 100 m long **Shelly Beach (756)**, a sand, rock and shell high tide beach, fronted by a mixture of rock and sand flats and backed by a foreshore reserve and car park. There is a boat ramp and more parking 200 m east of the beach. Further round the point and immediately below the large caravan park is an east facing, 100 m long beach **(757)** bordered by two 50 m long groynes and a fishing jetty. A large caravan park backs the northern end and the low energy beach is crossed by a slipway and boat ramp.

Porter Bay was converted from tidal flats to a large marina in the late 1980s. At the same time, the low energy beach and tidal flats that fronted the bay were formalised with training walls and terminal groynes, as well as being nourished with sand, while the creek was dredged to form the channel for the marina. The northern beach **(758)** is 120 m long, faces east and is fronted by 100 m wide tidal flats, and backed by a reserve and houses. The southern beach **(759)** is 300 m long and backed by a narrow reserve and beachfront houses, with car parking at either end. The southern shores of the bay extend eastward toward Billy Lights Point, with beach **760** occupying 300 m of the north facing shoreline. This is an irregular, low energy sand and cobble beach, fronted by 200 m wide tidal flats. A groyne and double boat ramp are located at the eastern end, with the beach building out against the groyne.

Beach **761** is a 600 m long sand beach that extends south of the low rocks of **Billy Lights Point** down to a low sand and rock foreland. The foreland is occupied by a large slipway and commercial boating facility (Fig. 4.124). On the south side of this facility is 300 m long beach **762**, wedged in between the slipway and low, rocky bluffs, 200 m south of which is a 400 m long jetty. The beach is also used for storing fish pens, which are located in the bay off the point, as part of a fish farm.

Figure 4.124 *Billy Lights Point in Port Lincoln houses a slipway and grain terminal.*

Swimming: None of these beaches are suitable for swimming, owing to either the presence of rocks, shallow

sand, rock flats and tidal flats or the boating and shipping activities.

Surfing: None.

Fishing: The Porter Bay training walls provide the best access to deeper water, as most of the large commercial jetties are not open to the public, while most of the shoreline is either too shallow or rocky.

Summary: A heavily developed section of shore with low energy, irregular sand and cobble beaches fronted by low sand and tidal flats, and crossed by some large jetties and several boat ramps.

Lincoln National Park

Established: 1941
Area: 29 060 ha
Beaches: 62 beaches (764-825)
Coast length: 109 km

Lincoln National Park occupies the entire 29 km^2 of Jussieu Peninsula, a large dune-covered, bedrock based peninsula which forms the southern tip of the Eyre Peninsula. The park consists of densely vegetated granite bedrock that rises up to 200 m high and which, along the southern shore, is blanketed by massive Pleistocene and Holocene dune deposits covering 19 km^2 and reaching heights of 200 m. The shoreline contains three environments: 34 km of low energy, north facing, predominantly sand-sandflat shore in Port Lincoln Proper, including Spalding Cove; the 35 km long, more exposed and bedrock-dominated, east facing gulf section between Cape Donington and Cape Catastrophe; and the 40 km long, exposed, high energy, southwest to south facing coast between Cape Catastrophe and Sleaford Mere.

Further information:
District Ranger, Port Lincoln, (08) 8688 3177

763-770   **PORT LINCOLN PROPER (W)**
           - Lincoln National Park

		Unpatrolled		
No.	Beach	Rating	Type	Length
763	Tulka	1	R+rocks/tidal flats	1.5 km
764	Tulka West	1	R+tidal flats	800 m
765	Tulka (E)	1	R+tidal flats	200 m
766	Horse Rock (W 2)	1	R+sand flats	1.1 km
767	Horse Rock (W 1)	1	R+sand flats	500 m
768	Horse Rock (E 1)	1	R+sand flats	1.2 km
769	Horse Rock (E 2)	1	R+sand flats	1 km
770	Horse Rock (E 3)	1	R+sand flats	1.5 km

Port Lincoln Proper is an almost landlocked extension of Port Lincoln that extends for 12 km to the southwest of Kirton Point. The V-shaped bay is dominated by low rock bluffs and tidal flats along its northern and westernmost shore, with low energy beaches and sand flats backed by low calcarenite bluffs backing most of the southern shore. Seagrass meadows lie off all the beaches. Proper Road parallels the western shore, providing access to the only settlement at Tulka, while the gravel road out through Lincoln National Park to Spalding Cove provides access to most of the southern shore beaches. Apart from about fifty shacks at Tulka, there is no development other than basic camping facilities at a few of the beaches.

The **Tulka** shack settlement is located on a rocky delta that protrudes 200 to 300 m into the bay. The small creek that has supplied the sediment crosses the northern end of the rock-sand beach. The crenulate, 1.5 km long beach (**763**) faces southeast and is fronted by a mixture of rock and sand flats.

The road entrance to Lincoln National Park is located at the road junction 1 km south of Tulka West, which is also the western apex of the bay. The shore here is occupied by a 500 m wide, 800 m long, low beach ridge plain (**764**), fronted by 500 m wide tidal flats. Immediately east is the first of the southern shore beaches, a 200 m long strip of high tide sand (**765**) backed by low bluffs and fronted by 100 m wide sand-tidal flats. Four kilometres to the east is beach **766**, a 1.1 km long, north to northwest facing beach with 50 m wide sand flats.

Beach **767** lies immediately west of Horse Rock. The Spalding Cove gravel road first reaches the coast here and runs along the back of the beach. It consists of a northwest facing, 500 m long high tide beach and 50 m wide sand flats, then seagrass meadows. On the east side of the rock is beach **768**, a scalloped, north facing, 1.2 km long beach and 50 m wide sand flats. The shore continues to the east with a series of similar beaches. Beach **769** is 1 km long and beach **770** is 1.5 km long. These four beaches are all backed by the road, with a camping area in lee of beach 769.

Swimming: These are all usually calm, low gradient, relatively safe beaches.

Surfing: None.

Fishing: Best out in the bay, with generally shallow conditions along the shore.

Summary: An accessible, though quiet, section of the bay offering a number of little used beaches.

771-782   PORT LINCOLN PROPER (E) – Lincoln National Park

Unpatrolled				
No.	Beach	Rating	Type	Length
771	Woodcutters (S 2)	1	R+sand flats	700 m
772	Woodcutters (S 1)	1	R+sand flats	800 m
773	Woodcutters	1	R+sand flats	1.2 km
774	Woodcutters (N)	1	R+sand flats	900 m
775	Stamford	1	R+sand flats	800 m
776	Surfleet Pt (W)	1	R+sand flats	900 m
777	Surfleet Cove	1	R+sand flats	700 m
778	Spalding Cove	1	R+sand flats	2.5 km
779	Spalding Cove (N 1)	1	R+sand flats	900 m
780	Spalding Cove (N 2)	1	R+sand flats	500 m
781	Fisherman Pt	1	R+sand flats	1.3 km
782	Engine Pt	1	R+sand flats	900 m

Stamford Hill is a 145 m high, north-south trending ridge, which was climbed by Matthew Flinders in 1802 when he surveyed the area. Today a monument to Flinders sits on the crest. The shore to the west of the hill trends to the southwest and contains four beaches. These beaches face almost due west and receive the full brunt of westerly wind waves generated within the bay. As a consequence, all are fronted by 200 to 300 m wide sand flats containing three to four low ridges, backed by a narrow high tide beach and usually stable low foredune, with low rocky bluffs to each end, except for 774 which is backed by continuous bluffs. Beaches **771, 772, 773** and **774** are 700 m, 800 m, 1.2 km and 900 m long respectively. There is a camping area behind Woodcutters Beach (773).

On the north side of Stamford Hill is **Stamford Beach (775)** which has a picnic and camping area and a walking track to the crest of the hill. The 800 m long beach faces north and is fronted by 300 m wide ridged sand flats. A small headland separates it from beach **776**, a similar, 900 m long beach, accessible via a vehicle track across the base of Surfleet Point to Spalding Cove.

Spalding Cove is a 2.5 km wide, 3 km long, U-shaped cove, bordered by Surfleet Point in the west and Fisherman Point to the east. The cove contains the main Spalding Cove beach (778) together with three beaches along the sides of the cove. Surfleet Cove is located immediately south of Surfleet Point and consists of an irregular, 700 m long, east facing beach (**777**) fronted by largely seagrass-covered sand flats. The vehicle track from Stamford Hill reaches the southern end of this beach, where a beachfront camping area is located.

The main cove beach (**778**) has road access to either end of the 2.5 km long beach, with a picnic and camping area at the western end. It is a low energy, very protected beach, backed by a few low beach-foredune ridges and a 300 m wide salt lake and fronted by 300 m wide ridged sand flats (Fig. 4.125). The eastern shore of the cove contains two west facing beaches, both accessible via a vehicle track that follows the eastern shore up to Engine Point. Beach **779** is a curving, 900 m long sand beach, with a low foredune and sand flats which widen to 100 m at the northern end and with seagrass growing to the shore. Beach **780** is a similar low energy beach that curves, with some undulations, for 500 m to the lee of Fisherman Point, the eastern boundary of the cove.

Figure 4.125 *Spalding Cove beach, shown here at low tide, is a protected, low energy beach fronted by wide sand flats.*

As well as the prominent Fisherman Point, two other points form the eastern shore of Port Lincoln Proper, namely Engine Point and Cape Colbert. Between these three points are the final two port beaches, both accessible by the gravel road up to Cape Donington. Beach **781** faces west and curves round for 1.3 km between Fisherman Point and Engine Point, with 100 m wide ridged sand flats and a 10 m high foredune. Beach **782** is 900 m long and curves around between Engine Point and Cape Colbert, and with slightly more exposure, has 200 m wide ridged sand flats and an equally high foredune. There are camping and picnic areas behind both beaches.

Swimming: The Stamford Hill and Spalding Cove picnic areas are the most popular locations and like all the beaches, are relatively safe for swimming, with deeper water over the seagrass meadows being the only hazard.

Surfing: None.

Fishing: Better out in the bay over the seagrass meadows.

Summary: Twelve pristine, low energy beaches all dominated by sand flats and seagrass meadows, with picnic and camping areas at six of the beaches.

783-792   **CAPE DONINGTON to MACLAREN POINT - Lincoln National Park**

Unpatrolled			
No.	Beach	Rating	Type	Length
783	Cape Colbert	2	R/LTT	400 m
784	Cape Donington	2	R/LTT	600 m
785	Cape Donington (S)	3	R/LTT	50 m
786	Beach 786	2	R/LTT	350 m
787	Beach 787	3	R/LTT	30 m
788	Carcase Rock (1)	2	R/LTT	900 m
789	Carcase Rock (2)	2	R/LTT	850 m
790	MacLaren Pt (N)	2	R/LTT	1.4 km
791	MacLaren Pt	2	R/LTT	1 km
792	MacLaren Pt (S)	2	R/LTT	350 m

Cape Donington is the northern tip of the Jussieu Peninsula. At the cape, the low energy shore to the west changes dramatically as the coast south of the cape faces due east, exposing it to low swell out of the south, as well as easterly wind waves, with waves usually averaging less than 1 m. Between Cape Donington and Cape Catastrophe, 30 km to the south, is 40 km of densely vegetated, bedrock dominated shoreline, containing 21 beaches, all within the national park. Apart from basic camping areas at a few beaches, there are no other facilities. Only five are accessible by car, another seven by 4WD, with nine having no vehicle access.

Cape Colbert beach (**783**) lies immediately west of Cape Donington. It is a 400 m long, north facing sand beach bordered by granite rocks, with a fringe of rocks along the base of the beach, then a narrow sand bar. The road to the Cape Donington lighthouse clips the eastern end of the beach. This is usually a calm beach, with the rocks posing the greatest risk.

On the eastern side of Cape Donington is the cape beach (**784**), a 600 m long, east to northeast facing white sand beach, usually fronted by a narrow, attached bar. There is a gravel road to the southern end, where camping is permitted. Three hundred metres to the south is a 50 m pocket of sand (beach **785**) in amongst the prominent granite rocks, that is accessible on foot or by 4WD.

Beach **786** is located 3 km south of the cape and is a more isolated, 350 m long sand beach, accessible only via a 4WD track from the south. The beach is bordered by granite rocks, which also front the northern 100 m of the beach. Beach **787** lies 300 m to the south and is a 30 m pocket of sand, backed by an essentially boulder beach.

Carcase Rock is a small granite outcrop lying 250 m off the northern end of beach **788**. The beach is 900 m long and faces east, with a slight protrusion in the beach owing to wave refraction around the rock. There is vehicle access to the northern end, with tracks leading north to beach 786 and south to beach 789, from which it is separated by a small bluff. Beach **789** is an 850 m long, straight white sand beach that terminates at a sloping, 300 m long granite headland, which separates it from beach 790. Both beaches have a usually narrow, attached low tide bar.

Beach **790** is located on the north side of **MacLaren Point**, a prominent, 20 m high granite headland that protrudes 1 km to the east. The beach is 1.4 km long and faces east, curving round in lee of the point. A few rocks outcrop along the centre of the beach and there is 4WD access to the southern end and the point. On the south side of the point is beach **791**, a 1 km long, southeast facing, slightly curving white sand beach, backed by a low foredune. The MacLaren Point track reaches the centre of the beach, with tracks leading to small informal parking areas at either end. There is a 500 m long walking track from the southern parking area round the edge of the 300 m wide southern headland to beach **792**. This beach faces southeast, is 350 m long, with a narrow attached bar and is backed by vegetated bedrock slopes.

Swimming: These beaches usually receive low swell and wind waves averaging 0.5 m high which break close to shore, usually across a narrow bar. Conditions are relatively safe away from the rocks and reefs.

Surfing: Usually too low, with only a shorebreak during occasional higher swell.

Fishing: Most campers and visitors come to fish both the numerous rocks, as well as off the beach.

Summary: Ten natural beaches with reasonable access, some camping areas and generally clean white sand and low waves, bordered by granite rocks throughout.

793-798   **POINT HASELGROVE to SHAG COVE - Lincoln National Park**

No.	Beach	Rating	Type	Length
Unpatrolled				
793	Point Haselgrove	3	R/LTT	30 m
794	Taylors Landing (N)	3	R/LTT	3.8 km
795	Taylors Landing	2	R/LTT	200 m
796	Taylors Landing (S 1)	3	R	50 m
797	Taylors Landing (S 2)	3	R	40 m
798	Shag Cove	3	R	400 m

Point Haselgrove is a 25 m high, rounded granite headland. On the north side of the point is a 30 m long pocket of high tide sand (beach **793**), entirely backed by sloping granite rocks and fronted by a 100 m long, narrow sand bar. A coastal walking track skirts the rear of the beach.

On the south side of the point is the longest and most exposed beach on the eastern side of the peninsula (**794**). It is a relatively straight, 3.8 km long, southeast facing white sand beach. It is fronted by a continuous narrow bar and backed by a continuous low foredune, with a small salt lake toward the southern end (Fig. 4.126). The only access to the beach is via the walking track that starts at Taylors Landing, 500 m south of the southern end of the beach.

Figure 4.126 *Point Haselgrove to beach 794, with seagrass meadows paralleling the beach.*

Taylors Landing beach (**795**) is accessible by car and is a relatively popular spot, with an informal camping area, beach boat launching and parking area. The landing was used to service Taylor Island, located 4 km to the east. The beach is 200 m long, with cliffed bluffs backing the northern half, while the southern end provides the area for access and boat launching. The beach receives low waves and seagrass grows almost to the shore.

South of the landing, all the way to Memory Cove, is the least accessible part of the coast. Beach **796** is located 1 km south and consists of a 50 m pocket of sand, surrounded by low bluffs and fronted by seagrass at the shore. It can only be accessed on foot along the end of the rocks and bluffs. Another 1.5 km to the south is another 40 m pocket of sand (**797**) in amongst a predominantly rocky shoreline.

Shag Cove lies 5 km south of the landing and can be accessed along a rough, poorly defined track off the Memory Cove track. The curving, east facing, 1 km long cove contains a discontinuous 400 m long strip of sand (**798**) along its northern shore. It is backed by low bluffs and fronted by seagrass.

Swimming: Taylors Landing is by far the most popular beach, with the others only accessible on foot or by boat. It is a protected and relatively safe beach, apart from the boat traffic.

Surfing: Usually too low, with high swell producing a shorebreak on beach 793 and at Taylors Landing.

Fishing: Most fishers tow their boats to Taylors Landing for deep water fishing.

Summary: Apart from Taylors Landing, an isolated and little used section of natural coast.

799-803   **MEMORY COVE - Lincoln National Park**

Unpatrolled				
No.	Beach	Rating	Type	Length
799	Beach 799	3	boulder	60 m
800	Beach 800	3	boulder	60 m
801	Memory Cove (N 2)	3	boulder	100 m
802	Memory Cove (N 1)	3	boulder	40 m
803	Memory Cove	2	R/LTT	220 m

The most isolated and difficult to access part of the park is located at the southeastern tip of the peninsula. A winding, 20 km long 4WD track leads out to Memory Cove, however the four beaches to the north can only be reached by boat, or with difficulty along the rocks.

Beach **799** is 3 km north of the cove and consists of a steep, 60 m long boulder beach, partly fronted by rock flats. Five hundred metres to the south is north facing, 60 m long beach **800**. It is a steep boulder beach fronted by relatively deep water, providing good protected anchorage.

Memory Cove has three beaches. On the northern shore is 100 m long beach **801**, an east facing boulder beach. Four hundred metres to the south is 40 m long beach **802**, set in a U-shaped cove and consisting of boulders with a veneer of sand. The main Memory Cove beach (**803**) lies in the southern corner of the cove and is a 220 m long, northeast facing white sand beach, backed by a tree-shaded beachfront camping area (Fig. 4.127). The usually low wave beach has a narrow, attached bar, with seagrass growing 50 m offshore. There is a monument to Matthew Flinders at the cove. He landed at the cove in 1802, when searching for six missing seamen. The surnames of the seamen live on in the islands along the shore, that he named in their memory.

Swimming: Memory Cove is a relatively safe swimming spot, with the deeper water off the beach being the main hazard.

Surfing: None.

Fishing: Memory Cove is popular for both beach and rock fishing.

Summary: It is a long but worthwhile drive to Memory Cove for the scenery, history and pristine camping area.

Figure 4.127 *The isolated Memory Cove beach during typical low wave conditions.*

Thistle Island
Thistle Island is a calcarenite-capped granite island lying 7 km east of Cape Catastrophe (Figure 4.121B). Its western shore is exposed to the full force of the Southern Ocean, while the eastern shore faces into quieter gulf waters. The island is 17 km long, between 1 and 5 km wide and averages 100 m in height. Much of the elevation is due to massive layers of dune calcarenite that have been deposited on the granite base at high stands of the sea over the past two million years. The island has two sides, a quieter northern and eastern side facing into the gulf waters, where the sole settlement at Cosy Cottage is located, and a high cliffed, exposed and dune-capped western shore, which is battered by the Southern Ocean waves and winds. The island has 47 km of shoreline and 25 beaches, which range in length from 50 m to 1.85 km.

T-1 – T-7   **THISTLE ISLAND (E) – OBSERVATORY PT to MARBLE BAY**

Unpatrolled				
No.	Beach	Rating	Type	Length
T-1	Observatory Pt (E)	4	R/LTT	1.85 km
T-2	Crawford (N)	4	R/LTT	1.7 km
T-3	Crawford (S)	4	R/LTT	450 m
T-4	Nautilus Beach	4	R/LTT	1.65 km
T-5	Trevally Point	4	R/LTT	400 m
T-6	Whalers Bay	6	LTT/TBR	1.85 km
T-7	Marble Bay	4	R/LTT	100 m

Observatory Point is the northernmost tip of Thistle Island and is composed of a series of up to 15 foredune ridges that have prograded northward to form a low, sandy foreland, with beaches to either side (Fig. 4.128). The eastern beach (**T-1**) faces northeast and runs straight from the low foreland for 1.85 km to the 20 m high rocks of Shag Point. It is backed by up to eight foredune ridges, which narrow to the south. The beach receives waves averaging about 1 m, which maintain a reflective high tide beach, usually fronted by a narrow bar, with seagrass 100 m offshore.

Figure 4.128 *Observatory Point at the northern tip of Thistle Island consists of up to 15 foredune ridges that have built out the northern end of the island up to 1 km and impounded salty Swan Lake.*

South of Shag Point is a 6 km section of high cliffs, formed in scarped calcarenite stacked between 100 m to 200 m high. Along the base of the cliffs are five beaches, with the central three being inaccessible.

The northern **Crawford Beach (T-2)** is 1.7 km long, faces northeast and is backed by bluffs which rise from 10 m in the north to 100 m in the south, with the main island vehicle track providing access to the northern end. It receives low waves and usually has a narrow, attached bar, with seagrass 100 m off the shore. The southern Crawford Beach **(T-3)** is a 450 m long strip of sand lying at the base of 150 to 200 m high cliffs and is accessible only by boat, with similar waves and bars as the northern beach. **Nautilus Beach (T-4)** occupies the southern section of high cliffs, terminating at Trevally Point. It is 1.65 km long and backed by cliffs which rise to the island crest of 224 m (Fig. 4.129). It has a 50 m wide attached bar and usually low surf, though occasional high swell can reach the beach. On the southern side of **Trevally Point** is a narrow, 400 m long, northeast facing beach **(T-5)**, wedged in at the base of steep 150 to 200 m high cliffs.

Whalers Bay (T-6) is bordered in the north by 150 m high bluffs, which decrease in elevation to 10 m at the southern end of the curving, northeast to north facing, 1.85 km long bay. The bay is protected in the south by Horny Point and is the site of the sole settlement on the island, Cosy Cottage, which consists of about twenty houses spread along the southern shores and point. There is a small jetty in lee of the point. The beach receives waves averaging about 1 m, but occasional higher waves can maintain a double bar system, with several rips cutting the inner bar. Even during intervening lower wave conditions, the rip channels usually remain.

On the southern side of Horny Point is **Marble Bay (T-7)**, an east facing bedrock bay containing a 100 m long strip of sand at the base of the rocks, with seagrass 50 m offshore.

Figure 4.129 *Nautilus Beach has received some sand that has been blown as dunes over the crest of the island and consequently slid down the backing 200 m high cliffs onto the low energy beach.*

Swimming: Whalers Bay is the most popular location and is relatively safe under normal low waves. However waves over 1 m will generate several rips along the beach.

Surfing: Only during periods of higher swell does surf break along the east coast beaches.

Fishing: There is good beach and rock fishing along these beaches, with the middle three being only accessible by boat.

T-8 – T-17   **THISTLE ISLAND – HECLA COVE to WATERHOUSE PT**

	Unpatrolled		
No.	Beach	Rating	Type	Length
---	---	---	---	---
T-8	Hecla Cove (N)	4	R+rocks/reef	700 m
T-9	Hecla Cove (S)	4	R+rocks/reef	300 m
T-10	Osprey Pt (N 2)	4	R+reef	200 m
T-11	Osprey Pt (N 1)	3	R	80 m
T-12	Loot Bay	4	R+reef	250 m
T-13	Waterhouse Bay (N)	3	R	70 m
T-14	Waterhouse Bay	3	R	1.2 km
T-15	Waterhouse Bay (S)	3	R	600 m
T-16	Waterhouse Pt (W 1)	5	LTT	250 m
T-17	Waterhouse Pt (W 2)	4	R+reef	100 m

The southeastern end of the island consists of lower calcarenite-capped granite slopes, reaching the shore as low headlands, bluffs and bedrock reefs, in amongst which are ten generally lower energy beaches. A vehicle track runs along the shore, providing access to all but the last beach.

Hecla Cove is an open, east facing, 1 km long bay, containing two near-continuous strips of high tide sand beach (**T-8** and **T-9**), bordered and separated by granite rock platforms and backed by a low climbing foredune. Waves are low to calm and seagrass and reefs dominate the bay floor. The bay is backed by the island's landing ground.

Osprey Point is one of three protruding granite rocks that form a series of small sand-filled embayments along the 700 m of shore between Hecla Cove and Loot Bay. In amongst the rocks are two small beaches (**T-10** and **T-11**). These are 200 m and 80 m long respectively and both are protected by the rocks and reefs, with low waves and seagrass growing to the shore.

Loot Bay (**T-12**) is a semicircular, 300 m wide bay containing a curving, 250 m long, low energy beach, which is backed by 20 m high bluffs and fronted by a seagrass covered bay floor. Past the southern end of Loot Bay is 2 km long, east facing **Waterhouse Bay**, which contains three beaches. Beach **T-13** is a 70 m pocket of sand located at the very northern end, just 100 m south of Loot Bay. It is bordered by low granite rocks in the north and a small tombolo to the south. The main beach (**T-14**) runs for 1.2 km to some low rocks that separate it from the southern beach (**T-15**), a 600 m long strip of sand that terminates at Peet Point. All three beaches usually receive low waves and have seagrass growing to within 50 m of the shore.

Five hundred metres west of the southern shores of Waterhouse Bay, on the western side of **Waterhouse Point**, is beach **T-16**, a south facing, more exposed, 250 m long beach. It receives waves averaging over 1 m, which maintain a 50 m wide attached bar, with rips flowing out against the boundary rocks during higher waves. It is backed by vegetated dunes rising to 30 m, with vehicle access to the eastern corner. Beach **T-17** lies 600 m further west in a deeper, more protected, southeast facing bay. The low energy, 100 m long sand beach is bordered by rock platforms.

Swimming: These are relatively safe beaches during normal low wave conditions, however beware during higher waves because of the numerous reefs and, on the more exposed beach, headland and reef-controlled rips.

Surfing: Best chance is the southern Waterhouse Point beach.

Fishing: The best fishing is also in the south where deep water is closer to shore.

Summary: An accessible but undeveloped section of the island, offering several low energy beaches and rocky coast.

T-18 – T-25   THISTLE ISLAND (W) – PINNACLE ROCK to SNUG COVE

No.	Beach	Rating	Type	Length
	Unpatrolled			
T-18	Pinnacle Rock	8	LTT+reef	850 m
T-19	Beach T-19	8	R+reef	50 m
T-20	False Creek	8	RBB	150 m
T-21	False Ck (N 1)	8	RBB	250 m
T-22	False Ck (N 2)	7	LTT/TBR	150 m
T-23	Mitlers Cove	3	R+reef	180 m
T-24	Snug Cove (W)	3	R+reef	250 m
T-25	Snug Cove	4	R-LTT	1.7 km

The western side of Thistle Island is one of the most wave and wind battered parts of the South Australian coast. The 25 km of shore is dominated by calcarenite cliffs rising between 100 and 200 m (Fig. 4.130), with eight beaches totalling only 3.5 km in length and much of this on the more protected northern Observatory Point. For the most part it is exposed to high southwest waves, which have long ago eroded the beaches that supplied the sand to the top of the cliffs. In some places the dunes moved 3 km across the top of the island to reach the eastern shore, where they would have cascaded down the 200 m high cliffs onto the eastern beaches.

Figure 4.130 *Fossil Point (foreground) looking into O'Loughlin Bay. Most of the island is composed of Pleistocene dune calcarenite overlying bedrock, visible here at the point. The calcarenite reaches a maximum height of 234 m and indicates the height to which climbing clifftop dunes can reach, which have then moved across and cascaded over onto the east side of the island, as indicated in Fig. 4.129.*

Pinnacle Rock is a solitary calcarenite sea stack which lies at the northern end of beach **T-18**, an 850 m long, curving high tide sand beach, backed by sloping bluffs and cliffs rising to 200 m and fronted by a 200 m wide, reef-strewn surf zone. For these reasons the beach is all but inaccessible. Two kilometres to the north is a 50 m long beach (**T-19**) almost entirely surrounded by 100 m high cliffs. Waves reach the beach through a 30 m gap in the cliffs, while reefs and surf extend 200 m further seaward.

Cliffs dominate the next 3 km of shore, before False Creek is reached. There is no creek in the small embayment, hence the name. There are, however, three beaches between the 'creek' and the northern Carrington Point. **False Creek** beach (**T-20**) is a 150 m long, exposed, west facing sand beach, backed by active dunes extending 1 km inland and fronted by a rip-dominated, 200 m wide surf zone. A 4WD track leads to the bluffs on the north side of the beach. On the northern side of the bluffs is beach **T-21**, a 250 m long sand beach backed by vegetated, sand-covered bluffs, with permanent rips at either end crossing the 200 m wide surf zone. Its northern neighbour (**T-22**), in lee of the point, also receives some protection from reefs 400 m off the beach. The lower waves produce an attached bar along the 150 m long, bluff-backed beach.

On the north side of Carrington Point is semicircular **Mitlers Cove**, which contains a 180 m long protected beach (**T-23**). The beach is backed by 10 m high bluffs, paralleled by a 4WD track. It receives low waves and has seagrass growing to the shore.

Along the 3.5 km northern side of the island between Nose Point and Observatory Point are two beaches. The first is a 250 m long, west facing, low energy beach (**T-24**) on the western side of Snug Cove. The beach is protected from the ocean swell by both Cape Catastrophe and Hopkins and Smith Islands. As a consequence, it receives low swell which surges up the steep beach face, with seagrass growing right off the beach and often littering the shore. A 4WD track reaches the southern end of the beach.

Snug Cove is a curving, north to northwest facing, 1.7 km long beach that forms the western side of the Observatory Point foreland. It commences in the south as a protected, calm bay with a narrow beach and seagrass growing up to the shore. Wave energy increases slightly toward the point, which has prograded almost 1 km into the gulf as a series of 15 foredune ridges. At the point it converges with the Observatory Point beach (**T-1**).

Swimming: The only safe beaches on the western side of the island are the northern Mitlers and Snug Coves. The exposed beaches are all extremely hazardous, with large waves, permanent rips and some reefs.

Surfing: The best chance is at False Creek, with some point breaks also off Carrington Point and Mitlers Cove. However, local knowledge is essential.

Fishing: There are good beach rips at False Creek, while rock fishing is extremely hazardous on the western side.

Summary: A dramatic, high energy section of shore, in places impossible to safely access.

804-813   CAPE CATASTROPHE to CAPE TOURNEFORT - Lincoln National Park

No.	Beach	Rating	Type	Length
Unpatrolled				
804	Cape Catastrophe (1)	4	R+reef	300 m
805	Cape Catastrophe (2)	5	R+platform	400 m
806	West Point (1)	5	R/LTT+reef	450 m
807	West Point (2)	4	R+reef	650 m
808	Termination	10	boulder	100 m
809	Jussieu Bay (E)	10	TBR+reef	700 m
810	Jussieu Bay (W)	10	TBR+reef	700 m
811	Curta Rocks (1)	7	TBR+reef	700 m
812	Curta Rocks (2)	7	TBR+reef	1 km
813	Cape Tournefort	7	TBR+reef	650 m

Cape Catastrophe is the southeastern tip of both the Jussieu and Eyre peninsulas and represents the demarcation between the protected gulf waters and shores and the exposed, high energy, wind and wave-dominated coast that extends all the way to Western Australia. Access to this part of the peninsula is via the Memory Bay 4WD track, which provides access to five of the ten beaches between Cape Tournefort and Cape Catastrophe.

The first two beaches, while facing south and exposed to periodic high waves, are moderately protected by West Point and inshore reefs, with waves averaging about 1 m. Beach **804** lies immediately west of the cape and is a 300 m long, steep reflective beach, backed by dune-covered, 20 to 30 m high cliffs, with calcarenite rocks and reefs scattered along the shore. Its neighbour, beach **805**, is 400 m long and protected by more extensive reefs extending up to 200 m off the beach (Fig. 4.131). It is backed by 50 to 70 m high cliffs. The Memory Bay track follows the cliff top above beach 805, however as a result of the cliffs, access to both beaches requires a steep climb.

West Point, 5 km southwest of Cape Catastrophe, is in fact the southernmost point of the peninsula. In lee of the 150 m high calcarenite headland are two partly protected beaches. Beach **806** is a 450 m long, south facing, coarse sand beach exposed to waves averaging over 1 m, which break heavily on a usually narrow bar and patchy inshore reefs (Fig. 4.132). Two hundred metres to the west is beach **807**, a 650 m long high tide sand beach, which is more protected by the point and fronted by near continuous calcarenite rocks and reefs, resulting in low waves at the shore. However, gaps in the reefs result in permanent rips when waves are breaking. A track off the Memory Bay track leads to a lookout above the eastern end of beach 806, providing an excellent view from the 40 m high bluffs.

Figure 4.131 *The two beaches (804 and 805) immediately west of Cape Catastrophe (far right). Note the calcarenite reefs in the surf, which lower waves at the shore, but also induce strong permanent rips (lower)*

Figure 4.132 *View along beach 806 showing waves breaking heavily across the usually narrow bar and inshore rocks.*

On the exposed western side of West Point, the coast consists of 8 km of high cliffed calcarenite coast extending up to Jussieu Bay. Three kilometres north of the point is a deep indentation containing a 100 m long, steep boulder beach (**808**), fronted by a 100 m wide sand and rock-dominated surf zone and backed by steep, 160 m high slopes (Fig. 4.133). It is possible to access this dangerous beach down the slopes from the Memory Bay track.

Jussieu Bay is an exposed, 5 km long, southwest facing section of cliffed calcarenite coast, capped by continuous clifftop dunes which have extended up to 6 km inland and to heights of 200 m. Along the base of the cliffs are three remnants of a once more continuous beach, all exposed to waves averaging 2 m. Beach **809** is located at the eastern end of the bay and is a 700 m long strip of high tide sand wedged in between 200 m high cliffs and near continuous calcarenite reefs, with permanent rips flowing out of five breaks in the reef across a 300 m wide surf zone. Beach **810** lies 1 km to the west and is similar at 700 m in length, but backed by steeper 180 m high cliffs, with less continuous reefs and four permanent rips across the wide

surf zone. Neither beach is accessible, owing to the cliffs, and both are highly dangerous (Fig. 4.133).

Figure 4.133 *Termination Beach (808), immediately west of West Point, is one of the most dangerous beaches on the coast, exposed to high waves breaking on a boulder beach backed by high cliffs.*

The western bay beach (**811**) lies in lee of **Curta Rocks**, which lower the waves to 1.5 m. The 700 m long, south facing beach has a predominantly sand beach and 200 m wide surf zone, with rocks and reef increasing toward the eastern end. It usually has two beach rips and three permanent rips. Its neighbour, beach **812**, faces southwest and receives some protection from the rocks and Cape Tournefort. It also has a 200 m wide surf zone, with two beach rips and permanent boundary rips. Both beaches are backed by 10 to 20 m high bluffs, with a bluff-top 4WD track from Cape Tournefort reaching the western end of Jussieu Bay (beach 811).

On the western side of beach 812 and in lee of **Cape Tournefort**, is beach **813**, a 650 m long sand beach dominated by intertidal calcarenite reef. While waves are lowered to average less than 1.5 m, three permanent rips (one central and two boundary) cross the 100 m wide surf zone. There is a parking area above the western side of the beach, providing a good view of the beach and access down the dune-covered slopes.

Swimming: All these beaches should be treated with respect, as they range from moderately to extremely dangerous, beside being isolated and dominated by rocks, reefs and in most cases, permanent rips. Only swim on the more protected beaches if waves are low.

Surfing: There is surf on the more exposed beaches, but most are very hazardous.

Fishing: There is excellent rock and beach fishing on the protected beaches, but be careful accessing and using the more exposed section of coast.

Summary: Ten dynamic beaches backed by high calcarenite cliffs, offering extensive scenery, but generally unsafe for water-based activities.

814-816   WANNA

No.	Beach	Unpatrolled Rating	Type	Length
814	Wanna (S)	5	R+reef	50 m
815	Wanna (N 1)	3	R	200 m
816	Wanna (N 2)	4	R+reef	150 m

Wanna is the name of the area where the gravel road terminates at a lookout above the 50 m high cliffs and the 4WD track begins. As a consequence of the road access, it is relatively popular and fortunately has one of the less hazardous beaches on this section of coast. Six hundred metres south of the lookout, at the base of 20 m high cliffs, is beach **814**. It consists of a 50 m long pocket of sand wedged into a break in the cliffs. It is fronted by 400 m of shallow reefs and an island, resulting in usually calm conditions at the shore.

Beach **815** is located 500 m north of the lookout, with a 4WD track leading to the slopes above the southern end. This beach is protected by headlands, an islet and reef and receives waves averaging less than 1 m, resulting in a usually steep reflective beach free of rips, though beachrock does dominate the northern half (Fig. 4.134). Its neighbour, beach **816**, is 150 m long and also protected, but more difficult to access owing to backing bluffs. It is fronted by rocks and reefs.

Figure 4.134 *The first beach at Wanna (815) is protected by headlands and reefs and usually has low waves (right). Waves rapidly increase in height along its northern neighbour (816), producing a wider surf and more hazardous conditions.*

Swimming: During the usual calm conditions, beach 815 offers a chance of low waves and rip-free waters.

Surfing: There are some point breaks off these beaches, however local knowledge and experience are required to surf these waters.

Fishing: This is an accessible and relatively popular fishing spot, offering relatively safe rock and beach fishing.

Summary: Wanna is worth visiting for the spectacular views, as well as offering a place to possibly swim and fish.

817-825   SLEAFORD BAY

No.	Beach	Unpatrolled Rating	Type Inner	Outer Bar	Length
817	Sleaford Bay (1)	6	R	RBB	600 m
818	Sleaford Bay (2)	8	TBR	RBB	5.5 km
819	Sleaford Bay (3)	8	TBR	RBB	1.4 km
820	Sleaford Bay (4)	8	TBR	RBB	2.3 km
821	Sleaford Bay (5)	8	TBR	RBB	1.5 km
822	Sleaford Bay (6)	8	TBR+reef	-	800 m
823	Sleaford Bay (7)	5	R+rocks	-	30 m
824	Sleaford Bay (8)	6	R+rocks	-	200 m
825	Sleaford Bay (9)	6	R+rocks	-	350 m

Sleaford Bay contains one of the more exposed, high energy beaches of the southern Australian coast. It is backed by massive Pleistocene dune systems, extending up to 9 km inland and to heights of 150 m, that are overlain by Holocene dunes extending up to 3 km inland. The bay and its beaches are accessible by car at the western Sleaford Mere end, at Wanna and by 4WD through the dunes and along the beaches. However apart from the car park, there are no facilities at the beach.

The bay beaches all share a common 13 km long, energetic surf zone. However the beach is cut by calcarenite bluffs and reefs, and bedrock in the west, into one main beach (818) and eight smaller beaches. Wave height peaks along the main beach and slowly decreases in height to the west. Beach **817** begins 1 km west of Wanna, as a 600 m long, southwest facing sand beach that is exposed to high waves, but protected at the shore by a near continuous beachrock reef lying 50 m offshore. As a result, it has a lower energy reflective to low tide terrace inner bar, then the reef and a high energy outer bar. Two permanent rips drain through the beachrock reef, while one large rip usually crosses the outer surf zone.

The main Sleaford Bay beach (**818**) faces south-southwest and extends for 5.5 km. Apart from a boundary calcarenite reef in the east, it is a continuous sand beach fronted by a 400 m wide surf zone, containing an energetic inner bar cut by rips every 500 m and a high energy outer surf zone with larger rips. It is backed by an unstable foredune area and then massive active dunes extending up to 2 km inland (Figs. 4.135 & 4.136). 4WD tracks lead through the dunes to the beach. At the western end it terminates at some low calcarenite bluffs which back the beach for 2 km before protruding across it to form the boundary with beach 819.

Figure 4.135 *The central section of Sleaford Bay showing the double bar, rip-dominated surf zone and backing wide, active sand dunes.*

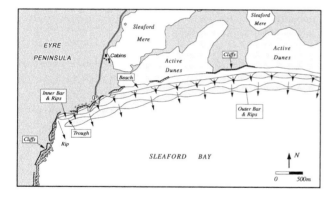

Figure 4.136 *The western half of Sleaford Bay, highlighting the inner and outer bar rips along this exposed, hazardous beach.*

Beach **819** extends for 1.4 km, backed by continuous irregular, 10 to 20 m high bluffs. The bluffs cut the high tide beach into several compartments, all of which are fronted by the continuous 300 m wide, double bar surf zone dominated by several large rips. After a 600 m long section of continuous bluff and no beach, beach **820** is a 2.3 km long section of open bluff and reef-free beach, backed by active dunes extending up to1 km inland. It is fronted by a 300 m wide surf zone containing a double rip-dominated bar system, with usually six rips crossing the inner bar.

Beach **821** faces south and is a 1.5 km long, bluff-dominated section, resulting in a narrower, discontinuous beach fronted by a 200 to 300 m wide surf zone, with prominent rips spaced every 400 m crossing the inner bar. Beach **822** is an 800 m long section immediately east

of the main car park. However, this relatively popular beach is dangerous, owing to the high waves and several permanent rips against the bluffs and fronting reefs that cross the beach.

The three small beaches west of the car park are free of dunes and dominated by bedrock. Beach **823** lies immediately west of the main car park and is a 30 m pocket of sand and boulders in the apex of a 150 m long bay. Beach **824**, also backed by a car park, is a 200 m long high tide sand beach fronted by continuous rock reefs, while beach **825** is 350 m long, with a crenulate high tide beach, also fronted by up to 200 m of reefs and rocks, and backed by the two western car parks.

Swimming: These are all hazardous beaches, even on the more moderate western beaches. Do not swim here unless waves are low and you know how to handle the conditions and rips, as people have drowned in the rips near the car park.

Surfing: This is the closest surfing beach to Port Lincoln and consequently moderately popular, with both beach and inshore reef breaks.

Fishing: The western car park beaches are very popular, while the more adventurous head through the dunes to fish the large rips that dominate the main beach.

Summary: Well worth the drive from Port Lincoln to visit Sleaford Mere for a glimpse of a high energy beach and dune system. However only enter the water if you really know how to handle the conditions.

826-836   **SLEAFORD BAY to D'ANVILLE BAY**

No.	Beach	Rating	Type	Length
Unpatrolled				
826	Beach 826	7	boulder	40 m
827	Beach 827	5	LTT/TBR	250 m
828	Beach 828	8	TBR+rocks	50 m
829	Fishery Bay	5	LTT/TBR	950 m
Whalers Way				
830	Cape Wiles (W 1)	8	R+reef	80 m
831	Cape Wiles (W 2)	8	R+reef	100 m
832	Black Lookout	10	R+reef	150 m
833	Groper Bay	8	TBR	70 m
834	Cowrie Beach	8	TBR+reef	100 m
835	Red Banks Beach	5	LTT	70 m
836	D'Anville Bay	9	R+reef+RBB	120 m

The western side of Sleaford Bay is bounded by a sloping bedrock shore that extends 10 km south to Cape Wiles. It then turns west and extends as 70 to 100 m high calcarenite cliffs for 6 km to Cape Carnot, from where another 20 km of high calcarenite cliffs extends all the way to Shoal Point. Between the Sleaford Bay car parks and D'Anville Bay is 30 km of bedrock and calcarenite-dominated shore, containing 11 beaches totalling only 2 km in length. Cliffed rocky shore and generally high waves dominate.

Beach **826** is backed by farmland 3.5 km south of the car parks. It is a 40 m long, east facing boulder beach, with a 50 m wide sand bar off the beach. Waves average 1 m and a permanent rip drains the small embayment. Beach **827** is located 2 km east of Fishery Bay. It is a 250 m long white sand beach that faces south and receives waves averaging 1 to 1.5 m, which usually maintain permanent rips at either end of the beach. It is bordered by 20 m high granite headlands and is backed by now stable dunes extending 300 m inland. Two hundred metres to the west is the pocket 50 m long beach **828**, a sand beach in a 50 m wide embayment, with a rock strewn surf zone drained by a permanent rip. The walking track from Fishery Bay reaches the top of the beach.

Fishery Bay (829) is a slightly curving, southeast facing, 950 m long, low gradient sand beach and bar, which is moderately sheltered by its orientation and prominent 20 to 40 m high headlands, resulting in waves averaging about 1 m (Fig. 4.137). There are usually two beach rips along the beach, with topographic rips forming against the headlands during periods of higher waves. There is a car park in lee of the beach, which is backed by a foredune, then older scarped calcarenite bluffs. The bay was the site of a whaling station early last century and some of the remains of this are still visible on the eastern rocks.

Figure 4.137 *Fishery Bay beach showing the typical wide, low gradient beach and usually low waves.*

Whalers Way is a privately owned and run coastal reserve. The reserve encompasses 10 km of spectacular high energy, cliffed coastline, with good gravel road access and numerous viewing points.

Cape Wiles forms the southeastern boundary of Whalers Way and is a 130 m high, calcarenite-capped granite headland. One kilometre west of the cape are two pockets of sand wedged in at the base of 130 m high, sheer cliffs (Fig. 4.138). They can be best viewed from Black Lookout. Beach **830** is 80 m long, while its neighbour (**831**) is 100 m long. Both beaches are fronted by 300 m of rocks and reefs, with surf over the reefs and a permanent rip draining each beach, while the lowered waves result in steep reflective beaches. Six hundred metres west of Black Lookout is another 150 m long sliver of sand (**832**), lying at the base of 100 m high cliffs and also fronted by 300 m wide shallow reefs.

Figure 4.138 *Beach 831 lies at the base of 130 m high calcarenite cliffs. This beach is only accessible from the sea and, while waves are lowered by reefs, rock falls are an additional hazard.*

Groper Bay is a cliffed bay, with a 70 m long high tide sand beach (**833**) fronted by continuous beachrock located in its northern corner. The beach is fronted by a sandy seabed with waves averaging about 1 m, but there is no access to the beach. **Cowrie Beach (834)** is located in lee of 50 m high Cape Carnot. It is backed by cliffs, with a steep track down to the beach that is used by adventurous fishers. The beach is composed of sand and rock and is fronted by a patchy reef surf zone.

At Cape Carnot the coast swings to face west and 3 km north of the cape is another small accessible beach at **Red Banks (835)**. The 70 m long beach faces south into a moderately protected, 500 m long embayment, resulting in waves averaging about 1 m which break across a 50 m wide bar. This beach offers the least hazardous swimming, if waves are low.

Three kilometres north of Red Banks is a small rocky embayment called **D'Anville Bay**. It contains a 120 m long high tide sand beach (**836**), backed by 60 m high cliffs and fronted by a reef-strewn, 300 m wide, high energy surf zone.

Swimming: This is a very hazardous section of shore that has claimed several lives, particularly of people being washed off the rocks. The only beaches to consider swimming at are Fishery Bay and possibly Red Banks, the rest are very dangerous and mostly inaccessible.

Surfing: The best surf is at Fishery Bay during moderate to higher swell, when both points provide left and right breaks.

Fishing: This is a popular fishing area, but fish the beach rips to stay safe, as rock fishing is extremely hazardous along this section of coast.

Summary: A spectacular section of high cliffed coast, containing a smattering of small beaches.

Coffin Bay National Park

Established: 1982
Area: 29 106 ha
Beaches: 86 beaches (837- 923)
Coast length: 127 km

Coffin Bay National Park encompasses the Coffin Bay Peninsula, bordered in the south by the exposed high energy beaches and massive dune systems of Gunyah Beach, Avoid Bay and Sensation Beach, in the west by the rocky, west facing shore between Point Whidbey and Point Sir Isaac, and in the east by the more quiescent shores of Coffin Bay and Port Douglas. In all, it has 127 km of shore, containing 86 low through high energy beach systems, with most of the land surface covered by vegetated Pleistocene calcarenite and both active and stable Holocene dunes. There is an access track to Sensation Beach and camping areas at Sensation and Mullalong beaches, and in Coffin Bay at Morgans Landing, Black Springs, The Brothers and Yang Bay. Camping in Coffin Bay National Park requires a permit from the Coffin Bay Ranger Station.

Further information:
Coffin Bay Ranger Station, (08) 8685 4047

837, 838   SHOAL POINT, GUNYAH - Coffin Bay National Park

No.	Beach	Rating Inner	Type	Outer Bar	Length
\multicolumn{6}{c}{**Unpatrolled**}					
837	Shoal Pt	8	TBR		500 m
838	Gunyah	8	TBR/RBB	RBB/LBT	15.5 km

The 40 km section of southwest facing coast between Cape Carnot and Point Avoid is one of the highest energy in Australia. It faces directly into the prevailing high southwest waves and winds and has been battered by these processes throughout the Holocene period. What remains today is 25 km of Pleistocene calcarenite cliffs, averaging over 100 m high and all capped by Holocene clifftop dunes. The cliffs then give way to 15 km long Gunyah

Beach. This high energy beach has a double bar, 500 m wide surf zone, containing massive beach rips averaging over 500 m in spacing. The largest rips reach 1 km in spacing, placing them amongst the biggest beach rips in the world.

Shoal Point is a slightly protruding point in the cliffed section, which on its eastern side has trapped the only remnant of a once massive beach system: a 500 m long pocket of sand (**837**) at the base of the 120 m high cliffs. The sand beach is backed by steep vegetated slopes and fronted by a 300 m wide surf zone, containing one central beach rip and a permanent rip against the western rocks. A 4WD track runs along the top of the cliffs, but there is no access to the beach.

Gunyah Beach (**838**) begins 4 km west of Shoal Point and runs for 15.5 km to Point Avoid. It faces southwest for most of its length, curving round in lee of Point Avoid to face southeast. There is a gravel road to a large parking area behind the beach at Point Avoid, otherwise the only access is by 4WD along the treacherous beach. The beach at Point Avoid is moderately protected, with waves averaging less than 1.5 m, however within 1 km of the point the beach is fully exposed and waves average about 2 m. The massive rips and wide surf dominate for the next 14 km (Fig. 4.139). The entire high energy section is also backed by some of the most massive active dunes in Australia, extending in places up to 10 km inland.

Figure 4.139 *The far western end of Gunyah Beach. Note the wide surf and well defined rip channel with backing embayment. This is an accessible but very hazardous beach that has drowned unwary swimmers.*

Swimming: The Point Avoid end is only moderately hazardous when lower waves prevail, but rips still persist. Most of the beach is extremely hazardous and not suitable for swimming.

Surfing: There are usually an abundance of beach breaks right along the beach, but conditions are often treacherous owing to the strong rips.

Fishing: This beach has the biggest rip holes in Australia, some 200 m across and is a relatively popular beach fishing location.

Summary: A high energy, dynamic section of coast, spectacular to see but usually unsafe for swimming.

839-848   **AVOID BAY - Coffin Bay National Park**

No.	Beach	Rating	Type	Length
Unpatrolled				
839	Point Avoid	4	R+reef	120 m
840	Point Avoid (N 1)	5	R+reef	250 m
841	Point Avoid (N 2)	5	R+reef	300 m
842	Point Avoid (N 3)	5	R+reef	200 m
843	Point Avoid (boat ramp)	5	R+reef	1.1 km
844	Avoid Bay (E 2)	4	R/LTT	160 m
845	Avoid Bay (E 1)	6	LTT-TBR	2.7 km
846	Avoid Bay (rocks)	8	TBR+reef	200 m
847	Avoid Bay (W)	7	TBR	5 km
848	Black Rocks (E)	8	TBR	300 m

Avoid Bay is a 7 km wide, relatively open, southwest facing bay, bordered in the east by Point Avoid and in the west by the Black Rocks islets, lying 1.5 km off the shore and the backing calcarenite headland. Between the two points is 15 km of rocky and sandy shoreline, including nine beaches. Three of these are backed by some massive Holocene dune transgression, extending up to 4 km across the Coffin Bay Peninsula to the quiet waters of Port Douglas. There is 2WD access to most of the Point Avoid beaches and the eastern end of the bay beaches, with a camping area on Point Avoid and a rather hazardous boat ramp at beach 843.

Point Avoid beach (839) is tucked in lee of some rocks and a substantial beachrock reef at the very western end of Gunyah Beach, where the coast changes orientation and faces southwest. However, the rocks and reefs lower waves at the shore of the 120 m long beach to about 0.5 m, producing a relatively quiet lagoon between the reef and the shore. Nevertheless, it is only safe to swim close to the beach, as currents flow through the reef. A car park is located immediately behind the beach.

To the north of Point Avoid is a series of rock and reef-controlled, west to north facing beaches, all of which receive waves less than 1 m high at the shore. The first **(840)** is a curving, 250 m long, west facing sand and beachrock beach, backed by 10 to 20 m high calcarenite bluffs and fronted by 200 m of scattered reefs. It can be accessed from the point lookout and car park. Beach **841** is a curving, 300 m long, north facing, relatively protected beach, backed by 20 m high bluffs and fronted by deeper reefs, with usually low waves at the shore. Beach **842** is a 200 m long, north facing beach that receives lower waves, which surge up the steep beach face. A car park just off the point road is located above the western end of the beach.

The boat ramp beach **(843)** is 1.1 km long and faces northwest. The steep boat ramp access is located at the quieter eastern end, while reefs lower waves along much of the beach, resulting in a continuous steep, reflective beach (Fig. 4.140). The point road runs along the top of the beach. There is a second car park above the western boundary bluffs, below which is 160 m long beach **844**. It has less reef and receives waves averaging over 1 m, which maintain a narrow bar and surf zone.

Figure 4.140 *The beach immediately east of the Point Avoid boat launching area (843) is well protected by headlands and reefs and usually has low waves surging up the steep beach, a marked contrast from nearby Gunyah Beach (Fig. 4.139).*

The main Avoid Bay beaches begin with the eastern 2.7 km long beach **(845)**, which faces essentially west. Waves increase up the beach from over 1 m to 1.5 m in the north, resulting in about ten beach rips dominating the central to northern section of beach. It terminates at some protruding calcarenite bluffs and reefs, which also contain 200 m long beach **846**. This beach receives waves averaging over 1.5 m, which break across a reef-strewn surf zone to produce hazardous conditions. There is also a permanent rip which drains the small embayment. Both beaches are backed by extensive, though vegetated, parabolic dunes.

Between these bluffs and the headlands in lee of Black Rocks is 5 km long beach **847**, a curving, west to southwest facing, moderate to high energy beach, backed by active sand dunes. Several large beach rips dominate the 300 m wide surf zone, together with a permanent rip at the eastern end. The beach terminates at bluffs which protrude part way across the beach. On the other side of the bluffs is 300 m long beach **848**, which is bordered by the prominent Black Rocks headland. It has a strong permanent rip against the headland rocks (Fig. 4.141).

Swimming: The two most popular spots are at the protected Point Avoid 'lagoon' (beach 839) and the boat ramp beach (843), with the other point beaches having low waves but rocks and reefs, while the bay beaches have higher waves and strong beach rips.

Surfing: The best beach breaks are along the more exposed western bay beaches.

Fishing: There are numerous spots to fish the rocks and reefs around the point, with a 4WD required to reach the better beach rips along the beach.

Summary: Point Avoid is a relatively popular picnic and swimming area, as well as providing basic camping facilities and an exposed boat ramp for experienced users.

Figure 4.141 *The western end of Avoid Bay (beaches 847 & 848) is exposed to moderate to high waves and has a wide, rip-dominated surf zone, with permanent rips against the western headland (bottom).*

849-858   BLACK ROCKS to SENSATION BEACH - Coffin Bay National Park

No.	Beach	Rating	Type	Length
Unpatrolled				
849	Black Rocks (W 1)	8	R+reef	350 m
850	Black Rocks (W 2)	8	R+reef	200 m
851	Black Rocks (W 3)	7	R+reef	100 m
852	Black Rocks (W 4)	8	R+reef	170 m
853	Black Rocks (W 5)	8	TBR+reef	950 m
854	Black Rocks (W 6)	8	R+reef	400 m
855	Black Rocks (W 7)	8	R+reef	450 m
856	Sensation Beach (E)	8	TBR+reef	450 m
857	Sensation Beach	7	TBR	4.5 km
858	Misery Bay (point)	3	R	60 m

Sensation Beach occupies the western half of a 10 km wide, southwest facing embayment, containing 12 km of shore, with the eastern half dominated by calcarenite cliffs and reefs. The only vehicle access to the entire section is in the west at Sensation Beach, where there is also a camping area. Most of the shore is, however, exposed to moderate to high waves and accessible only on foot.

Beach **849** lies immediately west of the protruding **Black Rocks** headland. It consists of a 350 m long, west facing

high tide sand beach fronted by reef along the eastern half, with a 150 m wide rip-dominated surf zone along the western half and 20 m high bluffs behind. Two kilometres to the west are two adjoining pockets of sand totalling 200 m in length (beach **850**), wedged in at the base of 40 m high bluffs and fronted by deeper reefs extending 200 m off the beach. Seven hundred metres to the west is a 100 m wide, semicircular gap in the cliffs containing 100 m long beach **851**. It consists of a curving high tide beach, surrounded by 50 m high cliffs and fronted by a reef-filled bay, with one permanent rip draining the entire system.

Beach **852** lies 600 m further west and is a straight, 170 m long sand beach, with a 100 m wide, reef-filled surf zone containing one large permanent rip. Its neighbour, beach **853**, is a straight, 950 m long sand beach, with reefs lining the western end, but with a sand bar and rip system along most of the beach, containing one beach rip, with a permanent rip at the eastern end. It is backed by 40 m high cliffs. It merges at its western end with beach **854**, a 400 m long high tide beach fronted by near continuous beachrock reef and a 100 m wide surf zone, containing a permanent rip toward the eastern end. Beach **855** lies 200 m to the west and consists of a 450 m long, discontinuous strip of sand wedged in at the base of 30 m high cliffs and fronted by continuous rocks and reefs and a 150 m wide surf zone.

The western boundary bluffs of beach 855 mark the beginning of Sensation Beach. The eastern section is a 450 m long beach (**856**), separated from the main beach by a small protruding bluff and fronted by a partly reef-filled, 150 m wide surf zone. The main **Sensation Beach** (**857**) is a curving, 4.5 km long beach that faces southwest, then south and finally southeast in the western corner. Much of the beach is fronted by a rip-dominated, 200 m wide surf zone with usually ten beach rips along this section. The western corner swings around in lee of the Misery Bay headland, reducing waves to about 1 m, which maintain a wide, low gradient beach and attached bar. Most of the beach is backed by active dunes, extending up to 3 km inland.

Tucked in on the north side of Misery Bay point is beach **858**, a 60 m long, northeast facing, headland-bound, low energy reflective beach.

Swimming: The least hazardous place to swim is at the western end of Sensation Beach, when waves are less than 1 m, as higher waves will maintain strong rips. All the other beaches are both difficult to access and very hazardous.

Surfing: Sensation Beach usually has a few kilometres of beach breaks.

Fishing: Most fishers head up Sensation Beach to fish the rip holes and the protected rocks of Misery Bay headland.

Summary: It is a long 4WD trip to reach Sensation Beach, but worth the effort to see the pristine white sand beach, surf and dunes.

859-870   **WHIDBEY BEACH WILDERNESS AREA (S) - Coffin Bay Peninsula**

Unpatrolled					
No.	Beach	Rating	Type	Length	
859	Whidbey (1)	9	R+reef/rocks	200 m	
860	Whidbey (2)	6	LTT/TBR	1 km	
861	Whidbey (3)	4	R+reef	200 m	
862	Whidbey (4)	4	R+reef	50 m	
863	Whidbey (5)	7	TBR+reef	100 m	
864	Whidbey (6)	7	TBR	100 m	
865	Whidbey (7)	6	TBR/LTT	350 m	
866	Whidbey (8)	6	R+reef	150 m	
867	Whidbey (9)	7	TBR+reef	350 m	
868	Whidbey (10)	4	R+reef	100 m	
869	Whidbey (11)	4	R+reef	100 m	
870	Whidbey (12)	4	R+reef	200 m	

Whidbey Beach is the name of a 6 km section of predominantly rock-dominated, south to southeast facing shore running west of the Misery Bay headland. The entire section of coast lies within the Coffin Bay Wilderness Area and can only be accessed on foot, with vehicles prohibited. A walking track commences at Sensation Beach and follows the old 4WD track along the coast to Point Whidbey and across to Boarding House Bay.

The twelve beaches occupy about half of the shoreline, with small headlands, bluffs, rocks and reefs forming four smaller 'bays' and dominating each of the beaches. All the beaches receive some protection from Point Whidbey and waves tend to average 1.5 m or less. Beaches 859 and 860 occupy the first bay west of the headland. Beach **859** is a 200 m long, southwest facing sand beach, fronted by a 100 m wide surf zone dominated by rocks and reefs. Its neighbour, beach **860**, is the longest of the Whidbey beaches, at 1 km in length. It faces south and is a continuous sandy beach, with a 100 m wide bar and surf zone, with usually two to three beach rips, as well as permanent rips against the rocks at each end. It is backed by a foredune and moderately active sand dunes.

Beach **861** lies at the western end of the second 'bay', a 500 m long rocky shoreline, with the 200 m long, southeast facing beach tucked into the corner, bordered by rock platforms and bluffs and fronted by reefs extending up to 200 m offshore. As a consequence, waves are low at the shore.

The third 'bay' is occupied by four beaches. Beach **862** is a 50 m long, west facing pocket of sand backed by 20 m high bluffs, with rocks and reefs running the length of the beach. Next door is 100 m long beach **863**, with a central rip-dominated sand bar and rocks and reefs to either side. Beach **864** is similar, with more sand and less rock, and is bordered by 10 m high bluffs. Beach **865** is the longest beach in the bay, at 350 m in length and has a more continuous, 50 m wide attached bar, which is cut by one or two rips during higher seas. Immediately west is beach **866**, a 150 m long strip of high tide sand at the base of 20 m high bluffs, fronted by 50 m wide rocks and reefs, with deeper reefs offshore.

Beach **867** lies 1 km to the west in the fourth 'bay' and is a solitary, 350 m long sand beach with a 100 m wide bar, usually containing two beach rips, with permanent rips against the boundary rocks and reefs. The walking track leads to the eastern end of the beach. Five hundred metres to the west is beach **868**, a protected, 100 m long, east facing pocket of high tide sand, bordered and fronted by rocks and reefs. Beach **869** is another 400 m west and is similar, with a 100 m long, east facing pocket of sand. A spur off the walking track reaches the western end of the beach. Beach **870** lies 300 m further west and is a straight, 200 m long sand beach, protected by islets and reefs extending 500 m offshore, resulting in usually calm conditions at the shore, with seagrass growing to the beach. The main walking track skirts the back of the beach.

Swimming: Be careful if swimming along this section, as wave height varies considerably between the beaches. While the sandy beaches are freer of rocks and reefs, they usually have rips. The least hazardous areas are some of the reef-protected, low energy beaches, so long as there is no surf. Beware of the exposed rip dominated beaches.

Surfing: There are a number of potential beach, reef and point breaks along here. However as it's a long walk, local knowledge is essential.

Fishing: This section offers rock, reef and beach fishing, but again it can be a long walk.

Summary: As this is a wilderness protection area, you can only access it on foot. However, it is well worth the walk along this very scenic section of bluffs, headlands and small beaches.

871-880   **POINT WHIDBEY WILDERNESS AREA - Coffin Bay Peninsula**

No.	Beach	Rating	Type	Length
	Unpatrolled			
871	Point Whidbey (E 1)	5	LTT+reef	80 m
872	Point Whidbey (E 2)	4	R+reef	180 m
873	Point Whidbey (N 1)	6	R+reef	300 m
874	Point Whidbey (N 2)	6	R+reef/rock	70 m
875	Point Whidbey (N 3)	6	R+reef/rock	200 m
876	Boarding House Bay (1)	4	R/LTT+reef	250 m
877	Boarding House Bay (2)	5	R/LTT+reef	1.2 km
878	Boarding House Bay (3)	7	LTT+reef	500 m
879	Boarding House Bay (4)	6	R+reef	250 m
880	Boarding House Bay (5)	5	R+reef	280 m

Point Whidbey is a calcarenite-capped granite headland that forms the westernmost point of the Coffin Bay Peninsula. The coast for several kilometres either side of the point is dominated by 10 to 30 m high, calcarenite-capped bedrock, with rocks and reefs lining the irregular shore. In amongst the 12 km of shore around the point are ten generally small beaches, most protected to some degree by their orientation, embaymentisation and offshore rocks and reefs. The entire point area lies in the wilderness zone and is closed to vehicles, with a 4WD track reaching the coast at the northern end of Boarding House Bay and forming the northern boundary of the wilderness area. There are walking tracks to the point and to Boarding House Bay.

Beaches 871 and 872 lie approximately 2 km east of the point. They face south and lie deep within adjoining small bays. Beach **871** is an 80 m pocket of sand partially protected by reefs and rocks, with a narrow clear access zone to deeper water. Waves break 200 m offshore on the reefs and are usually low at the beach. Beach **872** is a 180 m long strip of sand backed by 20 m high bluffs, with rocks crossing the middle of the beach and deeper reefs in the bay lowering waves at the shore to about 0.5 m.

The northern side of Point Whidbey is dominated by bedrock points and calcarenite bluffs for the first 4 km, within which are three small beaches in the deeper embayments. Beach **873** lies 1 km northeast of the point and is a 300 m long, northwest facing beach, backed by 20 m high bluffs and crossed by some large granite outcrops, with 100 m of broken bedrock reef off the beach producing a reef-strewn surf zone. Beach **874** is 1 km to the north and is a 70 m pocket of sand surrounded by bluffs and 200 m of bedrock reefs, resulting in low waves at the shore. Beach **875** is a 200 m long, west facing beach, also wedged in below 20 m high bluffs. Scattered rocks and reefs extend 400 m off the beach, resulting in waves averaging 1 m at the shore.

Boarding House Bay is an open, west facing, 3 km long bay, which contains four beaches. Beach **876** is located at the southern end, in lee of rocks that extend 300 m off the southern point. The 250 m long beach is usually reflective, with deeper water off the beach and reef further offshore. Beach **877** is the main bay beach. It is 1.2 km long and fronted by initially shallow, then deeper reefs, which permit waves to increase from 1 to 1.5 m up the beach. An energetic reflective beach dominates the southern half, with a narrow attached bar developing and rips increasing in occurrence and strength up the beach. The beach is backed by bluffs rising to 40 m in the north. Beach **878** is a segmented 500m long continuation of the main beach, but is crossed by some protruding bluffs, rocks and reefs which, combined with the higher waves, result in a dangerous reef-dominated, 100 m wide surf zone.

At the northern end of the bay the bluffs protrude slightly westward and rocks and reefs again dominate. Beach **879** is a 250 m long reflective beach backed by 20 m high bluffs, with waves breaking across a 50 m wide, reef-dominated surf zone. Its neighbour, beach **880**, is 280 m long and more protected by a line of shallow reefs 100 m offshore, forming a quieter 'lagoon' in their lee. The 4WD track reaches the bluffs above the northern end of this beach.

Swimming: All these beaches are potentially hazardous owing to persistent waves, rocks, reefs and strong permanent rips on some. Use extreme care if swimming along this coast, with beach 880 offering the least hazardous swimming in the 'lagoon' close inshore.

Surfing: There are a number of left and right reef breaks around the point, with local knowledge required to find and surf these spots.

Fishing: There are many excellent beach, rock and reef fishing locations right around the point, however it's a long walk to reach most.

Summary: A true wilderness area offering a relatively wild, rock-dominated coast, containing a dozen small, highly variable, rock and reef-dominated beaches.

881-893   REEF POINT to POINT SIR ISAAC - Coffin Bay Peninsula

Unpatrolled				
No.	Beach	Rating	Type	Length
881	Reef Point (S)	5	R+reef	920 m
882	Reef Point (N 1)	5	R+reef	920 m
883	Reef Point (N 2)	5	LTT/TBR	1.2 km
884	Reef Point (N 3)	5	R+reef	110 m
885	Reef Point (N 4)	5	R	600 m
886	Reef Point (N 5)	5	R/LTT	100 m
887	Reef Point (N 6)	5	LTT	550 m
888	Mullalong	4	R+reef	300 m
889	Mullalong (N 1)	4	R-LTT	1.8 km
890	Mullalong (N 2)	4	R/LTT	150 m
891	Mullalong (N 3)	4	R/LTT	80 m
892	Mullalong (N 4)	4	R/LTT	60 m
893	Mullalong (N 5)	4	R+reef	120 m

Reef Point is a prominent 40 m high, calcarenite-capped bedrock cliff located halfway up the western side of Coffin Bay Peninsula. Between the point and the northern tip of the peninsula is 11 km of bedrock-dominated, west to northwest facing coast, containing twelve beaches totalling 7 km in length, all bordered and often backed by steep cliffs and bluffs. There is vehicle access only to Mullalong Beach, where there is also a basic camping area. This is the only development on this section of coast. The remainder of the beaches can only be reached on foot.

Immediately south of Reef Point is beach **881**, a straight, 920 m long, west facing, steep, reflective sand beach, backed by climbing dunes in the south and 20 to 30 m high cliffs in the north, that are capped by clifftop dunes. Partially stable dunes extend 500 m in from the beach. Waves break on deeper reefs up to several hundred metres offshore, lowering wave height at the beach to about 1 m, maintaining the steep cusped beach and narrow surf zone. Higher waves produce a heavy shorebreak.

On the north side of the point are two straight sections of west-northwest facing shore, each containing three beaches. The first section is 2.5 km long and contains beaches 882-884. Beach **882** is a narrow, 920 m long, steep reflective beach backed by 10 to 20 m high bluffs and fronted by deeper reefs up to 1 km offshore. Wider bluffs separate it from beach **883**, a 1.2 km long sand beach which has waves increasing from 1 m in the south to 1.5 m in the north, producing occasional beach rips northward, including a permanent rip against the northern rocks. It is backed by bluffs, then climbing dunes extending up to 1.5 km inland. Beach **884** is located amongst the northern boundary bedrock cliffs and is a 110 m long, slightly curving pocket beach, bordered by bedrock platforms and backed by climbing dunes.

Bedrock reefs extend up to 500 m off the beach, lowering waves to about 0.5 m at the shore.

The second 1.5 km long section contains beaches 885-887. Beach **885** is a slightly curving, 600 m long, steep, cusped reflective beach, with rocks and reefs off the southern end and deeper reefs off the remainder, all lowering waves to about 1 m at the shore. It is backed by steep, vegetated, 40 m high bluffs. Beach **886** is a 100 m long pocket of sand bordered and backed by 40 m high cliffs and only accessible around the rocks at low tide. Beach **887** is a 550 m long sand beach, the southern half of which is shared with bedrock outcrops, while the northern half is a straight sand beach fronted by a 50 m wide attached bar, with permanent rips to either end. A 50 m high cliff and rugged rock platforms form the northern boundary, with access only down the steep, 40 m high backing bluffs.

Mullalong Beach (**888**) is the only readily accessible beach along this section of coast. A 4WD track leads to the rear of the foredune, where there is also a camping area. The 300 m long beach is bordered by 30 to 40 m high, calcarenite-capped bedrock headlands and rugged rock platforms, together with rock reefs occupying much of the 50 to 100 m wide surf zone. Waves are low in lee of the reefs, resulting in a narrow 'lagoon' between the reefs and steep shore. However, permanent rips drain the 'lagoon' out through the reefs, so beware if swimming.

North of Mullalong Beach is a 2.5 km long, west facing, open bay, containing four beaches. The first of these, **889**, is a curving, slightly crenulate, 1.8 km long beach, which begins in the south as a low energy reflective beach protected by the headland and reefs. There is 4WD access to the headland above the southern end of the beach. Waves increase up the beach, to average over 1 m. These maintain a heavy shorebreak along the narrow attached bar, with beach rips forming during higher waves and a permanent rip draining the southern reef area. The northern end of the 'bay' is dominated by 40 m high calcarenite cliffs, which contain three small pockets of sand. Beach **890** is a 150 m long sand pocket, which can only be accessed down the steep backing bluffs. Its neighbour, beach **891**, is similar but only 80 m long, while beach **892** is just 60 m long, with rock reefs extending 200 m off the northern end. These three beaches all receive waves averaging over 1 m, which maintain a steep beach and narrow attached bar, with rips occurring during higher waves.

Beach **893** lies 1.5 km south of the northern tip of the peninsula. It is a curving, 120 m long, west facing pocket of high tide sand, backed by 10 to 20 m high bluffs and fronted by a small reef-filled bay, resulting in low waves at the shore.

Swimming: Mullalong Beach and its northern neighbour (beach 889) are the only accessible beaches along this section and both are protected by reefs, resulting in low waves at the shore. They are relatively safe for swimming close in to shore, with surf, reefs and rips off both beaches.

Surfing: The beaches tend to have no surf or a heavy shorebreak, but there are some possible reef breaks at Reef Point and Mullalong Beach.

Fishing: Most fishers head for the accessible Mullalong Beach to fish the beach, reefs and rocks.

Summary: A rugged and difficult to access section of coast, with vehicle access to only two of the twelve beaches and most requiring a 1 to 2 km long walk from the track and a steep climb to reach.

894   SEASICK BAY

No.	Beach	Rating	Type	Length
		Unpatrolled		
894	Seasick Bay	4	R-LTT	800 m

Seasick Bay (894) is a curving, north facing, 800 m long sand beach, backed by a 5 to 10 m high foredune. It is bounded in the west by the lighthouse-capped, 20 m high, northernmost tip of the peninsula and in the west by 20 m high Point Sir Isaac (Fig. 4.142). The beach receives waves that are lowered by refraction around the northern tip and average about 1 m at the beach, maintaining a usually rip-free, narrow bar and low surf. There is 4WD access to the beach and headlands, with a lookout at the lighthouse and a camping area on Point Sir Isaac.

Figure 4.142 *Seasick Bay lies in lee of Point Sir Isaac and is protected from the dominant southwest swell, resulting in usually low wave conditions.*

Swimming: A relatively safe beach during normal low wave conditions, with the shorebreak increasing with higher waves.

Surfing: Usually just a low shorebreak.

Fishing: The rocks at either end are the most popular location.

Summary: It's a long, rough drive out to the point, which provides one of the more accessible and safer beaches on this exposed part of the coast.

895-901   COFFIN BAY (W) - Coffin Bay National Park

No.	Beach	Rating	Type	Length
		Unpatrolled		
895	Pt Sir Isaac (S 1)	1	R+sand flats	300 m
896	Pt Sir Isaac (S 2)	1	R+sand flats	1.2 km
897	Pt Burgess (N)	1	R+sand flats	250 m
898	Pt Burgess (S)	1	R+sand flats	250 m
899	Phantom Cove	1	R+sand flats	2.4 km
900	Seven Mile	1	R+sand flats	9.2 km
901	Pt Longnose (N)	2	R	4.8 km

Coffin Bay is a protected, U-shaped, north facing bay lying in lee of Point Sir Isaac. It is 12 km wide between the point and the eastern Frenchman Bluff and 10 km deep. It contains 30 km of bay shoreline, plus another 60 km of very low energy shore in Port Douglas, which enters the southern part of the bay. All the bay shore west of Point Longnose and all of Port Douglas west of Coffin Bay township lies within the national park.

The first 20 km of shore between Point Sir Isaac and Point Longnose face east to north and are protected from most ocean swell, being dominated more by local wind waves. Only immediately in lee of Point Isaac and out on Point Longnose does low, refracted swell regularly reach the shore. The beaches are all fronted by shallow sand flats, then deeper seagrass meadows.

On the southern side of **Point Sir Isaac** are two beaches, both accessible via 4WD tracks. The first (**895**) is a 300 m long, southeast facing strip of sand backed by 10 to 20 m high bluffs, with bedrock outcrops bordering each end. There are 4WD tracks along the bluffs, with the best access from the adjoining southern beach. This beach (**896**) runs due south for 1.2 km from the base of Point Sir Isaac. It is backed by some low foredune ridges and fronted by sand flats that widen to 300 m off the junction with beach 895. Low, refracted swell reaches both beaches, producing steep, reflective beach faces, but usually no surf.

Point Burgess is located 2 km south of Point Sir Isaac and is the last bedrock point on the western side of Coffin Bay, giving way to continuous sand beaches. Beach **897**

lies on the northern side of the point. It is a 250 m long, east facing beach bordered by low, rock-fronted headlands, backed by vegetated slopes and fronted by 100 m wide sand flats, then seagrass meadows. A 4WD track runs along the back of the beach. On the southern side of the point is southeast facing beach **898**, a 250 m long, crenulate sand beach, backed by grassy slopes, with 200 m wide sand flats off the beach.

Phantom Cove (899) begins immediately south of beach 898 and is a crenulate, 2.4 km long, east facing, low energy sand beach fronted by sand flats up to 300 m wide. The beach is backed by a mixture of vegetated recurved spits and foredunes that formed across the backing salt flats, prograding the shoreline up to 300 m into the bay. A 4WD track reaches the northern end of the beach, after skirting the back of the salt flats.

Seven Mile Beach (900) is a 9.2 km (6 mile) long, curving beach that initially faces east, then swings round to form the southern, north facing shore of the bay. It is fronted by sand flats up to 500 m wide, which are dominated by transverse sand ridges, then deeper seagrass meadows. Waves are usually low to calm, with swell rarely reaching the shore. Much of the southern section of beach is backed by generally stable transgressive dunes, which have moved across the peninsula from Sensation Beach, 6 km to the south. Because of the high and irregular dune topography, as well as a 500 m patch of active dunes, the main 4WD track runs along the beach, heading inland at Morgans Landing, where the dunes are absent and where the beach trends north. There is also a camping area near the landing.

Point Longnose beach (**901**) commences at the end of Seven Mile Beach, where there is a sand foreland marked by a sharp turn in the shore. The point is attached to the mainland for 2 km, before extending east as a 3.5 km long, 100 to 200 m wide, low, vegetated sand spit (Fig. 4.143). The spit has been slowly growing across the mouth of Port Douglas, leaving behind the remnants of up to 20 former spits and narrowing the once 6 km wide mouth to 2 km. The main 4WD track reaches the western end of the spit, however vehicles are prohibited on the spit itself. The more exposed location, together with deeper water off the spit, permits slightly higher waves to reach the shore, which maintain a reflective beach with no sand flats.

Swimming: These are all relatively safe beaches, especially those fronted by sand flats. Just beware of the deep tidal channel and strong tidal currents off the tip of Point Longnose.

Surfing: None.

Fishing: Best off the northern rocks, or out over the deeper seagrass meadows.

Summary: Vehicle access to all these beaches is via the main 4WD track. It is a long, often sandy drive, but worth the effort to visit this pristine part of the coast.

Figure 4.143 *Point Longnose is a low, vegetated sand spit that has built out for 3.5 km across the entrance to Port Douglas (right).*

902-914   **PORT DOUGLAS (W) - Coffin Bay National Park**

	Unpatrolled			
No.	Beach	Rating	Type	Length
902	Pt Longnose (S)	1	R+sand flats	2.8 km
903	Salt Waterhole	1	R+sand flats	1.4 km
904	Port Douglas (3)	1	R+sand flats	2.1 km
905	Port Douglas (4)	1	R+sand flats	180 m
906	Port Douglas (5)	1	R+sand flats	250 m
907	Port Douglas (6)	1	R+sand flats	50 m
908	Black Springs (1)	1	R+sand flats	250 m
909	Black Springs (2)	1	R+sand flats	300 m
910	Black Springs (3)	1	R+sand flats	250 m
911	Black Springs (4)	1	R+sand flats	150 m
912	Port Douglas (11)	1	R+sand flats	100 m
913	Port Douglas (12)	1	R+sand flats	50 m
914	Eely Point (N)	1	R+sand flats	100 m

Port Douglas is an almost landlocked bay that enters the southern side of Coffin Bay between Point Longnose and Farm Beach. The port has more than 60 km of shoreline located in the main 'port' area, as well as additional shoreline in the southern Yangie Bay, the eastern Kellide Bay and northern Mount Dutton Bay. The only settlement in the entire area is the Coffin Bay township, which lies along the southern entrance to Kellide Bay. All the beaches within the port receive only wind waves generated within the bay, as well as having a lower tide range owing to the restriction at the entrance. All are fronted by sand flats, with seagrass occupying the deeper parts of the bay.

Along the western side of the port between Point Longnose and Eely Point is 15 km of predominantly sandy shore containing 13 low energy beaches and sand flats. The southern side of **Point Longnose** consists of a narrow, 2.8 km long, south facing beach (**902**), which is fronted by the 2 to 4 km wide abandoned flood tide deltas of the former port entrance, resulting in usually calm conditions at the beach. The entire point, including this beach, is off limits to vehicles, with the best access being by boat.

Beach **903** is a 1.4 km long, east facing shore that is eroding and truncating Holocene beach ridges along its southern half, with the sediment being moved northward to built a low sand spit along the northern half. Samphire and tidal flats lie in lee of the spit. A spur off the main 4WD track reaches the southern end of the beach. Beach **904** is a 2.1 km long, relatively straight, east facing beach, backed by vegetated transgressive dunes that have blown up to 5 km from Avoid Bay. The beach terminates at a low bedrock headland and is fronted by 200 to 300 m wide ridged sand flats. At the headland the shore turns and trends more toward the southeast, with bedrock and calcarenite dominating the shore and beaches to the east. The main track parallels the beach 100 to 200 m inland.

Beach **905** is a 180 m long, east facing sand beach, bordered by low, sand-covered bedrock headlands. Its neighbour, beach **906**, is similar except 250 m in length. There is no vehicle access to either beach.

A short spur off the main vehicle track does however reach the **Black Springs** area, which provides access to beaches 907, 908 and 909. Beach **907** is a 50 m long pocket of north facing sand, with the 4WD track leading to a small parking area right above the rocky, low bluff-bound beach. Black Springs beach (**908**) is 250 m long, faces north and has vehicle access to the grassy slopes behind the beach. The beach is bordered by low granite headlands and backed by partially cleared slopes, which are used for camping. The vehicle track continues east to the bluffs above the western end of beach **909**, a 300 m long, northeast facing sand beach, backed by partially cleared slopes. Its neighbours, **910** and **911**, are 250 m and 150 m long respectively, with low, crumbling calcarenite bluffs separating the three beaches and no vehicle access.

Beaches **912** and **913** are two strips of sand, 100 m and 50 m long respectively, located 2 km northwest of Eely Point. They face north and are backed by low calcarenite bluffs, with no vehicle access. Finally, beach **914** is a narrow, curving, east facing, 100 m long strip of sand on the northern side of Eely Point, with no vehicle access.

Swimming: All these beaches receive only wind waves generated in the bay and are calm when winds are low or offshore. They are all fronted by shallow sand flats of varying width.

Surfing: None.

Fishing: Better out in the port over the seagrass meadows.

Summary: The Black Springs beaches are the only ones regularly accessed by vehicle, the others are easier to access by boat.

915-923   **PORT DOUGLAS (S) - Coffin Bay National Park**

No.	Name	Rating	Type	Length
		Unpatrolled		
915	Eely Point (S)	1	R+sand flats	3.5 km
916	Port Douglas (15)	1	R+sand flats	2 km
917	Port Douglas (16)	1	R+sand flats	100 m
918	Port Douglas (17)	1	R+sand flats	200 m
919	Port Douglas (18)	1	R+sand flats	300 m
920	Port Douglas (19)	1	R+sand flats	150 m
921	Yangie Bay	1	R+sand flats	550 m
922	Port Douglas (21)	1	R+sand flats	1.5 km
923	Port Douglas (22)	1	R+sand flats	2.8 km

The southern shore of Port Douglas runs for 15 km between Eely Point and the town of Coffin Bay. The shoreline contains both sand and rock-dominated sections, with small Yangie Bay entering the port at its southern extremity. The main 4WD track skirts the back of the shore, with vehicle access restricted to the Eely Point beach (915) and the first beach in the national park (923). The others are only accessible by boat. All the beaches receive only wind waves during onshore winds and calms often prevail. They all generally consist of a narrow high tide beach fronted by shallow sand flats.

Eely Point is a low calcarenite bluff, with a 3.5 km long, east facing, curving beach (**915**) extending immediately south of the point. The main access track parallels the back of the entire beach, providing access at a few locations. The beach is backed by low, vegetated transgressive dunes and fronted by 200 to 300 m wide sand flats, containing transverse sand ridges (Fig. 4.144). The beach terminates at a recurved spit, that is enclosing part of the former sand flat and is separated from beach 916 by a 50 m wide entrance. Beach **916** continues on for 2 km as a slightly curving, north facing sand beach, backed by vegetated dunes then calcarenite and fronted by sand flats that narrow from 300 m in the west to 150 m in the east, where it terminates at low calcarenite bluffs. The main access track passes the western end of the beach.

Beaches **917** and **918** are two neighbouring beaches on the north side of Yangie Bay. They are 100 m and 200 m long respectively, bordered by low calcarenite bluffs and fronted by 100 m wide sand flats, with no vehicle access. Beaches **919** and **920** lie immediately north of the entrance to Yangie Bay. They are 300 m and 150 m in length, and are bordered and separated by low calcarenite bluffs, with access only by boat.

Figure 4.144 *The low energy beach east of Eely Point, showing the transverse intertidal sand ridges and the high backing dune that has blown over from the ocean beach.*

On the south side of the Yangie Bay entrance is 550 m long beach **921**, a north facing, very low energy beach fronted by 200 m wide sand flats and bordered in the east by calcarenite bluffs, with no vehicle access.

To the east of Yangie Bay are two longer sand beaches. Beach **922** is 1.5 km long and faces northwest across 300 to 400 m wide ridged sand flats. It is backed by vegetated transgressive dunes that originated from Gunyah Beach, 8 km to the southwest. Beach **923** is the easternmost beach in the national park, with the park boundary crossing toward the eastern end of the 2.8 km long beach. It is also backed by largely vegetated transgressive dunes, some of which have moved 10 km inland from the Gunyah Beach source. Toward the eastern end of the beach, near the ranger station, the dunes are replaced by a 500 m wide, low foredune ridge system. This beach is fronted by sand flats reaching up to 1 km in width.

Swimming: These are all relatively safe, with calm to low wave beaches fronted by shallow sand flats.

Surfing: None.

Fishing: Generally too shallow off the beaches, with deeper water out over the seagrass meadows.

Summary: Most of these remote beaches can only be accessed by boat, with the vehicle track only touching three of the beaches.

924-930   COFFIN BAY TOWN BEACHES

No.	Beach	Rating	Type	Length
		Unpatrolled		
924	Coffin Bay (1)	1	R+sand flats	200 m
925	Coffin Bay (2)	1	R+sand flats	150 m
926	Coffin Bay (3)	1	R+sand flats	50 m
927	Coffin Bay (4)	1	R+sand flats	50 m
928	Coffin Bay (5)	1	R+sand flats	100 m
929	Coffin Bay (6)	1	R+sand flats	100 m
930	Coffin Bay (7)	1	R+tidal flats	550 m

Coffin Bay is a small but growing settlement located along the southern entrance of Kellide Bay where it joins Port Douglas. The town extends for 3 km along the shore, with fishing boats moored in the 100 m wide tidal channel that links the two bays. The town is bordered by Coffin Bay National Park to the south, Kellide Bay Conservation Park to the east and the bay to the north. It is consequently a focus not only for fishing, but also for visitors to the surrounding parks, bays and beaches.

Six small adjacent, west facing beaches are located along the western shore of the town, each bordered by calcarenite bluffs and backed by a foreshore reserve, access roads and houses. They are all fronted by sand flats, which are 200 m wide in the south, narrowing to 50 m in the north.

The first of the beaches (**924**) is 200 m long and backed by the access road to the national park. The second (**925**) is 150 m long, with a narrow reserve then houses. Beaches **926** and **927** are both 50 m long pockets of sand, bordered by low calcarenite bluffs and backed by the reserve and houses. Beach **928** is 100 m long, while the northern beach, **929**, is 100 m long and terminates at a low bluff that forms the southern point of the entrance to Kellide Bay. The sand flats narrow to less than 50 m off this beach. At the point the shore turns east and beach **930** is a curving, crenulate, 550 m long, narrow beach, that faces north across 200 m wide sand flats and the deeper Kellide Bay entrance channel.

Swimming: These beaches receive only low wind waves and are fronted by shallow sand flats.

Surfing: None.

Fishing: Best in the channel.

Summary: One of the few developed sections of coast in this part of the peninsula, with good access to all beaches and a foreshore reserve between the houses, roads and beaches.

931-941   PORT DOUGLAS (E)

Unpatrolled				
No.	Beach	Rating	Type	Length
931	Koromoonah	1	R+tidal flats	150 m
932	Horse Peninsula (1)	1	R+tidal flats	300 m
933	Horse Peninsula (2)	1	R+sand flats	250 m
934	Horse Peninsula (3)	1	R+sand flats	450 m
935	Horse Peninsula (4)	1	R+tidal flats	100 m
936	Horse Peninsula (5)	1	R+sand flats	800 m
937	Horse Peninsula (6)	1	R+tidal flats	700 m
938	Horse Peninsula (7)	1	R+tidal flats	200 m
939	Horse Peninsula (8)	1	R+tidal flats	450 m
940	Horse Peninsula (9)	1	R+sand flats	1.5 km
941	Little Douglas	1	R+sand flats	900 m

The **Horse Peninsula** is a 10 km long, up to 1 km wide, north-south trending, low calcarenite ridge surrounded by Mount Dutton Bay to the east and the northern section of Port Douglas to the west. It is in fact a lithified Pleistocene (i.e. at least 100 000 years old) sand spit formed in the same manner as the new modern spit (beaches 935-941). A gravel road from Farm Beach terminates at the Little Douglas shacks in the northern corner, the only occupation on the peninsula. From there a 4WD track runs south, following the rear of the low dunes that back the beaches, along the length of the peninsula and provides rough access to the beaches.

Koromoonah beach (931) is a 150 m long, crenulate strip of sand on the south facing 'foot' of the rocky peninsula. It is bordered by low calcarenite bluffs, with a clump of rocks forming a small cuspate foreland in the centre. Koromoonah homestead lies 300 m west of the beach.

Beach **932** lies in a curving, 300 m long bay at the 'heel' of the peninsula. It faces southwest and has stunted mangroves in the western corner, and 200 m wide sand flats off the beach. The main track passes the back of the beach.

The remaining beaches are all located on the west facing shore and can potentially receive higher wind waves, however shallow rock flats significantly lower wave height on some of the beaches. Beach **933** lies 2 km north of the 'heel' of the peninsula. It is a 250 m long, southwest facing strip of sand on the southern side of a small cuspate foreland, formed in lee of a low calcarenite outcrop. It is backed by a low foredune and fronted by 150 m wide sand flats. The northern side of the foreland is occupied by 450 m long beach **934**, which is fronted by a mixture of sand and low rock flats, with rock reefs bordering each end. Beach **935** is a curving, 100 m long beach fronted by 50 m wide rock-tidal flats, then 400 m wide sand flats.

Beach 935 is the beginning of a continuous chain of beaches all the way to the Little Douglas shacks, with rock outcrops and forelands defining the boundaries. Beach **936** is an 800 m long, continuous sandy beach that faces west and is fronted by 300 m wide sand flats, while low 200 m wide dunes back the beach. It terminates at a protruding sandy foreland. Seven hundred metre long beach **937** continues to the north. It is fronted by shallow rock flats, which form a tidal flat along the more crenulate beach. It merges in the north with 200 m long beach **938**, which is also fronted by rock flats and terminates at a foreland composed of low, vegetated rock flats.

Beach **939** begins amongst rock flats, which give way to sand flats halfway along the 450 m long beach. To the north is 1.5 km long beach **940**, the longest of the peninsula beaches. Sand moving south from Little Douglas is building the beach out in a series of elongate spits, impounding salt and samphire flats. Finally, beach **941** begins at the rocks forming the eastern entrance to Little Douglas Bay (Fig. 4.145) and runs south for 900 m to an inflection that separates it from beach 940. The shacks lie behind the northern end of the beach, which is also used for launching boats. Sand flats up to 1 km wide extend off the beaches, with only the tidal channel into Little Douglas Bay at the very northern end providing deeper water for boating.

Figure 4.145 *Little Douglas Bay has been formed as the spit from Farm Beach has built southward, enclosing the bay. Beach 941 commences on the right, while the tip of beach 942 is seen to the left.*

Swimming: Most of these beaches are too shallow for swimming, except at high tide. Conditions are often calm with only low wind waves providing low surf.

Surfing: None.

Fishing: Best at Little Douglas in the tidal channel.

Summary: Apart from Little Douglas, this is a little used or visited section of coast.

942-952   **COFFIN BAY (E)**

No.	Beach	Unpatrolled Rating	Type	Length
942	Farm Beach (S)	2	R+sand flats	2.3 km
943	Farm Beach	3	R	2 km
944	Farm Beach (N)	3	R	200 m
945	Beach 945	5	R+rocks	50 m
946	Beach 946	5	R+rocks	70 m
947	Gallipoli Beach	3	R	250 m
948	Mena Hill (1)	4	R	400 m
949	Mena Hill (2)	4	R	180 m
950	Mena Hill (3)	4	R	180 m
951	Mena Hill (4)	4	R	1.1 km
952	Mena Hill (5)	4	R	150 m

Farm Beach is a small shack and freehold settlement located at the northern end of the beach of the same name. A gravel road runs out to the settlement and the beach is a popular boat launching area. This beach also marks a transition between the very protected beaches of Port Douglas to the south and the increasingly exposed beaches to the north.

Farm Beach consists of three sections. Farm Beach south (**942**) is a 2.3 km long, up to 300 m wide sand spit, that has been slowly extending south into Port Douglas, in the process impounding the backing Little Douglas Bay. The beach forms the eastern side of the entrance to Port Douglas and as a result, strong tidal currents flow in the deep channel 100 m off the beach. The beach usually receives only very low swell or wind waves and is fronted by 100 m wide sand flats, then the tidal channel. The only access is by 4WD down the beach.

The main Farm Beach (**943**) extends north for 2 km from a sandy inflection point adjoining the southern beach, up to the northern boundary bluffs next to the boat ramp area, with the shacks and houses located on the backing bedrock slopes. The beach is exposed to low refracted ocean swell, which averages less than 0.5 m. Calm conditions are also common. Apart from the access road and shacks, there are no facilities at the beach. Beach **944** lies below the northern bluffs and can be reached on foot at low tide from the main beach. It is 200 m in length, faces west and also receives low waves.

North of Farm Beach, steep bluffs and cliffs composed of metasedimentray rocks dominate the shore for the first 15 km, with usually small beaches wedged into gaps in the bluffs. The shore trends north-south, with all the beaches facing west. Swell refracted around Point Sir Isaac, 15 km to the west, gradually increases northwards. A 6 km long 4WD track runs along the top of the bluffs, becoming poorly defined north of Gallipoli Beach but providing access to the bluffs above the beaches as far as

beach 952. However a steep descent is usually required to reach the beaches.

Beaches **945** and **946** are two pockets of sand, 50 m and 70 m long respectively, located 2 km north of the boat ramp. They are backed by steep, 20 m high bluffs, with rock debris covering much of both beaches. **Gallipoli Beach (947)**, 3 km north of the ramp, was used for the famous landing site in making the 1981 movie 'Gallipoli' and the name has stuck. Like its Turkish namesake, the 250 m long beach is backed by steep, 30 m high bluffs. Waves average about 0.5 m and regular beach cusps are a feature of the beach (Fig. 4.146).

Figure 4.146 *A view south along the usually low energy Gallipoli Beach, with a series of active and higher, larger inactive beach cusps apparent. The higher ones, outlined by the vegetation, represent the limit of past storm erosion.*

Immediately north of Gallipoli Beach are three small bluff-bound beaches. The bluffs rise 170 m to the crest of **Mena Hill** within a kilometre of the coast. Beach **948** is 400 m long and is not only bordered by 30 m high bluffs, but cut into three sections by bluffs protruding onto the beach. Beach **949** is a 180 m long pocket of sand wedged below 40 m high bluffs, while beach **950** is also 180 m long, with similar steep bluffs bordered by bedrock platforms to either end.

On the northern side of beach 950 is 1.1 km long beach **951**, the longest beach in this section. It is a relatively straight, west facing, steep sand beach that receives waves averaging about 0.5 m, which surge up the beach face. It is backed by steep, 50 m high bluffs, requiring a steep climb down to the beach at the southern end. Its neighbour, beach **952**, is a 150 m long strip of sand at the base of 30 m high cliffs and is only accessible by boat or around the rocks at low tide from beach 951.

Swimming: Farm Beach is the most accessible and popular location, while Gallipoli is the more scenic. All beaches are relatively safe under normal low wave to calm conditions, though access and rocks are a hazard on the northern beaches.

Surfing: Usually too small, with only shorebreaks in larger swell.

Fishing: Most fishers launch their boats at Farm Beach to head out to the deeper reefs.

Summary: An accessible, relatively low energy section of coast, mainly frequented by boat fishers.

953-960   **FRENCHMAN BLUFF to COLES POINT**

No.	Beach	Rating	Type	Length
Unpatrolled				
953	Frenchman Bluff (S 3)	4	R	1 km
954	Frenchman Bluff (S 2)	3	R	130 m
955	Frenchman Bluff (S 1)	4	R+cliffs+reef	200 m
956	Frenchman Bluff (N)	4	R/LTT	550 m
957	Coles Point (S 4)	5	R/LTT+rocks	400 m
958	Coles Point (S 3)	5	R/LTT+rocks	600 m
959	Coles Point (S 2)	4	R+reef	200 m
960	Coles Point (S 1)	3	R+reef	500 m

Between the Frenchman Bluff area and Coles Point is 8 km of more irregular bedrock-dominated coast, with usually steep bluffs and cliffs backing eight beaches. The bluffs above all the beaches are accessible from the gravel road out to Frenchman Bluff in the south (beaches 953-956) and from the Coles Point road in the north (beaches 957-960). Apart from the access road, tracks and lookouts, there is no development or facilities. Two kilometres north of the bluff and adjacent to the shoreline is the prominent 50 m high Frenchman Lookout.

This section of coast faces generally west and receives lower swell that slowly increases in size northwards, with some of the beaches receiving protection from headlands and reefs. Beach **953** is a curving, 1 km long sand beach, which lies immediately below the point where the Frenchman Bluff road reaches the shore. It receives waves averaging about 0.5 m in the south, which decrease along the more protected, south facing northern section. There is a good view of the beach from a northern car park, however the best access is at the southern end. On the west side of the car park, at the base of 20 m high cliffs, is beach **954**. This is a 130 m long, south facing reflective pocket beach, backed by the bluffs and bordered by two small headlands that both contain car parks.

Beach **955** lies below 40 m high cliffs on the southern side of Frenchman Bluff. It is a 200 m long sliver of sand fronted by deeper reefs, with only low waves at the shore. While the road runs along the top of the bluffs, the beach can only be accessed down the steep eastern bluffs. On the north side of the bluff is 550 m long, west facing beach **956**. This is a more exposed beach receiving waves between 0.5 and 1 m on average, which maintain a narrow

attached bar and surf zone. However the beach is backed by steep, 50 m high bluffs, with the best access down a gully at the northern end.

Two and a half kilometres to the north is the first of the **Coles Point** beaches. Beach **957** is a 400 m long, west facing sand beach backed by 20 m high bluffs, with a vehicle track leading to the southern bluffs. The beach receives waves averaging up to 1 m, which break across a 50 m wide, reef and rock-filled surf zone, with a permanent rip draining the surf. On the northern side of the rocks is 600 m long beach **958**, a straight, west facing beach that receives waves averaging 1 m. These maintain a 50 m wide bar which is usually cut by two beach rips, together with some rocks along the beach. The vehicle track runs along the 20 m high bluffs behind the beach, with no formal foot access. It is bordered by a protruding bluff at the northern end, on the northern side of which is 200 m long beach **959**, a pocket of sand backed by the bluffs and partially fronted by reefs. Bluffs and a rock platform separate it from beach **960**, which lies on the southern side of Coles Point. The beach is 500 m long, faces south and curves round slightly under 20 m high bluffs. There are bluff-top car parks at either end, with the best access down the steep southern bluff.

Swimming: While all these beaches receive relatively low waves, their difficult foot access and the presence of rocks and reef make them moderately hazardous. Use care in accessing and swimming on any of these beaches.

Surfing: Usually only a heavy shorebreak on the more exposed beaches during higher waves.

Fishing: There are many good rock fishing spots, but again access can be difficult and watch for larger waves.

Summary: Most people visit this section of the coast for the views from the many lookouts, with only a few keen fishers using some of the beaches.

961-967   **GREENLY BEACH, MOUNT GREENLY**

No.	Beach	Rating	Type	Length
Unpatrolled				
961	Greenly Beach	5	LTT/TBR	1.4 km
962	Greenly Beach (N)	5	LTT/TBR	600 m
963	Mount Greenly (1)	5	LTT+reef	150 m
964	Mount Greenly (2)	5	LTT+reef	450 m
965	Mount Greenly (3)	6	TBR+reef	1.9 km
966	Mount Greenly (4)	6	TBR	300 m
967	Mount Greenly (5)	6	TBR	300 m

The first 6 km of coast north of Coles Point is a relatively straight section of bedrock and calcarenite-dominated coast containing six beaches, all bounded and dominated to varying degrees by the rocks and reefs. Wave height continues to increase slowly up the coast and the first beach rips and permanent rips north of Coffin Bay are encountered. Access to Greenly Beach and beaches 962 to 967 is via the Greenly Beach road and tracks to the north.

Greenly Beach (961) is located 10 km west of the highway and is a relatively popular swimming and surfing beach, particularly as it offers some of the first consistent surf north of Coffin Bay (Fig. 4.147). The beach is afforded some protection in the southern corner by Coles Point, with the remainder trends due north and is exposed to westerly waves, which average about 1 m but can range from calm to higher conditions. Rips are usually present up the beach, with strong rips forming during higher waves. The beach terminates at a 10 m bluff fronted by a rock platform, on the north side of which is beach **962**. This is a slightly crenulate, 600 m long beach bordered by bluffs, with some reef toward the centre and permanent rips at either end. Both beaches are backed by continuous active sand dunes extending up to 1 km inland. The access road runs out to the southern end of Greenly Beach and parallels the rear of the sand dunes, with 250 m high Mount Greenly immediately to the east.

Figure 4.147 *Greenly Beach and Mount Greenly. Waves and surf usually increase up the beach.*

Beach **963** lies 3 km north of Greenly Beach and just below the access road, with a small car park overlooking the southern end of the 150 m long beach. The sand beach has a surf zone dominated by rocks and reefs, and is drained by a permanent rip. The backing bluffs, rocks and reefs continue north and dominate its neighbours, 450 m long beach **964** and 1.9 km long beach 965. The main vehicle track terminates on the bare, 40 m high bluffs behind beach **965**, with access via a gully to the beach. This beach usually has two to three beach rips, as well as permanent reef-controlled rips. There are also some quieter rock pools in the reefs.

Immediately north are beaches **966** and **967**, two similar 300 m long, bluff-bordered and backed beaches, with jagged rock platforms off the headlands. A 4WD track winds through the backing bluff-top dunes to provide access to the rear of the beaches. Both have 100 m wide surf zones dominated by permanent headland-controlled rips.

Swimming: Greenly Beach is by far the most accessible and popular of these beaches. It is relatively safe during normal low waves, but will have strong rips and at times a heavy shorebreak when waves exceed 1 m. Most of these beaches are relatively hazardous, owing to the rocks, reefs and rips.

Surfing: Only during higher swell at Greenly and the other more open beaches.

Fishing: This is a popular section of coast, offering permanent rip holes which can be fished from the beaches or rock platforms.

Summary: An accessible and relatively scenic section of coast, however be careful if swimming or rock fishing.

968-972   KAPUNTA

No.	Beach	Rating	Type	Length
		Unpatrolled		
968	Kapunta (1)	7	R+rock&reef	50 m
969	Kapunta (2)	7	R+rock&reef	50 m
970	Kapunta (3)	8	TBR+rock&reef	100 m
971	Kapunta (4)	7	LTT+rock&reef	80 m
972	Kapunta (5)	7	R+rock&reef	50 m

Calcarenite cliffs continue to dominate the shore between 6 to 9 km north of Coles Point. The cliff line is irregular in places, with a few small pocket sand beaches wedged in the deeper crenulations. Five beaches, with a combined length of 300 m, occupy parts of this 3 km section of rocky shore. All are backed by 40 to 50 m high calcarenite cliffs that are capped by deflated cliff tops and scattered dunes, with cleared farmland a few hundred metres inland. Access to the cliffs above the beaches is via 4WD tracks running along the tops of the cliffs.

Beaches **968** and **969** lie immediately north of beach 967. They are two 50 m pockets of sand, wedged in at the base of 40 m high bluffs and fronted by a mixture of rocks, reefs and surf, with a permanent rip draining each beach.

Beach **970** is a straight, 100 m long beach fronted by a 100 m wide, partially reef-controlled surf zone, with a permanent rip flowing out from the southern end of the beach. Beach **971** is an 80 m long strip of sand below steep bluffs, with boulder debris on the beach. A 150 m wide, reef-dominated surf zone lowers waves at the shore. Beach **972** is a 50 m pocket of sand wedged deep in a small embayment and completely backed by steep cliffs. A farm boundary fence and 4WD track terminate above the beach.

Summary: These are six difficult to find and access beaches, none are suitable for safe swimming. There is potential for surfing breaks amongst the reefs and point, as well as rip and rock holes for fishing, but local knowledge is essential and waves can be a hazard on the rocks.

973-983   CONVENTION BEACH

No.	Beach	Unpatrolled Rating	Type	Length
973	Convention (S 3)	8	LTT+rocks/reef	60 m
974	Convention (S 2)	8	TBR+rocks	50 m
975	Convention (S 1)	8	TBR	40 m
976	Convention	7	TBR/RBB	2.2 km
977	Convention (N 1)	7	TBR/RBB	500 m
978	Convention (N 2)	7	TBR/RBB	100 m
979	Convention (N 3)	8	TBR+rocks/reef	350 m
980	Convention (N 4)	8	TBR+rocks/reef	120 m
981	Convention (N 5)	8	TBR+rocks/reef	210 m
982	Convention (N 6)	8	TBR+rocks/reef	70 m
983	Convention (N 7)	8	RBB+rocks/reef	600 m

Convention Beach is a 7 km long beach that is broken into eleven small beaches by protruding calcarenite bluffs, with all the beaches sharing a continuous high energy, 200 to 300 m wide, rip-dominated surf zone. The entire beach is backed by generally active dunes, averaging 1 km in width, then farmland. The highway lies 7 km inland, with access in the south to beaches 973-976 via a series of 4WD tracks running in from the highway and in the north via farm tracks and off the northern Mount Drummond road to beaches 977-982. Local knowledge is needed to locate the tracks and permission may be required to cross farmland. There are no facilities and no development along the beach.

The entire beach system faces almost due west and receives moderately high waves, averaging 1.5 m in height, with higher waves common. These maintain the rip-dominated surf zone and, together with the bluffs, rocks and reefs, produce generally hazardous surf conditions.

Beach **973** lies at the southern end of the surf zone and 1.5 km south of the main beach. It is a 60 m pocket of northwest facing sand bordered by 30 m high bluffs, together with rocks and reefs in the 200 m wide surf zone and a permanent rip draining the beach. One kilometre to the north is beach **974**, a 50 m strip of sand bordered and

backed by 50 m high cliffs, with foot access via a steep, vegetated gully. Beach **975**, 200 m to the north, is a similar 40 m pocket of sand, but has no safe access.

The main Convention Beach (**976**) is a straight, west facing, 2.2 km long sand beach, backed by active dunes extending up to 1 km inland and fronted by a rip-dominated, 300 m wide surf zone, usually containing ten strong beach rips. It is bordered at either end by small calcarenite bluffs.

Beaches 977 and 978 are a continuation of the main beach, with small bluffs truncating the beach at the shoreline, while the surf zone continues on uninterrupted. Beach **977** is 500 m long, while **978** is just 100 m long. Both are bordered and backed by the 30 m high bluffs, with some rock debris on the beaches and reef in the surf zone.

Beaches 979 to 982 occupy a 1 km long section of bluff-dominated shore, capped by unstable clifftop dunes extending several hundred metres inland. Each of the beaches lies in an indentation in the 30 m high bluffs and is fronted by the continuous surf zone (Fig. 4.148). Beach **979** is a discontinuous 350 m long, narrow strip of sand broken by some major rock falls. Beach **980** is a 120 m long, wider beach bordered and backed by the bluffs. Beach **981** is 210 m in length, with scattered reef in the surf, while beach **982** is a 70 m pocket beach, also with reef in the inner surf zone.

Figure 4.148 *The northern Convention beaches (beaches 979 to 982) occupy indentations in the 30 m high, dune-capped calcarenite bluffs, with a continuous outer bar linking the beaches.*

Beach **983** is the northernmost segment of Convention Beach and is a straight, 600 m long, exposed beach backed by low vegetated bluffs and partially active, 800 m wide sand dunes. The surf zone averages 300 m in width and usually contains three to four strong beach rips.

Swimming: These are eleven hazardous beaches, particularly those dominated by the rocks and reefs, with strong persistent rips on all beaches.

Surfing: Best chance on the more open beach breaks, however access can be difficult.

Fishing: There are deep permanent rips and holes on all these beaches.

Summary: An energetic, generally bluff-dominated section of coast, with limited and difficult access, resulting in little usage other than by local farmers and knowledgeable fishers.

984-986   **PICNIC BEACH**

		Unpatrolled			
No.	Beach	Rating	Type		Length
			Inner	Outer Bar	
984	Picnic	7	TBR	RBB	7 km
985	Picnic (W 1)	4	LTT+rocks/reef		150 m
986	Picnic (W 2)	4	R+rocks/reef		250 m

Picnic Beach (984) is a 7 km long beach that begins on the south side of Point Drummond as a south facing, partially protected white sand beach, that curves around to face southwest and finally west as it runs down to link with the northernmost Convention Beach. There is vehicle access via Point Drummond to a car park in lee of the northern end of the beach, where waves are generally less than 1 m in height. However, south of the car park wave height increases rapidly and a 300 m wide, rip-dominated surf zone can contain up to 20 rips in the inner surf zone, with larger rips in the outer surf zone. Generally active sand dunes, up to 2 km wide, back the entire beach, with farmland further inland and 160 m high Mount Drummond in the south.

West of the car park are two small, more protected beaches that become increasingly dominated by the rugged metasedimentary rocks of the point. Both can be accessed from the main beach, with a car park located above the southernmost beach. Beach **985** is 150 m long and usually has a narrow, shallow bar and low waves, with rocks and reefs scattered across the bar. Beach **986** is a more irregular, 250 m long, curving sand and boulder beach, which is broken by more prominent rocks and reefs. A steep gravel track has been cut in the 20 m high red bluffs to provide access to the beach for launching small boats.

Swimming: The northern, more protected section of the main and western beaches usually have lower waves and are relatively safe. However rips usually form from the car park south, so care is needed if swimming down the beach.

Surfing: This is a popular spot to check the swell and surf, with numerous beach breaks down the beach.

Fishing: A popular fishing beach with both persistent rip holes, as well as more sheltered rock fishing in lee of the point.

Summary: A traditionally popular beach, as the name implies, but with no facilities.

987-1003   **POINT DRUMMOND**

	Unpatrolled			
No.	Beach	Rating	Type	Length
987	Pt Drummond (S 14)	4	R/LTT+rocks/reef	100 m
988	Pt Drummond (S 13)	4	R/LTT+rocks/reef	50 m
989	Pt Drummond (S 12)	6	R+rocks/reef	60 m
990	Pt Drummond (S 11)	7	R+platform	200 m
991	Pt Drummond (S 10)	7	R (boulder)	100 m
992	Pt Drummond (S 9)	7	R (boulder)	150 m
993	Pt Drummond (S 8)	5	R/LTT+reef	50 m
994	Pt Drummond (S 7)	4	R+rocks/reef	250 m
995	Pt Drummond (S 6)	4	R+reef	350 m
996	Pt Drummond (S 5)	5	R+platform	250 m
997	Pt Drummond (S 4)	6	R+rocks/reef	40 m
998	Pt Drummond (S 3)	6	R+rocks/reef	40 m
999	Pt Drummond (S 2)	6	R+rocks/reef	100 m
1000	Pt Drummond (S 1)	5	R+rocks/reef	100 m
1001	Pt Drummond	6	R+platform	150 m
1002	Pt Drummond (N 1)	6	R+rocks/reef	250 m
1003	Pt Drummond (N 2)	5	R+reef	160 m

Point Drummond is a prominent 40 m high bedrock headland composed of deeply weathered granite and gneiss, with the weathering producing the red capping. It protrudes 4 km west of the general trend of the coast. On the south side of the point are 4 km of irregular bedrock cliffs, bluffs and rock platforms, together with 14 small rock and reef-dominated beaches (987-1000), including some boulder beaches. While the beaches face generally west into the swell, the fronting rocks and reefs tend to lower waves at the shore. On the north side is the small point beach (1001) and two reef-dominated beaches (1002-1003). The Point Drummond-Picnic Beach road runs around the perimeter of the point and provides good access to the cliffs and bluffs above all the beaches, with foot access usually down steep slopes.

Immediately north of the southern tip of the point is beach **987**, a 100 m long sand beach fronted by 150 m of rocks and reefs. Its neighbour **988** is a 50 m pocket of sand, also fronted by 150 m wide reefs and rocks. One hundred metres to the north is beach **989**, a 60 m strip of sand in lee of 50 m of shallow reefs.

Five hundred metres to the north is 200 m long beach **990**, a more irregular, rock-dominated beach fronted by both a platform and rocks and reefs. Beaches **991** and **992** are both steep boulder beaches, 100 m and 150 m long respectively.

Beach **993** lies 2 km south of the point and has a vehicle track leading to a lookout above the 50 m long beach. The beach lies in a small, narrow, reef-filled embayment, resulting in low waves at the shore. On the north side of the lookout is a 400 m wide embayment containing beaches **994** and **995,** two curving 250 m and 350 m long beaches, both backed by steep, 40 m high bluffs and fronted by deeper reefs, with low waves at the shore.

Beaches 996, 997 and 998 all lie within 1 km of the point. They face west and are more exposed, with waves breaking across shallow reefs and rocks up to 200 to 300 m off the beaches. Beach **996** is 250 m long, backed by steep bluffs and fronted by a platform, then the rocks and reefs. Beaches **997** and **998** are two 40 m long pockets of sand in semicircular breaches in the bluffs. Both are fronted by the wider reef-strewn surf zones.

Beach **999** lies just south of the point and consists of a 100 m long pocket of sand located in a 200 m long V-shaped bay, bordered by prominent rock platforms and bluffs. Beach **1000** is located on the western side of the point and is a curving, 100 m long sand beach facing into a 200 m long, narrow embayment and is surrounded on three sides by 20 m high bluffs. Beach **1001** is located at Point Drummond. It is a slightly curving, 150 m long sand beach fronted by a 40 m wide intertidal platform, with reef further offshore and prominent rocky bluffs to either end.

On the northern side of the point are beaches 1002 and 1003. Beach **1002** is an irregular, 250 m long beach containing three sand arcs formed in lee of rock outcrops, while beach **1003** is a slightly curving, 160 m long sand beach, backed by steep, 40 m high bluffs and fronted by shallow reefs extending 300 m offshore.

Swimming: While all the beaches are protected to varying degrees by the rocks and reefs, with often low waves at the shore, be careful if swimming here as rocks abound and some beaches are drained by strong permanent rips.

Surfing: There is potential for waves amongst the many reefs and points around Point Drummond, with the access road providing a good view of all the surf around the point.

Fishing: A relatively popular fishing area with generally good access to a wide range of beach, rock and reef sites.

Summary: An accessible section of rugged coast, mainly offering views to the passing tourist, with only a few locals actively using the beaches for fishing and occasionally surfing.

1004-1005 HILL BAY

| No. | Beach | Unpatrolled | | |
		Rating	Type	Length
1004	Hill Bay	2	R+reef	200 m
1005	Hill Bay (N)	4	R+reef	200 m

Hill Bay lies 3 km northeast of Point Drummond and is an open, west facing bay, backed by 30 m high bluffs, but protected by a line of granite reefs running off the southern point. The reef provides a rare safe anchorage on this exposed section of coast, which is used by professional fishermen to temporarily moor their boats (Fig. 4.149).

Figure 4.149 *Hill Bay offers safe, if difficult to access, anchorage in lee of a granite reef. Clifftop dunes blanket the southern bluffs.*

Inside the 1.5 km wide bay are two 200 m long beaches. The southern beach (**1004**) faces north toward the moorings and consists of a protected narrow strip of sand at the base of sand-draped, 30 m high bluffs. There is a car park on top of the bluff providing a view of the bay. On the north side of the bay is beach **1005**, which faces west but is afforded protection by the reef-filled bay, with low waves breaking across a shallow reef immediately off the beach. It is backed by the bluffs, with a steep climb required to reach the beach.

Swimming: Both beaches are relatively protected, however access is difficult and neither is recommended for swimming.

Surfing: There are some low reef breaks over the northern beach.

Fishing: Most fishers come in here to moor their boats, with good fishing inside the protecting reefs.

Summary: One of the few 'working' beaches on this section of coast.

Regional Map 10: Western Eyre Peninsula

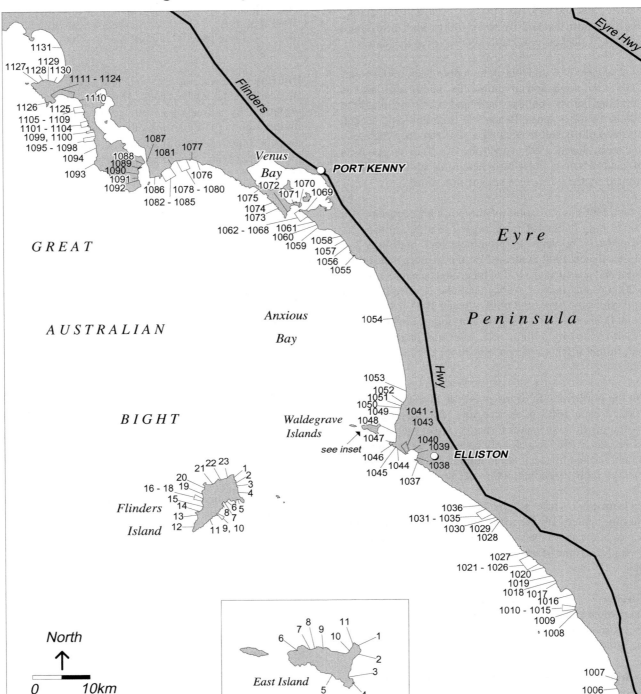

Figure 4.150 *Regional Map 10 – Western Eyre Peninsula*

1006-1007 KIANA BEACH

No.	Beach	Rating	Type Inner	Outer Bar	Length
			Unpatrolled		
1006	Kiana	8	TBR/RBB	RBB	1.1 km
1007	Kiana (N)	3	R		80 m

Kiana Beach (1006) is an isolated strip of sand bordered by predominantly high calcarenite cliffs. It lies 8 km north of Hill Bay and 17 km south of Sheringa, the nearest neighbouring beaches. The beach is located just 1.5 km west of the highway and there is a vehicle track out to the northern and southern ends, with a steep climb down the 30 m high backing bluffs to reach the beach. The beach is 1.1 km long, relatively straight and faces due west. The 1013 and bluffs impinge on the beach in two locations, almost cutting it into three. It receives high westerly swell which maintains a 300 m wide, double bar system, with usually five beach rips cutting the inner bar.

One kilometre to the north is beach **1007**, an 80 m long pocket of sand wedged in at the base of a 300 m deep embayment and encircled by steep, 30 m high bluffs. It is located just 500 m west of the highway, with a vehicle track winding out to the southern side of the bluffs.

Swimming: Kiana is a very hazardous, rip-dominated beach, while the northern beach is safer but difficult to access on foot.

Surfing: There are usually beach breaks at Kiana, with a right point break at the northern beach.

Fishing: Kiana is relatively popular for beach fishing in the deep rip holes and off the southern rocks.

Summary: Two beaches close to the highway offering excellent views of the rugged coast from the overlooking car parks. However the beaches are only suitable for experienced surfers and fishers.

1008-1016 SHERINGA BEACH

No.	Beach	Rating	Type	Length
			Unpatrolled	
1008	Sheringa (S 8)	3	R+reef	30 m
1009	Sheringa (S 7)	8	R+rocks/reef	110 m
1010	Sheringa (S 6)	8	R+rocks/reef	60 m
1011	Sheringa (S 5)	8	R+rocks/reef	50 m
1012	Sheringa (S 4)	8	R+rocks/reef	40 m
1013	Sheringa (S 3)	8	R+rocks/reef	30 m
1014	Sheringa (S 2)	8	R+rocks/reef	30 m
1015	Sheringa (S 1)	8	LTT+rocks/reef	150 m
1016	Sheringa Beach	4	R/LTT+rocks/reef	3.9 km

Sheringa Beach is one of the more accessible beaches on this section of the west coast, lying 6 km off the highway, with a good gravel road right to the northern end of the beach. The beach is bordered by prominent calcarenite headlands, with the first kilometre of the southern headland containing eight small pocket beaches, all backed by 10 to 20 m high calcarenite bluffs and cliffs.

Beach **1008** is a 30 m long pocket of sand lying inside a 200 m deep, V-shaped breach in the cliffs. It is protected by outer reefs and waves are usually low at the shore. There is a 4WD track right to the top of the beach, with a steep climb down the bluffs to reach the shore. The track then runs north along the cliffs toward Sheringa Beach, past the seven more exposed neighbouring beaches. Beach **1009** is 110 m long, with rock debris on the beach and a 100 m wide, reef-dominated surf zone. Beaches **1010, 1011, 1013, 1014** are four neighbouring pockets of sand, each separated by small protruding calcarenite cliffs and sharing a common reef-dominated, 100 m to 200 m wide, higher energy surf zone. Beach **1015** is a 150 m long sand beach fronted by a 200 m wide reef, which has a left hand surf break running into the beach.

Sheringa Beach (**1016**) commences at the northern end of the bluffs and sweeps in a slight curve for 3.9 km to the northern calcarenite bluffs, which extend 1 km west of the beach. The entire bay area seaward of the beach is dominated by reefs, which cause wave breaking during high waves and lower waves at the shore to usually less than 1 m, to maintain a narrow, often barless, surf zone against the beach. A foredune backs the beach, then 1 to 2 km of active white sand dunes. The main access, car park and beach boat launching area are all located at the very northern end of the beach, which is also backed by a high tide boulder beach.

Swimming: The main Sheringa Beach is a relatively safe location owing to the protection of the reefs. Rips do form during higher waves and rocks and reefs line much of the seabed.

Surfing: Sheringa offers both low beach and outer reef breaks, both along the beach and off the southern headland.

Fishing: A popular area both for launching boats and to fish the inner reefs from the beach.

Summary: A relatively popular spot for sightseers, while the local surfers and fishers use the beach.

1017-1027 BACK SHERINGA, TUNGKETTA

		Unpatrolled			
No.	Beach	Rating	Type		Length
			Inner	Outer Bar	
1017	Back Sheringa	7	R/LTT		300 m
1018	Back Sheringa (N 1)	7	R/LTT+reef	RBB	600 m
1019	Back Sheringa (N 2)	7	LTT+reef	RBB	600 m
1020	Back Sheringa (N 3)	7	TBR+reef	RBB	1.1 km
1021	Back Sheringa (N 4)	7	LTT/TBR+reef		250 m
1022	Back Sheringa (N 5)	7	R+reef		100 m
1023	Back Sheringa (N 6)	7	R+reef		30 m
1024	Back Sheringa (N 7)	7	R+reef		160 m
1025	Back Sheringa (N 8)	7	R+reef		210 m
1026	Back Sheringa (N 9)	7	R+reef		80 m
1027	Tungketta	7	R+reef		250 m

The gravel road to Sheringa Beach runs out to the end of the northern headland, then turns and runs north for 3 km as far as beach 1019, beyond which is a 6 km long bluff-top 4WD track to beach 1027. Apart from the access tracks, there are no facilities at any of these beaches. All the beaches face west to southwest and are exposed to waves averaging 1.5 to 2 m, which break across outer reefs and, on some beaches, over outer bars.

The first back beach (**1017**) lies 2 km north of the headland and is one of three west facing beaches bordered by 10 m high calcarenite bluffs and separated by lower bluffs (Fig. 4.151). The gravel road runs along the bluffs behind the beaches, with a parking area at the northern end of beach 1019. Beach 1017 is 300 m long and is fronted by 300 m wide deeper reefs, with permanent rips at either end. The reef diminishes north of the beach and beaches **1018** and **1019** are adjoining 600 m long beaches, with usually an attached inner bar containing permanent rips at either end and an outer bar 300 m off the beach. The northern ends of both beaches become increasingly dominated by low protruding bluffs and rocks.

Figure 4.151 *Sheringa Beach's northern beaches begin at beach 1017 (lower) and continue north in lee of an energetic reef and rip-dominated surf zone.*

Beach **1020** is the longest beach along this section of coast. The 1.1 km long white sand beach faces west and is exposed to most swell, with waves breaking over both outer bars and scattered reefs, particularly toward either end. Three permanent rips drain the 300 m wide surf zone, while active dunes extend 1.5 km inland from the beach.

The remaining beaches are all fronted by shallow reef-platforms, resulting in lower waves at the shore, with waves breaking over deeper reefs further out. Beach **1021** is 250 m long and backed by collapsing 10 m high calcarenite bluffs and fronted by a 100 m wide reef-dominated surf zone. Beach **1022** is a 100 m long narrow strip of sand below 15 m high bluffs, while beach **1023** is a 30 m pocket of sand encased in bluffs and fronted by extensive reefs.

Beach **1024** is a straight, 160 m long beach, bounded by prominent protruding northern bluffs and fronted by 200 m to 300 m wide reefs, with a right hand surf break straight off the beach. Beach **1025** lies on the northern side of the bluffs and is a 210 m long sand beach, with a slight foreland in the centre in lee of shallow, 150 m wide reefs.

Three hundred metres to the north is 80 m long beach **1026**. It is a west facing beach protected by protruding bluffs and reefs, which result in usually calm conditions at the shore. Beach **1027** is the northernmost beach in this section and lies 2 km due east of **Tungketta Reef**. It is 250 m long, backed by steep, 20 m high bluffs and fronted by continuous shallow, 50 m wide reefs, then deeper reefs.

Swimming: Apart from beach 1026 these are all potentially hazardous beaches, dominated by reefs and fronted by strong permanent rips. Be very careful if entering the water anywhere along this section of coast.

Surfing: There are a number of reef and beach breaks along this section, with a drive along the bluff top providing a good view of the surf.

Fishing: This is a popular area to fish the deep permanent rip holes, especially on the first three beaches.

Summary: A partly accessible and relatively scenic section of coast, offering natural beaches with potential for surf and fishing, though hazardous for swimming.

1028-1036 LOCH WELL BEACH

		Unpatrolled			
No.	Beach	Rating	Type		Length
			Inner	Outer Bar	
1028	Loch Well (S 4)	8	RBB	RBB	450 m
1029	Loch Well (S 3)	8	RBB	RBB	200 m

1030	Loch Well (S 2)	8	RBB	RBB	750 m
1031	Loch Well (S 1)	8	RBB	RBB	400 m
1032	Loch Well Beach	8	RBB	RBB	500 m
1033	Loch Well (N 1)	8	RBB	RBB	200 m
1034	Loch Well (N 2)	8	RBB	RBB	200 m
1035	Loch Well (N 3)	8	RBB	RBB	150 m
1036	Loch Well (N 4)	8	RBB	RBB	400 m

Loch Well Beach is an accessible fishing spot located 3 km off the highway. It is a popular spot for viewing the cliff-dominated shore. The main beach lies immediately below the car park and access stairs. It is part of a 5 km series of nine, southwest facing beaches that share common continuous inner and outer bars containing strong rips spaced approximately every 350 m, with waves averaging about 2 m. The beach is truncated by 100 m high calcarenite spurs which break it into the nine sections, four to the south of the main beach and four to the north. Only the main beach is accessible by car, with the others only reached by climbing down steep gullies, while the northernmost beach (1036) is backed by 90 m high cliffs and is inaccessible.

Beach **1028** lies 2.5 km south of the car park. It is a 450 m long beach backed by steep cliffs, with a bare gully providing foot access. Two large rips usually drain the 300 m wide surf zone. Beach **1029** is 200 m long and more confined by the backing cliffs, with a gully also providing foot access. Beach **1030** is the longest continuous section of the beach at 750 m in length, however it is relatively narrow and backed by high cliffs, with access via two gullies toward the northern end. Beach **1031** is also a relatively narrow, 400 m long strip of sand below the cliffs and lies 300 m south of the car park. It can be reached during low waves and at low tide around the rocks from the main beach.

The main Loch Well Beach (**1032**) lies 100 m below the car park, from which there are dramatic views of the beach. Wooden stairs provide safe access to the beach. The beach is usually dominated by two inner rips, with larger rips in the outer 300 m wide surf zone (Fig. 4.152).

Three hundred metres north of the car park is beach **1033**, a 200 m long beach bordered by protruding 80 m high bluffs. It can also be accessed during low waves around the headland from the main beach. Immediately north is beach **1034**, a small, 200 m long pocket of sand wedged in a steep gully. Beach **1035** is a V-shaped, 150 m long section of beach backed by a steep V-shaped valley. One kilometre to the north is the final beach (**1036**), a straight, 400 m long, narrow strip of sand at the base of 80 m high cliffs, with no safe access.

Swimming: All these beaches including the main beach are very hazardous, with strong, deep rips always present right off the shore. Only enter the surf if you are experienced and can handle such conditions.

Surfing: The car park provides an excellent spot to check the swell and surf below. There are normally good beach breaks under low to moderate swell.

Fishing: A popular beach for fishing the deep rip holes right off the beach.

Summary: An accessible viewing spot and beach, worth the short drive out just for the view.

Figure 4.152 *The main Loch Well beach lies below the clifftop car park (centre). The car park and access have been upgraded since this photo was taken. Beware, however, as this is an exposed and very dangerous rip-dominated beach, is shown in this view.*

1037-1041 ELLISTON (WATERLOO BAY)

No.	Beach	Unpatrolled Rating	Type	Length
1037	Back	5	R+reef	120 m
1038	Boord	3	R	700 m
1039	Main	3	R/LTT	1 km
1040	Jetty	3	LTT	1 km
1041	Wellesley	3	R	500 m

Elliston is a small town located on the southern shores of Waterloo Bay. The bay forms an almost circular break in the 50 m high calcarenite cliffs that dominate the coast to the north and south. The entrance to the bay is 1.5 km wide, with Wellington Point forming the southern head and Salmon Point the northern (Fig. 4.153). A number of rock reefs lie off each point, with a 300 m gap between the reefs permitting waves and boats into the protected bay.

The **Back Beach** (**1037**) is located on the west facing, ocean side of the town, with one of the main streets running straight to the northern end of the 120 m long beach. The beach is set in a semicircular gap in the calcarenite bluffs and is fronted by shallow 100 m wide reefs.

The semicircular shoreline of Waterloo Bay runs for 5 km and faces initially west, then south and finally almost east against Wellesley Point. Most of the shore consists of low energy sandy beach, interspersed with low calcarenite bluffs. The bluffs divide the bay into five beaches. All receive low waves averaging between 0.5 and 1 m. As the waves squeeze through the entrance reefs, they are not only reduced in size, but they spread out (refract) around the bay shore, causing the shoreline and the orientation of the four beaches to range from west to east.

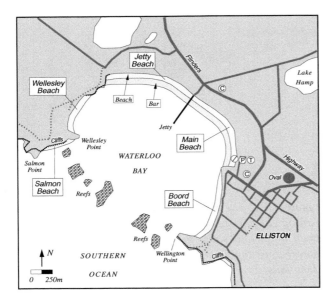

Figure 4.153 *Waterloo Bay is a semicircular bay formed in lee of a breach in the calcarenite cliffs to either side. The entrance reefs represent the original alignment of the rocky shore. The small town of Elliston is located on the southern shores of the bay.*

Boord Beach (1038) lies just inside Wellington Point. It is a 700 m long, narrow, west facing beach, which is backed by low calcarenite bluffs and has calcarenite rocks and reefs outcropping along the beach. The low waves and medium sand produce a relatively steep beach with no bar and deep water close inshore. A road runs along the top of the bluffs, providing good access to the length of the beach.

Main Beach (1039) is the focus of most recreational activities in Elliston. The road into town terminates at the bluff between Boord Beach and Main Beach, with a car park behind the southern end of Main Beach, fronted by a concrete-sand boat ramp. The 1 km long beach is composed of finer sand and has a low gradient, firm surface and shallow water off the beach (Fig. 4.154). It faces west, then southwest and terminates at the northern bluffs, which are the site of the 300 m long Elliston jetty.

Jetty Beach (1040) extends for 1 km to the west of the jetty bluffs. Like Main Beach it is composed of fine sand, however it receives slightly higher waves and has a wide, low beach fronted by a shallow, attached bar. Waves

usually average less than 1 m. There is road access to the beach from the jetty and at the western bluffs.

Figure 4.154 *Main Beach at Elliston, running north of the boat ramp (lower). The protected beach usually has calm to low wave conditions.*

Wellesley Beach (1041) lies tucked in behind Wellesley Point, which protrudes for 200 m into the bay from the western end of the 500 m long beach. The beach faces southeast and is well protected from waves entering the bay. As a result waves average 0.5 m at the beach, increasing slightly in size toward the east. The beach is usually moderately steep, with deeper seagrass-covered sand immediately offshore.

Swimming: The most suitable and popular beach at Elliston is Main Beach. It has the best access and toilet facilities. It is also relatively safe, with usually low waves and no rips. The other beaches are also relatively safe, though more remote, with Boord also having some rock reefs on and off the beach.

Surfing: There is usually no surf in the bay. During a big outside swell there is a dumping shorebreak on Jetty Beach.

Fishing: The jetty and bluffs around the beach are the best spots to reach deeper water, together with the more exposed and hazardous outer points.

Summary: Elliston is a nice little town on the shores of a usually quiet and attractive bay.

1042-1046 SALMON BEACH to CAPE FINNISS

Unpatrolled			
No.	Beach	Rating	Type	Length
1042	Salmon Beach	4	R+reef	150 m
1043	Salmon Point	4	R+reef	50 m
1044	Cape Finniss (S 3)	4	R+reef	40 m
1045	Cape Finniss (S 2)	3	R+reef	80 m
1046	Cape Finniss (S 1)	4	R+reef	60 m

Salmon Point forms the northern entrance to Waterloo Bay. Between the point and Cape Finniss, 4 km to the northwest, is a straight section of 40 to 50 m high calcarenite cliffs. Located in some of the indentations in the cliffs are five small beaches, all protected by extensive shallow reefs. A gravel road runs along the top of the cliffs to Cape Finniss, providing access to the bluffs above all the beaches, with beach access only via a steep climb down the bluffs.

Salmon Beach (1042) is a small 150 m long beach extending between Wellesley Point and Salmon Point. It faces south and is fronted by continuous reefs over which high waves break, dissipating into a shallow 'lagoon' between the reefs and the low energy beach. The beach is narrow and backed by 20 to 30 m high calcarenite bluffs. One hundred metres to the west is its smaller neighbour, 50 m long beach **1043**. It is also fronted by wide, shallow reefs and backed by steep, 30 m high bluffs.

Beach **1044** lies 2 km south of the cape and is a 40 m pocket of sand wedged below steep, 50 m high bluffs and fronted by 100 m wide, shallow reefs. Three hundred metres to the north is beach **1045**. It lies in a 300 m deep embayment and consists of two small pockets of sand totalling 80 m in length, with rocks crossing the centre. Reefs lower waves across the entrance to the bay and conditions are often calm at the shore. Beach **1046** is located just 500 m south of the cape and lies in a semicircular embayment at the base of steep, 50 m high bluffs. The 60 m long beach is fronted by 200 m of reef, which also substantially lower waves at the shore.

Summary: Most people drive along the Cape Finniss tourist drive to view this coast. Only a few fishers venture down to these small beaches, none of which are suitable for swimming.

Waldegrave Islands Conservation Park	
Established:	1967
Area:	434 ha
West Island:	approx. 30 ha & 2 km of rocky shoreline
East Island:	approx. 400 ha & 10 km of shoreline
East Island:	Beaches 1-10

E-1–E-11   EAST ISLAND - Waldegrave Islands Conservation Park

Unpatrolled				
No.	Beach	Rating	Type	Length
E-1	McLachlan Pt (E)	4	R/LTT	420 m
E-2	McLachlan Pt (S)	3	R+reef	150 m
E-3	Point Watson (N)	3	R+reef	200 m
E-4	Point Watson	5	R+reefs	40 m
E-5	Goose Hill	7	LTT+reefs	250 m
E-6	west beach	7	R+rocks/reef	120 m
E-7	north beach 1	6	R+rocks/reef	100 m
E-8	north beach 2	6	R+rocks/reef	150 m
E-9	north beach 3	6	R+reef	60 m
E-10	McLachlan Pt (W 2)	5	LTT	330 m
E-11	McLachlan Pt (W 1)	5	LTT	60 m

East Island is the larger of the two granite Waldegrave Islands. Its southeastern tip is located 2 km off Cape Finniss and 8 km northwest of Elliston. The island is relatively flat, with a maximum height of 40 m. It has 10 km of shore and 11 beaches totalling just under 2 km in length. Most of the shore consists of bedrock forming generally low bluffs fronted by jagged rock platforms. The smaller West Island has 2 km of rocky shore with just one 30 m wide sand-boulder beach toward the southeast end, which provides the only landing on the island.

The East Island beaches are all small and half are dominated by rocks and boulders. On the east side of the northern tip of the island, **McLachlan Point**, is a 420 m long, northeast facing beach **(E-1)** which receives low refracted waves and usually has a reflective beach with a narrow bar. On the southern side of the point is a narrow, 150 m long strip of sand **(E-2)** lying below 10 m high bluffs and facing southwest toward Cape Finniss. It is sheltered by the island, as well as reefs lying between the island and cape, and is usually calm.

Point Watson forms the southeast tip of the island. On its north side is protected, northeast facing beach **(E-3)**. It is 200 m in length and, while fronted by reefs, provides the best anchorage on the island. Right on the tip of the point is a small 40 m pocket of southeast facing sand **(E-4)** surrounded by 10 m high bluffs, with waves breaking over the fronting rocks and reefs.

There is only one beach **(E-5)** on the 3 km long southern side of the island. It is a 250 m long, south facing sand beach, backed by 30 m high bluffs rising to 40 m high Goose Hill, the peak of the island. The beach is fronted by surf breaking across 100 m wide reefs, the only surf on the island.

The north side of the island has 4 km of rocky shoreline containing six small beaches. Toward the western tip of the island is a curving, 120 m long strip of high tide sand **(E-6)** fronted by boulders and reef flats, located in a small 100 m wide bay. Waves are usually less than 1 m high. One kilometre east of the western tip are beaches **E-7** and **E-8**, two adjoining sand and boulder beaches, 100 m and 150 m long respectively. They both face north and are fronted by 100 m wide rock and reef flats. Seven hundred metres to the east is a 60 m pocket of sand **(E-9)** contained in a small bay, fronted by scattered reefs, with jagged granite rock platforms to either side.

On the west side of McLachlan Point are the only two sand beaches with any beach surf on the island. **E-10** and **E-11** are two adjoining 330 m and 60 m long beaches, which receive waves averaging just over 1 m that break across a continuous narrow sand bar.

Swimming: The only beaches suitable for safe swimming are the northern beaches on either side of McLachlan Point, the remainder are generally too rocky.

Surfing: The only surf worth travelling out to the island for is along the central section of the south coast, around the Goose Hill area.

Fishing: Most fishing is done by boat over the many reefs around the islands.

Summary: An island conservation park offering rocky shore, a few beaches and no development.

Flinders Island

Flinders Island lies approximately 40 km due west of Elliston. It is the largest island of the Investigator Group, with a maximum length of 12 km and an area of 25 km². It has a shoreline length of 32 km, including 23 beaches. The island is composed of resilient granite and gneiss, which make up most of the shoreline, producing often jagged rock platforms and reefs. The island rises to a maximum height of 60 m and for the most part has a relatively flat, gently undulating surface, much of which has been cleared for grazing. The western side of the island has the most extensive beach systems, most backed by stable dune systems extending a few hundred metres inland. There are only two properties on the privately owned island, both located on the more protected eastern side. Accommodation is available on the island, with adventurous surfers heading out to try the waves on the exposed western shore.

F-1 – F-4   **FLINDERS ISLAND (northeast)**

	Unpatrolled			
No.	Beach	Rating	Type	Length
F-1	Pt Malcolm (S 1)	5	R/LTT	40 m
F-2	Pt Malcolm (S 2)	4	R/LTT	300 m
F-3	Pt Malcolm (S 3)	4	R/LTT	140 m
F-4	Front Beach	4	R/LTT	1.7 km

Point Malcolm is a 20 m high, dune-capped granite headland that forms the northern tip of the island. From the point the shore trends south for 5 km to the prominent 50 m high Bobs Nose headland. In between are four relatively protected, east facing beaches, all receiving

waves usually less than 1 m high. Beach **F-1** is a 40 m pocket of sand bordered by rocks and rock platforms, located 600 m south of the point and backed by steep, 30 m high bluffs.

Just over 1 km south of the point is the first of three more extensive sand beaches, all receiving low waves and usually fronted by a narrow, attached sand bar and low surf. The first beach, **F-2**, is 300 m long and bordered by high granite boulders, with boulders also outcropping toward the northern end of the beach. It is backed by dune-draped bedrock slopes, scrub, then farmland. Two hundred metres to the south is beach **F-3**, a 140 m long pocket of sand also bordered by granite slopes and backed by vegetated bedrock slopes. Front Beach (**F-4**) is one of the longer island beaches. It is 1.7 km long and backed by 200 m wide, low foredunes, then farmland. One of the two island farms is located behind the southern end of the beach.

Summary: Four relatively safe beaches, apart from the difficult access to the first beach. The southern beach near the farm is the easiest to access and usually free of surf.

F-5 – F-11   **FLINDERS ISLAND (southeast)**

	Unpatrolled			
No.	Beach	Rating	Type	Length
F-5	Bobs Nose (E)	5	LTT	250 m
F-6	Bobs Nose (W)	3	R+reef	150 m
F-7	Groper Bay (N)	4	R+reef	100 m
F-8	Groper Bay (S)	3	R+reef	650 m
F-9	Kapara (N 2)	4	R+reef	60 m
F-10	Kapara (N 1)	3	R+reef	120 m
F-11	Kapara (S)	6	R-boulder	100 m

Bobs Nose is a prominent 50 m high cliff that divides the northeast and southeast sections of the island. Between Bobs Nose and the southern tip is approximately 14 km of bedrock-dominated shoreline containing six small sand beaches and one boulder beach.

Beach **F-5** is located at the base of 20 m high cliffs and consists of a 250 m long strip of sand, facing south but receiving relatively low waves averaging about 1 m. These break across the narrow beach and up to the base of the cliffs at high tide. The northern farm lies to the north of this beach.

Beach **F-6** is a 150 m long, narrow strip of sand located 1.5 km north of the second island farm. It faces east and is relatively protected by its location and deeper reefs offshore. Beaches F-7 and F-8 occupy part of Groper Bay location of the main island settlement and landing.

Beach **F-7** is a 100 m long strip of sand on the northern side of the bay, fronted by continuous bedrock reefs. Beach **F-8** fronts the farm and is a curving, 650 m long sandy beach bordered by bedrock boulders and reefs, with some outcropping on the beach. Conditions are often calm at both beaches.

Two kilometres south of the farm is a small, 200 m wide, north facing bay containing two protected pocket beaches, located just north of the *Kapara* wreck. Beach **F-9** is a 60 m pocket of high tide sand fronted by continuous bedrock, while beach **F-10** is similar, but 120 m in length with some open sand in the centre. Another 2 km further south is beach **F-11**, a more exposed, 100 m long boulder beach exposed to waves averaging over 1 m. It is backed by 20 m high vegetated bluffs.

Summary: The only beach suitable for safe swimming is in Groper Bay (F-8). The others, while receiving low waves, are remote and dominated by rocks and reefs.

F-12 – F-20 **FLINDERS ISLAND (west)**

Unpatrolled			
No.	Beach	Rating	Type	Length
F-12	Dunstan Pt (S 2)	10	R-boulder	400 m
F-13	Dunstan Pt (S 1)	7	TBR	1.1 km
F-14	Dunstan Pt (N 1)	4	LTT	300 m
F-15	Dunstan Pt (N 2)	6	LTT/TBR	1.9 m
F-16	Gem Pt (S 5)	3	R	100 m
F-17	Gem Pt (S 4)	3	R	250 m
F-18	Gem Pt (S 3)	4	R+reef	250 m
F-19	Gem Pt (S 2)	4	R/LTT	1.3 km
F-20	Gem Pt (S 1)	4	R+reef	800 m

The western side of the island extends for 9 km north of the southern point. It contains nine predominantly sandy beaches, some exposed to relatively high waves and dominated by rips. Apart from the backing cleared higher land there is no development on this part of the island.

Beach **F-12** is located 500 m north of the southern tip of the island and consists of an exposed, 400 m long, west facing boulder beach, covered by a 100 m long veneer of sand toward the northern end. The entire beach is fronted by a 200 m wide reef-strewn surf zone, with a strong permanent rip in front of the sand section. This is the most hazardous beach on the island. It is also the site of the island's best surfing wave, a heavy righthander know as *Kitchenview*.

Beach **F-13** is the first of the sand beaches that dominate much of the western side of the island and begins 3 km north of the southern point. It is a 1.1 km long, southwest facing, moderately exposed beach, containing a 100 m wide surf zone usually cut by five rips. It is backed by

stable dunes, including clifftop dunes behind the northern half of the beach that extend up to 400 m inland. It is bordered by granite platforms and boulders.

Beach **F-14** lies on the northern side of the granite Dunstan Point. It is 300 m in length, faces northwest and receives refracted waves averaging about 1 m, which break across a narrow continuous bar. It terminates at a section of rocks and boulders, on the northern side of which is 1.9 km long beach **F-15**. This curving, west facing beach is bordered by granite headlands and receives waves averaging 1.5 m, which break across a 100 m wide, rip-dominated surf zone. It is backed by stable dunes extending up to 300 m inland.

On the northern headland is a 100 m wide, U-shaped, west facing bay, within which is 100 m long beach **F-16**. It is protected by the protruding arms and reefs and receives only low waves at the shore, which consists of the small sand beach together with rocks and bedrock outcrops. On the north side of the point is 250 m long beach **F-17**, a northwest facing reflective beach receiving waves averaging under 1 m. It terminates at a low bluff, on the northern side of which is a second 250 m long reflective beach (**F-18**), with both beaches protected by the southern point and deeper reefs offshore.

Beach **F-19** lies immediately to the north and is a 1.3 km long, northwest facing, moderately protected beach, backed by low bluffs and some active dunes. It is fronted by deeper reefs which lower the waves to about 1 m at the shore, where they surge up a steep reflective beach. The northern boundary is a curving granite point, on the northern side of which is beach **F-20**, an 800 m long, west facing reflective beach bordered by the prominent northern Gem Point and granite rock platforms, with bedrock reefs both outcropping along parts of the beach and extending 500 m off the southern point. The reefs lower the waves to about 1 m at the shore.

Summary: The west coast has the most exposed and hazardous beaches, as well as a few smaller, more protected beaches. Use care if swimming anywhere along here, as the more exposed beaches have rips, while rocks and reefs are common along most beaches.

F-21 – F-23 **FLINDERS ISLAND (northwest)**

Unpatrolled			
No.	Beach	Rating	Type	Length
F-21	Gem Beach	5	LTT/TBR	2.4 km
F-22	Flinders Bay (W)	4	R/LTT	550 m
F-23	Flinders Bay (E)	4	LTT	1.7 km

The northwestern section of the island is 5 km in length, terminating at Point Malcolm, the northern tip of the island. It is the most sand-dominated section of the island, containing three beaches totalling 4.6 km in length.

Gem Beach (**F-21**) is the longest beach on the island. The curving, 2.4 km long beach faces northwest and receives waves lowered by refraction to average 1 to 1.5 m at the shore, where they maintain a continuous 50 m wide bar often cut by several rips, especially along the eastern half of the beach. It is bordered by granite headlands and backed by continuous, largely vegetated sand dunes extending up to 500 m inland, with an active sand blow toward the western end.

Beach **F-22** is a more protected, 550 m long, curving beach bordered by a granite headland to the west and a sand foreland in lee of a prominent inshore reefs at the eastern end. On the east side of the foreland, beach **F-23** extends for 1.7 km to within 500 m of the northern tip at Point Malcolm. It receives waves averaging about 1 m, which break over a narrow continuous bar. Both beaches are backed by low, vegetated dunes up to 400 m wide, then farmland.

Summary: These are moderately safe beaches during normal low wave conditions, with rips developing when waves exceed 1 m.

1047-1052 **ANXIOUS BAY (S)**

No.	Beach	Rating	Type	Length
Unpatrolled				
1047	Cape Finniss	2	R	150 m
1048	Cape Finniss (N)	2	R+sand flats	3.9 km
1049	Walkers Rock (S 3)	3	R/LTT	2 km
1050	Walkers Rock (S 2)	5	R/LTT+reef	300 m
1051	Walkers Rock (S 1)	5	LTT	300 m
1052	Walkers Rock	2	R/LTT	600 m

Cape Finniss is a prominent 40 m high, calcarenite-capped granite headland, which marks an abrupt change in the trend of the coast and, together with the Waldegrave Islands, forms the southern boundary of 30 km long Anxious Bay, an exposed, high energy, west facing shore. The southern end of the bay is afforded protection by the cape, the islands and some reefs, resulting in generally lower energy beaches. Between the cape and Walkers Rock, 8 km to the north, are six beaches, with vehicle access to the coast at the cape and Walkers Rock, and beach boat launching also at both locations.

Beach **1047** is a 150 m long, northwest facing reflective beach located on the north side of Cape Finniss, immediately west of the main beach (1048). It is bordered and backed by 20 m high calcarenite bluffs, with the cape tourist road providing access to the top of the bluffs. The beach receives low waves which surge up the steep beach face.

The main cape beach (**1048**) commences on the west side of the cape and runs north for 3 km to a protruding sand foreland formed in lee of the Waldegrave Islands and then on for another 1 km to a small calcarenite bluff. The cape road runs out to a large car park and boat launching area located 1 km north of the cape, with only 4WD access to the northern beach. While stable dunes back the southern half of the beach, active dunes extend up to 400 m in from the northern half. The beach commences with a calm shoreline fronted by seagrass-covered sand flats, which extend all the way to East Island, 800 m offshore and along the beach as far as the foreland. Beyond the foreland waves slowly increase and break across a narrow attached bar.

Beach **1049** is a straight, 2 km long, west facing beach which is backed by active dunes between 200 and 700 m in width, which are in turn backed by farmland. The beach is moderately protected by the islands and receives waves averaging about 1 m, increasing slightly to the north. The waves break across a narrow attached bar, with rips forming when waves exceed 1 m. The only access is via 4WD tracks.

Beaches 1050 and 1051 are two 300 m long neighbouring beaches which lie immediately south of Walkers Rock. They are both bordered by small, 10 m high calcarenite bluffs. Beach **1050** has reef off each bluff and a permanent central rip, while **1051** has a sand beach and surf, with permanent rips against the bluffs at each end. Waves average between 1 and 1.5 m.

Walkers Rock is a protected, curving, 600 m long, northwest to west facing 'bay' (**1052**) sheltered by reefs extending north of the southern calcarenite point (Fig. 4.155). The reefs lower waves to about 0.5 m at the shore and calm conditions are common. There is a car park and basic camping area behind the southern end of the beach, with the beach used for launching boats. The beach terminates at a small northern sand foreland, formed in lee of the reefs and a small islet lying 200 m offshore.

Swimming: The safest swimming is at the access points at the main cape beach and Walkers Rock. The other beaches, while receiving lowered waves, can have rips, especially the two on the southern side of Walkers Rock.

Surfing: There are breaks on Cape Finniss over the calcarenite reefs, one a left hander called *Blackfellows*. These are dangerous and require local knowledge and experience to surf. This was also the site of a fatal shark attack in 2000. There are also beach breaks at the two more exposed beaches just south of Walkers Rock, with low refracted swell breaking at Walkers Rock during big outside wave conditions.

Fishing: Most fishers launch their boats at the cape and Walkers Rock to fish the many inshore reefs, with good beach fishing on the south side of the rocks.

Summary: Two accessible parts of the coast, offering generally lower wave conditions and natural beaches.

Figure 4.155 *Walkers Rock beach (centre) showing road access and bluff-top car park. While the southern beach (1051) receives moderate energy waves, the main beach (1052) is protected by a continuous outer reef, resulting in usually calm conditions at the shore.*

Lake Newland Conservation Park

Established: 1991
Area: 8448 ha
Beach: part of 1054
Coast length: 15 km

Lake Newland Conservation Park encompasses both Lake Newland and the fronting 1 to 2 km wide beach and dune system. There is no public access from the highway, with only 4WD access along the beach from Talia or Walkers Rock.

1053, 1054 ANXIOUS BAY

No.	Beach	Rating	**Unpatrolled** Type Inner	Outer Bar	Length
1053	Walkers Rock (N)	6	TBR		3.9 km
1054	Anxious Bay	7	TBR/RBB	LBT	23.5 km

Anxious Bay contains one of the more exposed, higher energy beach systems on the western Eyre Peninsula. While the southern end of the bay is protected by the Waldegrave Islands, the central and northern 38 km of shore is fully exposed to the persistent high southwest winds and waves. There are two beaches in this section of the bay, with vehicle access only to the very northern end of the main beach at Talia. The bay owes its origin to a geological depression called the Ponda Basin, which crosses the coast at the bay and extends offshore across the continental shelf.

The southern beach (**1053**) extends from immediately north of Walkers Rock beach for 3.9 km to a small calcarenite bluff. It is relatively straight, faces west and receives moderate waves that increase up the beach to average over 1.5 m at the bluff. These break across a low gradient, 200 m wide surf zone containing widely spaced beach rips, with a permanent rip against the bluff. The beach is backed by moderately active dunes up to 2 km in width and then the southern salt flats of Lake Newland. The only access to the beach is via 4WD through the dunes or along the beach.

The main Anxious Bay beach (**1054**) is 23.5 km long and faces almost due west into the high prevailing swell. It has a 200 to 300 m wide surf zone, which has a low gradient in the south, then a 10 km section dominated by parallel beachrock reefs in the centre, which lie between 50 and 200 m off the beach (Fig. 4.156). Several strong permanent rips run out through gaps in the rock. North of the rocks, strong beach rips dominate up to Talia, with a permanent rip against the northern headland. The beach is backed by the massive dune systems, then Lake Newland and farmland, with the only public access at Talia, or along the beach via 4WD. This is a relatively remote and hazardous beach, great to explore but hazardous for swimming. There is a lonely monument to a 1928 drowning victim overlooking the northern end of the beach.

Figure 4.156 *Part of the central section of Anxious Bay beach showing the offshore parallel calcarenite reefs, with sections of protected beach in their lee, but permanent rips draining out through gaps in the reef (centre) resulting in hazardous swimming conditions.*

Swimming: Two exposed, rip-dominated, potentially hazardous beaches, especially in lee of the beachrock reefs and at Talia.

Surfing: There are many beach breaks and some reef breaks along the beaches, with a 4WD required to access them.

Fishing: Relatively popular for fishing the many beach rips.

Summary: A long, exposed, natural beach, much now preserved in the conservation park.

1055-1061   TALIA to BEACH 1061

No.	Beach	Unpatrolled Rating	Type	Length
1055	Talia (N)	3	R+rock&reef	200 m
1056	Mt Carmel (S)	4	R+rock&reef	80 m
1057	Mt Carmel	7	TBR	1.3 km
1058	Mt Carmel (N)	7	TBR+reef	600 m
1059	Beach 1059	7	R+reef	250 m
1060	Beach 1060	7	LTT/TBR+reef	800 m
1061	Beach 1061	6	LTT+reef	360 m

Talia is famous for its spectacular coastal scenery associated with the red sandstone and capping calcarenite. Waves have eroded along the joint lines in the sandstone and undercut the calcarenite to form sea caves and blowholes. A gravel road from the highway runs along the top of the cliffs, providing good access to and views of the coast. The sandstone section extends for about 5 km north of the beach, before calcarenite cliffs dominate the shore and continue on for another 10 km to Venus Bay.

The first beach (**1055**) is located 4 km north of the main beach. It lies in a small, partly protected, west facing bay which lowers waves to about 1 m at the beach. The beach is a 200 m curving strip of sand partly fronted by a rock platform, with a 50 m opening to the deeper water of the bay. While waves are lower at the beach, the swell breaks heavily on either side of the bay. Two kilometres to the north is the second beach (**1056**), which is an 80 m strip of sand bordered by protruding sandstone rock platforms and fronted by a small, boulder-filled gap in the platforms. Waves break 200 m offshore on bedrock reefs, resulting in low waves at the shore.

Mount Carmel beach lies immediately west of 60 m high Mount Carmel. There are two beaches, the main beach (**1057**) is a straight, 1.3 km long, west facing, exposed beach that receives high waves and is dominated by strong permanent rips to either end, as well as usually two more central beach rips. The car park on the bluffs above the beach provides a good view of the surf and rips. A small, 30 m high point forms the northern boundary, on the northern side of which is the 200 m long beach **1058** (Fig. 4.157). This beach can only be accessed at low tide around the point or down the steep backing 40 m high bluffs.

Steep cliffs continue for another 4 km before a series of three near continuous, open, southwest facing beaches are reached. The three beaches are backed by continuous 10 to 20 m high calcarenite bluffs, with an assortment of

4WD tracks providing access to each. The first (**1059**) lies in lee of a calcarenite-capped bedrock point, with reef extending 200 m off the point and providing a left-hand surf break. The 250 m long beach is protected by the reef, with low waves at the shore and an easy paddle out to the break. Immediately north is beach **1060**, an 800 m long strip of sand wedged in at the base of the bluffs. It is fronted by deeper reefs and more sand towards the north, with both permanent reef-controlled and beach rips dominating the 100 m wide surf zone. The northern beach (**1061**) is 360 m long, straight and fronted by deeper reefs, which tend to lower waves at the shore to less than 1.5 m, but still sufficient to maintain permanent rips at either end.

Figure 4.157 The northern end of Mount Carmel beach (1057) showing the rip-dominated surf zone and backing active clifftop dunes.

Swimming: This is a hazardous section of coast with persistent high waves. The more protected beaches are dominated by rocks and reefs, while the more open beaches have strong rips.

Surfing: There are beach and reef breaks at the Mount Carmel beaches and the point break at beach 1059.

Fishing: Mount Carmel beach is a popular spot to fish the deep rip holes. Be careful rock fishing as large waves can wash over the platforms.

Summary: A reasonably accessible and scenic section of coast.

1062-1069   NEEDLE EYE to VENUS BAY

No.	Beach	Unpatrolled Rating	Type	Length
1062	Needle Eye (S 2)	8	TBR+reef	150 m
1063	Needle Eye (S 1)	8	TBR+reef	200 m
1064	Needle Eye (N)	8	TBR+reef	130 m
1065	Venus Bay (back bch 1)	8	TBR+reef	70 m
1066	Venus Bay (back bch 2)	8	TBR+reef	150 m
1067	Venus Bay (back bch 3)	8	TBR+reef	30 m
1068	South Head	6	LTT+reef	30 m
1069	Venus Bay (jetty)	1	R+sand flats	500 m

Venus Bay is a 75 km² embayment formed in a depression behind high calcarenite cliffs, with a deep, 300 m wide entrance connecting it to the sea. The small town of Venus Bay rests on the southern side of the entrance and backs on to the 30 m high calcarenite cliffs. Extending south of South Head, the southern entrance to the bay, is 1.5 km of cliffed coast containing seven small cliff-bound beaches (Fig. 4.158). A town street leads to the top of three of the beaches, with walking tracks leading to the tops of the others.

Figure 4.158 *The exposed calcarenite cliffs south of Venus Bay looking north to the entrance, with the small town (centre) spreading from the lee of the cliffs down to the shores of the bay.*

Needle Eye is a calcarenite sea stack located 1 km south of South Head. There are two small beaches to the south (**1062** and **1063**), both backed by steep cliffs and consisting of 150 m and 200 m long strips of high tide sand, fronted by rip-dominated, 150 m wide surf zones. Beach **1064** lies immediately north of the stack and is a similar, 130 m long, cliff-bound beach, with reefs off the southern end and a permanent rip toward the northern end. Beaches **1065** and **1066** are neighbouring 70 m and 150 m long beaches, located just 100 m from the northern houses of the town. They are also backed by 20 m high cliffs, with 200 m wide, rip-dominated surf zones.

Out on South Head are two small beaches, both located in gaps in the cliffs. Beach **1067** is 30 m long and faces south, while beach **1068** is located at the very tip of the point and faces west. Both are backed by steep bluffs and have reef-dominated surf zones.

Inside the bay at the centre of the town is a 500 m long sand foreland (**1069**), which protrudes out close to the deep tidal channel. The boat ramp is located at the northern end, a 100 m long, curving jetty in the centre, with a foreshore reserve and caravan park backing the beach, then the town centre. Conditions are usually calm at the beach, however there are strong tidal currents off the jetty.

Swimming: The only safe beach is the town beach, away from the jetty and deeper channel water.

Surfing: There are beach and reef breaks along the base of the cliffs, but they are difficult to access.

Fishing: The jetty is the most popular location.

Summary: Venus Bay is a popular holiday and fishing centre offering both the bay and nearby beaches, as well as spectacular rocky coastal scenery.

Venus Bay Conservation Park

Established:	1976
Area:	1423 ha
Beaches:	1070-1075 (6 beaches)
Coast length:	bay shore approx. 20 km, ocean shore approx. 12 km

Venus Bay Conservation Park occupies the southern 7 km of a 12 km long calcarenite peninsula that terminates at 100 m high Cape Weyland. The peninsula is 500 m wide at its narrowest and widens to about 2 km at the cape. The eastern side of the peninsula is bordered by the quiet waters of Venus Bay and has about 20 km of bay shoreline, while the western side faces the Southern Ocean and has about 12 km of low through very high energy, predominantly cliffed shoreline, including five beaches. The entrance to Venus Bay runs along the southern side of the cape.

1070-1075 CAPE WEYLAND - Venus Bay Conservation Park

Unpatrolled				
No.	Beach	Rating	Type	Length
1070	Cape Weyland (1)	1	R+reef	50 m
1071	Cape Weyland (2)	2	R	1.9 km
1072	Cape Weyland (3)	5	R+platform	100 m
1073	Cape Weyland (N 1)	10	LTT+reef	100 m
1074	Cape Weyland (N 2)	10	LTT+reef	150 m
1075	Cape Weyland (N 3)	10	TBR+reef	70 m

Cape Weyland is a prominent calcarenite headland that partly protects the entrance to Venus Bay, located 3 km to the northeast. In between the cape and the entrance are three south facing beaches protected by the cape, with extensive reefs extending up to several hundred metres off the beaches.

Beach **1070** is a protected, low wave, 50 m long, west facing pocket of sand, backed by 10 m high calcarenite bluffs and fronted by shallow reefs. The backing bluffs form the northern entrance to Venus Bay. The beach faces west along its neighbour, 1.9 km long, curving, south facing beach **1071**, which is protected by extensive deeper offshore reefs, with waves averaging less than 1 m at the shore. The waves surge up the steep sandy beach face and usually maintain up to 30 regularly spaced beach cusps. The beach is bordered by calcarenite-capped headlands and backed by 200 to 600 m wide, generally stable dunes, which have blown up onto the backing calcarenite and over into Venus Bay, where the sand has been reworked to form low energy spits fringed by mangroves. The main 4WD track terminates at the northern corner of the beach.

Beach **1072** is located 1 km north and east of the cape. It is a 100 m long, south facing strip of sand at the base of 60 m high cliffs. The beach is fronted by an intertidal calcarenite platform and then a 100 m wide, reef-dominated surf zone, drained by a permanent rip. There is no foot access to the beach.

Beaches 1073, 1074 and 1075 are three exposed, high energy, west facing pocket beaches located at the base of high cliffs, with no foot access. Beach **1073** is 100 m long, set in a semicircular gap in the cliffs and fronted by a 200 m wide reef and rip-dominated surf zone. Beach **1074** is a 150 m long strip of sand at the base of 100 m high cliffs capped by a small clifftop dune. The beach is littered with cliff debris and fronted by a reef-strewn, 200 m wide surf zone. Beach **1075** is a 70 m long sand beach set in a small, V-shaped gap in the 40 m high cliffs. It is fronted by two reefs, between which flows a strong permanent rip, often flowing 400 m seaward (Fig. 4.159).

Figure 4.159 *Beach 1075, located 7 km north of Cape Weyland, is an inaccessible pocket of sand, with an energetic surf zone drained by a strong permanent rip, clearly visible in this view.*

Swimming: The longer, protected beach 1071 is the only accessible beach and the only one suitable for swimming. Just watch the heavy shorebreak when waves exceed 0.5 m.

Surfing: None.

Fishing: Only at beach 1071, as the rest of the coast consists of high cliffs and inaccessible beaches.

Summary: A spectacular section of cliffed coast exposed to high waves on the west side, with the quiescent Venus Bay on the eastern side.

1076-1085   **ANXIOUS BAY (N)**

No.	Beach	Unpatrolled Rating	Type	Length
1076	Horseshoe Bay (E)	8	TBR/RBB+reefs	1.4 km
1077	Horseshoe Bay	7	TBR+headlands	200 m
1078	Horseshoe Bay (W)	7	R/LTT+reefs	900 m
1079	Tyringa (E 4)	8	R/LTT+platform	500 m
1080	Tyringa (E 3)	7	R+platform/reefs	1.2 km
1081	Tyringa (E 2)	8	R+platform/reefs	900 m
1082	Tyringa (E 1)	6	R+reefs	50 m
1083	Tyringa	7	R+reefs	1 km
1084	Tyringa (W 1)	6	R+reefs	300 m
1085	Tyringa (W 2)	5	R+reefs	120 m

The northern end of Anxious Bay is dominated by calcarenite-capped bedrock, that extends from Cape Weyland for 30 km to Baird Bay. Ten kilometres east of Baird Bay, the cliffs decrease in height and rock and reef-dominated sand beaches lie along their base. All the beaches are accessible via 4WD tracks from Baird Bay and the backing farmland.

Beach **1076** is the easternmost of these beaches, beginning 10 km due east of Baird Bay settlement. It is a 1.4 km long, south facing, exposed beach, receiving some protection from deeper reefs off the beach and bordered by 20 m high calcarenite bluffs. Waves average less than 2 m and maintain usually two beach rips, as well as reef-controlled rips and a steep beach face. It is backed by unstable then stable dunes extending up to 1 km inland, then farmland. Vehicle tracks are located at both ends and the centre of the beach.

The western bluffs form the boundary of **Horseshoe Bay** beach **(1077)**, a curving, 200 m long, steep sand beach bordered and backed by 20 m high calcarenite bluffs. Although waves are lower at the beach (Fig. 4.160), a deep permanent rip drains the small bay.

West of Horseshoe Bay beach, bedrock-based, calcarenite cliffs dominate, with the beaches wedged in at the base of the bluffs. Beach **1078** is a crenulate, 900 m long high tide sand beach lying at the base of 20 to 30 m high cliffs and fronted by a mixture of bedrock and deeper reefs. Beach **1079** is similar, 500 m in length, with the western half fronted by 50 to 100 m wide, irregular bedrock platforms. High waves break on reefs up to 500 m off the beach. Beach **1080** lies at the base of a 4WD access track and is backed by vegetated, 20 m high bluffs and fronted by scattered reefs extending up to 300 m offshore, which lower waves to about 1 m at the shore.

Beach **1081** is a crenulate high tide beach fronted by irregular bedrock platforms up to 200 m wide, with rock pools in amongst the bedrock. It is backed by vegetated bluffs that decrease in height to the west, with a 4WD track running along the top of the bluffs. Beach **1082** is a 50 m pocket of sand located on a small bluff-backed beach. It provides access to a left hand surfing break on a reef immediately east of the beach.

Figure 4.160 *The usually low energy beach at Horseshoe Bay (1077).*

Tyringa Beach (1083) is named after the backing farm. It is a 1 km sand beach, with the surf zone a mixture of exposed and deeper reefs, some extending 500 m off the beach. Wave height is variable along the beach and maintains about five permanent rips. Immediately west is 300 m long beach **1084**, which is moderately protected by bedrock reefs, with bedrock also exposed along the beach and in the surf. Its neighbour (**1085**) is a more protected, steep, 120 m long beach, with usually low waves against the beach.

Swimming: These are all accessible, but isolated and potentially hazardous beaches, with the safest swimming close inshore at Horseshoe Bay beach, when there is no surf in the bay and at beach 1084. However strong rips drain all beaches, so be careful.

Surfing: There are reef breaks off a few of the beaches, with local knowledge and experience required to find and surf them.

Fishing: This is a popular beach and rock fishing area, with good access and numerous deep holes.

Summary: A scenic section of coast accessible via the bluff-top tracks.

1086, 1087 SILICA BEACH, BAIRD BAY

Unpatrolled			
No.	Beach	Rating	Type	Length
1086	Silica Beach	4	R	1.5 km
1087	Baird Bay	1	R+sand flats	4.9 km

The eastern entrance to Baird Bay is marked by a low, right-angle shaped headland, to the east of which extends straight, 1.5 km long **Silica Beach (1086)**, with its eastern boundary formed by a straight, 600 m long calcarenite headland. This steep, south facing beach is famous for its pure coarse silica sand, which has been mined in the past. While exposed to southern winds, the beach is protected by Jones Island and deeper reefs extending over 1 km off the beach, resulting in waves averaging less than 1 m at the shore, where they surge up the steep, narrow, cusped beach face. Most of the beach is backed by low calcarenite bluffs and a 4WD track.

The main **Baird Bay** beach (**1087**), is a 4.9 km, west facing, protected beach, which begins 600 m north of the eastern bay entrance and extends 2 km to Baird Bay settlement, and another 2 km into the protected bay. The beach receives no swell and is fronted by seagrass-covered sand flats, which narrow to 100 m off the settlement (Fig. 4.161), where the 200 m wide tidal channel for the bay is located.

Figure 4.161 *The small Baird Bay settlement fronts the low energy bay beach, permitting boats to moor off the beach.*

Baird Bay is an elongate, 15 km long, 1 to 3 km wide, shallow bay, with a 300 m wide entrance adjacent to the small settlement. Most of the bay is surrounded by farmland, with one bay island and Jones Island forming part of the Baird Bay Islands Conservation Park.

Swimming: Both beaches are relatively safe, just beware of the strong tidal flows off the bay beach.

Surfing: None.

Fishing: Baird Bay is a focus for bay, beach and rock fishing on the adjoining beaches.

Summary: A quiet settlement with about 15 houses and a basic camping area, but no services.

1088-1092   BAIRD BAY ENTRANCE (W)

		Unpatrolled		
No.	Beach	Rating	Type	Length
1088	Baird Bay (W 1)	1	R+sand flats	900 m
1089	Baird Bay (W 2)	1	R+sand flats	1 km
1090	Baird Bay (W 3)	1	R+sand flats	1.2 km
1091	Baird Bay (W 4)	1	R+sand flats	400 m
1092	Baird Bay (W 5)	1	R+sand flats	60 m

The western side of the entrance to Baird Bay lies at the base of the 20 km long Calca Peninsula, with vehicle access via the peninsula. It consists of a series of low energy recurved spits, formed from sand moving into the bay and partly narrowing the entrance to 300 m opposite the settlement.

Beach **1088** lies opposite the Baird Bay settlement and its growth as a series of recurved spits has narrowed the inlet. Today the southeast facing, 900 m long beach is a relatively stable, low wave beach. However there are strong tidal currents in the deep channel off the eastern end, while to the west, seagrass-covered sand flats widen to 400 m. Beach **1089** is a more active, 1 km long, low energy recurved spit, that is slowly building across beach 1088. Tidal flats back the beach, with no vehicle access. Beach **1090** is a third recurved spit, which is attached to the coast and runs for 1.2 km to the western end of its neighbour (1091). Both beaches are fronted by extensive seagrass-covered sand flats.

Beach **1091** is a 400 m long beach backed by low dunes over bedrock, with a farm located at its southern end. Sand flats extend about 400 m off the beach. Two hundred metres to the south of the farm is a 60 m long pocket of sand (**1092**) backed by low dunes and cleared farmland.

Summary: Five low energy 'bay' beaches lying at the tip of the Calca Peninsula, all fronted by shallow sand flats and backed by private farmland.

1093-1100   SEARCY BAY (S)

		Unpatrolled		
No.	Beach	Rating	Type	Length
1093	Point Labatt	5	R+platform	100 m
1094	Searcy Bay (1)	6	R+reef	1.9 km
1095	Searcy Bay (2)	6	R+rocks/reef	100 m
1096	Searcy Bay (3)	6	R+reef	50 m
1097	Searcy Bay (4)	6	R+rocks/reef	200 m
1098	Searcy Bay (5)	6	TBR+rocks/reef	600 m
1099	Searcy Bay (6)	6	R+reef	100 m
1100	Searcy Bay (7)	6	LTT+reef	450 m

Searcy Bay is an exposed, rocky, west to southwest facing embayment bordered by Point Labatt in the south and Slade Point 13 km to the north. The shoreline consists of a mixture of calcarenite-capped bedrock bluffs and cliffs, with a scattering of beaches, most lying in lee of reefs. A gravel road provides access via the Baird Bay shore to Point Labatt, with 4WD tracks leading to the beaches.

Point Labatt is the site of a sea lion colony and was declared a conservation park in 1973. The small 39 ha park occupies the area around the point, including the small beach. The 40 m high point is composed of calcarenite-capped granite and has granite reefs extending up to 600 m off the shore. The sea lions reside on the northern side of the point, where the 100 m long sand beach (**1093**) is located in amongst the rocks and reefs. There is a gravel road to viewing platforms above the beach, however public access to the beach is prohibited.

One and a half kilometres north of the point is the first of a series of seven reef-protected sand beaches that occupy the next 5 km of west facing shore. Beach **1094** is the longest of the beaches, at 1.9 km and is protected by an offshore reef in the south, with deeper northern reefs permitting higher waves to the shore. The beach is steep and usually cusped, with surf breaking over rocks and reefs and a strong permanent rip toward the northern end. Beaches **1095** and **1096** lie out on an irregular granite point and consist of 100 m and 50 m long pockets of sand backed by bluffs and fronted by extensive bedrock reefs and rocks, resulting in generally low waves at the shore.

On the north side of the point is a 1.6 km long, open embayment containing four near continuous beaches. Beach **1097** lies in the more protected southern corner and is a 200 m long, northwest facing, sandy reflective beach, with a rock-dominated shoreline. Beach **1098** is a 600 m long, more exposed, west facing beach, with a sand high tide beach and bar, but with rocks outcropping along the shore. Two to three rips cross the bar, together with permanent rips at either end. Beach **1099** is a curving, southwest facing, 100 m long beach, lying in lee of a shallow reef, with jagged calcarenite bluffs to either end. Beach **1100** occupies the northern end of the 'bay'. It is a southwest facing, curving, 450 m long, steep sand beach, with a narrow attached bar and deeper reefs offshore.

Swimming: These are all potentially hazardous beaches owing to the high outer waves, rocks, reefs and rips. Only the protected southern end of beach 1094 offers relatively safe swimming close inshore.

Surfing: There are many reef and point breaks along this section that would be worth checking out during low to moderate swell.

Fishing: Rip holes and deeper gutters are found along all of the beaches.

Summary: Most visitors drive out to see the sea lions at the point. For those with a 4WD, there are additional scenic attractions along the exposed bay shore.

1101-1110   SEARCY BAY (centre)

No.	Beach	Rating	Type	Length
		Unpatrolled		
1101	Searcy Bay (8)	7	R+headland&reef	40 m
1102	Searcy Bay (9)	7	R+headland&reef	110 m
1103	Searcy Bay (10)	7	R+headland&reef	60 m
1104	Searcy Bay (11)	7	R+reef	230 m
1105	Searcy Bay (12)	7	R+reef	40 m
1106	Searcy Bay (13)	7	R+reef	130 m
1107	Searcy Bay (14)	7	R+reef	240 m
1108	Searcy Bay (15)	7	R+reef	200 m
1109	Searcy Bay (16)	7	R+reef	150 m
1110	Searcy Bay (17)	7	R/LTT+reef	2.2 km

The central section of Searcy Bay is an isolated 6 km stretch of coast, dominated by dune-capped calcarenite and containing ten beaches, with nine of them barely totalling 1 km in length. While the Point Labatt road runs 1 to 2 km to the east, there is only one formed track leading to the coast, as well as 4WD access tracks across the backing dunes. Several beaches are not accessible by vehicle and a few are backed by high cliffs, with no foot access.

Beach **1101** lies 500 m south of the southern access track. It is a 40 m pocket of sand at the base of 40 m high, dune-capped calcarenite cliffs that extend up to 300 m seaward, forming a V-shaped bay. The high tide sand beach is fronted by 100 m of shallow reefs. Beach **1102** is located below the end of the access track and is a 110 m long sand beach fronted by 200 m of shallow reefs and surf and backed by 30 m high cliffs. Its northern neighbour (**1103**) is a 60 m long, northwest facing pocket of sand, also located below the end of the access track. A sea stack lies 50 m off the beach, surrounded by shallow reefs and

surf. Beach **1104** is 230 m long, backed by 20 m high cliffs and fronted by 200 m wide surf that breaks over shallow reefs. There is no safe foot access to any of these beaches.

Beaches **1105, 1106** and **1107** are three neighbouring pockets of sand, each backed and separated by 30 m high calcarenite cliffs and headlands. They are 40 m, 130 m and 240 m long respectively, with no vehicle access to the backing bluffs and no safe foot access to the beaches. Each is fronted by shallow reefs, widening to 200 m off beach 1107 and drained by permanent rips.

Beaches **1108** and **1109** are two narrow high tide sand beaches lying in lee of reefs extending up to 500 m seaward. Immediately to the north is the longest beach in the bay, 2.2 km long beach **1110** (Fig. 4.162). This beach lies in lee of reefs up to 800 m wide off the southern third, with deeper reefs to the north. The reefs both lower waves in their lee as well as forming a sand foreland 500 m up the beach. The central-northern section has a narrow attached bar, with permanent rips against the foreland and northern headland. The entire beach is backed by active dunes extending up to 1 km inland.

Figure 4.162 *The central section of Searcy Bay showing reef-fronted pocket beaches 1108 and 1109 (right) and the longer, more exposed, high energy beach 1110, backed by active sand dunes.*

Swimming: These are ten often difficult to access, reef and rip-dominated beaches. While some have low waves at the shore, be very careful as permanent rips drain all the beaches.

Surfing: There are several reef breaks, some well offshore, with local knowledge essential to find the best locations.

Fishing: The long beach 1110 offers excellent holes and gutters in amongst the reefs.

Summary: The southern section is accessible via tracks off the Point Labatt road, with two viewing points offering excellent views of this rugged coastline.

1111-1124   SEARCY BAY (N)

No.	Beach	Rating	Type	Length
Unpatrolled				
1111	Searcy Bay (18)	5	R+reef	100 m
1112	Searcy Bay (19)	5	R+reef	40 m
1113	Searcy Bay (20)	5	R+reef	150 m
1114	Searcy Bay (21)	6	R+reef	80 m
1115	Searcy Bay (22)	6	R+reef	50 m
1116	Searcy Bay (23)	6	R+reef	60 m
1117	Searcy Bay (24)	7	R+reef	300 m
1118	Searcy Bay (25)	7	R/LTT+reef	150 m
1119	Searcy Bay (26)	7	R+reef	260 m
1120	Searcy Bay (27)	6	R/LTT+reef	100 m
1121	Searcy Bay (28)	5	R+reef	130 m
1122	Searcy Bay (29)	4	R/LTT+reef	400 m
1123	Searcy Bay (30)	4	LTT	70 m
1124	Searcy Bay (31)	4	R	200 m

The northern end of Searcy Bay consists of 3.5 km of gradually curving, southwest to finally south facing shore, terminating in lee of Slade Point. Along this section are 13 generally small, headland-bound, reef-fronted beaches, five are less than 100 m long and the longest only 400 m. Access to all the beaches is via 4WD tracks, apart from the very northern beach which has a gravel road from Sceale Bay, 3 km to the north.

Beaches **1111** to **1115** are five neighbouring beaches located along the base of 900 m of calcarenite cliffs and bluffs, each separated by small protruding headlands. They are all fronted by 100 to 200 m wide shallow reefs and rock flats, with waves averaging 1.5 m breaking on the outer edge of the reefs and lower wave conditions at the shore. Only 1112 is readily accessible, the others are backed by steep cliffs.

Beach 1115 terminates at a more prominent headland, on the northern side of which begins a series of eight beaches extending 2.5 km to Slade Point. Beach **1116** is a 60 m pocket beach fronted by deeper reefs and beach **1117** a 300 m long sand beach fronted by a continuous shallow reef. Beach **1118** consists of two small pockets of sand totalling 150 m in length, with reefs and rocks off the beach. Beach **1119** is 260 m long, backed by 30 to 40 m high cliffs, with vehicle access to the top of the cliffs but foot access only around the rocks from beach 1118. Beaches **1120**, **1121** and **1122** are three neighbouring beaches, 100 m, 130 m and 400 m long respectively, all backed by 40 m high cliffs and are difficult to access on foot. They are fronted by shallow reefs which diminish toward the west.

Beach **1123** is a 70 m pocket of sand wedged into a V-shaped gap in the 30 m high cliffs, with access via a steep gully. It has a narrow bar and usually low waves. Beach

1124 lies in lee of Slade Point and is a curving, east facing, 200 m long, steep reflective beach, usually containing several well developed beach cusps. A right hand surf breaks over the western point.

Swimming: The northern beach in lee of Slade Point is the only readily accessible and relatively safe beach, with usually a low surging break against the shore. All the other beaches are both difficult to access and dominated by rocks, reefs and permanent rips.

Surfing: The Slade Point break is the best along this section, with other opportunities amongst the many reefs.

Fishing: Most fishers head for the more accessible Slade Point area, though there are many deep holes along the beaches in amongst the many reefs and rocks.

Summary: A generally difficult to access section of cliff, rock and reef-dominated small beaches.

1125-1130   SLADE POINT to CAPE BLANCHE

No.	Beach	Rating	Type	Length
Unpatrolled				
1125	Slade Point (1)	2	R	250 m
1126	Slade Point (2)	3	R	80 m
1127	Cape Blanche (1)	4	R+rocks	1.1 km
1128	Cape Blanche (2)	4	R+rocks	50 m
1129	Cape Blanche (3)	4	R	60 m
1130	Sceale Bay	4	R+rocks	300 m

Slade Point and Cape Blanche form the boundaries of a 5 km long, calcarenite-capped bedrock headland that extends 5 km seaward and separates the southern Searcy Bay from Sceale Bay. The headland rises to 70 m at Slade Point and 100 m at the cape, with the intervening cliffs averaging 70 to 80 m in height and capped by a thin layer of clifftop dune extending up to 1 km inland. There is road access to Slade Point from Sceale Bay settlement and 4WD tracks from the settlement out along the northern shore to Cape Blanche. While rocky shore and cliffs dominate much of the 14 km of shoreline between the beaches of Searcy Bay and Sceale Bay, there are six small beaches on the north and south sides of the headland.

In lee of **Slade Point** is a small, 300 m wide bay containing two protected, south facing reflective beaches (Fig. 4.163). Beach **1125** is a 250 m long sand beach that forms a small tombolo in the centre, attached to a bedrock reef. It is backed by a narrow foredune and 30 m high bluffs. A vehicle track leads to the top of the beach, with a steep track down to the beach. Its neighbour is an 80 m strip of sand (**1126**) lying at the base of the cliffs, accessible around the rocks from the main beach. Waves average less than 1 m in the bay and surge up the steep beach faces, as well as breaking on reefs off the beaches.

Figure 4.163 *The two Slade Point beaches are well protected by the point and reefs, and are linked to each other by a small tombolo (centre).*

On the north side of **Cape Blanche** are four protected, low energy, rock-dominated beaches, all lying at the base of steep slopes. The first beach (**1127**) begins 500 m east of the cape and extends for 1.1 km along a curving, north facing embayment. It consists of a narrow strip of sand backed by 50 to 70 m high bluffs and is partially covered by rock debris from the bluffs, with rocks along and off the beach. Waves average less than 1 m and maintain a steep beach face. There is no vehicle access and no easy foot access to the beach. Beach **1128** is a 50 m long pocket of sand at the base of two small sloping gullies, with rocks littering the beach and shoreline. The cape 4WD track runs 200 m inland along the top of the slopes. One kilometre further into the bay is beach **1129**, a 60 m long pocket beach fringed by small headlands, with a 4WD track terminating on the eastern headland. Beach **1130** is a 300 m long, crenulate high tide sand beach, largely fronted by rocks and reefs, which lies immediately below the western end of the Sceale Bay settlement. Waves are usually low to calm at the beach, which was once the site of a jetty.

Swimming: The main Slade Point beach (1125) is the best and safest beach for swimming under normal low wave conditions. All the others are either difficult to access or dominated by rocks and reefs.

Surfing: None, apart from reef breaks at Slade Point during higher swell conditions.

Fishing: There are numerous sites around the cape, with the safest sites along the more protected southern and northern shores.

Summary: A high, dramatic headland and rocky section of coast, containing a mixture of small rock and reef-dominated beaches, with a caravan park at Sceale Bay.

1131   SCEALE BAY - YANERBIE

No.	Beach	Unpatrolled Rating	Type	Length
1131	Sceale Bay-Yanerbie	2→6→2	LTT→TBR→LTT	14.7 km

Sceale Bay is a curving, southwest facing bay bordered by Cape Blanche and Speed Point. There is a 7.5 km wide entrance between the points, with 20 km of bay shoreline, including the main 14.7 km long beach (**1131**) which begins at Sceale Bay settlement and runs north, then west and finally south at Yanerbie shack settlement. The southern end of the bay is backed by low stable foredunes, the northern end has the massive white Yanerbie sand dunes which extend up to 4.5 km inland. There is road access in the south to the small Sceale Bay settlement and in the north to the beachfront Yanerbie shacks.

The southern end of the beach is protected by Cape Blanche and shore-parallel reefs and is usually calm, with the low gradient beach used for launching small boats. A few hundred metres up the beach waves begin to slowly increase, averaging 1.5 to 2 m north of the settlement. For the next 10 km, high waves, a 200 to 400 m wide, low gradient beach and surf zone, and large rips dominate, with dunes also increasing in size and instability. The final 2 km of beach swing around to face south, then west in lee of Speed Point. The waves decrease and calm conditions usually prevail in front of the Yanerbie shacks, with the beach fronted by shallow seagrass beds and used for launching and mooring small boats.

Swimming: The safest swimming is at either end, with hazardous conditions along the central higher energy section.

Surfing: There are beach breaks all the way along the more exposed central section, with the best locations on the southern reefs, with a right hander known as *The Island* and a left called *Squirrels*. During big seas there is also a small right up at Yanerbie.

Fishing: The small settlements at either end are both occupied by fishers, with most using boats to fish the bay and reefs.

Summary: An accessible bay with two small settlements, including a caravan park at Sceale Bay and the magnificent dunes at the northern end.

Regional Map 11: Streaky Bay to Ceduna

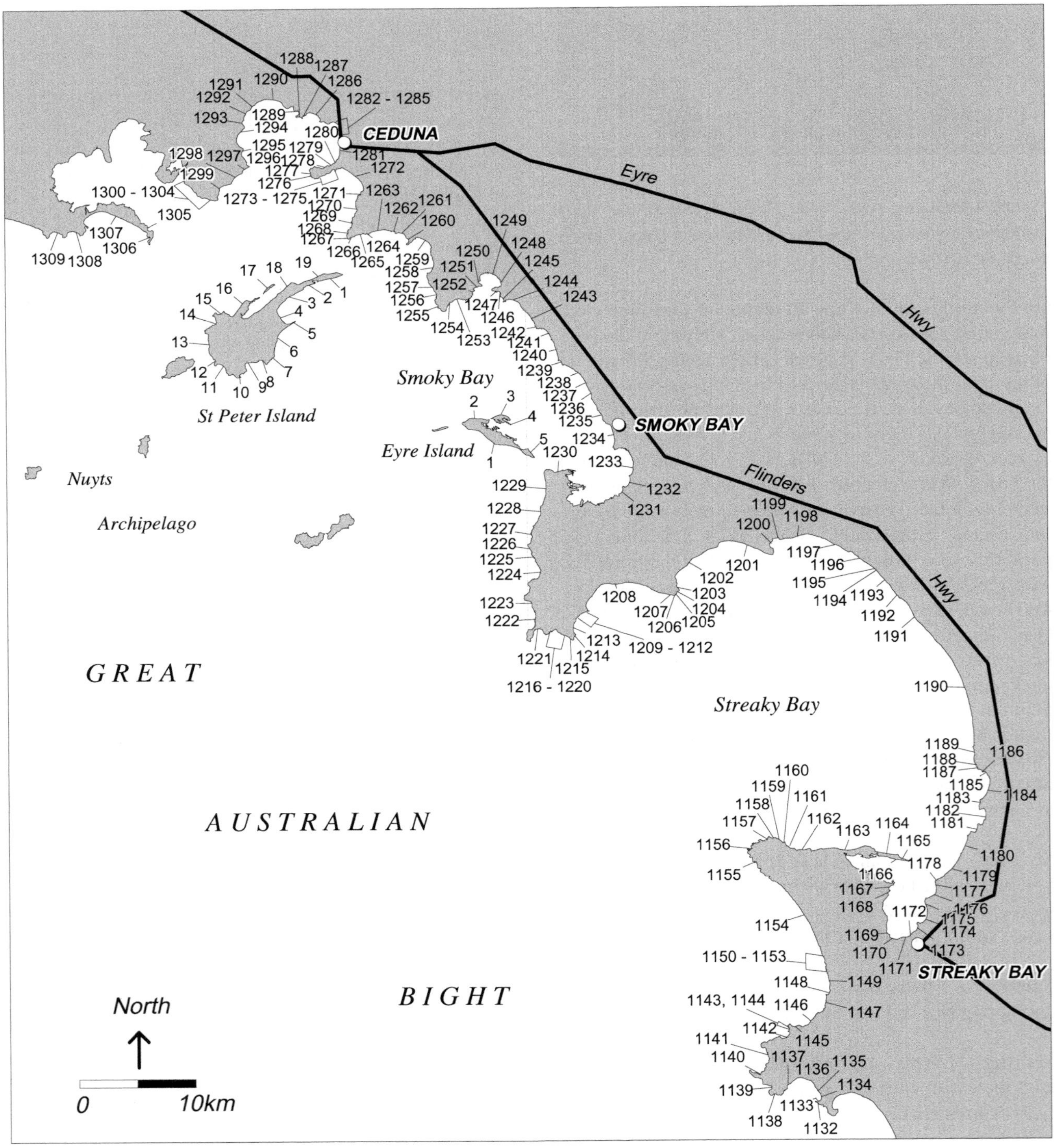

Figure 4.164 *Regional Map 11 – Streaky Bay to Ceduna*

1132-1138 SPEED POINT to SMOOTH POOL

No.	Beach	Rating	Type	Length
Unpatrolled				
1132	Speed Pt	7	R+platform	240 m
1133	Speed Pt (N 1)	4	R+reef	50 m
1134	Speed Pt (N 2)	4	R+reef	900m
1135	Speed Pt (N 3)	5	R-LTT	100 m
1136	Speed Pt-Smooth Pool	4	R+reef	2.4 km
1137	Smooth Pool (E 2)	3	R+reef	120 m
1138	Smooth Pool (E 1)	5	R+reef	100 m

Speed Point and **Smooth Pool** are two low granite headlands, both capped by calcarenite. They are located 4 km apart and protrude seaward to form a southwest to south facing, predominantly sandy embayment. In addition to the headlands, reefs extend for 1 km into the bay north of Speed Point and scattered reefs extend 1.5 km south of Smooth Pool, resulting in substantial wave breaking and, together with wave refraction, lower waves to less than 1 m along much of the 4 km long shoreline. There is vehicle access to both headlands, with the centre of the bay shore backed by moderately active sand dunes extending up to 1 km inland.

On the tip of Speed Point is a straight, 240 m long high tide sand beach (**1132**) fronted by a 100 m wide intertidal rock platform, across which waves break heavily. Immediately in lee of the point is a 50 m long, north facing pocket beach (**1133**) fronted by shallow reefs. The gravel road terminates on a low headland separating this beach from the curving, 900 m long beach **1134**. There is a car park at the protected southern end of this beach, with waves increasing slightly along the beach and reefs dominating off the northern end.

Beach **1135** is a 100 m long, west facing beach, bordered by small calcarenite bluffs and fronted by 600 m wide shallow reefs. The reefs continue north for 500 m along beach **1136**, a curving, 2.4 km long, west and finally south facing beach. It receives low waves, which surge up a usually cusped beach face. The main active dunes back the central to northern section of this beach.

Beach **1137** is a 120 m long, protected, south facing beach, the first encountered on the Smooth Pool tourist drive. A track off the road runs 100 m out to the rear of the usually calm beach, which is protected by reefs extending 1 km into the bay. One and a half kilometres to the west is beach **1138**, a 100 m long, south facing beach, bordered by irregular granite rocks and fronted by reefs to either side extending up to 1 km offshore. These lower waves at the beach to less than 1 m. There is a small dune climbing up onto the low bluffs behind the beach, with a gravel road running along the top of the bluff. The road

continues on to Smooth Pool, a large tidal pool located on the headland.

Swimming: The most popular beach is 1134, which is both readily accessible by car and offers usually calm waters near the car park. Most of the other beaches have low waves, but are dominated by rocks and reefs.

Surfing: The only surf is on the many reefs off most of the beaches, with the best being a left over the rocks at *Smooth Pool*.

Fishing: The best fishing is off the rocks and reefs, however beware of waves in the more exposed locations.

Summary: A relatively small, lower energy embayment, protected by the headlands and reefs, with good access to each end but no facilities.

1139-1145 POINT WESTALL to THE DREADNOUGHTS

No.	Beach	Rating	Type	Length
Unpatrolled				
1139	Smooth Pool (N)	6	R+platform	280 m
1140	Point Westall	4	R+rocks/reef	160 m
1141	Granite Rock	6	R+platform	1.4 km
1142	The Dreadnoughts (W 3)	4	R+reef	1.2 km
1143	The Dreadnoughts (W 2)	5	R+reef	160 m
1144	The Dreadnoughts (W 1)	5	R+reef	80 m
1145	High Cliff	3	R	700 m

The eight kilometres of irregular coast from Smooth Pool round to The Dreadnoughts consists of 50 to 60 m high calcarenite cliffs, sitting on a granite basement. In amongst the rock-dominated indentations are several beaches, with varying degrees of protection. The Westall Way scenic drive follows the top of the cliffs from the pool around to The Dreadnoughts, before returning to the main road. However there are no facilities on the coast.

Immediately north of Smooth Pool is a 300 m wide, 500 m long bay, containing a 280 m long high tide beach (**1139**) at its eastern extremity. The beach is fronted by a continuous bedrock platform, with waves breaking along the reefs on either side of the bay.

Point Westall is a 30 m high, 1 km long, finger-like point, on the eastern side of which is a protected, 160 m long, north facing pocket of sand (**1140**), backed by 50 m high bluffs and fronted by shallow reefs. Between the point and the northern Granite Rock headland is a second 1.4 km long, west facing, more exposed beach (**1141**), consisting of a disjointed high tide sand beach fronted by a mixture of elevated Pleistocene and intertidal Holocene beachrock. Strong permanent rips flow out of gaps in the beachrock.

To the east of Granite Rock, the coast faces northwest and is more protected by the headlands and reefs. Beach **1142** is the longest of three beaches that lie immediately west of the 60 m high Dreadnought cliffs. It is a curving, 1.2 km long beach backed by dune-draped, 20 m high bluffs, with a steep reflective beach usually receiving waves less than 1 m (Fig. 4.165). Some rock debris from the bluffs litters the beach. In amongst the Dreadnought cliffs are beaches **1143** and **1144**, two 160 m and 80 m long pocket beaches, both bordered and backed by 50 m high cliffs and fronted by shallow reefs.

Figure 4.165 Beach 1142 lies along the base of the 60 m high Dreadnought cliffs. The protected, low energy reflective beach is subject to rock falls off the backing cliffs.

On the eastern side of the point is beach **1145**, a narrow, curving 700 m long, low energy beach backed by 60 m High Cliff, that decreases in height to the west. It is fronted by shallow reefs, particularly to the east, where there is a left hand surfing break called *Indicators*.

Swimming: The only beaches suitable for relatively safe swimming are 1142 and 1145, but even here beware of reefs and rocks. The others are either too exposed, difficult to access or dominated by rocks and reefs.

Surfing: The best looking break is *Indicators,* which is located in front of the car park at High Cliff.

Fishing: There are numerous sites around this section of rocky coast offering moderately protected deep gutters in amongst the rocks and reefs.

Summary: A very scenic and accessible section of clifftop, with some of the beaches difficult to access down the steep bluffs and cliffs.

1146-1154   CORVISART BAY

No.	Beach	Rating	Type	Length
Unpatrolled				
1146	Corvisart Bay (1)	4	R+reef	2 km
1147	Corvisart Bay (2)	4	R	1.2 km
1148	Corvisart Bay (3)	5	R+reef	560 m
1149	Corvisart Bay (4)	7	TBR+reef	1.4 km
1150	Corvisart Bay (5)	7	TBR+reef	100 m
1151	Corvisart Bay (6)	7	TBR+reef	300 m
1152	Corvisart Bay (7)	7	TBR+reef	400 m
1153	Corvisart Bay (8)	7	TBR+reef	50 m
1154	Back Beach	8	TBR/RBB	7.6 km

Corvisart Bay is an 18 km long, exposed, west to southwest facing embayment, bounded by Cape Bauer in the north and The Dreadnoughts in the south. The bay and its shore have been battered by strong westerly waves and winds, leaving generally depleted narrow beaches, interspersed with beach and dune rock. Because of the persistent westerly swell, reef controlled rips and, in the north, beach rips, abound. The eight beaches are all backed by eroded calcarenite bluffs, partly covered by clifftop dunes extending up to 2 km inland. The only vehicle access to the bay beaches is in the south at the High Cliff car park and in the centre at the Streaky Bay Back Beach (1154). The remainder require a 4WD to access via the deflated backing clifftop and dunes.

The southernmost bay beach (**1146**) is a crenulate, 2 km long, northwest facing beach, moderately protected by High Cliff headland. It consists of a narrow high tide beach, fronted by a near continuous calcarenite reef and backed by low bluffs and a foredune. Wave height increases east along the beach, averaging just over 1 m at the eastern end, where the waves provide a left hand surf break over the reef, called *Granites*. Beach **1147** is a straight, 1.2 km long, northwest facing reflective beach, which receives waves averaging about 1 m that surge up the steep beach face, but increase in height up the beach, with a narrow bar forming during higher waves and a permanent rip against the small northern point. On the other side of the point is beach **1148**, a narrow, indented, 560 m long, moderate energy beach, which has a strong permanent rip against the northern reef. Beach **1149** is a crenulate, 1.4 km long, west facing beach that receives waves averaging over 1.5 m, which break across a reef-dominated surf zone, with a few strong rips flowing out through gaps in the reefs.

The next 2 km are taken up by calcarenite bluffs, including three bluff-bound beaches. Beach **1150** is a 100 m long pocket of sand fronted by bedrock reefs, beach **1151** is 300 m long with rocks outcropping on the beach and a strong permanent rip, beach **1152** is 400 m long and backed by steep, 20 m high bluffs with a strong central rip, while beach **1153** is a 50 m pocket of sand fronted by a wide surf zone.

Back Beach (1154) begins on the north side of the bluff, and extends for 7.6 km to Cape Bauer. Beachrock reefs dominate the beach for one kilometre north of the car park, beyond which is a more open beach with an offshore bar. High waves and rips dominate the entire beach. Along the southern section there are permanent reef-controlled rips, while north of the car park beach rips are spaced approximately every 400 m and flow out through a 300 m wide surf zone.

Swimming: The only safe beaches are the southern two or three during normal low wave conditions. The remainder are exposed to persistent higher waves and are dominated by strong permanent reef-controlled rips, and beach rips in the north.

Surfing: The best surf is on the southern reef break at *Granites* and along the reef breaks in front of and north of the *Back Beach* car park.

Fishing: Back Beach is a very popular location for beach fishing, with deep holes dominating the beach.

Summary: An exposed section of calcarenite coast with good access at either end, but generally hazardous swimming conditions.

Streaky Bay

Streaky Bay is a large west facing bay, bordered by Cape Bauer in the south and Point De Mole 25 km to the north. The 600 km² semicircular bay has a 100 km long shoreline containing 50 low energy beaches, all fronted by sand flats. The bay is protected from ocean waves by the headlands and the massive sand shoals that fill much of the bay. The town of Streaky Bay is located at the southernmost extremity of the bay in the protected Blanche Port, while the very small Haslam settlement is located on the eastern shore. Otherwise it is backed by farmland and Acraman Creek Conservation Park in the north.

1155-1160   CAPE BAUER (STREAKY BAY)

		Unpatrolled		
No.	Beach	Rating	Type	Length
1155	Cape Bauer (1)	5	R+reef	80 m
1156	Cape Bauer (2)	4	R+reef	120 m
1157	Cape Bauer (3)	3	R+reef	250 m
1158	Cape Bauer (4)	1	R+sand flats	240 m
1159	Cape Bauer (5)	1	R+sand flats	600 m
1160	Cape Bauer (6)	1	R+sand flats	200 m

Cape Bauer forms the northern boundary to the energetic Corvisart Bay, as well as the southern headland for the

larger and quieter Streaky Bay. The cape consists of a calcarenite-covered bedrock point, reaching a maximum height of 80 m near the northwestern tip, where a lighthouse is located. There is a gravel road out to the lighthouse, providing access to the bluffs above the few beaches that occupy indentations in the northern rocky coastline.

Beach **1155** is an 80 m long pocket of sand located 200 m south of the end of the cape road. It is backed by 40 m high cliffs and fronted by a surf-covered reef. Beach **1156** lies at the northwestern tip of the cape and is a moderately protected, 120 m long, northwest facing pocket beach, backed by a 20 m high bluff, then the road. Beach **1157** lies at the northern tip of the cape and is protected from most ocean swell, resulting in a 250 m long, low energy reflective beach, with the road skirting its western end.

Beaches 1158, 1159 and 1160 are three sheltered, low energy beaches fronted by sand flats, on the northwestern side of the cape. They are backed and bordered by grassy slopes that rise toward the crest of the cape, with the cape road running along the base of the slope just behind the beaches. Beach **1158** is a 240 m long pocket of sand fronted by 100 m wide sand flats. Its neighbour (**1159**) is a 600 m long, more crenulate, bedrock-controlled strip of sand and sand flats, while **1160** is a 200 m long pocket of sand bordered by low, grassy headlands, with sand flats widening to 400 m.

Swimming: Apart from beach 1155, these are low energy, relatively safe beaches, especially those fronted by sand flats.

Surfing: None at the beaches.

Fishing: The best spots are off the cape, however beware of waves on the more exposed western side.

Summary: There is a gravel road out to the lighthouse which passes close to all of these beaches.

1161-1166   GIBSON SPIT (STREAKY BAY)

		Unpatrolled		
No.	Beach	Rating	Type	Length
1161	The Spit (1)	1	R+sand flats	600 m
1162	The Spit (2)	1	R+sand flats	1.5 km
1163	The Spit (3)	1	R+sand flats	6.2 km
1164	The Spit (4)	1	R+sand flats	1.7 km
1165	The Spit (5)	1	R+sand flats	1.1 km
1166	The Spit (6)	1	R+sand flats	2 km

The Spit is a 5 km long spit of sand that protrudes into Streaky Bay, partly blocking Blanche Port (Fig. 4.166). The sediment that originates on the eastern side of Cape

Bauer moves alongshore to supply the slowly growing spit. It manifests itself as a series of dynamic onlapping smaller spits terminating in the main spit at Point Gibson. All the beaches face north and are fronted by 500 to 1000 m wide sand flats, then seagrass meadows. The cape road backs the western three beaches, with the eastern spit beaches only accessible by crossing two tidal creeks.

Figure 4.166 *View east along The Spit, which terminates at Point Gibson and partly blocks the entrance to Blanche Port.*

Beach **1161** is a 600 m long series of recurved spits, backed by an infilled lagoon and fronted by 1 km wide sand flats. Beach **1162** is a 1.5 km long spit with up to ten abandoned recurved spits attesting to its eastward migration, in the process enclosing an elongate, 1 km long, mangrove-filled lagoon between the beach and the road. Beach **1163** is the longest section and consists of a 6.2 km long, crenulate, north facing series of abandoned spits. The eastern 2 km of spits have enclosed a narrow, 2 km long lagoon between them and the neighbouring beach (**1164**), with a small, narrow tidal creek now separating the two beaches. Beach 1164 is a relatively straight, 1.7 km long part of the spit between the tidal creek and a small bedrock headland, that separates it from the most easterly part of the spit.

The eastern tip of the spit consists of the northern beach (**1165**), a 1.1 km long, low strip of sand terminating at narrow Point Gibson. Sand flats extend another 1 km east of the point, narrowing Blanche Port to 1.2 km, a third its width before being narrowed by the spit. On the south side of the spit is 2 km long beach **1166**, which forms a continuous south facing, very low energy beach facing into Blanche Port and backing both beaches 1164 and 1165, that terminates amongst the extensive mangroves that occupy the southern side of the spit.

Swimming: These are all very low energy beaches fronted by shallow sand flats, with swimming only possible at high tide.

Surfing: None.

Fishing: Generally too shallow off the beaches.

Summary: An interesting low energy, yet dynamic section of coast, readily accessible in the west along the cape road.

1167-1177   BLANCHE PORT (STREAKY BAY)

No.	Beach	Rating	Type	Length
Unpatrolled				
1167	Shag Point (N)	1	R+sand flats	300 m
1168	Shag Point (S)	1	R+sand flats	100 m
1169	Doctors Beach (W)	1	R+sand flats	400 m
1170	Doctors Beach	1	R+sand flats	800 m
1171	Jetty Beach	1	R+sand flats	200 m
1172	Jetty Beach (E)	1	R+sand flats	150 m
1173	Yacht Club Beach	1	R+sand flats	600 m
1174	Crawfords Landing (S)	1	R+sand flats	1.6 km
1175	Crawfords Landing	1	R+sand flats	1.2 km
1176	Perforated Rocks (S 2)	1	R+sand flats	1.5 km
1177	Perforated Rocks (S 1)	1	R+sand flats	300 m

Blanche Port is a north facing, U-shaped bay at the base of which is the town of Streaky Bay. The long Gibson Spit narrows the bay entrance to just over 1 km in the north, otherwise the bay averages about 3 km in width and 7 km in length. Around its 23 km of protected shoreline are the southern side of the spit (beach 1166), extensive mangroves in the northwest corner and a mixture of generally low calcarenite bluffs bordered by twelve beaches and sand flats totalling 9 km in length. Apart from the spit all the beaches are accessible by vehicle, with farmland and the Streaky Bay township backing the beaches.

Shag Point is a low, protruding calcarenite bluff with two narrow, low energy beaches located on either side. Beach **1167** lies on the north side of the point. It is a curving, 300 m long, east facing beach, fronted by shallow sand flats and seagrass meadows and backed by the cape road and farmland. Its neighbour (beach **1168**) lies on the south side of the point and is 100 m long.

Doctors Beach lies at the base of the bay and consists of two parts. The first is beach **1169**, a 400 m long, northeast facing high tide beach fronted by 500 m wide sand flats. A small tidal creek drains the backing samphire flats and separates it from the main 800 m long Doctors Beach (**1170**), which is backed by the Streaky Bay caravan park. Both beaches have built northward into the bay as several low beach ridges.

Streaky Bay township is built on sloping calcarenite, with 10 m high bluffs along the north facing waterfront. Below the bluffs and town centre are two narrow, low energy beaches, consisting of a mixture of sand and rocks and both partly covered by seawalls. The main **Jetty Beach** (**1171**) is 200 m long, while the eastern beach (**1172**) is 150 m long. A 300 m long jetty crosses the main beach, with boats moored off the beach (Fig. 4.167).

Figure 4.167 *The Streaky Bay township waterfront is dominated by the long jetty, with two usually calm bay beaches to either side.*

The eastern shore of the bay receives higher westerly wind waves and consists of five longer and more continuous beaches. The yacht club beach (**1173**) is located immediately northeast of the town. It is a 600 m long, curving, west facing, low energy beach fronted by 200 m wide sand flats. The yacht club and boat ramp are located toward the centre of the beach. It terminates at a 10 m high knoll, on the north side of which is a 2.8 km long, curving beach which sweeps up to **Crawfords Landing** at the northern tip. A rock outcrop toward the centre breaks the beach into a 1.6 km long southern beach (**1174**) and the 1.2 km long northern beach (**1175**). Low dunes and a vehicle track parallel the back of the beach and lead to the landing.

North of the landing is another 1.8 km long, curving, west facing beach, consisting of a 1.5 km long southern section (**1176**) backed by low dunes and a 300 m long northern segment (**1177**) backed by low grassy bluffs. **Perforated Rocks** (also known as the Little Islands) lie 200 m off the northern beach.

Swimming: These are all usually calm to low wind wave beaches fronted by shallow sand flats, with best swimming toward high tide.

Surfing: None.

Fishing: The Streaky Bay jetty is one of the more popular locations on the whole coast.

Summary: A series of accessible, low energy beaches surrounding the relatively quiet Blanche Port.

1178-1184   PERFORATED ROCKS to PERLUBIE BEACH (STREAKY BAY)

No.	Beach	Rating	Type	Length
	Unpatrolled			
1178	Perforated Rocks (N 1)	1	R+sand flats	200 m
1179	Perforated Rocks (N 2)	1	R+sand flats	3 km
1180	Perforated Rocks (N 3)	1	R+sand flats	1.8 km
1181	Perlubie (S 3)	1	R+sand flats	800 m
1182	Perlubie (S 2)	1	R+sand flats	1.6 km
1183	Perlubie (S 1)	1	R+sand flats	300 m
1184	Perlubie Beach	1	R+sand flats	2.4 km

The southeastern shore of Streaky Bay consists of a gently curving 10 km section of low energy, west facing shore that extends from Perforated Rocks to the popular Perlubie Beach. All the beaches are protected by Cape Bauer and the shallow shoals that fill much of the entrance to the bay, as well as by the rocks and Eba Island. The highway runs a few kilometres inland, almost reaching the coast at Perlubie Beach. Low, narrow dunes and farmland back most of the beaches, with the best public access at Perlubie Beach.

Perforated Rocks (Little Islands) consist of two small rock islets lying 100 m off the shore and 200 m apart. In lee of the rocks is 200 m long beach **1178**. A vehicle track leads to the beach, which is backed by a low grassy bluff. Beach **1179** continues north from the small sand foreland that separates it from 1178. The beach is 3 km long and faces west toward Gibson Point. It is backed by a strip of low, 50 m wide foredunes, then farmland and terminates at a low, slightly protruding bluff.

Beach **1180** continues on the north side of the bluff for 1.8 km to a more prominent northern bluff. It is a straight, west facing, low energy beach, fronted by narrow sand flats and backed by low dunes and farmland. On the northern side of the bluff is beach **1181**, an 800 m long, west facing beach fronted by deeper sand flats and some calcarenite reef toward the northern end. A 2 km long vehicle track runs out from the highway to the southern end.

Beach **1182** is a 1.6 km long beach lying in lee of Pigface Island. The southern half has extensive reef flats, while the northern half has 300 m wide sand flats capped by 13 low sand ridges, while up to 30 low beach ridges back the beach all the way to the highway.

Beach **1183** lies on the south side of the sand foreland formed in lee of Eba Island. The 300 m long beach is fronted by deeper sand flats and some scattered reefs. It is bordered in the south by some low bluffs, with several shacks located on the bluffs and an access track from the highway to the shacks.

Perlubie Beach (1184) is the most popular beach in Streaky Bay. It is located just off the highway, 15 km north of Streaky Bay township. A road runs to a bluff-top car park overlooking the northern end of the beach, with vehicle access also onto the beach, which is used for launching boats at high tide. The northern end of the beach also provides a second car park, toilet block, water and some shade shelters on the beach, which is also used for camping (Figs. 4.168 & 4.169).

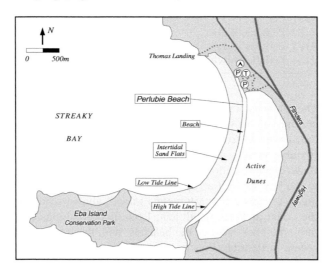

Figure 4.168 *Perlubie Beach.*

Figure 4.169 *The northern end of Perlubie Beach, a popular swimming and camping site.*

The beach is 2.4 km long and faces west. Waves entering the wide bay entrance 20 km to the west are greatly reduced by the time they reach the beach, with additional protection to the south from Eba Island. The beach is composed of fine sand which, with the low waves, produces a wide, flat, firm beach suitable for driving cars on. It is fronted by 200 m wide, ridged, shallow sand flats, with a long wade required to reach deep water. The southern sand flats extend all the way to Eba Island. Seagrass is common on the beach and extensive bare sand dunes extend up to 400 m inland.

The rear of the wide beach becomes a major camping area over the Christmas holidays, with numerous caravans, tents and shelters erected along the northern end of the

beach. The campers use the beach to launch their boats, with farmers using old tractors to tow their boats out across the sand flats. There is also a New Years Day horse race along the beach.

Swimming: These are all relatively safe beaches inshore and on the sand flats. Deep water lies off the sand flats.

Surfing: There is usually no surf, with strong westerlies only producing a low wind wave.

Fishing: The best fishing is from a boat in the bay. The northern bluffs at Thomas Landing provide the only shore access to deep water.

Summary: Perlubie is a popular spot with few facilities, but a beautiful long beach good for camping, boating and relatively safe swimming. The remainder are only used by locals.

1185-1188   THOMAS LANDING (STREAKY BAY)

No.	Beach	Rating	Type	Length
Unpatrolled				
1185	Perlubie (N)	2	R+sand flats	150 m
1186	Thomas Landing	2	R+sand flats	50 m
1187	Thomas Landing (N 1)	2	R+sand flats	50 m
1188	Thomas Landing (N 2)	2	R+sand flats	50 m

The northern end of Perlubie Beach is bordered by low calcarenite bluffs, which extend north for 2 km along the base of 25 m high Perlubie Hill. In four indentations in the bluffs are four small, low energy beaches fronted by sand flats. All four are accessible by vehicle tracks running north from Perlubie Beach and in from the highway.

The first beach (**1185**) is 150 m long and shares its 200 m wide ridged sand flats with the main Perlubie Beach. The now abandoned Thomas Landing was located on the bluffs at the northern end of this beach, providing reasonable access to deeper water beyond the sand flats. The other three (**1186, 1187, 1188**) are all 50 m long pockets of sand wedged beneath the 5 m high bluffs running north of the landing. They are all fronted by 100 m wide sand flats and seagrass meadows.

Summary: Four low energy pockets of sand offering a little seclusion and usually low wave to calm conditions.

1189-1191　PERLUBIE HILL to HASLAM (STREAKY BAY)

		Unpatrolled		
No.	Beach	Rating	Type	Length
1189	Perlubie Hill	1	R+sand flats	800 m
1190	Aldergrove	1	R+sand flats	11.5 km
1191	Haslam	1	R+sand flats	4 km

Perlubie Hill is a low calcarenite-capped rise that provides a view up the long sandy section of bay shore toward Haslam. On the north side of the hill is an 800 m long sand beach (**1189**), bordered at both ends by beachrock reefs and low bluffs, and fronted by 200 m wide ridged sand flats. Active dunes extend 500 m in from the beach. On the north side of the bluffs is 11.5 km long **Aldergrove Beach** (**1190**), a west facing beach exposed to wind waves formed within the 20 m wide bay. It is fronted by sand flats averaging 150 m in width and backed by generally low, unstable dunes extending on average about 500 m inland, then cleared farmland. Because of the farmland there is no public access to the beaches.

Aldergrove Beach terminates at a protruding, 1 km long section of low bluffs, with **Haslam Beach** (**1191**) extending from the bluffs up to Haslam. This 4 km long beach is part cuspate foreland, protruding 200 m bayward in lee of a large sand shoal, while at Haslam it narrows as it becomes backed by low bluffs, terminating at a calcarenite reef. A 200 m long jetty crosses the high tide beach, reef and sand flats, to provide access to deeper water.

Swimming: Haslam offers the best access and, as with the others, usually has low waves breaking over the shallow sand flats.

Surfing: None.

Fishing: Best off the jetty at Haslam.

Summary: Haslam offers good beach access but no facilities other than the jetty and picnic area.

1192-1195　HASLAM NORTH (STREAKY BAY)

		Unpatrolled		
No.	Beach	Rating	Type	Length
1192	Haslam (N 1)	2	R+sand flats	300 m
1193	Haslam (N 2)	2	R+sand flats	400 m
1194	Haslam (N 3)	2	R+sand flats	200 m
1195	Haslam (N 4)	2	R+sand flats	150 m

North of Haslam, 10 m high calcarenite bluffs dominate the southwest facing bay shore for 6 km. At the base of

the bluffs are four small, narrow beaches located between 1 and 4 km north of the jetty. Beach **1192** is a crenulate, 300 m long beach, while beach **1193** is 400 m long, with outcrops of beachrock. Beach **1194** is a protruding high tide beach in lee of a calcarenite reef, with fringing beachrock, while beach **1195** is a 150 m long pocket beach also dominated by reefs and beachrock.

Summary: Four difficult to access, low energy beaches, dominated by beachrock and all backed by farmland.

1196, 1197　FLAGSTAFF LANDING (STREAKY BAY)

		Unpatrolled		
No.	Beach	Rating	Type	Length
1196	Haslam (N 5)	2	R+sand flats&reef	1.4 km
1197	Flagstaff Landing	2	R+sand flats&reef	2.6 km

The low bluffs terminate about 7 km northwest of Haslam, beyond which is a highly crenulate, 1.4 km long, southwest facing beach (**1196**) that is dominated by three reef sections, with the beach widening in lee of each of the reefs. The beach is backed by farmland, with no public access.

Flagstaff Landing is located 2.5 km south of Flagstaff Corner. The landing is today only used for launching small boats, with no facilities. The beach (**1197**) extends for 2.6 km to the east of the landing, with reefs bordering either end. The landing is located in a small open area next to the reef at the very western end of the beach, right below the end of the access road.

Summary: Two long, low energy beaches, with the landing used to launch boats and good fishing off the bordering reefs.

Acraman Creek Conservation Park	
Established:	1991
Area:	3960 ha
Beaches:	4 beaches (1198-1201)
Coast length:	11 km

Acraman Creek Conservation Park incorporates the present 5 km long tidal creek and its network of smaller tributaries and few hundred hectares of mangrove woodlands, as well as an adjoining dammed and dry former neighbouring creek, which preserves a record of the former creek channels and bed.

1198-1203 ACRAMAN CREEK to POINT DE MOLE (STREAKY BAY)

No.	Beach	Rating	Type	Length
	Unpatrolled			
1198	Acraman Ck (E)	2	R+sand&tidal flats	3.4 km
1199	Acraman Ck (inlet 1)	2	R+tidal flats	400 m
1200	Acraman Ck (inlet 2)	2	R+tidal flats	700 m
1201	Pt Lindsay	2	R	5.3 km
1202	Pt Lindsay (W)	1	R+sand flats	6 km
1203	Pt De Mole (E 3)	1	R+sand flats	600 m

Acraman Creek is the largest tidal creek in Streaky Bay and an important mangrove habitat, with the entire system classified as a conservation park. The creek has been formed by the development of two barrier-spit systems which, during the past few thousand years, have slowly impounded much of the creek area. The eastern barrier consists of beach **1198**, a 3.4 km long beach backed in part by up to 12 beach ridges, with a low, narrow, 1.8 km long recurved spit extending west to form the eastern side of the inlet. The Flagstaff Landing road provides access to the eastern end of the beach.

The 2 km wide creek mouth contains two low barrier islands, fronted by 1 km wide sandy tidal flats and surrounded by tidal creeks and mangroves. The eastern island (**1199**) is 400 m long and backed by mangroves, with tributary creeks to either side. The western island (**1200**) is 700 m long, with the main creek forming the boundary between it and Point Lindsay.

Point Lindsay is a low, prominent recurved spit that forms the western side of Acraman Creek inlet. The 5.3 km long beach (**1201**) and spit receives occasional low swell which has helped build the beach up to 600 m into the bay as a series of low beach-foredune ridges, as well as more than 1 km east to partly enclose the backing creek and mangroves. A vehicle track reaches the western end of the beach, from there, a 4WD track runs along the back of the beach to the point, where the inlet beach is used to launch boats.

Immediately west of the beach is a 6 km long, low energy, crenulate, southeast facing beach (**1202**) that runs round to the lee of **Point De Mole**, where it faces east. The beach receives increasing protection toward the west, where it is fronted by 400 m wide sand flats and seagrass meadows. Beach **1203** lies immediately west of the point and is a 600 m long, northeast facing, protected beach fronted by sand flats which narrow toward the point. A 4WD track runs out along the back of the beach to the point. The point is the northern boundary of Streaky Bay and also marks the boundary between the low energy bay beaches and the higher energy ocean beaches to the west.

Swimming: The best location is at the road access at the western end of Point Lindsay beach. Be careful in the creek, as there are deep tidal channels and currents, while shallow tidal flats front the other beaches.

Surfing: None.

Fishing: Acraman Creek is a popular location for boat and shore fishing.

Summary: A natural coastal system containing the creek, mangroves and the beaches, all preserved in the conservation park.

1204-1208 POINT DE MOLE, GASCOIGNE BAY

No.	Beach	Rating	Type	Length
	Unpatrolled			
1204	Pt De Mole (E 2)	3	R+reef	150 m
1205	Pt De Mole (E 1)	3	R+reef	100 m
1206	Pt De Mole	4	R+beachrock	100 m
1207	Pt De Mole (W)	4	R+reef	600 m
1208	Gascoigne Bay	3	R	9.5 km

Gascoigne Bay is an open, south to southeast facing, 9 km wide bay located immediately northwest of Streaky Bay and bordered by Point De Mole and Point Collinson. All the beaches on the point and in the bay receive usually low ocean swell and tend to be reflective, with steep beach faces. There is 4WD access to Point De Mole, with a gravel road running along the western side of the bay beach out to Edward Bay.

Beaches **1204** and **1205** lie on the eastern side of the point and face southeast into Streaky Bay. They are 150 m and 100 m long respectively, backed by low, sand-draped bluffs. They receive low ocean swell, which breaks over shallow reefs extending 50 m off the beaches.

On the south facing side of the point is beach **1206**, a crenulate, 100 m long beach, protected to the east by a beachrock reef. On the western side of the reef is 600 m long beach **1207**, which has rocks and reefs extending up to 800 m offshore, lowering waves to about 0.5 m at the shore (Fig. 4.170).

The main beach (**1208**) is a 9.5 km long, crenulate, south facing beach, partly protected by extensive reefs in the bay. Between the point and the vehicle access track, it faces south and is backed by dunes up to 1 km wide. West of the access track it faces southeast and is protected by Point Collinson, with shallow reefs off the low energy beach, while it is backed by low, stable beach-foredune ridges.

Figure 4.170 *View west of Point De Mole showing reef-protected beaches 1206 (foreground) and 1207.*

Swimming: The best location is at the Gascoigne Bay access track, where waves are usually low. Beware of the rocks and reefs out on Point De Mole.

Surfing: There are some reef breaks off Point De Mole and in lee of Point Collinson.

Fishing: Most fishers head for the reefs around the points.

Summary: An open, lower energy, ocean bay with limited access and no facilities.

1209-1221 POINT COLLINSON, EDWARD BAY

No.	Beach	Rating	Type	Length
	Unpatrolled			
1209	Pt Collinson (E 5)	3	R+reef flats	800 m
1210	Pt Collinson (E 4)	3	R+reef flats	100 m
1211	Pt Collinson (E 3)	3	R+reef flats	50 m
1212	Pt Collinson (E 2)	3	R+reef flats	50 m
1213	Pt Collinson (E 1)	2	R+reef flats	1.2 km
1214	Pt Collinson	4	R+reef flats	250 m
1215	Pt Collinson (W 1)	5	R+platform	450 m
1216	Pt Collinson (W 2)	5	R+platform	100 m
1217	Pt Collinson (W 3)	5	R+platform	150 m
1218	Pt Collinson (W 4)	5	R+platform	100 m
1219	Edward Bay (E 2)	4	R+reef	300 m
1220	Edward Bay (E 1)	5	R	100 m
1221	Edward Bay	4	R	850 m

Between Point Collinson and a low, flat, rocky point 2.5 km to the northeast is an open, southeast facing bay containing five low energy beaches, all fronted by shallow reef and some sand flats. Four-wheel drive tracks provide access to most of the beaches, which are backed by low, vegetated dunes.

Beaches **1209, 1210, 1211** and **1212** are four southeast facing sand beaches, backed by 10 m high foredunes and

fronted by shallow reef flats extending up to 200 m into the bay. As a consequence, only low waves reach the beach at high tide, with calm conditions at low tide. Beach **1213** is a 1.2 km long, curving, east facing, low energy beach that terminates in lee of the point. It is fronted by reef and sand flats up to 200 m wide, with usually calm conditions at the shore. The Point Collinson track runs along the back of the beach.

Beach **1214** lies on the eastern tip of Point Collinson. It is a 250 m long, southeast facing high tide beach, fronted by 400 m wide intertidal calcarenite reef flats. On the western side of the point the coast faces southeast and the first 2.5 km of shore is dominated by intertidal 50 to 100 m wide calcarenite reefs and backing low bluffs. Along the base of the bluffs are four beaches. The first (**1215**) begins at the tip of the point and is backed by a low foredune and 400 m wide unstable dunes. The 450 m long beach protrudes seaward in lee of the rock flats and reefs, with surf breaking on the outer reefs 400 m offshore. Small protruding bluffs separate the next three beaches (**1216, 1217, 1218**), which are 100 m, 150 m and 100 m long respectively. Moderate waves break heavily on the reefs but are usually calm against the shore, while dunes cap the backing low bluffs.

Beach **1219** is a curving, 300 m long, south facing reflective beach bounded by 10 m high dune-capped bluffs and bordered by a 200 m wide calcarenite platform in the east and jagged granite bedrock platform in the west. Its smaller neighbour (**1220**) is a 100 m long pocket beach bordered and backed by 20 m high bluffs, with access to both beaches requiring a climb down the bluffs.

Edward Bay beach (**1221**) is located in lee of Point Brown. It is a curving, 850 m long, south facing beach composed of coarse quartz sand derived from the eroding point. Combined with waves lowered to about 1 m, the coarse sand produces a steep, cusped beach face (Fig. 4.171), with a heavy shorebreak during periods of higher waves. There is vehicle access to the rear of the beach, where there are some informal camp sites, but no facilities.

Figure 4.171 *Edward Bay beach, with waves surging up the steep, cusped beach face.*

Swimming: Edward Bay is the most accessible and relatively safe under normal low waves. Most of the other beaches are dominated by shallow rock, reef and sand flats.

Surfing: The only surf is on the reef breaks off Point Collinson and Edward Bay.

Fishing: The point and bay areas are relatively popular sites for both rock and beach fishing.

Summary: An out of the way but accessible section of rock and reef-dominated coast that becomes increasingly exposed to higher waves toward the west.

1222-1230 POINT BROWN to CAPE MISSIESSY

No.	Beach	Unpatrolled Rating	Type	Length
1222	St Mary Bay	4	R-LTT	2.2 km
1223	St Mary Bay (N)	5	R+reef	350 m
1224	Pt Dillon (N 1)	5	R+reef	800 m
1225	Pt Dillon (N 2)	5	R+reef flats	700 m
1226	Pt Dillon (N 3)	5	R+reef flats	800 m
1227	Pt Dillon (N 4)	5	R+reef	200 m
1228	Cape Missiessy (S)	6	R+reef	800 m
1229	Cape Missiessy	2	R+reef&sand flats	2.8 km
1230	Cape Missiessy (E)	3	R+tidal flats	2.4 km

The coast between Point Brown and Cape Missiessy consists of 16 km of exposed, west facing calcarenite shore composed of low calcarenite bluffs, some capped by clifftop dunes, together with nine beaches, all afforded substantial protection from fronting reefs. There is vehicle access only to St Mary Bay in the south, with only 4WD tracks to the other beaches and bluffs.

St Mary Bay is an open, west facing, 2 km long bay bordered by 30 m high Point Brown and Point Dillon, which rises inland to 40 m. There are two beaches in the bay, the main 2.2 km long beach (**1222**) which is backed by active dunes and a 350 m long northern beach (**1223**). The points and reefs lower waves at both beaches to about 1 m which, along the main beach, maintain a steep reflective beach, which forms a narrow attached bar cut by rips during higher waves, particularly toward the northern end. The northern beach is fringed and fronted by reefs.

Three kilometres north of Point Dillon is the first of four reef-controlled beaches. Beach **1224** is a curving, 800 m long beach protected by the point and fringing rock reefs, particularly along the southern half of the beach, with low waves breaking across the reefs. One kilometre to the north are the straight neighbouring beaches **1225** and **1226**, 700 m and 800 m long respectively. Both are fronted by shallow reef flats up to 500 m wide, which

form a sheltered 'lagoon' in their lee, resulting in low energy beaches (Fig. 4.172). Beach **1227** lies another 1 km to the north. It is a more crenulate, 200 m long beach fronted by irregular reefs, with two permanent rips flowing through gaps in the reefs.

Figure 4.172 *Beaches 1225 and 1226 are two of a series of reef-protected beaches that extend north of Point Dillon.*

Beach **1228** is a straight, 800 m long beach fronted by shallow reef and sand flats that decrease waves to less than 0.5 m at the shore and calms often prevail. The beach terminates at a low calcarenite bluff, beyond which extends the 8 km² Cape Missiessy barrier system, comprising beaches 1229 and 1230. The barrier consists of a series of northward and eastward prograding beach ridges and recurved spits, the latter extending over 2 km into Smoky Bay. More recently, grazing stock have destabilised the ridges and a 6 km² active dune system covers the ridges and is blowing into the backing mangrove system. Beach **1229** is a relatively straight, 2.8 km long, low energy beach fronted by sand flats that widen considerably off the cape, where the southern entrance for Smoky Bay is located. At the cape the beach turns and faces north. The low, 2.4 km long beach (**1230**) grades into mangroves at the eastern end. It is fronted by 100 to 300 m wide sand tidal flats, then a deep, 200 m wide tidal channel, with Eyre Island on the north side of the channel.

Swimming: The southern end of St Mary Bay is the most accessible and relatively safe when waves are less than 1 m. A heavy shorebreak develops during higher waves. Most of the other beaches are difficult to access and dominated by reefs.

Surfing: Only on the reefs off Point Brown.

Fishing: St Mary Bay and the rocks around Point Brown are the most accessible and popular locations.

Summary: Apart from the southern St Mary Bay, this is a relatively inaccessible section of coast.

EY-1 – EY-5   EYRE ISLAND

Unpatrolled				
No.	Beach	Rating	Type	Length
EY-1	Eyre Is (S)	1	R+sand flats	8 km
EY-2	Eyre Is (N 1)	1	R+sand flats	2.5 km
EY-3	Eyre Is (N 2)	1	R+sand flats	1.8 km
EY-4	Eyre Is (N 3)	1	R+sand flats	1 km
EY-5	Eyre Is (N 4)	1	R+sand flats	700 m

Eyre Island is a low, 8 km long sand island that, together with Little Eyre Island and surrounding sand flats, forms an 11 km long, low barrier across the southern half of Smoky Bay. The islands themselves are sheltered by calcarenite reefs further offshore and consequently, despite their seemingly exposed location and southwest orientation, they receive only low wind waves along their southern shores, with lower energy conditions on the northern shores. Eyre Island is composed of multiple low beach ridges and recurved spits, which link to form a continuous 8 km long southern beach and four discontinuous northern beaches. It has 16 km of shoreline and an area of 15 km², including a central 5 km² mangrove stand. The island is one of several that comprise the Nuyts Archipelago Conservation Park.

The main southern beach (**EY-1**) is a double scalloped, 8 km long, low energy beach, fronted by 100 m wide sand flats and then more than 1 km of seagrass meadows. Sediment is moving along the beach, eroding from the western end while building spits at the eastern tip. The north side of the island consists of four beaches (EY-2 – EY-5) as well as exposed mangroves between beaches EY-3 and EY-4. Beach **EY-2** is a curving, 2.5 km long beach that begins at the western tip and terminates in a narrow spit and tidal creek, with backing mangroves (Fig. 4.173). Beach **EY-3** commences on the other side of the creek and is a 1.8 km long series of up to 40 truncated recurved spits. It is slowly overlapping beach **EY-4**, which is a narrow, 1 km long active spit growing to the east, while in the west it is fronted by mangroves. To the east of the beach are two tidal creeks and 2 km of mangrove shoreline. On the eastern side is beach **EY-5**, a 700 m

long, low beach that links with the eastern end of beach EY-1.

Figure 4.173 *The northwestern side of Eyre Island consists of a series of multiple beach ridges fronted by wide sand flats. As the 30 ridges have built eastward, they have afforded increasing protection to the backing 5 km² of mangroves.*

Summary: A low-lying nature reserve accessible by boat from Smoky Bay and surrounded by low energy beaches, sand flats and seagrass meadows.

1231-1244   SMOKY BAY

Unpatrolled				
No.	Beach	Rating	Type	Length
1231	Smoky Bay (S 3)	1	R+tidal flats	700 m
1232	Smoky Bay (S 2)	1	R+tidal flats	1.2 km
1233	Smoky Bay (S 1)	1	R+tidal/sand flats	1.2 km
1234	Smoky Bay	1	R+sand flats	2.4 km
1235	Smoky Bay (N)	1	R+sand flats	1 km
1236	Koppi Tucka (1)	1	R+sand flats	1.2 km
1237	Koppi Tucka (2)	1	R+sand flats	200 m
1238	Koppi Tucka (3)	1	R+sand flats	800 m
1239	Saddle Peak (S)	1	R+sand flats	400 m
1240	Saddle Peak	1	R+sand flats	2 km
1241	Saddle Peak (N)	1	R	1.2 km
1242	Pines Way (1)	1	R+LTT	150 m
1243	Pines Way (2)	1	R+LTT	1.3 km
1244	Pines Way (3)	1	R+LTT	1.4 km

Smoky Bay is a 20 km long, 10 km wide, southwest facing bay that is bordered by sandy Cape Missiessy in the south and the calcarenite Cape D'Estrees to the north. In addition, the Eyre Islands and associated sand flats block the southern half of the bay, narrowing the northern entrance to 9 km, while the Franklin Islands, lying 20 km offshore, also reduce waves from the southwest. Only in the northern half does low swell enter the bay, with all the bay beaches dominated by low wave conditions and most fronted by sand flats and seagrass meadows. The bay's 35 km shoreline contains 23 beaches, with 16 dominated by tidal and/or sand flats. The small town of Smoky Bay is located just off the highway toward the southern end of the bay, with no other development other than farmland. Laura Bay Conservation Park is located in the north.

Cape Missiessy shelters the southernmost section of the bay, with about 5 km² of mangroves in lee of the cape. As the shore swings round to face west, low wind waves maintain low energy beaches and sandy tidal flats. Beaches 1231, 1232 and 1233 are three low energy, northwest facing beaches lying between Smoky Bay township and the mangroves in lee of the cape. Beach **1231** is a curving, 700 m long, high tide sand and shell beach fronted by 1 km wide tidal flats, with a small tidal creek at the western end, while low calcarenite reefs border the eastern end. Beach **1232** is a curving, 1.2 km long beach also fronted by wide tidal flats, bordered by low bluffs and backed by a few beach ridges, which widen to the north. Beach **1233** has a 1.2 km long high tide beach with tidal flats, grading to vegetated tidal flats in the north. A gravel road parallels the northern half of the beach.

Smoky Bay township is fronted by beach **1234**. It is a 2.4 km long, west facing beach, protected by Eyre Island 6 km across the bay. The beach has sand flats up to 600 m wide, which narrow to 150 m in the south where a 300 m long jetty is located, backed by the town centre (Fig. 4.174). Waves average less than 0.3 m and boats are moored off the beach.

Figure 4.174 *Smoky Bay jetty and township. Note the wide sand flats that necessitate the long jetty.*

North of the town, the shoreline faces west toward Eyre Island and consists of low calcarenite bluffs fronted by a series of low energy beaches, which are backed by farmland all the way to Laura Bay. While the highway parallels the coast 1 to 2 km inland, public access to the coast is restricted owing to the farmland.

Beach **1235** lies 2 km north of the town. It is a 1 km long, low energy beach with 400 m wide sand flats. Beaches **1236, 1237** and **1238** are similar, at 1.2 km, 200 m and 800 m in length respectively, each bordered by low bluffs and backed by generally dune-covered bluffs. Beach **1239** is located on a small headland below **Saddle Peak**, a 20 m high rise. The 400 m long beach is bordered by more

prominent bluffs, fronted by 400 m wide ridged sand flats and backed by an unstable 20 m high foredune.

North of the headland the beaches face across the 10 km wide bay toward the open ocean and low swell and higher wind waves reach the beaches. As a consequence, the sand flats narrow to more wave-dominated low tide terraces, with larger 20 m high foredunes and some blowouts backing the beaches.

Beach **1240** is a straight, west facing, 2 km long beach, fronted by 100 m wide sand flats and backed by low dunes up to 600 m wide. Its northern neighbour (**1241**) is the highest energy beach in the central bay and comprises a 1.2 km long reflective beach with no sand flats. Beach **1242** lies on a small protruding headland and is a 150 m long pocket beach bordered by low protruding bluffs, with a 50 m wide low tide terrace. Immediately north of the headland is beach **1243**, a more irregular, 1.3 km long beach backed by a crenulate bluff and fronted by a 100 m wide low tide terrace. In the northeast corner of the bay is southwest facing beach **1244**, a 1.4 km long, lower energy beach fronted by ridged sand flats which widen to 200 m in the west. It is however backed by the largest and most active dunes inside the bay, with one 20 m high blowout extending 400 m inland.

Swimming: The best location is at Smoky Bay beach, which offers good access and parking as well as the facilities of Smoky Bay.

Surfing: None in the bay.

Fishing: The Smoky Bay jetty is the most popular location, with others heading out in boats to fish the bay and tidal channels.

Summary: A large protected bay offering quieter waters for boating.

Laura Bay Conservation Park	
Established:	1973
Area:	267 km
Beaches:	3 beaches (1245-1247)
Coast length:	2 km

Laura Bay Conservation Park encompasses a 2 km long strip of coast and the backing dunes and vegetated land extending up to 2.5 km inland. The western side of the southern headland was also the site of a timber wharf, constructed in 1911 to load grain and a well preserved water tank built in 1914.

1245-1254   LAURA BAY to CAPE D'ESTREES - Laura Bay Conservation Park

No.	Beach	Unpatrolled Rating	Type	Length
1245	Dog Point (W)	2	R	300 m
1246	Sandy Cove	2	R	400 m
1247	Laura Bay (S)	2	R	100 m
1248	Laura Bay (E)	1	R+sand/tidal flats	700 m
1249	Laura Bay	1	R+sand/tidal flats	650 m
1250	Laura Bay (W)	1	R+sand/tidal flats	200 m
1251	Cape D'Estrees (E 4)	1	R+sand flats	800 m
1252	Cape D'Estrees (E 3)	1	R+sand flats	100 m
1253	Cape D'Estrees (E 2)	1	R+sand flats	800 m
1254	Cape D'Estrees (E 1)	1	R+sand flats	150 m

Laura Bay is located at the northern end of Smoky Bay, 3 km east of Cape D'Estrees. A gravel road from Ceduna crosses the back of the bay and provides access to the small conservation park, the old water storage tanks and some of the beaches. There are ten low to very low energy beaches within and adjacent to the bay and Cape D'Estrees.

On the southeastern side of the bay are three exposed, southwest facing beaches all located in the park. They receive waves averaging about 0.5 m and are fronted by 50 m wide low tide bars. Beach **1245** is a narrow, 300 m long beach lying at the base of low calcarenite bluffs, which also border the beach. The eastern bluffs protrude 200 m into Smoky Bay and terminate at Dog Point. Sandy Cove (beach **1246**) is a 400 m long, slightly curving beach contained between two 20 m high headlands, with active 20 m high dunes extending 300 m in from the beach. Beach **1247** lies on the southern side of the 30 m high, conical eastern headland for Laura Bay and consists of a 100 m strip of high tide sand backed by steep, 10 m high bluffs.

Laura Bay is a small, semicircular, south facing bay, 1 km wide at the mouth, widening to 2 km inside. It is very protected from ocean waves and usually calm conditions prevail at the three shelly beaches, each fronted by sand and tidal flats a few hundred metres wide and bordered and backed by stands of low mangroves. Beach **1248** is a low gradient, 700 m long high tide beach fronted by 400 m wide sand flats, with mangroves fringing the western end. Its neighbour, beach **1249**, is 650 m long, with small tidal creeks and mangroves to each end and 500 m wide sand flats. These two beaches are part of a low beach ridge-samphire system that has prograded 1 km into the bay. Beach **1250** is tucked inside the western headland and consists of a 200 m long, east facing strip of high tide sand fronted by 400 m wide sandy tidal flats.

Outside the bay, between the western headland and Cape D'Estrees are four more exposed beaches. The first two,

1251 and **1252**, face southeast and lie immediately west of the headland. The first links the small headland to the mainland. It is 800 m long and backed by a low dune system, while it is fronted by 100 m wide ridged sand flats. Its neighbour is similar, only 100 m in length and partially backed by low bluffs. One kilometre to the west is beach **1253**, an 800 m long, south facing, narrow high tide beach backed by low calcarenite bluffs, with a narrow low tide bar and usually low waves. Immediately west and just inside the cape is 150 m long beach **1254**, a similar bluff-bound beach.

Swimming: The most accessible and suitable beach for swimming is 1246, located immediately south of the Laura Bay water tank. The remainder are too shallow or more difficult to access.

Surfing: None.

Fishing: The several headlands are surrounded by slightly deeper water and seagrass meadows.

Summary: A reasonably accessible section of coast containing the small, low energy Laura Bay, the conservation park and a few more exposed small beaches.

1255-1266   DECRES BAY

No.	Beach	Unpatrolled Rating	Type	Length
1255	Cape D'Estrees (W 1)	1	R+LTT	100 m
1256	Cape D'Estrees (W 2)	1	R+LTT	300 m
1257	Cape D'Estrees (W 3)	1	R+LTT	200 m
1258	Decres Bay (W)	1	R+sand flats	1.2 km
Wittelbee Conservation Park				
1259	Decres Bay	1	R+sand flats	1.8 km
1260	Wittelbee Point (W 1)	1	R+sand flats	250 m
1261	Wittelbee Point (W 2)	1	R+sand flats	650 m
1262	Wittelbee Point (W 3)	1	R+sand flats	250 m
1263	Cape Vivonne (E 4)	1	R+sand flats	1.6 km
1264	Cape Vivonne (E 3)	1	R+sand flats	150 m
1265	Cape Vivonne (E 4)	1	R+sand flats	650 m
1266	Cape Vivonne (E 1)	1	R+sand flats	200 m

Decres Bay is an open, southwest facing, 10 km long bay, bordered by Cape D'Estrees and Cape Vivonne. It is afforded considerable protection by St Peter Island and its extensive sand shoals, and by the Franklin Islands. As a consequence, the 15 km of low calcarenite shoreline receives low waves and most of the 11 bay beaches are fronted by sand flats and seagrass meadows. The small Wittelbee Conservation Park is located in the centre of the bay, with road access only to the park and the adjoining main Decres Bay beach (1259).

Cape D'Estrees is a southward protruding, 20 m high calcarenite headland. On the west side of the cape, 10 to 20 m high bluffs trend north for 6 km. Along the base of the bluffs are four irregular bluff-dominated beaches. Beach **1255** lies 600 m north of the cape and is a 100 m long, west facing, narrow pocket beach with a bar then seagrass 50 m offshore. Eight hundred metres to the north is beach **1256**, a discontinuous 300 m long beach backed and cut in parts by the bluffs. Immediately north is beach **1257**, a similar 200 m long beach. Bluffs then dominate for 2 km to beach **1258**, a crenulate, bluff and reef-dominated high tide beach fronted by 100 m wide sand flats.

The main **Decres Bay** beach (**1259**) is a 1.8 km long, southwest facing sand beach, backed by generally vegetated dunes rising to 20 m and extending 400 m inland, and fronted by 400 m wide ridged sand flats. Decres Bay road runs past the beach, with vehicle access at either end. Wittelbee Point protrudes 1 km into the bay at the western end. Wittelbee Conservation Park includes the western half of this beach, as well as the point and two western beaches.

On the western side of the point is 250 m long beach **1260**. It is backed by calcarenite bluffs and fronted by 100 m wide sand flats. The bluffs continue from the point for 5 km to **Cape Vivonne**. In between are another six narrow beaches, all lying at the base of the 10 to 15 m high bluffs. Beach **1261** is 650 m long and fronted by 200 m wide ridged sand flats. Beach **1262** is a 250 m long pocket beach, backed and bordered by the bluffs. Beach **1263** is a 1.6 km long, narrow beach backed by dune-draped bluffs, with the sand flats widening to 200 m in the west. Beach **1264** is a 150 m long pocket beach, while beach **1265** a 650 m long, narrow high tide beach, with the bluffs protruding onto the beach in places. Beach **1266** lies immediately east of 20 m high Cape Vivonne. It is a 200 m long high tide beach fronted by 200 m wide sand flats. A 4WD track runs out to the cape and terminates at this beach.

Swimming: The most accessible and attractive beach is the main Decres Bay beach (1259), which usually has low waves, with shallow sand flats off the beach.

Surfing: None.

Fishing: Better off the points, which provide access to deeper water.

Summary: A little visited section of coast, close to Ceduna, but with limited access.

SP-1 – SP-19 ST PETER ISLAND

No.	Beach	Rating	Type	Length
\multicolumn{5}{c}{**Unpatrolled**}				
SP-1	St Peter spit (S 1)	1	R+sand flats	2.4 km
SP-2	St Peter spit (S 2)	1	R+sand flats	1.2 km
SP-3	Mt Younghusband (S)	1	R+sand flats	3 km
SP-4	St Peter Is 4	1	R+tidal flat	1.5 km
SP-5	St Peter Is 5	1	R+tidal flats	400 m
SP-6	St Peter Is 6	1	R+sand flats	4.2 km
SP-7	Hawks Nest (E)	2	R+reef	200 m
SP-8	Hawks Nest	2	R+reef	300 m
SP-9	Hawks Nest (W)	2	R+reef	1.2 km
SP-10	St Peter Is 10	2	R+reef	100 m
SP-11	St Peter Is 11	2	R+reef	1.2 km
SP-12	St Peter Is 12	3	R+reef	300 m
SP-13	St Peter Is 13	3	R+reef	2.8 km
SP-14	St Peter Is 14	3	R+reef	1.1 km
SP-15	St Peter Is 15	2	R+reef	150 m
SP-16	Bob Bay	1	R+sand flats	3.5 km
SP-17	Bob Bay (N)	1	R+sand flats	2 km
SP-18	Mt Younghusband (N)	1	R+sand flats	2 km
SP-19	St Peter spit (N)	1	R+sand flats	3.7 km

St Peter Island is the largest of the western Eyre Peninsula islands and the largest in the Nuyts Archipelago Conservation Park. It has an area of 42 km² and 40 km of shoreline. The island lies 12 km due south of Ceduna and is privately owned. Despite its exposed location, the island shoreline receives low to very low waves owing to its southwest-northeast orientation, the protection of Goat Island and bedrock reefs off the exposed southwest side of the island. The island's 19 beaches are all relatively low energy, many fronted by sand and tidal flats. The island consists of a calcarenite-capped bedrock core rising to a maximum elevation of 48 m at Mt Younghusband. East of the mount is a 4 km long sand spit, while to either side are sheltered bays with mangrove stands. The eastern tip of the spit lies 3 km due south of Cape Vivonne.

The eastern 4 km of the island consists of a low, 300 to 500 m wide spit composed of up to 30 multiple, truncated recurved spits, with sediment moving along both the north and south sides of the spit. The spit comprises three beaches: a continuous northern 3.7 km long beach (**SP-19**) fronted by 300 to 600 m wide sand flats and two similar beaches on the south side, 2.4 km and 1.2 km in length (**SP-1** and **SP-2**), also fronted by sand flats, but separated by a mangrove-filled tidal creek.

The spits attach to the base of Mt Younghusband, the southern side of which forms the northern boundary of a U-shaped, east facing bay containing three very low energy beaches. Beach **SP-3** is a south facing, 3 km long, narrow beach, backed by the slopes of the mount and fronted by 400 m wide tidal flats. It terminates at the base of the bay where the shore turns and faces east. This section consists of a 1.5 km long, narrow strip of sand (**SP-4**) backed by samphire flats and fronted by 500 m wide tidal flats. The southern side of the bay consists of a 1 km long spit, containing mangrove shore and a 400 m long, low energy beach and tidal flats (**SP-5**). The sand shoals off the spit continue east for 5 km, paralleling the main northern spit.

The southeast shore of the island consists of calcarenite-covered slopes. The first beach, **SP-6**, is the longest on the island at 4.2 km in length. It is backed by the slopes, and fronted by sand flats which are 800 m wide in the east, narrowing to 100 m in the west. At the southern end of the beach the coast trends west, exposing the shore to higher offshore waves. However rock reefs lower the waves at the shore and reflective beaches dominated by reefs prevail. Beach **SP-7** is a 200 m long pocket of sand enclosed in reefs extending 200 m offshore. Beach **SP-8**, known as Hawks Nest, is a 300 m long, south facing beach contained in a small, 500 m wide embayment. It is backed by low dune-draped bluffs, with reefs extending across the bay entrance and low waves at the shore. Its neighbour, beach **SP-9**, is a narrow, 1.2 km long, south facing beach backed by low grassy bluffs and also fronted by scattered shallow reefs.

Beach **SP-10** is the southernmost beach on the island. It is a 100 m long pocket of sand wedged in a bedrock embayment and protected by a small headland and deeper reefs. Two kilometres to the west is beach **SP-11**, a curving, southwest facing, 1.2 km long, narrow reflective beach, occupying an open bay with reefs lying up to 500 m offshore and low waves at the beach. The beach terminates in the west at a prominent 30 m high headland. On the western side of the headland is a curving, 300 m long pocket beach (**SP-12**), bordered by the headland and reefs extending 600 m off the western side. On the western side of the reefs is the 2.8 km long, west facing beach **SP-13**, which is protected by Goat Island and its backing shoals, that extend the 1.5 km from the island to the beach. Wave height and beach width increase north along the beach, which is backed by a low foredune and vegetated dunes extending up to 600 m inland.

The northern side of the island begins at the northern end of beach SP-14, where a prominent 30 m high headland trends northeast. At the end of the headland is the curving, 1.1 km long beach **SP-14**. The beach faces west and is free of inshore reefs, but still receives low swell and usually has a steep reflective beach face (Fig. 4.175). The beach terminates at a low point, on the north side of which is a lower energy, 150 m long, northwest facing, narrow beach (**SP-15**), fronted by scattered rocks, reefs and seagrass meadows.

To the west of beach SP-15 the shoreline decreases in energy, with only the occasional higher southwest swell reaching the north side of the island. Such events during the past few thousand years have slowly developed a crenulate, 3 km long sand spit. As the spit has grown east it has enclosed Bob Bay, a 5 km long, partly mangrove-

filled, shallow bay. The beach (**SP-16**) consists of multiple, truncated recurved spits, with sand flats up to 400 m wide. The spit has subsequently been breached, forming a detached 1.1 km long barrier island-spit (**SP-17**). This barrier island also partially encloses the backing bay, with the tidal inlets located either end of the island (Fig. 4.176).

Figure 4.175 *Beach SP-14 on the midwestern side of St Peter Island faces southwest but is sufficiently protected to have a lower energy reflective to low tide terrace beach.*

Figure 4.176 *Bob Bay has been formed by the growth of western (foreground) and longer eastern (background) sand spits, which form two 'tails' to St Peter Island.*

As the spits have been prograded to the west, they have increasingly sheltered the backing bedrock shore. The once more exposed beach has become fronted by sand flats and mangroves as the protection increases. The remaining section of unprotected shore contains a 2 km long beach (**SP-18**), a narrow beach that hugs the base of the northern slopes of Mt Younghusband and is fronted by sand flats.

Swimming: While most of the island beaches are relatively safe owing to the normally low waves and lack of beach rips, the best for swimming are some of the southeast and southwest beaches which have deeper water close inshore. However during periods of higher waves, these beaches will have a heavy shorebreak and some reef-controlled rips.

Surfing: There are a number of potential reef breaks along the southern shore.

Fishing: The island has numerous good locations for rock, reef and beach fishing.

Summary: A privately owned island that has been cleared for grazing and is surrounded by 19 natural, undeveloped beaches.

1267-1272 CAPE VIVONNE, BOSANQUET BAY

Unpatrolled				
No.	Beach	Rating	Type	Length
1267	Cape Vivonne	1	R+sand flats	50 m
1268	Cape Vivonne (N 1)	1	R+sand flats	60 m
1269	Cape Vivonne (N 2)	1	R+sand flats	200 m
1270	Cape Vivonne (N 3)	1	R+sand flats	200 m
1271	Cape Vivonne (N 4)	1	R+sand flats	250 m
1272	Bosanquet Bay	1	R+sand flats	3 km

Bosanquet Bay is a 4 km wide bay located between Cape Vivonne and Cape Thevenard. The centre of the bay consists of a long beach, with small bluff and rock-bound beaches along each of the capes, all receiving low waves and fronted by sand flats. A 4WD track runs out to Cape Vivonne, while Decres Road runs along the back of the main beach and the town of Thevenard is located on Cape Thevenard.

Cape Vivonne extends for 5 km in a southerly direction and consists of 10 to 20 m high calcarenite bluffs, overlying granite bedrock. In amongst the bluffs are five small beaches. Beach **1267** lies on the tip of the cape and is a 50 m pocket of shelly high tide sand fronted by intertidal bedrock reefs. Immediately to the north is beach **1268**, a 60 m long high tide beach. It is backed by 10 m high bluffs and fronted by sand flats, with granite boulders to either end and bluff debris on the beach. One kilometre to the north is beach **1269**, a 200 m long beach with backing bluffs and 100 m wide sand flats. Beach **1270** lies another 1 km to the north and is similar, 200 m long with 100 m wide sand flats and dunes draping the backing 10 m high bluffs. At the north end of the cape is beach **1271**, a curving, 250 m long beach, backed by low dunes and fronted by 200 m wide sand flats. It lies immediately east of the main beach and is accessible via a rough vehicle track.

The main Bosanquet Bay beach (**1272**) is a 3 km long, southwest facing sand beach, backed by older vegetated dunes extending up to 400 m inland and fronted by ridged sand flats which widen from 200 m in the east to 600 m in the west (Fig. 4.177). There is vehicle access at either end.

Figure 4.177 *View east across Bosanquet Bay, with the 600 m wide ridged sand flats in foreground.*

Swimming: Bosanquet Bay beach is one of the more attractive and popular beaches in the Ceduna-Thevenard area, offering clean sand, usually low wind waves to calm conditions and shallow sand flats. The cape beaches are more difficult to access and dominated by the bluffs and rocks.

Surfing: None.

Fishing: Best out on the rocks off the cape beaches.

Summary: An undeveloped section of natural coast right next to Ceduna, offering both the long bay beach and the rocky cape beaches.

1273-1280 THEVENARD

Unpatrolled				
No.	Beach	Rating	Type	Length
1273	Cape Thevenard (E 4)	2	R+rocks+flats	150 m
1274	Cape Thevenard (E 3)	2	R+sand flats	150 m
1275	Cape Thevenard (E 2)	2	R+sand flats	300 m
1276	Cape Thevenard (E 1)	2	R+sand flats	150 m
1277	Cape Thevenard	2	R+sand flats	250 m
1278	Thevenard (N 1)	1	R+sand flats	100 m
1279	Thevenard boat ramp	1	R+sand flats	100 m
1280	Thevenard (N 3)	1	R+sand flats	100 m

Cape Thevenard is a distinctive narrow, 3 km long, 10 to 20 m high, calcarenite-capped headland that separates Bosanquet and Murat bays. The western tip of the cape reaches relatively deep water and was chosen last century as a port. Today it remains the port for Ceduna and is the last commercial port until Esperance, 1100 km to the west. The small town of Thevenard and the port facilities are located along the centre to western end of the cape. Calcarenite bluffs and low energy beaches and sand flats surround the cape, with vehicle access to all the shore.

The western corner on the southern side of the cape is filled with mangroves. Once free of the mangroves, there are four small beaches. Beach **1273** is a 150 m long high tide beach fronted by rock and sand flats extending 300 m off the beach. Beach **1274** is a straight, 150 m long, south facing strip of high tide sand, backed by a low bluff and fronted by 200 m wide sand flats. Beach **1275** is backed by commercial activities and crossed in the centre by a concrete boat ramp and jetty. The once 300 m long beach now consists of an eastern 100 m long section backed by pine trees and storage sheds, and a 100 m beach on the south side of the jetty, with grain storage sheds in lee of the bluffs. On the western tip of the cape, beach **1276** is located on the southern side of the main port jetty, with port facilities backing and surrounding the 150 m long beach. On the north side of the jetty is beach **1277**, a 250 m long beach also backed by port facilities, with the beach also used as a boat launching area. There are rocks on the beach and sand flats, while it is backed by 10 m high bluffs.

The northern side of Cape Thevenard forms the southern boundary of Murat Bay. Along the 2.5 km, north facing cape shore are three low energy beaches. Beach **1278** is located 1 km east of the cape and consists of a high tide sand and shell beach fronted by 400 m wide sand flats, with salt marsh fringing the western end of the beach. Seaview Terrace runs along the back of the beach. Beach **1279** is a similar 100 m long, low energy high tide beach, bordered by salt marsh in the west, while the Thevenard boat ramp extends 200 m across the eastern end of the beach to reach deeper water out past the sand flats. On the northern side of the boat ramp is beach **1280**, another 100 m long, low energy high tide beach with 200 m wide sand flats and the main Thevenard road behind.

Swimming: None of these beaches are good for swimming owing to the shallow sand flats and rocks on some, as well as commercial and boating activity.

Surfing: None.

Fishing: The jetties and boat ramp are the favoured locations.

Summary: Cape Thevenard is a low, protruding headland surrounded by calcarenite bluffs and eight small, low energy beaches and sand flats, as well as the main port and boating facilities.

1281-1285   CEDUNA (MURAT BAY)

Unpatrolled				
No.	Beach	Rating	Type	Length
1281	Ceduna	1	R+sand flats	1.5 km
1282	Ceduna (N 1)	1	R+sand flats	200 m
1283	Ceduna (N 2)	1	R+sand flats	250 m
1284	Ceduna (N 3)	1	R+sand flats	300 m
1285	Ceduna (N 4)	1	R+sand flats	600 m

Ceduna is the first location after Port Augusta where it is possible to see the sea, on the long drive across southern Australia. It is also the last town in South Australia, before crossing the Nullarbor to Western Australia. Consequently it is a popular spot to stop. The town is located on the eastern shores of Murat Bay, a circular, low energy, south facing bay. The bay is 5 km wide at its entrance between Cape Thevenard and Matts Point, with a shoreline length of 22 km containing 18 very low energy beaches. The shoreline is protected both by the bay's enclosed nature as well as the blocking of swell by St Peter and Goat Islands. As a consequence, local wind waves dominate over ocean swell in the bay and sand and tidal flats ring the shoreline.

The main Ceduna beach (**1281**) begins at the base of Cape Thevenard and runs due north for 1.5 km to 10 m high bluffs. The beach is backed, from south to north, by the sailing club, a foreshore reserve, a caravan park and the town centre, as well as being crossed by two boat ramps and the 400 m long jetty (Fig. 4.178). The low energy beach is fronted by 150 m wide sand flats and backed by low crumbing bluffs, which deposit rock debris on the sand.

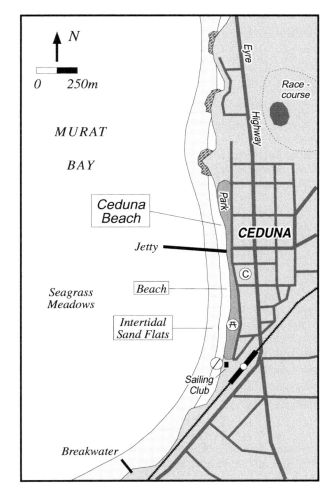

Figure 4.178 *Ceduna is located on the shores of Murat Bay.*

Beaches **1282** and **1283** are two neighbouring beaches and sand flats, 200 m and 250 m long respectively. They continue on along the base of the bluffs and are backed by the northern part of the town, including the hospital and the highway.

Beach **1284** lies at the northern end of town and right next to the highway. It is a low, 300 m long strip of sand backed by the highway and salt flats, with a truck depot behind the southern end. Its neighbour, beach **1285**, commences where the Denial Bay road leaves the highway and runs along the back of the 600 m long beach. It faces southwest and is fronted by 600 m wide ridged sand flats.

Swimming: The only beach suitable for swimming is the main town beach and then only at high tide when the sand flats are covered.

Surfing: None.

Fishing: The long jetty is the most popular fishing spot in Ceduna.

Summary: A very popular location, with the beaches mainly used for yachting and boating activities.

1286-1296 MURAT BAY (N)

Unpatrolled				
No.	Beach	Rating	Type	Length
1286	Round Point (1)	1	R+sand flats	1.1 km
1287	Round Point (2)	1	R+tidal flats	200 m
1288	Low Point	1	R+tidal flats	300 m
1289	Rocky Point	1	R+tidal flats	200 m
1290	Rocky Point (W)	1	R+tidal flats	1.6 km
1291	McKenzie Landing	1	R+tidal flats	200 m
1292	Denial Bay (N)	1	R+tidal flats	1 km
1293	Denial Bay	1	R+tidal flats	600 m
1294	Denial Bay (S)	1	R+tidal flats	1.2 km
1295	Matts Point (N 2)	1	R+tidal flats	400 m
1296	Matts Point (N 1)	1	R+tidal flats	400 m

The northern shore of Murat Bay extends for 14 km from where the Denial Bay road leaves the highway, around past Denial Bay to Matts Point, the western boundary of Murat Bay. This is a very low energy shore consisting of low calcarenite bluffs and tidal flats. The Denial Bay road runs along the back of the beaches, providing access to most.

Beach **1286** is a curving, southwest facing, 1.1 km long, low sand beach fronted by 400 m wide sand flats and bordered on the west by the low, rocky **Round Point**. A small aboriginal community consisting of several houses is located at the western end of the beach. Immediately north of Round Point is beach **1287**, a curving, 200 m long, east facing high tide beach and tidal flats. The

western end of the point is called **Low Point** and between the two points is 300 m long beach **1288**, which is fronted by 100 m wide tidal flats. On the north side of Low Point is beach **1289**, a curving, 200 m long beach and tidal flats that run north to the low **Rocky Point**. The point area and the three beaches are accessible along rough vehicle tracks.

West of Rocky Point is a 1.6 km long, low sand beach (**1290**), backed by the road and low beach ridges and samphire flats, and fronted by 600 m wide tidal flats, with mangroves fringing the western end of the beach. The mangroves extend for 800 m, then give way to the small, 200 m long beach **1291**, which is fronted by the same tidal flats.

The small settlement of **Denial Bay** is dominated by the central jetty, which extends across the shallow tidal flats 400 m into the bay (Fig. 4.179). To either side are the northern beach **1292**, a slightly crenulate, 1 km long, narrow high tide beach, the central 600 m long beach, **1293**, that fronts the town and is crossed by the jetty toward the northern end, while to the south, beach **1294** faces east, curving round for 1.2 km to some low bluffs. All three beaches have a few scattered mangroves and are fronted by tidal flats. At the end of the bluffs is a point, beyond which is 400 m long beach **1295**, which faces southeast across 800 m wide tidal flats. Scattered mangroves fringe the beach, while an oyster farm is located on the tidal flats, with the processing plant on the point. Finally on the north side of **Matts Point** is a curving, 400 m long, east facing beach (**1296**) fronted by similar tidal flats and backed by low, dune-fronted bluffs.

Figure 4.179 *The small Denial Bay settlement and its 400 m long jetty is located on the very low energy western shores of Murat Bay, with seagrass growing right to the shoreline.*

Swimming: None of these low beaches and tidal flats are suitable for swimming, owing to the usually shallow water.

Surfing: None.

Fishing: The end of the Denial Bay jetty has the only deep water in this part of the bay.

Summary: A very low energy, quiet section of the bay, with reasonable access to all the beaches.

1297-1304   MATTS POINT to ROCKY POINT

Unpatrolled				
No.	Beach	Rating	Type	Length
1297	Matts Pt (S 1)	1	R+sand flats	700 m
1298	Matts Pt (S 2)	1	R+sand flats	1.3 km
1299	Cape Beaufort	1	R+sand flats	2.4 km
1300	Rocky Pt	1	R+sand flats	1 km
1301	Rocky Pt (W 1)	1	R+sand flats	100 m
1302	Rocky Pt (W 2)	1	R+sand flats	150 m
1303	Rocky Pt (W 3)	1	R+sand flats	700 m
1304	Rocky Pt (W 4)	1	R+tidal flats	100 m

Matts Point forms the western boundary to Murat Bay and once round the point, the shore is oriented initially southeast for over 4 km to Cape Beaufort, then south for another 3 km into the mouth of Tourville Bay. Along this 8 km section are eight beaches which, while still protected by Goat and St Peter Islands, do receive low swell and higher wind waves. However all remain fronted by sand flats, while they are backed by low, scrub-covered calcarenite, then farmland, with access via informal 4WD tracks.

Beach **1297** is a straight, 700 m long, southeast facing beach located on the southern side of Matts Point, with rocks and low bluffs to either end. The beach has built out 50 m from the backing low bluffs and is fronted by 300 m wide sand flats. Around the next point is 1.3 km long beach **1298**, which has prograded 300 m into the bay. It is fronted by 300 m wide ridged sand flats in the north, widening to 800 m in the south. A low rock-tipped foreland separates it from beach **1299**. This beach initially faces southeast, where it is straight and has 800 m wide sand flats, then turns to face east as it fronts crenulate, 10 m high bluffs. It terminates 2.4 km to the south at Cape Beaufort, where the tidal flats narrow to 50 m. There is a 4WD track along the back of the bluffs to the cape.

On the western side of the cape the beaches face into more open ocean, but are still protected by sand flats which widen considerably into adjoining Tourville Bay. Beach **1300** is a straight, south facing, 1 km long beach, bordered in the west by low Rocky Point. The sand flats widen from 50 to 500 m along the beach. On Rocky Point is a 100 m long pocket beach (**1301**) bordered by rocks, followed by more rocks then a similar 150 m long beach (**1302**), and finally a 700 m long beach (**1303**) that

terminates when the shore trends north. As the shore turns, there is one last beach (**1304**), a 100 m long pocket beach, beyond which mangroves dominate the shore.

Summary: An isolated, low energy section of low coast, dominated by usually vegetated, low calcarenite bluffs, and fronted by sandy-shelly beaches and wide sand flats.

1305-1310   POINT PETER to BIELAMAH SANDHILLS

Unpatrolled				
No.	Beach	Rating	Type	Length
1305	Point Peter (N)	1	R+sand flats	400 m
1306	Point Peter	8	TBR+headlands	150 m
1307	Davenport	8	TBR	6.5 km
1308	Point James (E)	8	LTT+reef	600 m
1309	Point James	8	TBR+reef	1.05 km
1310	Bielamah sandhills	8	TBR+beachrock	5.4 km

Point Peter marks the southern headland for Tourville Bay and represents the first high energy headland and mainland coast since Point Brown, 50 km to the southeast. It is also the first in a series of granite and metasedimentary headlands all linked by generally high energy beaches and dune systems.

There are two beaches on the 1 km long granite Point Peter. On the northern protected side is a 400 m long strip of sand (**1305**) fronted by sand flats up to 800 m wide. It terminates in a dynamic spit of sand that represents movement of marine sand into Tourville Bay. On the exposed southern side of the point is a 150 m long pocket beach (**1306**) bordered by jagged granite headlands and backed by steep, vegetated, 20 m high bluffs. Waves average over 1.5 m and a strong rip always flows out against one of the headlands.

The point itself is tied to the mainland by a 6.5 km long, southwest facing beach (**1307**). The beach maintains a high energy surf zone for 4 km, before decreasing in energy in lee of the western headland, with sand flats in front of the far western, northeast facing end of the beach. The Davenport Creek road terminates at the low energy end of the beach, while active dunes extend, on average, 1 km in from the main beach, in places moving over the backing mangroves of Davenport Creek (Fig. 4.180).

There are three 20 m high metasedimentary headlands located within 3.5 km of the western end of the beach, the westernmost headland is called Point James. The three contain two exposed beaches between them and are capped by low, vegetated clifftop dunes. Four-wheel drive tracks lead to each of the headlands. The first beach (**1308**) is a moderately steep, 600 m long beach receiving waves averaging less than 1.5 m, which maintain a beach and narrow bar, with rips usually restricted to either end against the rocks. Steep, 20 m high, vegetated bluffs back the beach, which occupies half of a 1 km long embayment, the western half of the shore comprising an irregular, 50 m wide, metasedimentary rock platform. A 4WD track reaches the bluffs above the centre of the beach, with a steep sandy descent to the beach. The second beach (**1309**) is just over 1 km long and is backed by some active dunes climbing over the bluffs. It receives slightly higher waves and has a wider surf zone, together with beachrock reefs in the surf, resulting in five to six rips along the beach. There is 4WD access to the eastern headland and the bluffs above the western end of the beach.

Figure 4.180 *Davenport beach, showing the moderate energy surf and active dunes that are blowing across the backing mangroves and into Davenport Creek.*

West of **Point James** is a straight, south-southwest facing, 5.4 km long, high energy beach (**1310**) backed by a foredune system, then the active **Bielamah sandhills**, which extend up to 1.5 km inland. The beach has strong beach rips located about every 400 m along the beach, with a deep trough off the bar and a beachrock reef paralleling the beach between 300 m offshore in the east, to 500 m offshore in the west. Waves break over the reef and a few strong reef-controlled rips flow well seaward. Four-wheel drive tracks reach each end of the beach and skirt the rear of the dune system.

Swimming: The only area suitable for swimming is in front of the car park and to the west, at the Davenport Creek beach, where waves are usually low to calm. All the other beaches, apart from 1305, are exposed to high waves and dominated by rips and, in places, reefs.

Surfing: The best surf is east of the car park on the beach breaks. The other beaches are more difficult to access and some are dominated by reefs.

Fishing: There are good beach rips on all the exposed beaches.

Summary: There is good access to Davenport Creek and the main beach, with 4WD tracks reaching the other beaches.

Regional Map 12: Great Australian Bight

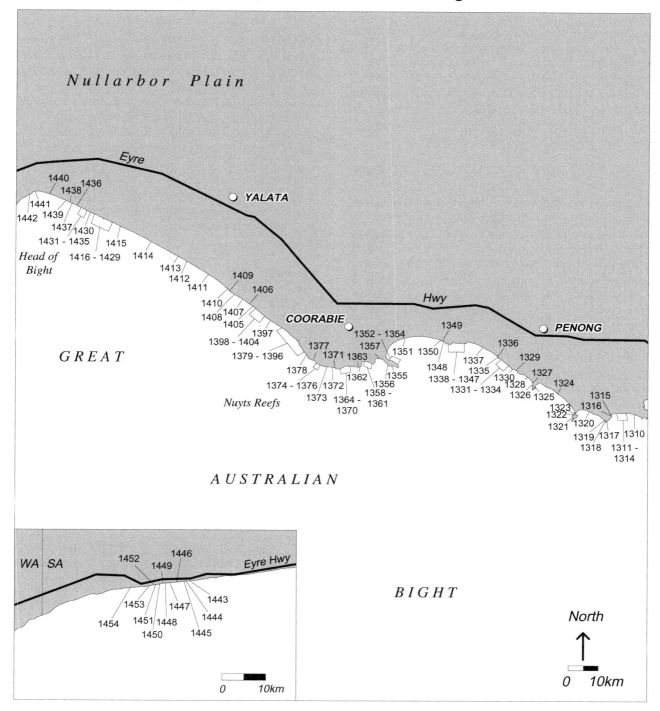

Figure 4.181 *Regional Map 12 – Great Australian Bight*

1311-1319 ROCKY POINT AREA

The coast west of the **Bielamah sand dunes** is dominated by 20 m high calcarenite bluffs for 10 km to Rocky Point, a calcarenite-capped metasedimentary headland. Below most of the bluffs are narrow strips and pockets of sand, most fronted by calcarenite reefs and beachrock, with reef further offshore also decreasing wave height. As a consequence, there are nine moderate to low energy beaches along this section. A gravel road reaches the rear of beach 1314 and follows the bluffs for 3 km out to Rocky Point, with 4WD access along the bluff top to all other beaches. Apart from the tracks and farmland further inland, there is no development at the coast.

The first four beaches (1311-1314) all face due south. Beach **1311** is a 120 m long pocket of sand lying immediately west of the Bielamah sand dunes. It is backed and bordered by 10 to 15 m high calcarenite bluffs, with reefs along the beach extending 200 m offshore, resulting in waves averaging about 1 m at the beach. Two hundred metres to the west is beach **1312**, a 350 m long, narrow beach at the base of steep bluffs which rise to 30 m in the west. There is no foot access. Offshore reefs lower waves to less than 1 m at the shore, maintaining a steep reflective beach.

Beach **1313** is a curving, 300 m long reflective beach, also protected by offshore reefs and with bedrock reefs to either side. The beach fills a small gap in the bluffs, with a foredune almost climbing to the top of the 20 m high bluffs providing a steep, sandy access track. Beach **1314** begins immediately to the west. It is a discontinuous, 3 km long, narrow beach, backed by 10 m high, partially dune-draped bluffs. Bedrock and calcarenite reefs, as well as bedrock, outcrop along the beach, both lowering waves to under 1.5 m and inducing several permanent rips (Fig. 4.182).

Figure 4.182 *The shoreline east of Rocky Point contains a reef-bound, low energy high tide sand beach (1314), with 'lagoons' forming in the reef.*

At the western end of beach 1314, the coast turns and trends southwest for 3 km to Rocky Point. Along this section are two protected pocket beaches. Beaches **1315** and **1316** are 50 m and 150 m long respectively, face east and lie below steep, 20 m high bluffs. They are both fronted by 200 m wide shallow reefs, resulting in low wave to calm conditions at the shore.

Surrounding the tip of **Rocky Point** are three protected beaches. To the east is an east facing, 200 m long, narrow reflective beach (**1317**), backed by low bluffs and broken by bedrock rocks and reefs. On the western side of the point are two reflective pockets of sand (beaches **1318** and **1319**). They are 100 m and 80 m long respectively and are bordered and backed by 10 m high bluffs, with shallow reefs off each beach reducing waves to less than 1 m at the shore.

Swimming: Most of these beaches receive reduced to low waves, however rocks, reefs and permanent rips in the more exposed sections result in moderate beach hazards.

Surfing: There are many reef breaks right along this section, including out on Rocky Point.

Fishing: All the beaches offer good rock and reef fishing, with beach 1314 having the best rip holes.

Summary: A remote but accessible section of shore, with some beaches requiring a climb down the bluffs to reach.

1320-1323 ROCKY POINT to POINT BELL

Unpatrolled				
No.	Beach	Rating	Type	Length
1320	Rocky Pt-Pt Bell	6	R-TBR-R	12 km
1321	Point Bell (W 1)	4	R/LTT+reef	600 m
1322	Point Bell (W 2)	4	R/LTT+reef	250 m
1323	Point Bell (W 3)	4	R/LTT+reef	200 m

Rocky Point and Point Bell are two protruding metasedimentary headlands, both capped by calcarenite. Between the two is a 12 km long beach (**1320**) backed by active dunes extending from 500 m to 1.5 km inland. The beach faces south-southwest for most of its length, swinging round in lee of Point Bell to face east. There is vehicle access to Rocky Point and the rear of Point Bell, with 4WD access along the low gradient, white sand beach and out onto the point.

While the beach faces into the dominant swell, it is protected in the east by shore parallel beachrock reefs and by Finders Rocks. The rocks lie 3 km offshore, with a slight cuspate foreland in their lee. Finally in the west, the 5 km long southern protrusion of Point Bell, together with reefs extending another 3.5 km off the point, all lower waves at the shore. As a result waves peak in the centre at about 1.5 m, where they maintain a low gradient transverse bar and rip system with widely spaced rips, decreasing in energy to the east in lee of the beachrock reefs and to the west in lee of Point Bell. The sand on the eastern side of Point Bell is very fine and white, in contrast to the coarser brown sand on the western side of the point. Where the two sands merge on the beach and in the dunes, they produce some distinctive size and colour-sorted dune features.

Point Bell Conservation Reserve	
Established:	1993
Area:	602 ha
Beaches:	5 beaches (1320-1324)
Coast length:	11 km

Point Bell has a metasedimentary basement capped by calcarenite and clifftop dunes, which reach a maximum height of 50 m. There are no beaches along the 3 km rocky eastern foreshore of the point. On the more exposed western side are three small pocket beaches, each backed and bordered by calcarenite-capped bedrock. Beach **1321** is located on the western side of the 1.5 km long, narrow tip of the point. It is a 600 m long, southwest facing, steep beach, often reflective, but exposed to higher waves during high seas. While it is backed by steep, 40 m high bluffs, a sand dune ramp toward the western end provides access down to the beach. Beaches **1322** and **1323** are located at the eastern end of the main beach and are separated by two protruding bluffs. They are 250 m and 200 m long respectively and face northwest along the main beach. Refraction around the point and shallow reefs off both beaches lower waves to less than 1 m at the shore, maintaining steep reflective beaches.

Swimming: The safest swimming is in the calm waters in lee of Point Bell, as rips dominate the more open central beach, with permanent rips toward Rocky Point. The Point Bell beaches, while moderately protected, are dominated by reefs and strong rips during bigger seas.

Surfing: There are reef breaks off both points and beach breaks along the main beach.

Fishing: Both points attract rock fishers, with the beach providing access to some of the central and eastern rip holes.

Summary: An exposed, moderate energy beach backed by active dunes, bordered in the west by the very prominent Point Bell.

1324-1328 SHELLY, PORT LE HUNTE, CACTUS, CASTLES, CHADINGA

		Unpatrolled		
No.	Beach	Rating	Type	Length
			Inner Outer Bar	
1324	Cantaby-Shelly	7	TBR-RBB	19 km
1325	Port Le Hunte	3	R+reef	650 m
1326	Cactus	5	R+reef	250 m
1327	Castles	5	R+reef	400 m
1328	Chadinga Sandhills	7	TBR RBB	4.3 km

Point Bell forms the eastern boundary of a 19 km long, high energy, southwest facing beach. The western boundary is the equally prominent Point Sinclair, which protrudes 3 km to the south. While access to Point Bell is limited, there is a gravel road out to Point Sinclair and the adjacent Port Le Hunte and Cactus beaches.

The main beach (**1324**) between Point Bell and Point Sinclair is known both as **Cantaby** Beach after the backing sandhills (Fig. 4.183), and in the west as **Shelly** Beach (Fig. 4.184). The 19 km long beach begins in the east with relatively coarse, brown, shelly sands, which in the west are replaced by fine, white, carbonate sands. The sands and the prevailing high waves maintain a 200 m wide, rip-dominated surf zone in the east-centre, with a wider, lower gradient and lower energy surf zone at the central cuspate foreland in lee of Pudding Rock, a series of reefs 3 km offshore. Between the foreland and Shelly Beach, the gradient stays low and the surf zone averages 400 m in width, with waves slowly decreasing toward Point Sinclair. There is 4WD access to Shelly Beach, which is a popular swimming and surfing spot. Active dunes, extending up to 4 km inland, back the entire beach.

Figure 4.183 *View across the base of Point Bell and along Cantaby Beach. Note the wide, high energy surf zone and backing active sand dunes.*

Point Sinclair is a 40 m high dune and calcarenite-capped granite headland that has small beaches to either side (Fig. 4.185). On the east is the curving, southeast facing, protected, 650 m long Port Le Hunte beach

(**1325**). The gravel road from Penong terminates on the 30 m high bluffs overlooking Port Le Hunte. A vehicle track leads down to the jetty, where there is a small informal camping area between the bluffs and a seawall. The low energy beach runs northeast from the jetty, with intertidal rock flats fronting the eastern half. There is a toilet block and fresh water at the jetty, but no other facilities. The 200 m long jetty is no longer used to ship wheat and a shark net has been strung along the jetty and back to the beach to provide a shark-proof swimming enclosure. This is necessary as there have been fatal shark attacks at Port Le Hunte in 1975 and at adjoining Cactus Beach in 2000. The beach is usually calm, with a high tide beach and a narrow bar, then deeper water. During summer fishers launch their boats from the beach and moor them just offshore.

Figure 4.184 *Shelly Beach lies at the western end of Cantaby-Shelly beach, and has a wide, white sand, low gradient beach and surf zone.*

On the western side of the point are the world famous **Cactus** (**1326**) and **Castles** (**1327**) beaches and their surrounding surfing breaks (Fig. 4.186). The Penong Road runs along the back of Cactus Beach, with a large camping area set amongst the dune scrub between the road and beach, and good vehicle and foot access to the back of the beach. There is a small camp store, which provides the only commercial activity in the area. The beaches are 250 m and 400 m long respectively. They face west and are backed by a low foredune, bordered by calcarenite bluffs and fronted by exposed beachrock and shallow calcarenite reefs. In lee of the reefs is a narrow high tide sand beach and, while waves can be large on the outer reefs, they are usually less than 0.5 m when they finally reach the beach. However both beaches are drained by strong permanent rips, particularly off Castles.

In addition to the *Cactus* left and *Castles* right surf breaks off the beaches, to the south of Cactus, out on Point Sinclair, are *Witzigs*, *Backdoors* and *Cunns*, while off the north Castles bluff are *Caves*, *Crushers* and *Supertubes*. All the breaks are over calcarenite reefs and receive slight protection and cleaner waves owing to refraction around the point and over outer deeper reefs.

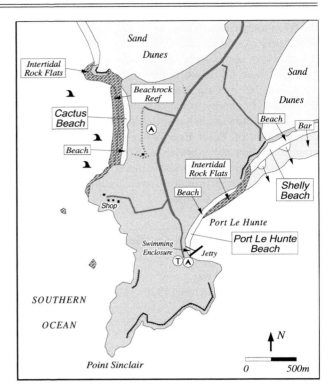

Figure 4.185 *The Point Sinclair area with Port Le Hunte and Shelly Beach to the east and Cactus Beach and surf breaks to the west.*

Figure 4.186 *Cactus Beach (left) showing the reef-dominated, surf filled bay, with the camping area between the road and the beach. To the right is Chadinga beach and its massive sand dunes.*

Immediately west of Castles are the massive **Chadinga sandhills**, fronted by a 4.3 km long, high energy beach (**1328**). The beach faces southwest and receives persistent high swell, which maintains an inner rip-dominated bar and an outer bar cut by larger rips, the whole surf zone averaging 500 m in width. In addition, there are strong permanent rips against the bordering calcarenite bluffs. The active dunes extend up to 3 km inland and are entirely contained in the Chadinga Conservation Reserve.

Chadinga Conservation Reserve

Established: 1993
Area: 8125 ha
Beaches: 2 beaches (1328, 1337)
Coast length: 17 km

Chadinga Conservation Reserve encompasses 80 km² of the largely active sand dunes in lee of Chadinga beach and the extensive, more vegetated dunes west of Eyre Bluff.

Swimming: It is difficult to swim at Cactus owing to the extensive rock flats and reefs. The best locations are inshore at Shelly Beach for surf and in the swimming net at Port Le Hunte for calm, shark-free conditions. Rocks, reef and rips dominate all other locations and the area has been the site of two fatal shark attacks.

Surfing: Cactus has several excellent breaks which have made this remote part of the Great Australian Bight a mecca for surfers, some of whom have stayed on to live in the dunes.

Fishing: Point Sinclair offers beach, rock and boat fishing and is a popular year-round fishing location with locals and travellers.

Summary: While Point Sinclair has been a popular location and a port for 100 years, it is the surf that has placed it on the map and led to the development of the camping area, with its open-air facilities built from the local rocks. It is worth the easy 20 km drive out to Cactus, with the last few kilometres crossing the salt flats of Lake McDonnell. It is an arid and interesting area, with massive sand dunes bordering the western entrance.

1329-1337 CHADINGA SANDHILLS to EYRE BLUFF

No.	Beach	Rating	Type Inner	Outer	Length Bar
		Unpatrolled			
1329	Beach 1329	8	RBB+reef		1.65 km
1330	Beach 1330	8	RBB+reef		150 m
1331	Beach 1331	8	RBB+reef		150 m
1332	Beach 1332	8	RBB+reef		1 km
1333	Beach 1333	8	RBB+reef		600 m
1334	Beach 1334	8	RBB+reef		1.6 km
1335	Beach 1335	8	RBB+reef		80 m
1336	Beach 1336	8	RBB+reef		200 m
1337	Eyre Bluff (E 3)	8	RBB	LBT	12.2 km

At the western end of the Chadinga sandhills, the beach terminates at 10 m high calcarenite bluffs which dominate the coast for the next 11 km, beyond which is a 12 km long beach running west to Eyre Bluff. This 23 km long section of coast is relatively isolated, with only rough 4WD access and no facilities or development. There are eight beaches along the bluff section and then the longer western beach, all facing southwest and exposed to persistent waves averaging 2 m in height.

Beach **1329** begins 2.5 km west of the dunes. It is bordered by bluffs and backed by both dunes, extending a few hundred metres inland, and in the west by 10 m high bluffs, with 4WD tracks leading to the top of the bluffs. In addition, beachrock and calcarenite reefs outcrop along the beach and in the 300 m wide, high energy surf zone dominated by strong permanent rips. Three hundred metres to the west is 150 m long beach **1330**, a bluff-bordered and backed beach, with a continuous, 150 m wide reef fronting the beach. Waves break across the reef and one large rip drains the surf. A track off the main 4WD track runs out to the top of the bluffs and along them to the west.

Beach **1331** lies another 2 km further west and is a pocket, 150 m long, bluff-bound beach fronted by a wide, rip-dominated surf zone. Immediately west is the 1 km long beach **1332**, which is partially broken by three small protruding bluffs and is backed by a few hundred metres of active dunes along most of its length. A 300 m wide surf zone and strong rips front the beach. Three hundred metres to the west is beach **1333**, a straight, 600 m long beach, backed by vegetated dunes extending 1 km inland and fronted by a high energy surf zone, with a strong, permanent, reef-controlled rip in the centre of the beach.

Beach **1334** is a 1.6 km long, crenulate beach that is bordered by bluffs and impacted by a discontinuous reef lying 200 m off the beach. Waves break heavily on the reef and three strong rips flow out through gaps in the reef, with a very large rip draining the western 'embayment'. Largely stable dunes extend up to 1.5 km inland.

Beaches 1335 and 1336 are two bluff and reef-bound beaches located at the eastern end of the longer beach that runs up to Eyre Bluff. Beach **1335** is an 80 m pocket of sand separated from 200 m long beach **1336** by a small bluff. Both beaches are fronted by shallow reefs and a 300 m wide surf zone and drained by strong permanent rips. A 4WD track reaches the dunes above beach 1336.

Beach **1337** extends from the end of the 4WD track for 12.2 km up to the easternmost of the Eyre Bluffs. It faces southwest, exposing it to persistent high waves which maintain a continuous inner bar, which expands to a double bar system to the west. Both bars are dominated by large rips, spaced at approximately 500 m intervals on the inner bar. Stable, 1.5 km wide dunes back the first 7 km, with active dunes along the western section of the beach. Salt lakes also back the central section of the dunes. Most of the dunes and lakes are part of the Chadinga Conservation Reserve.

Swimming: This is a remote, high energy, rip and reef-dominated section of coast and is unsuitable for safe swimming.

Surfing: There are numerous reef breaks along the bluff section and beach breaks in the west, which merge into a prominent reef break against Eyre Bluff, however local knowledge is required to find the right spots.

Fishing: There are deep beach, rip and reef holes the whole length of this section of coast. However beware of large waves breaking over the rocks.

Summary: A little used section of coast only accessible by 4WD, offering high energy beaches backed by a mixture of low bluffs and dune fields.

1338-1344 EYRE BLUFF to CLARE BAY

		Unpatrolled		
No.	Beach	Rating	Type	Length
1338	Eyre Bluff (E 2)	4	LTT	950 m
1339	Eyre Bluff (E 1)	3	R+reef	650 m
1340	Eyre Bluff (W 1)	4	R+reef	450 m
1341	Eyre Bluff (W 2)	4	R+reef	600 m
1342	Eyre Bluff (W 3)	4	R+reef	100 m
1343	Clare Bay (E)	5	R+reef	100 m
1344	Clare Bay	3	R+reef	800 m

Eyre Bluff marks the eastern end of a 9 km long section of exposed, south facing coast, dominated by 20 m high calcarenite bluffs and for the most part fronted by calcarenite reefs, extending up to 500 m offshore. While high waves break on the outer reefs, they are for the most part lowered, often substantially, before they reach the shore, resulting in lower energy beaches between the bluff and Clare Bay. The only public access to this section is to the shacks at Clare Bay, with 4WD access along the bluffs and clifftop dunes required to reach the tops of the other beaches.

Beach **1338** is a slightly curving, 950 m long, south facing beach protected by deeper offshore reefs. It is backed by steep, 20 m high bluffs and bordered by a prominent eastern point, and a foreland in lee of a reef to the west, neither of which provide foot access to the beach. Waves average about 1 m at the shore and break across a narrow attached bar. On the western side of the foreland is a curving, 650 m long reflective beach (**1339**) backed by a foredune and then sand-draped bluffs. There is a steep, sandy 4WD track down the dunes. Prominent reefs and surf lie 100 to 300 m off the beach, while Eyre Bluff extends 400 m seaward to form the western boundary.

On the western side of Eyre Bluff is a straight, 500 m long cliff, then a narrow, curving, 450 m long beach

(**1340**) located at the base of steep, 30 m high bluffs. Reefs extend a few hundred metres seaward of the beach and, although waves are low at the beach, a large rip drains the inner surf. Its neighbour (beach **1341**) is a similar, 600 m long, narrow, curving beach, backed by bluffs which decrease in height to the west, where a 4WD track reaches the top of the beach. A climb down the western end of the beach provides the only access to both beaches. Offshore reefs front much of the beach, with a rip running out through a gap in the reefs off the eastern half of the beach.

Beaches 1342 and 1343 are two 100 m long pockets of sand located immediately east of Clare Bay. Beach **1342** is backed and bordered by 20 m high bluffs and fronted by offshore reefs, while beach **1343** lies on the eastern headland of the bay and has surf breaking across 150 m wide reefs all the way to the beach.

Clare Bay (1344) is one of the few locations where there is vehicle access, via a 'public' farm track to the coast and two shacks. The curving, 800 m long beach is protected by continuous beachrock reefs lying 100-200 m off the beach, which result in a relatively calm lagoon and beach in their lee. However there is heavy surf over the reef and adjoining headlands (Fig. 4.187).

Figure 4.187 *Clare Bay has vehicle access to this remote bay. While heavy surf breaks over the reefs, a calm lagoon lies off the beach.*

Swimming: The only spot suitable for swimming is in the Clare Bay 'lagoon', all the other beaches are dominated by high outer surf, reefs and rips.

Surfing: There is potential for surf on a number of the reef breaks, with both left and right breaks off Clare Bay, however local knowledge is required to find and access many of these breaks.

Fishing: Clare Bay is the focus for fishing along this section, with most of the beaches offering access to deep reef and rip holes.

Summary: Bluffs and reefs dominate this section of coast, with generally lower energy beaches fronted by wide, reef-dominated surf.

1345-1350   CLARE BAY (W)

No.	Beach	Unpatrolled		
		Rating	Type	Length
1345	Clare Bay (W 1)	5	R+reef	400 m
1346	Clare Bay (W 2)	5	R+reef	150 m
1347	Clare Bay (W 3)	5	R+reef	150 m
1348	Clare Bay (W 4)	8	TBR+reef	150 m
1349	Clare Bay (W 5)	8	TBR+reef	350 m
1350	Clare Bay (W 6)	8	TBR+reef	350 m

To the west of Clare Bay, the calcarenite bluffs continue for another 4 km to the beginning of the long Fowlers Bay beach. In between are six small bluff and reef-dominated beaches.

On the far western side of the bay is southeast facing beach **1345**, a narrow, 400 m long beach located below 20 m high bluffs, with a reef extending 1 km east of the western headland, which substantially lowers waves at the shore. To the west of the headland, the bluffs rise to 30 m and become more irregular. Beach **1346** is a narrow, 150 m long high tide beach located at the base of the steep bluffs and fronted by a 100 to 200 m wide, shallow reef system, with a permanent rip flowing through the reefs. Beach **1347** is contained in an amphitheatre of 30 m high bluffs. It is just 150 m long and bordered by bluffs and reefs, with a strong rip exiting through a narrow gap in the reefs. There is no safe foot access to either beach (Fig. 4.188).

Figure 4.188 *Beach 1347 is typical of the bluff-bound beaches west of Clare Bay. While rocks and reefs lower waves at the shore, note the strong permanent rip running out the left side of the bay.*

Beach **1348** lies almost 2 km further west. It is the first of three adjoining beaches all dominated by high waves

and surf. The beach is a narrow, 150 m long strip of sand lying at the base of 30 m high bluffs. A 4WD track reaches the eastern end of the bluffs, however there is no safe foot access to the beach. The beach is fronted by deeper reefs and a 200 m wide, high energy surf zone drained by a strong rip system. Beach **1349** is an exposed, 350 m long beach backed by 15 m high, dune-capped bluffs, with active dunes on the western bluffs. Beach **1350** is a similar 350 m long beach, backed by bluffs covered with more active dunes.

Summary: An exposed, high energy, bluff and reef-dominated section of coast, unsuitable for swimming, but offering difficult access to surfing reefs and fishing holes.

1351   FOWLERS BAY

No.	Beach	Unpatrolled		
		Rating	Type	Length
1351	Fowlers Bay	7→1	TBR→R	24.9 km

Fowlers Bay beach (**1351**) begins in lee of the prominent Point Fowler and extends east for 16 km to the western end of the Eyre Bluffs. The 40 m high, up to 1 km wide point extends southeast for 5 km, affording considerable protection to the east facing western end of the beach, so much so that Port Eyre was established here as a wheat jetty last century, with the small settlement and 350 m long jetty still the only development on this section of coast. The spiraling, 25 km long beach initially consists of a western 5 km long, low energy section on either side of the jetty. The beach here is reflective, grading south into sand flats and north into a low tide terrace, with seagrass growing to the shore and often piled high on the beach. One kilometre south of the jetty is a sandy protrusion in the shoreline, built from dune sands blowing across the base of the point from Scott Bay. This form of sand transport, called 'headland bypassing', is continuing to supply sand to Fowlers Bay. As the beach swings around it becomes increasingly exposed to higher waves, as it faces southeast and finally south.

North of the jetty, the beach slowly increases in exposure and wave energy. After 5 km it becomes an energetic, south facing, 300 m wide transverse bar and rip system, with strong rips located every few hundred metres and permanent rips amongst the beachrock reefs at the eastern end of the beach. The exposed section is backed by the generally active Nantiby sand dunes, which extend on average about 1 km inland and reach a maximum height of 50 m at Nantiby Hill. There is 4WD access to the rear of the dunes at Nantiby Well, with road access to the small Port Eyre settlement.

The jetty was constructed in 1896 and the small settlement reached its peak in the 1890s, followed by a gradual decline and final abandonment in the 1950s. However since the 1980s, a few people have reoccupied the old houses and today it has a caravan park and kiosk. The Scott Bay sand dune has also posed a problem, prograding over and burying a few of the houses on the southern edge of the settlement (Fig. 4.189).

Figure 4.189 *The small Fowler Bay settlement and jetty occupy the protected eastern side of Point Fowler. However active dunes from Scott Bay have buried some of the old buildings.*

Swimming: Port Eyre has low waves but a lot of seagrass to contend with (Fig. 4.190), while the exposed beach is dominated by strong rips. The low wave area about 2 km north of the jetty offers the cleanest and usually safest section of beach.

Surfing: There are about 20 km of beach breaks, with conditions being wave and wind dependent.

Fishing: The jetty is the most popular location and attracts many of the visitors to this remote settlement.

Summary: Fowlers Bay and historic Port Eyre are easily accessible by gravel road and worth the visit to see the old, and now rejuvenated, settlement, as well as the surrounding beaches and dunes.

Figure 4.190 *Seagrass piled high on the beach at Fowlers Bay.*

Fowlers Bay Conservation Reserve	
Established:	1993
Area:	8649 ha
Beaches:	13 beaches (1351-1363)
Coast length:	23 km

Fowlers Bay Conservation Reserve incorporates Point Fowler and the adjoining Scott Bay and coast to Cape Nuyts, as well as some of the backing active dunes, Pleistocene barriers and salt lakes as far inland as Coorabie.

1352-1356 POINT FOWLER

Unpatrolled				
No.	Beach	Rating	Type	Length
1352	Fowlers Bay spit	1	R+sand flats	1.3 km
1353	Pt Fowler (E 2)	3	R	200 m
1354	Pt Fowler (E 1)	3	R	50 m
1355	Pt Fowler (W 1)	8	R+reef	100 m
1356	Pt Fowler (W 2)	8	R+reef	50 m

Point Fowler is one of the longer and more prominent headlands on the western Eyre Peninsula, and the last major projection before the more continuous coast to the west. The entire headland and coast for 20 km to the west is part of the Fowlers Bay Conservation Reserve. The 5 km long, southeast trending headland has a metasedimentary basement, capped by calcarenite reaching 45 m in height, with 10 to 30 m high bluffs bordering most of the point. The point has 10 km of predominantly rocky shoreline, with one northern sand spit and four small bluff-dominated beaches. A 4WD track from Port Eyre provides access to the headland and the bluffs above the beaches.

Beach **1352** is a migrating, 1.3 km long sand spit that represents a large pulse of sand that is slowly migrating toward the jetty, with the sand being moved by the low waves that run along the beach during high outside swell. An old seacliff behind the beach is now stranded up to 400 m inland, with the remains of a 19th century whaling site now located 100 m from the sea (Fig. 4.191). The northeast facing beach receives very low waves and is usually calm and littered with seagrass debris.

One kilometre to the east on the point proper is beach **1353**, a 200 m long, east facing reflective beach. It receives waves averaging about 0.5 m and is backed and bordered by steep, 10 m high bluffs. Out on the northern tip of the point is a sheltered, northeast facing, 50 m pocket of sand (beach **1354**) backed by 15 m high bluffs, with usually calm conditions at the beach.

The south facing western side of the point is exposed to high southwest waves and has an energetic shoreline. Beach **1355** is located 1 km west of the point and is a 100 m long strip of high tide sand, backed by steep, 30 m high bluffs capped by clifftop dunes and fronted by a rock and reef-strewn, 100 m wide surf zone. One kilometre further west is beach **1356**, a 50 m pocket of sand located in a gap in the 30 m high bluffs and fronted by a 150 m wide reef and surf zone. A 4WD track runs along the top of the bluffs, however there is no safe foot access to the beach.

Summary: The point provides a variety of coastal scenery and access to the top of the bluff-bound beaches. The spit is the only safe spot to swim, though it is fronted by shallow water. The other beaches are difficult, if not impossible, to safely access and the western ones are very dangerous.

Figure 4.191 *The migratory sand spit on the northern side of Point Fowler. The main track runs along below the old seacliff.*

1357, 1358 SCOTT BAY

| | **Unpatrolled** | | | |
No.	Beach	Rating	Type	Length
1357	Scott Bay	4→8→4	R→TBR→R	4.5 km
1358	Scott Pt (E)	3	R+reef	150 m

Scott Bay is a 4.5 km long, exposed, southwest facing bay, containing a predominantly energetic beach (**1357**) and massive backing sand dunes, some of which are spilling over into Fowlers Bay, 2 km to the northeast. The bay, while exposed in the central 3 km, is protected at both ends by prominent beachrock reefs extending up to 1 km into the bay from Point Fowler in the east and Scott Point in the west. In lee of the reefs, waves are low and the beach reflective, while along the central section waves average 1.5 to 2 m and several large rips cross the 300 m wide surf zone.

The beach terminates in the west at 20 m high calcarenite bluffs, which continue on south for 1 km to Scott Point where they reach 30 m in height. Along the base of the bluffs is 150 m long beach **1358**, a low energy, east facing beach protected by the point and reefs extending 400 m off the beach.

Swimming: The only safe areas are in lee of the reefs on the main beach and out on the point, though the point beach requires a climb down from the bluffs.

Surfing: There is surf over the eastern reefs, with beach breaks along the centre.

Fishing: This is a relatively popular beach for rock, reef and beach fishing.

Summary: An exposed bay, fronted by usually high seas and backed by massive active dunes.

1359-1362 SCOTT POINT to MEXICAN HAT

| | **Unpatrolled** | | | |
No.	Beach	Rating	Type	Length
1359	Scott Pt (W 1)	5	R+reef	150 m
1360	Scott Pt (W 2)	7	R+reef	200 m
1361	Mexican Hat	8	TBR+reef	2.5 km
1362	Mexican Hat (W)	5	LTT	300 m

West of **Scott Point** the coast runs relatively straight for 4 km to the lee of a sea stack, called Mexican Hat, then turns and trends south for 1.5 km to Cape Nuyts. The entire section is composed of 10 to 20 m high calcarenite bluffs backed by clifftop dunes. There is one long and three small beaches along the base of the bluffs, with vehicle tracks providing access along the top of the bluffs.

Beach **1359** lies immediately west of Scott Point and is a bluff-bound, 150 m long high tide beach fronted by shallow reefs, resulting in low waves at the shore. Access is down the steep, 20 m high bluffs. Beach **1360** lies 1 km west of the point and is a 200 m long beach, backed by lower dune-draped bluffs and fronted by scattered reefs extending 200 m offshore. The reefs provide calmer conditions at the shore, however strong currents and a permanent rip flow out through a gap in the reef.

The main beach (**1361**) is 2.5 km long and slightly crenulate in lee of deeper reefs, with a combination of reef-controlled and beach rips flowing out across the 300 m wide, high energy surf zone. The beach is backed by low bluffs, a deflation surface and then active dunes extending 1 km inland. At the western end of the beach, a beachrock reef lowers waves at the shore. A solitary tin shack is located on top of the bluffs. Beach **1362** lies just south of the shack and is a lower energy, 300 m long, east facing beach lying in lee of the cape, reefs and sea stack of **Mexican Hat**. The beach is narrow and awash at high tide, with a continuous, 50 m wide, shallow bar.

Summary: This is a bluff and reef-dominated section of exposed coast, with vehicle access to the top of the bluffs and onto the main beach. There are reef and beach breaks along this section, however it is mainly used for fishing the deep rip holes amongst the reefs.

1363-1369 CAPE NUYTS, WANDILLA

No.	Beach	Rating	Type	Length
Unpatrolled				
1363	Cape Nuyts	5	R+reef	100 m
1364	Cape Nuyts (W 1)	8	LTT+reef	100 m
1365	Cape Nuyts (W 2)	8	TBR+reef	80 m
1366	Cape Nuyts (W 3)	8	TBR+reef	150 m
1367	Cape Nuyts (W 4)	8	TBR+reef	250 m
1368	Cape Nuyts (W 5)	8	TBR+reef	1.1 km
1369	Wandilla (E)	6	R+reef	80 m

Cape Nuyts is a 50 m high calcarenite headland, to the west of which are 5 km of irregular, south facing, 30 to 60 m high, dune-capped calcarenite bluffs. The bluffs and the fronting reefs are exposed to high southerly swell and most of the beaches that supplied the bluff-top dunes have since been eroded. There are, however, six small beaches and one longer beach tucked in small embayments along the base of the bluffs.

Beach **1363** lies out on Cape Nuyts and is located in a 100 m wide amphitheater cut in the 40 m high bluffs. The curving, 100 m long beach is partially protected by reefs extending across the 60 m wide mouth, however a strong rip also flows through the gap. The beach can only be accessed down the steep debris slopes.

Beach **1364** lies 2.5 km to the west of the cape and is the first of a series of small bluff and reef-bound beaches. The 100 m long beach is backed by 40 m high bluffs, with no safe access. It is partly covered by beachrock reefs, with a 200 m wide surf zone and a strong permanent rip draining the small embayment. Beach **1365** lies just to the west and is similar, only 80 m long and backed by partially vegetated bluffs, with a strong rip draining the 200 m wide surf and prominent rock platforms to either side. Its neighbour is beach **1366**, an almost identical 150 m long beach, with bluffs, beachrock, a rock platform and a strong rip-controlled surf zone. Beach **1367** lies immediately west and is a 250 m long beach, backed by 50 m high bluffs and bordered by beachrock, with a strong rip exiting the western end of the beach.

The main beach in this section is **1368**, a 1.1 km long beach that sweeps in two south facing arcs, 900 m then 400 m long, formed in lee of a central reef. The beach lies at the base of bluffs, which decrease in height from 50 m in the east to 20 m in the west, where they become increasingly dune-draped. The main access road to this section of coast terminates behind the centre of the beach.

Beach **1369** lies in lee of a 500 m long, south trending headland. It is an 80 m long strip of sand at the base of steep, 20 m high bluffs, with the headland and reefs lowering waves at the shore.

Summary: An exposed, bluff dominated section of high energy shore, containing six small, difficult to access and hazardous beaches, with a longer central beach. All are dominated by reefs, high surf and strong rips.

1370-1377 WANDILLA BAY to CHEETIMA BEACH

No.	Beach	Rating	Type	Length
Unpatrolled				
1370	Wandilla Bay	8	TBR+reef	250 m
1371	Wandilla Bay (W 1)	7	R+reef	500 m
1372	Wandilla Bay (W 2)	6	R+reef	750 m
1373	Cabbots Well	7	R/LTT+reef	1.65 km
1374	Cabbots Beach	7	LTT/TBR+reef	300 m
1375	Cabbots Beach (W 1)	7	LTT/TBR+reef	400 m
1376	Cabbots Beach (W 2)	6	LTT/TBR+reef	300 m
1377	Cheetima Beach	7	R+reef	1.7 km

Wandilla Bay is a small, exposed bay located 7 km west of Cape Nuyts. West of the bay, the coast continues on due west for 6 km, then turns to face southwest between Cabbots and Cheetima beaches. In between are eight generally small beaches, all backed by low calcarenite bluffs up to 20 m high and fronted by scattered reefs. While high waves break on the reefs, the beaches are afforded varying degrees of protection. A gravel road provides vehicle access to Wandilla Bay, with 4WD tracks generally following the deflated bluff tops providing access to all the other beaches.

Wandilla Bay (**1370**) is a 250 m long, south facing, exposed beach, backed and bordered by low calcarenite bluffs, the eastern headland consisting of a narrow, 300 m long calcarenite spur. The beach and swash zone are a mixture of sand and calcarenite outcrops, while in the surf zone strong rips run out either end of the bay (Fig. 4.192).

Irregular bluffs dominate the shore for 2 km to the west, beyond which are two strips of high tide sand (beaches 1371 and 1372). Beach **1371** is 500 m long and fronted by continuous low tide beachrock, then a 150 m wide, rip-dominated surf zone. Beach **1372** is a crenulate, 750 m strip of high tide sand, with reef lying just off the shore producing lower waves against the beach. There is usually high surf over the reefs and a strong rip exiting the western end of the 'lagoon'.

Figure 4.192 *Wandilla Bay beach consists of a high tide sand beach fronted by near continuous reefs, with two permanent rips flowing out through gaps in the reef.*

Cabbots Well beach (**1373**) is the longest beach in this section. The south facing beach is exposed to high waves, which break over and around a discontinuous linear beachrock reef up to 400 m offshore. This lowers and refracts the waves, resulting in two small cuspate protrusions along the 1.65 km long beach. Wave height, while lower at the beach, is sufficient to maintain a series of five strong permanent rips. Active dunes extend up to 1 km in from the beach, which is also the site of a wind-powered well.

West of the beach are 2 km of highly irregular, 20 m high bluffs (containing an excellent left break in the second bay), then the coast turns to face southwest along the Cabbots to Cheetima beaches section. **Cabbots Beach** (**1374**) is a crenulate, 300 m long beach, with some beachrock outcrops. There is a left hand break into the beach and a strong rip at the western end. Immediately west is beach **1375**, a discontinuous, 400 m long beach which has low protruding bluffs impinging on the high tide beach, as well as beachrock in the inner surf zone. One kilometre to the west is beach **1376**, a more protected, southeast facing, 300 m long beach, with scattered reefs lying up to 1 km offshore and usually lower waves at the shore. It is backed by low bluffs and a vehicle track leads to the western end of the beach.

Cheetima beach (**1377**) is an exposed, southwest facing, crenulate, 1.7 km long high tide sand and cobble beach. The low tide zone and out into the surf are dominated by calcarenite reef flats and beachrock, which produce two cuspate protrusions, as well as an irregular surf zone. During moderate to high waves, the surf is 400 m wide, with a large rip exiting the eastern end of the beach. At the eastern end of the bluff is a monument to the two fishermen who lost their lives in the 1992 wreck of the Tritan Star at this site.

Swimming: These are generally accessible and lower energy beaches close inshore. However all are dominated by reefs and exposed to high waves and strong rips. So use caution if swimming anywhere along this section.

Surfing: There are several good reef breaks off both the beaches and some of the bluffs, with reasonably good access to most. However beware of the strong rips.

Fishing: This is a relatively popular area for beach and rock fishing, with numerous deep inshore holes.

Summary: An accessible section of coast containing a mixture of usually low bluffs, beaches, deflated surfaces and dunes.

1378-1384   **CAPE ADIEU (W)**

Unpatrolled				
No.	Beach	Rating	Type	Length
1378	Cape Adieu	8	R+platform	1.1 km
1379	Cape Adieu (W 1)	8	R+platform	60 m
1380	Cape Adieu (W 2)	8	R+platform	100 m
1381	Cape Adieu (W 3)	8	R+platform	200 m
1382	Cape Adieu (W 4)	8	R+platform	100 m
1383	Cape Adieu (W 5)	8	R+platform	200 m
1384	Cape Adieu (W 6)	8	R+platform	100 m

Cape Adieu is a 20 m high calcarenite bluff that marks a change in coastal orientation from a southerly to a more southwesterly aspect. It is also the last prominent shoreline inflection before the Head of Bight, 110 km to the northwest. Calcarenite bluffs dominate the first 15 km of coast west to the Tchalingaby dunefield. The first 6 km consists of 20 to 30 m high, generally steep bluffs, backed by deflated clifftops, then generally unstable clifftop dunes extending a few hundred metres inland. A 4WD track runs along the rear of the dunes, with tracks leading out to the cliffs. Below the bluffs are a series of seven generally small, bluff and reef-dominated beaches totalling only 1.8 km in length, with bluffs, intertidal rock platforms and reefs dominating in between. All are exposed to high southerly waves, with usually heavy surf seaward of and across the platform, and hazardous beach conditions.

Beach **1378** commences 1 km west of the cape and is a 1.1 km long, southwest facing, crenulate high tide sand beach, with the beach broken by long calcarenite outcrops and a continuous intertidal platform fronting the shore (Fig. 4.193). There are higher surf breaks seaward of the platform. A mixture of active dunes and low, dune-topped bluffs back the beach.

Figure 4.193 *Cape Adieu beach (1378) is typical of the reef and platform-controlled beaches along this high energy section of coast.*

The next six beaches are scattered along 4 km of steep, bluff-backed coast. Beaches **1379** and **1380** are neigbouring 60 m and 100 m long beaches, both backed by 30 m high bluffs and fronted by intertidal calcarenite platforms. Beach **1381** consists of two pockets of sand totalling 200 m in length. The beach is almost enclosed in 20 m high calcarenite bluffs, while it is exposed to high waves breaking across an intertidal platform. A permanent rip drains the embayment. Beach **1382** lies in a small cove, with the curving, 100 m long beach at the head of the cove fronted by a broken platform and deeper water. Beach **1383** is a 200 m long strip of crenulate high tide sand fronted by an irregular intertidal platform which includes some tidal pools, with usually a heavy surf breaking across the rocks. Beach **1384** is a 100 m strip of sand at the base of 20 m high bluffs and is fronted by an irregular, 100 m wide platform.

Summary: These are six hazardous beaches all fronted by 50 to 100 m wide intertidal rock platforms, with surf usually breaking seaward of the platforms and maintaining permanent rips. The western five are inaccessible on foot, while all are unsuitable for swimming and even difficult for surfers to negotiate. Fishers are the only ones who might productively use these beaches.

1385-1391 BEACHES 1385 to 1391

No.	Beach	Unpatrolled		
		Rating	Type	Length
1385	Beach 1385	8	R+platform	150 m
1386	Beach 1386	8	R+platform	100 m
1387	Beach 1387	8	R+platform	800 m
1388	Beach 1388	8	R+platform	450 m
1389	Beach 1389	8	R+platform	100 m
1390	Beach 1390	8	R+platform	100 m
1391	Beach 1391	8	R+platform	100 m

Beaches 1385 to 1391 occupy a southwest facing, 5 km section of shore backed by 20 m high calcarenite bluffs, with all the beaches fronted by calcarenite intertidal platforms and some reefs and all exposed to usually high southerly waves and winds. The clifftops are for the most part vegetated and stable, permitting 4WD tracks to run along the top of the cliffs providing good access to the top of the beaches, with foot access down to the beaches only possible where sand ramps are present. Low scrub and sheep country backs the beaches, with three fences terminating at the edge of the cliffs.

Beach **1385** is a boulder-covered, 150 m long strip of high tide sand at the base of steep bluffs, with no foot access. It is fronted by a flat, 50 m wide platform. Beach **1386** is a more embayed, 100 m long beach, with a tidal pool against the beach and a permanent rip to the outer surf zone. Beach **1387** is a crenulate, 800 m long high tide beach, backed by vegetated dunes which have climbed up onto the backing bluffs. It is fronted by a 100 m wide platform and wider surf zone (Fig. 4.194). Fence lines terminate above beaches 1386 and 1387. Beach **1388** is a highly irregular, 450 m long sand beach broken by outcrops. It is generally backed by dune-capped calcarenite bluffs and fronted by an irregular rock platform and wide surf. Beach **1389** is a 100 m long pocket of sand located in a small gap in the bluffs.

Figure 4.194 *Beach 1387 is a high energy beach fronted by a wide intertidal platform, with the narrow high tide beach sandwiched in between the active platform and a backing raised Pleistocene platform.*

Beaches **1390** and **1391** are two 100 m long pockets of high tide sand lying at the base of 20 m high bluffs, fronted by 50 to 100 m wide platforms. A main track and fence line terminate above beach 1390.

Summary: A section of coast accessible by 4WD, with variable foot access down to the calcarenite dominated beaches. All are dominated by the intertidal platforms and high surf, with strong rips seaward of the platforms. None are suitable for swimming, with even the occasional calmer inner tidal pool usually connected to seaward-flowing rips.

Wahgunyah Conservation Reserve

Established:	1993
Area:	29 163 ha
Beaches:	23 beaches (1391-1413)
Coast length:	48 km

The Wahgunyah Conservation Reserve is spread along 48 km of coast and extends up to 12 km inland, encompassing a section of coast dominated by high energy beaches and backed by several extensive sand dune systems, including the Tchalingaby, Chalgonippi, Wahgunyah and Ocock sandhills.

1392-1397 TCHALINGABY SANDHILLS

No.	Beach	Rating	Type Inner	Outer Bar	Length
		Unpatrolled			
1392	Tchalingaby (E 5)	8	R+reef	LBT	150 m
1393	Tchalingaby (E 4)	8	RBB+reef	LBT	800 m
1394	Tchalingaby (E 3)	8	RBB+reef	LBT	300 m
1395	Tchalingaby (E 2)	8	RBB+reef	LBT	500 m
1396	Tchalingaby (E 1)	8	RBB	D	1.9 km
1397	Tchalingaby	8	D	D	7.3 km

The **Tchalingaby Sandhills** are a 10 km² area of active sand dunes, extending for 5 km along the coast and up to 2.5 km inland, and reaching heights of 70 m. They also represent the first in a series of similar sized dune fields that lie between Cape Adieu and the Head of Bight. All are a product of their southwest orientation into the high southerly waves and winds. Four-wheel drive tracks provide access to the beaches east of the dunes, with the main track skirting the rear of the dunes then running up to 3 km inland. Much of the dune system is part of the Wahgunyah Conservation Reserve.

West of beach 1392 the shoreline runs relatively straight up to the main beach. However, protruding bluffs and fronting rocks and reefs separate the first 4 km into five beaches, that share a continuous outer bar and surf zone. All five are composed of fine sand which, in combination with waves averaging 2 m, maintains a double bar, 300 to 400 m wide surf zone, with the rocks and reefs inducing strong permanent rips off each beach. Beach **1392** is a 150 m long pocket of sand fronted by a 100 m wide platform, then a rip-dominated, 200 m wide surf zone in lee of the outer bar. Beach **1393** is an 800 m long beach backed by low bluffs, with four substantial rock outcrops along the shore, each resulting in strong rips, then a 300 m wide surf zone. Beach **1394** is a 300 m long, relatively wide beach, bordered by protruding bluffs and platforms, with strong rips at each end and wide surf. Beach **1395** is a 500 m long, bluff-backed beach, with the western end marked by a small stack at the tip of the protruding bluff.

A permanent rip lies at the eastern end, with a 400 m wide, double bar surf zone off the beach.

Beach **1396** extends for 1.9 km and for the most part consists of a 200 to 300 m wide, two to three bar surf zone - the product of both the high waves and fine sand. Calcarenite outcrops to either end induce permanent rips around the rocks. A 4WD track runs along the top of the western bluffs, which is also the site of a solitary fishing shack.

The main **Tchalingaby** beach (**1397**) is a relatively straight, 7.3 km long, southwest facing, low gradient, high energy beach, backed for the most part by its large and active dune field. Bluffs and a 4WD track overlook the eastern end, with access via the Wahgunyah Well track at the western end, but no formal access through the dunes. The beach is fully dissipative, with a 500 m wide, triple bar surf zone. However, the presence of a few scattered rocks and reefs along the beach and inner surf zone maintain eight permanent rips close inshore, with a very large rip immediately west of a central inflection in the shoreline.

Swimming: These are six high energy beaches, all dominated by a high outer surf zone and strong reef-controlled rips in the inner surf zone. Be very careful if swimming here as high wave set-up and set-down at the shoreline can trap the unwary.

Surfing: There are numerous beach and reef breaks along this section and generally good access.

Fishing: The permanent rips provide several relatively safe spots for fishing this energetic coast.

Summary: A relatively accessible stretch of exposed, high energy shoreline, with good 4WD access to all the beaches.

1398-1408 CHALGONIPPI to OCOCK SANDHILLS

No.	Beach	Rating	Type Inner	Outer Bar	Length
1398	Chalgonippi dunes	9	D+reef	D	1.4 km
1399	Chalgonippi dunes (W)	9	D+reef	D	150 m
1400	Wahgunyah Well (E)	9	TBR+reef	D	600 m
1401	Wahgunyah dunes	9	D+reef	D	1.4 km
1402	Wahgunyah dunes (W 1)	9	TBR+reef	D	500 m
1403	Wahgunyah dunes (W 2)	9	TBR+reef	D	60 m
1404	Wahgunyah dunes (W 3)	9	D	D	300 m
1405	Wahgunyah dunes (W 4)	9	D	D	2.1 km

1406	Coroma Well	9	D	D	900 m
1407	Ocock dunes	9	D	D	4.4 km
1408	Ocock dunes (W)	9	RBB	D	1.9 km

The coastline between the Chalgonippi and Ocock sandhills consists of a straight, 14.5 km stretch of predominantly sandy shoreline. It faces southwest and is exposed to persistent high waves and onshore winds. As a consequence it is dominated by wide, fine sand, low gradient, high energy beaches and surf, and for the most part backed by dunes extending up to 2.5 km inland and to 100 m in elevation. While the beaches are separated by bluffs, rocks and reefs, they share a common, continuous, 400 m wide surf zone. Two main access tracks reach the coast, at Wahgunyah Well in the east and the main Dog Fence track to the west, with no defined tracks to most of the shore. Apart from the tracks and two wells there is no development along the coast. All of this coastal section and most of the backing dunes are part of the Wahgunyah Conservation Reserve.

The **Chalgonippi dunes** are a 1 km long patch of active dunes extending up to 1 km inland. They are fronted by beach **1398**, a 1.4 km long, high energy beach bordered by 10 m high calcarenite bluffs and reefs. A dissipative double bar system dominates the main beach, with reef-induced permanent rips to either end. In amongst the western bluffs and reefs is an irregular, 150 m long, high tide pocket beach (**1399**), fronted by a 100 m wide intertidal platform and then a 200 m wide, double bar surf zone. Immediately west is beach **1400**, a similar 600 m long beach lying at the base of the bluffs, with a more irregular fronting reef and then the wide barred surf zone.

Wahgunyah Well is located toward the eastern end of beach **1401**, in the deflation basin between the beach and the Wahgunyah dunes, which extend another 2 km inland (Fig. 4.195). The 1.4 km long beach is bordered by calcarenite bluffs and reefs, with two reefs also located along the beach, both of which induce strong permanent rips. Beyond the reefs is a 400 m wide dissipative surf zone, with usually two to three bars. Amongst the bluffs at the western end of the beach are three bluff-bound beaches. Beach **1402** is an open, 500 m long dissipative beach, with strong permanent rips against the reefs at either end. Beach **1403** is a 60 m pocket of sand and reef fronted by the same surf zone, while beach **1404** is a 300 m long, more open beach, bordered by protruding bluffs and fronted by the wide surf.

To the west of these bluffs is an open, 2.1 km long sand beach (**1405**) that is also bordered in the west by low calcarenite bluffs. It has a 400 m wide, two to three bar surf zone, with increasing reefs toward the western end maintaining two strong permanent rips. In lee of the western bluffs is the **Coroma Well**, with a narrow, 900 m long beach (**1406**) lying along the base of the bluffs. The beach is awash at high tide and in high surf, while the

wide surf zone continues uninterrupted past the bluffs.

Figure 4.195 *View from the beach back toward Wahgunyah Well and massive dunes.*

The **Ocock Sandhills** lie between Coroma Well and the bluffs at the end of the Dog Fence track (Fig. 4.196). These dunes extend for 7 km along the coast, up to 1.5 km inland and reach heights of 100 m. They are fronted by two high energy beaches. Beach **1407** is a 4.4 km long beach with calcarenite bluffs and reefs to either end. Its neighbour (**1408**) is a 1.9 km long continuation which terminates at the more dominant Dog Fence bluffs. Reefs to either end induce three strong permanent rips in the inner surf zone.

Figure 4.196 *The Ocock Sandhills are typical of this section of the Eyre Peninsula. The beach is fronted by a 400 m wide, high energy surf zone, while massive active sand dunes extend 1.5 km inland.*

Swimming: While low gradient, sandy beaches dominate this section, they are exposed to persistent high waves, which induce strong set-up and set-down at the shore, together with strong rips adjacent to any inshore reefs. Use extreme care if swimming here and stay well clear of the rips.

Surfing: The best surf is usually found around the reefs, with many kilometres of beach breaks in between.

Fishing: The inner reefs and their deep holes offer the best fishing locations.

Summary: A section of coast requiring a good 4WD to negotiate and offering high energy beaches backed by massive dune systems.

1409-1412 DOG FENCE

No.	Beach	Rating	Type Inner	Outer Bar	Length
1409	Dog Fence bluff (1)	9	R+platform/reef		50 m
1410	Dog Fence bluff (2)	9	TBR+platform/reef		800 m
1411	Dog Fence Beach	9	D	D	14.9 km
1412	Dog Fence (W)	9	D+reef	D	4.5 km

Unpatrolled

A dog fence and maintenance track run due south from the Eyre Highway and reach the coast at a prominent 3.5 km section of 20 m high calcarenite bluffs, where the fence terminates at the edge of the bluffs. The track provides 4WD access to the coast and beaches and tracks to either side, including the western end of the Ocock beach and dunes. The four beaches along this 20 km long section of coast and their backing dune systems are all contained in the Wahgunyah Conservation Reserve. In amongst the bluffs are two bluff and reef-dominated beaches. Beach **1409** lies immediately west of the end of the dog fence. It consists of a 50 m pocket of high tide sand surrounded by steep, vegetated bluffs and fronted by an intertidal platform, then a 400 m wide surf zone, with waves breaking on outer shore-parallel calcarenite (beachrock) reefs. Four hundred metres to the west is a second bluff-backed beach (**1410**), an irregular beach consisting of a few pockets of sand in indentations in the bluffs and a reef and rip-dominated inner surf zone, with the same reef and surf offshore.

The western end of the bluffs marks the beginning of the 15 km long Dog Fence Beach (**1411**), one of the longer beaches west of Fowlers Bay. This is an exposed, southwest facing, high energy beach. The combination of high waves, fine sand and high winds maintains a triple bar, 400 to 500 m wide, dissipative surf zone, backed by continuous active dunes extending 1 to 2 km inland. In addition, the calcarenite reef off the bluffs parallels the first 5 km of shore, inducing a slightly lower energy, rip-dominated inner surf zone (Fig. 4.197). Reefs also occur along the western 3.5 km of shore resulting in several large permanent rip systems. The most prominent reef is backed by a low bluff, which forms the boundary with beach **1412**. This high energy, 4.5 km long beach has a wide, two to three bar surf zone, with near continuous inner reefs permitting a rip-dominated inner surf zone, including two large permanent rip embayments. The beach terminates at a low bluff and reefs that mark the boundary of the Yalata Aboriginal Land.

Figure 4.197 *The very eastern end of Dog Fence Beach is protected by shallow reefs. However wave energy quickly increases to the west and hazardous rips dominate the surf.*

Swimming: Dog Fence Beach is an accessible beach, however use care if swimming here as rips dominate the inner surf. Be very careful if swimming near any of the permanent reef-controlled rips, usually identifiable by their large embayments and deeper water off the shore.

Surfing: There is rideable surf on some of the reefs, with local knowledge and experience necessary to surf this remote section of coast.

Fishing: The best fishing location is in the rip holes at the eastern end of the beach.

Summary: An easy to find beach if you follow the dog fence, offering good views from the bluffs and many kilometres of high energy beach and surf.

1413-1415 YALATA

No.	Beach	Rating	Type Inner	Outer Bar	Length
1413	Shoulder Hill	9	TBR+reef	D	2.2 km
1414	Yalata Beach	9	D/RBB+reef	D	19.8 km
1415	Yalata (W)	9	TBR+reef	D	5.8 km

Unpatrolled

The 25 km long 4WD track from Yalata reaches the coast at a section of low calcarenite bluffs, fronted by a wide, high energy surf zone that dominates this section of coast. The bluffs also form the boundary between two beaches that mark the southern boundary of the Yalata Aboriginal Land (Fig. 4.198). To the east is beach **1413**, a 2.2 km long sand beach bordered by low bluffs and reefs, with patchy reef also occurring along the beach. These maintain strong rips in the inner surf, with a second then third bar located 500 m seaward of the shore. To the west of the bluffs begins the 20 km long **Yalata Beach (1414)**, a continuous, high energy, fine sand beach which, toward the western end, becomes increasingly interrupted by calcarenite outcrops at the shore that induce permanent rips in the inner surf zone. The outer surf zone gradually narrows from 500 m and three bars in the east to 300 m and two bars in the west, a response to both slightly lower waves and more medium sand. The beach is backed by massive dune systems, with stable, vegetated dunes extending up to 7 km inland and to heights of 90 m, and active dunes closer to shore moving sand up to 2 km inland.

Figure 4.198 *Yalata Beach (top) begins at the low bluffs which lie at the end of the 4WD track from Yalata. This is an exposed, rip-dominated, hazardous beach, especially around the bluffs where permanent rips prevail.*

The main beach terminates at a 200 m long reef, which separates it from the western beach (**1415**) which continues on for another 5.8 km. This beach is similar, with high waves, a 300 m wide double bar surf zone and an inner surf interrupted by several reefs, each of which induces strong permanent rips. Active dunes extending up to 1 km inland back the entire beach.

Swimming: The beach at the end of the Yalata track is a relatively popular swimming and fishing beach with the local community. It is, however, very hazardous owing to the pronounced set-up and set-down at the shore, reefs and strong rips adjacent to the reefs and rocks.

Surfing: The are many kilometres of beach and reef breaks along the shore.

Fishing: An excellent stretch of coast for fishing the wide surf and deep inner rip holes.

Summary: A remote section of coast, for which permission is required to access, consequently it is used by few outside the Yalata community.

1416-1429   COYMBRA

No.	Beach	Unpatrolled Rating	Type	Length
1416	Coymbra (1)	8	TBR+reef	400 m
1417	Coymbra (2)	7	R+reef	350 m
1418	Coymbra (3)	6	R+reef	150 m
1419	Coymbra (4)	6	R+reef	300 m
1420	Coymbra (5)	6	R+reef	100 m
1421	Coymbra (6)	6	R+reef	300 m
1422	Coymbra (7)	7	R+reef	100 m
1423	Coymbra (8)	7	R/LTT+reef	60 m
1424	Coymbra (1)	6	R+reef	300 m
1425	Coymbra (10)	6	R/LTT+reef	850 m
1426	Coymbra (camp)	6	R/LTT+reef	500 m
1427	Coymbra (12)	6	R/LTT+reef	1 km
1428	Coymbra (13)	7	LTT+reef	550 m
1429	Coymbra (14)	8	TBR+reef	600 m

Coymbra is the name of an 8 km section of coast that contains a total of 14 bluff and reef-bound beaches, all backed by deflated surfaces then active dunes extending up to 2 km inland. While the coast has a relatively straight trend and is exposed to waves averaging over 1.5 m, the reefs lower waves at the shore and cause considerable refraction, resulting in a series of arcuate, lower energy beaches, fronted by a mixture of platforms, reefs and reef-fronted lagoons. The main 4WD track runs in behind the active dunes, with a few informal access tracks across the dunes to the beaches and a solitary 'camping' site in lee of beach 1426.

The first four beaches are located amongst 2.5 km of undulating calcarenite bluffs and reefs. Beach **1416** forms the eastern boundary of the system and is an exposed, 400 m long, high energy beach drained by a strong rip through the 300 m wide surf, with waves breaking on reefs further seaward. The reefs become more continuous to the west and beach **1417** has a 200 m wide, reef-dominated surf zone, resulting in lower waves at the shore and a strong permanent rip exiting the eastern end of the beach. Beach **1418** is a 150 m long, reef and platform-bound beach, with only low waves reaching the shore at high tide and calm conditions at the shore at low tide. Beach **1419** is fronted by a relatively straight beachrock reef, which forms a small lagoon in its lee, which is free of waves and calm at low tide. The 300 m long, curving beach is backed by a vegetated foredune then active dunes.

The remaining Coymbra systems are contained in four slightly longer 'bays', each fronted by straight beachrock reefs, with lagoons between the reefs and the arcurate shorelines. The first 'bay' is 1 km long and contains four beaches (1420-1423) (Fig. 4.199). Beach **1420** is a protected, 100 m long, west facing high tide sand beach fronted by shallow reef flats, with bluffs and stacks forming the boundaries. Its 300 m long neighbour (**1421**) emerges from a series of calcarenite outcrops to form a sandy beach, backed by a low foredune and fronted by a deeper, 50 m wide lagoon, lying in lee of a continuous beachrock reef. Calm conditions prevail in the lagoon at low tide. The western end of the 'bay' is backed by 10 m high bluffs, in amongst which are two pockets of high tide sand (beaches **1422** and **1423**), 100 m and 60 m long respectively and fronted by 100 m wide shallow reef flats.

Figure 4.199 *The first of the Coymbra 'bays' (beach 1422) formed in lee of shore parallel beachrock reefs.*

The next bay contains solitary beach **1424**, a 300 m long, curving beach, impounded by protruding bluffs and a straight beachrock reef running between the bluffs and enclosing a calmer lagoon.

The main Coymbra 'bay' is located in front of the camping site. It is a 1.4 km long embayment fronted by a straight, but discontinuous, beachrock reef. The reef is linked to two sand forelands and impounds a 100 to 250 m wide lagoon. Within the lagoon are two curving beaches (**1425** and **1426**) that join at a central foreland. Waves break heavily on the reef, while gaps in the reef permit some wave energy to reach the shore, maintaining a shorebreak at the beach (Fig. 4.200). This water returns seaward through the gaps as two permanent rips, which increase in strength toward the reefs (Fig. 4.201).

Figure 4.200 *The main Coymbra beach (1425) is a curving, reef protected beach, typical of this 8 km long section of reef-dominated shore.*

Figure 4.201 *Beach 1426 highlights the impact of the beachrock reef in protecting the shore, while the gap allows waves to both shape the curved bay, as well as drain the surf via a permanent rip.*

To the west of the camp site is the longest 'bay', a 2.2 km long system containing three arcuate beaches. Beach **1427** at 1 km long is the longest in the Coymbra section. It is fronted by a near continuous beachrock reef and a 200 m wide lagoon and usually has low waves at the shore. Beaches **1428** and **1429** are fronted by a more broken reef which permits higher waves to reach the shore, resulting in a surf-filled lagoon and strong rips draining the 550 and 600 m long beach-lagoon systems respectively. Beach 1429 terminates at a more prominent calcarenite bluff littered with slabs of beachrock.

Swimming: Some of the 'lagoons' are relatively safe at low tide close inshore. However be careful as high set-up and set-down occurs within the lagoon, especially at high tide and some drain seaward by strong rips through the reefs.

Surfing: There is a lot of potential on some of the reefs.

Fishing: The few people who know this site mainly come to fish the many deep lagoons, holes and gutters.

Summary: A scenic and interesting section of reefs, lagoons, beaches and dunes.

1430-1439   **BEACHES 1430 to 1439**

		Unpatrolled		
No.	Beach	Rating	Type	Length
1430	Beach 1430	7	R+reef	2.6 km
1431	Beach 1431	7	R+reef	300 m
1432	Beach 1432	7	R+reef	300 m
1433	Beach 1433	7	LTT+reef	900 m
1434	Beach 1434	7	R+platform/reef	300 m
1435	Beach 1435	8	TBR+reef	1 km
1436	Beach 1436	8	TBR+reef	3 km
1437	Beach 1437	8	TBR+reef	500 m
1438	Beach 1438	8	TBR+reef	900 m
1439	Beach 1439	8	TBR+reef	2.9 km

To the west of the Coymbra beaches the coast continues straight toward the Head of Bight. The first 12 km is dominated by beachrock reef, which both outcrops on the beaches and, toward the east, parallels the shoreline, forming lagoons in its lee with more arcuate beaches. Toward the western end, the reefs begin to diminish and more exposed, higher energy systems result. The entire section is backed by generally vegetated dunes extending up to 1 km inland, with the main access track paralleling the back of the dunes and occasional informal tracks crossing the dunes to the ten beaches.

Beach **1430** is a 2.6 km long, crenulate beach fronted by a near continuous beachrock reef lying approximately 100 m off the beach, with a linear lagoon between the reef and shoreline. Waves break heavily on the reef resulting in a lower energy shorebreak and reflective beach face, with water returning seaward in permanent rips flowing through gaps in the reefs.

Beach **1431** is a curving, 300 m long beach bounded by sand forelands extending out to beachrock reefs. There is a 100 m wide gap in the reef which permits a high shorebreak at the beach and a permanent rip to exit through the gap. Its western neighbour (beach **1432**) is 300 m long and almost identical, with beachrock also outcropping along the shoreline. Beach **1433** is a 900 m long, more open beach fronted by a broken reef, together with reef along the central shoreline section. The net result is a more energetic, rip-dominated shoreline and surf zone.

Beach **1434** is a reef-controlled, 300 m long, curving beach, with intertidal reef running the length of the shoreline, as well as reef paralleling each end of the beach. Beach **1435** begins immediately to the west and is a 1 km long beach containing a mixture of shoreline reef and sand, together with scattered reef offshore. This permits higher waves to penetrate the reef and maintain a 300 m wide, rip-dominated surf zone. Beach **1436** is a similar 3 km long beach, with reef dominating the shoreline and a 300 m wide surf zone containing an inner rip-dominated

bar, and outer surf breaking on both a discontinuous bar and reefs. This beach is backed by more active dunes extending 1 km inland.

Beach **1437** is an arcuate, 500 m long beach bordered by prominent reefs extending 100 m seaward of each end, with permanent rips flowing out against each reef and a 200 m wide surf zone. Beach **1438** commences immediately to the west and is a 900 m long beach containing linear shoreline reefs to either end and a central sandy section dominated by a large permanent rip system. Beach **1439** runs for 2.9 km from the western reef to a prominent sand foreland in lee of a reef extending 500 m seaward. The beach is relatively straight, with some shoreline reef and a 200 m wide, rip-dominated surf zone.

Swimming: These are ten high energy, reef and rip-dominated beaches. The shoreline inside the beachrock lagoons can become relatively calm at low tide and during low to moderate waves, however beware of strong currents and rips flowing out of each lagoon, and the presence of rocks and reefs along some of the shorelines.

Surfing: There are numerous reef-controlled beaches along this section, with local knowledge required to find the better spots.

Fishing: This whole section offers numerous permanent rip holes, reefs and lagoons.

Summary: A beachrock-dominated, sandy coastline, backed by stable through unstable dunes, with access to the beaches requiring a trip over the intervening dunes.

1440-1442   **HEAD OF BIGHT**

		Unpatrolled		
No.	Beach	Rating	Type		Length
			Inner	Outer Bar	
1440	Head of Bight	8	TBR	LBT	9.5 km
1441	Twin Rocks	7	LTT/TBR		600 m
1442	Beach 1442	8	R+rocks/reef		50 m

The **Head of Bight** marks the apex of the Great Australian Bight and the boundary between the predominantly sandy beaches to the east and the 180 km of Nullarbor cliffs to the west. The head has also been a sink for sand driven onshore by the persistent high waves and winds, with dunes extending up to 8 km in lee of the beach (Fig. 4.202). A 4WD track descends the eastern end of the cliffs to provide access to the rear of beaches 1440 and 1441, with the main track then wandering across the deflation basin in lee of beach 1440 (Fig. 4.203).

The main beach (**1440**) commences at the sandy foreland that separates it from beach 1439 and, facing southwest, runs relatively straight for 9.5 km to the lee of Twin

Rocks. This is an exposed, high energy beach consisting of a 3 km long section of reef-controlled rips, then a 4 km section with an inner bar cut by rips spaced every 500 m and finally a lower gradient, 400 m wide, more dissipative surf zone at the head. It is backed by the widest active dune field on the Eyre Peninsula.

Figure 4.202 *The Head of Bight where the beach ends (right) and the Nullarbor cliffs begin (left), with massive sand dunes extending up to 8 km inland.*

Figure 4.203 *The track from the 40 m high cliffs down across the dunes to the Head of Bight, with the massive Yalata dunes behind.*

Twin Rocks are a 20 m high sea stack lying 400 m off the beach. To the west of the rocks is a 600 m long beach (**1441**), that is backed by calcarenite-capped Nullarbor limestone and terminates at the first of the exposed Nullarbor cliffs. The moderate energy, low gradient beach is usually covered in dense seagrass, with a 100 m wide, low gradient surf zone.

Beach **1442** is located 2 km west of Twin Rocks. It consists of a 50 m pocket of sand lying in an indentation in the Nullarbor Cliffs. A tourist drive to whale viewing points runs along the cliffs, with a car park located 100 m in from the beach. The beach, however, can only be accessed via a steep climb down the cliff debris. It has a rock-strewn inner surf zone that faces west across the 200 m wide outer surf zone.

Swimming: The Twin Rocks beach is moderately safe close inshore on the low gradient bar, while the main beach

is exposed, high energy and dominated by large inshore rips.

Surfing: The main beach offers some energetic beach and reef breaks.

Fishing: There are large rip holes along the main beach, as well as scattered shoreline reef.

Summary: Many visitors now view these beaches from the Twin Rocks tourist drive, with only a very few venturing down to the beaches and barely a soul venturing into the surf.

Great Australian Bight Marine Park

Established: 1998
Area: 1 975 900 ha
Coast area: 200 km west of Ceduna to WA border
                      (beaches 1414 to 1454)
Coast length: 242 km

The Great Australian Bight Marine Park was established to protect the southern right whale and Australian sea lion, both of which breed in the coast waters of the park. In addition it protects a representative area of the unique benthic sea floor flora, fauna and sediments.

Nullarbor National Park

Established: 1979
Area: 588 283 ha
Beaches: 12 beaches (1443 to 1454)
Coast: 187 km

Nullarbor National Park covers over 588 000 ha and was gazetted to protect part of the Nullarbor Plain and bordering cliffline. The southern boundary of the park includes 187 km of cliffline, while it extends up to 32 km inland.

The Nullarbor Cliffs

The Nullarbor Cliffs begin at Head of Bight and extend west for 209 km to Wilson Bluff, the border with Western Australia. The cliffs average 70 m in height, reaching a maximum of 90 m in places. They run uninterrupted for the first 179 km until the Merdayerrah Sandpatch. This sandpatch consists of both lithified Pleistocene and active Holocene dunes and sand ramps, that have run up against the cliffs, and in places blown over the top onto the plain.

Apart from the sandpatch there are no beaches along the 179 km long cliffed section. However this was not always the case, as there are a number of places where Holocene sand dunes sit on top of the cliff edge, indicating the presence of past sand ramps, fed by beaches at the base of the cliffs. All these beaches have since been eroded, leaving only the clifftop dunes as evidence.

The cliffs are composed of Tertiary limestone deposited in a shallow sea between 50 to 20 million years ago, then uplifted to form the flat Nullarbor Plain. The sea has since scarped the edge of the plain forming the cliffs. Three limestone units compose the cliffs. The lower units are a chalky white marine deposit and known as the Wilson Bluff Limestone. Overlying these are the Abrakurrie units which contain impurities, and finally the prominent thicker reddish Nullarbor Limestone.

The cliff edge is usually located within a kilometre or two of the highway and there are a number of sealed access points and viewing areas from which to see the cliffs, ocean and occasionally whales. No drive across the Nullarbor is complete without at least seeing a few of these spectacular sights.

1443-1454 MERDAYERRAH

Unpatrolled				
No.	Beach	Rating	Type	Length
1443	Merdayerrah 1	7	R+platform	200 m
1444	Merdayerrah 2	7	R+platform	450 m
1445	Merdayerrah 3	7	R+platform	150 m
1446	Merdayerrah 4	7	R+platform	300 m
1447	Merdayerrah 5	7	R+platform	500 m
1448	Merdayerrah 6	7	R+platform	600 m
1449	Merdayerrah 7	7	R+platform	600 m
1450	Merdayerrah 8	7	R+platform	800 m
1451	Merdayerrah 9	7	R+platform	400 m
1452	Merdayerrah 10	7	R+platform	600 m
1453	Merdayerrah 11	7	LTT+platform	1.8 km
1454	Merdayerrah Sandpatch	7	TBR+reef	8.5 km

The **Merdayerrah Sandpatch** is part of a 21 km long section of the Nullarbor Cliff which has been blanketed in Pleistocene and Holocene dunes (Figs. 4.204 & 4.205). The steep Pleistocene dunes have been lithified to calcarenite and drape the entire section of the cliffs to within 5 km of the Western Australian border at Wilson Bluff. More recent active Holocene dunes have partially covered the western 8 km of Pleistocene deposits and in a few locations have reached the top of the 90 m high cliffs to form active clifftop dunes. This entire section of coast lies within 500 m to 2 km of the Eyre Highway and toward the central section are three lookouts providing views of the cliffs, dunes and shoreline. The only vehicle access down to the shore is via two steep 4WD tracks. One is located above the eastern beaches (1443-1445) and a second is above beaches 1447-1449.

Steep gullies cutting through the calcarenite make it difficult to impossible to drive along some of the intervening slopes.

Most of the beaches consist of a strip of high tide sand wedged in between intertidal calcarenite platforms and backed by scarped calcarenite and/or limestone bluffs. Only the main beach is backed by dunes and has a more normal surf zone.

Beaches **1443**, **1444** and **1445** are three adjoining beaches, 200, 450 and 150 m long respectively, located 3 km from the eastern end of the calcarenite. Each lies in a slight indentation in the backing calcarenite slopes and is fronted by 50 to 100 m wide platforms, with waves breaking both over the platforms and further seaward. They are accessible via a steep vehicle track. Beach **1446** lies 2 km further west and is a similar solitary, 300 m long high tide beach and platform.

Figure 4.204 *View west along the Merdayerrah Sandpatch, with the dunes climbing the 90 m high backing limestone bluffs, forming active clifftop dunes.*

Figure 4.205 *View east from Wilson Bluff on the SA-WA border showing the protruding beaches and dunes of the Merdayerrah Sandpatch, South Australia's westernmost beach.*

Beaches **1447**, **1448** and **1449** are all accessible from the second rough track. They are 500, 600 and 600 m long respectively, each backed by low, scarped calcarenite bluffs and fronted by intertidal platforms up to 100 m wide. There is a rough fishing shack behind beach 1447, the only structure on this entire section of coast.

Beaches **1450, 1451** and **1452** are near continuous, straighter, 800, 400 and 600 m long high tide beaches, with similar backing bluffs and fronting platforms. Beach **1453** begins at an inflection in the calcarenite-dominated shoreline. The 1.8 km long beach represents a transition, with some platform sections and some open sections and a sandy surf zone dominated by reef-controlled rips.

The Merdayerrah Sandpatch backs the main beach. The beach (**1454**) is 8.5 km long and entirely backed by the active white, calcareous dunes, that in places extend 1 km inland, climbing the backing 90 m high slopes to reach the crest of the Nullarbor Plain. Where this occurs, it provides an easy, though steep, walking track down to the shore. The beach, while partly controlled by deeper reefs, is dominated by a series of rips across the inner bar.

Swimming: All the platform beaches are unsuitable owing to the dominance of the rock platforms, while the main beach is both difficult to access and dominated by rips.

Surfing: There are beach and reef breaks on the main beach, however you will have to walk down and back up the sand dune to reach the shore.

Fishing: Most local fishers head down the tracks to fish from the rock platforms.

Summary: The most western beach system in South Australia and a fitting backdrop to the massive Nullarbor Cliffs and the border to Wilson Bluff.

5. KANGAROO ISLAND

Kangaroo Island is the second largest island on the Australian coast. It has a 458 km long coastline containing 218 beaches, which occupy 156 km or 34% of the predominantly rocky coast. The beaches average only 700 m in length, which is indicative of the overriding control of headland rocks and reefs on beach location and length. There are no surf clubs or patrolled beaches on the island and consequently care must be taken in choosing a beach for surfing or swimming, particularly as most beaches are isolated and little used. The best beaches, close to the major settlements, are Emu Bay north of Kingscote and Hog Bay at Penneshaw. The most popular, though hazardous, surfing beach is Pennington Bay. Be very careful if visiting any of the exposed southern, western and some north coast beaches, as high waves and rips predominate.

The following chapter describes each beach on the island, beginning with the northernmost beach at Cape Rouge, and continuing clockwise around the island. Figure 5.1 indicates the location of each beach.

KI-1 – KI-5 CAPE ROUGE, BAY OF SHOALS

Unpatrolled				
No.	Name	Rating	Type	Length
KI-1	Cape Rouge	2	R+sand ridges	3.5 km
KI-2	Bay of Shoals (N)	1	R+tidal flats	9.25 km
KI-3	Bay of Shoals (S)	1	R+tidal flats	4.5 km
KI-4	The Bluff	2	R+sand ridges	750 m
KI-5	Beatrice Pt (W)	1	R+sand ridges	2 km

The **Bay of Shoals** is an east facing bay located on the northeastern tip of the island. The bay has a 5 km wide mouth and 16 km of very low energy shoreline. It is bordered in the north by Cape Rouge and in the south by Beatrice Point, both of which terminate in narrow sand spits. The North Cape Road runs along the back of the northern section of the bay, while the southern shores are accessible off The Bluff Road.

Regional Map 13: Kangaroo Island

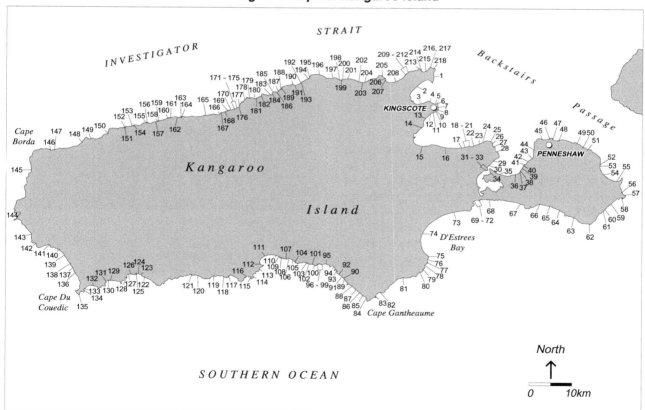

Figure 5.1 *Kangaroo Island regional map, indicating the location of the island's 218 beaches.*

Cape Rouge beach (**KI-1**) is a crenulate, east facing, 3.5 km long beach, that runs essentially due south from the northern 30 m high Point Marsden. The only access to the beach is from the North Cape Road, which runs 800 m west of the northern end of the beach. Otherwise there is no public access, as grazing land backs the beach. The beach consists of a narrow high tide beach, fronted by 100 m wide sand flats, containing two to three low ridges, then seagrass meadows. The sand flats widen off the southern tip, extending south as part of the 9 km long submerged spit which blocks the entrance to the bay.

The Bay of Shoals is circular in shape, with the northern and southern shoreline separated by a small tidal creek. The 9.2 km long northern shore faces southeast and consists of a narrow, low strip of high tide sand (**KI-2**), fronted by intertidal sand flats, which are vegetated in places and grade bayward into seagrass meadows, with seagrass debris littering the beach. The North Cape Road backs 5 km of this beach. The beach terminates at a small, 50 m wide tidal creek, on the south side of which begins the 4.5 km long, north facing southern beach (**KI-3**). This beach is similar in form, with the sand flats extending up to 200 m off the beach, before the water is deep enough for the seagrass meadows. It terminates at The Bluff; a 50 m high conical hill. The only access to this beach is on foot from The Bluff car park.

The Bluff beach (**KI-4**) is a 750 m long, northeast facing, crenulate strip of high tide sand fronted by partially vegetated, 100 m wide, ridged sand flats, then the deep tidal channel for the bay. There is a large car park and a popular boat ramp forming the eastern boundary. It protrudes out to the edge of the tidal channel, where several boats are usually moored.

To the east of the car park is a 2 km long, curving, north facing beach (**KI-5**) which terminates at the northern tip of the **Beatrice Point** sand spit. A gravel road runs from The Bluff along the back of this beach to the spit and then down to Kingscote. This protected beach is usually calm and consists of a narrow high tide beach fronted by 100 m wide sand ridges and flats, then the seagrass meadows and finally the tidal channel for the bay, which clips the northern tip of the spit.

Swimming: These beaches are generally unsuitable for swimming, owing to the shallow sand flats.

Surfing: None.

Fishing: Most fishing is done out in the bay, as the shore is too shallow.

Summary: A relatively large, low energy bay, mainly used for boating.

KI-6 – KI-13 BEATRICE POINT, KINGSCOTE, BROWNLOW

Unpatrolled				
No.	Name	Rating	Type	Length
KI-6	Beatrice Point (S)	1	R+sand flats	1.05 km
KI-7	Kingscote Jetty (N)	1	R+sand flats	900 m
KI-8	Kingscote Jetty (S)	1	R+sand flats	300 m
KI-9	Kingscote pool	1	R+pool	50 m
KI-10	Rolls Point	2	R+sand flats	300 m
KI-11	Yacht Club	1	R+sand flats	500 m
KI-12	Yacht Club (S)	1	R+sand flats	450 m
KI-13	Brownlow	1	R+sand flats	2.6 km

Kingscote is the main town on Kangaroo Island. It spreads along 2 km of shoreline that rises inland to an elevation of 50 m. The shoreline is dominated by low bluffs, along the base of which are several narrow high tide beaches fronted by wider sand flats. The entire shoreline faces east and is well protected from ocean waves, both by its orientation and by the Bay of Shoals sand spit, which extends south to end in line with the Kingscote jetty (Fig. 5.2). All the shoreline and beaches are readily accessible by vehicle.

Figure 5.2 *Kingscote is the largest town on Kangaroo Island and the main port of entry for tourists and cargo, both of which alight at the jetty. Low energy beaches lie to either side.*

The Kingscote shoreline starts at Beatrice Point 1.5 km north of the town centre. The point area was the site of the first settlement in South Australia, before Adelaide was selected. The beach (**KI-6**) runs due south for 1 km, first as a sandy spit, then below shaley, 20 m high bluffs. A road runs along the top of the bluffs, providing an excellent view of the beaches and point. The narrow high tide beach is fronted by 150 m wide sand flats. It terminates at a protrusion in the bluffs fronted by a rock groyne, which was the site of a jetty when the basalt bluffs were quarried for road base. On the other side of the groyne, beach **KI-7** continues on for 900 m to the Kingscote jetty. This is an irregular, narrow beach, backed by a seawall and also fronted by sand flats. There is a park and picnic area behind the beach, immediately north of the jetty.

On the south side of **Kingscote jetty** is a 300 m long, narrow, sand and gravel high tide beach (**KI-8**), fronted by shallow, 50 to 100 m wide sand flats, then seagrass meadows. A terraced park backs the centre of the beach, then a road that runs south to Rolls Point. The beach terminates at the Kingscote pool, which consists of a 50 m long sand beach (**KI-9**) enclosed in a pool structure. The pool is backed by a sloping grassy reserve and has dressing sheds and toilet facilities.

Immediately south of the pool is a narrow, 300 m long sand beach (**KI-10**) tucked beneath the low bluffs of Rolls Point. A road backs the bluffs and there is a walking track along the edge of the bluffs overlooking the beach. The bluffs contain remnants of a Pleistocene sand and cobble beach deposited at a former higher sea level.

Kingscote Yacht Club is located on the south side of Rolls Point. The club house and boat shed occupy part of a grassy foreshore reserve, fronted by the moderately steep beach (**KI-11**), then 200 m wide sand flats and seagrass meadows. There is a car park right behind the beach, which is used for launching boats across the beach. The 500 m long beach terminates at a low rock outcrop, beyond which a second beach (**KI-12**) continues on in a similar manner for 450 m to a second low rock outcrop. This beach is backed by a strip of more natural vegetation and a road.

Brownlow is a small beachfront settlement located 2 km south of Kingscote. The 1.5 km long settlement is fronted by a narrow high tide beach (**KI-13**), then sand flats that widen from 300 m in the north to 900 m in the south. Past the houses, the Kingscote golf course backs the beach for another 1.2 km. The beach terminates in a series of 500 m long spits that form the northern side of the Cygnet River mouth.

Swimming: The only spot suitable for swimming is the Kingscote pool, as all the beaches are fronted by shallow sand flats.

Surfing: None.

Fishing: The jetty is the most popular location, the beaches generally being too shallow.

Summary: These seven beaches border the island's major settlement. While they are all very accessible and some highly visible, they are little used for recreation, other than fishing and sailing at the yacht club beach.

KI-14   CYGNET RIVER DELTA

No.	Name	Rating	Type	Length
	Unpatrolled			
KI-14	Cygnet R delta	1	R+sand flats	1.7 km

The **Cygnet River** is the largest river on Kangaroo Island and reaches the coast on the sheltered western shore of Western Cove, resulting in a very low energy shoreline. While much of the 5 km long delta shore is tidal flats, south of the 50 m wide river mouth at Brownlow is a series of four shell beach ridges (**KI-14**) totalling 1.7 km in length, each separated by small tidal creeks. They are fronted by 400 m wide vegetated sand flats and backed by several hundred metres of scrubby high tide flats. This whole area is an important wetland and is totally undeveloped and not utilised apart from fishing in the river.

KI-15, KI-16 NEPEAN BAY, MORRISON

No.	Name	Rating	Type	Length
	Unpatrolled			
KI-15	Nepean Bay	1	R+sand flats	4.9 km
KI-16	Morrison	1	R+sand ridges	6.4 km

The southern shore of Western Cove is an 11 km long, continuous sandy beach, divided in two by a cuspate foreland, which is also the site of the 2 km long, 30 ha **Nepean Bay Conservation Park**. The beaches are backed by a series of low beach ridges, 1.5 km wide in the centre, with the entire system representing a massive regressive barrier system which has built out into the cove during the past 6 000 years. There is a settlement called Western Cove, which extends for 1.5 km along Nepean Bay, and a gravel road backing 2 km of Morrison Beach, but little other development and no facilities.

Nepean Bay Conservation Park
Established: 1974
Area: 30 ha
Coast length: 2 km
Beaches: part of KI-16 & 17
Nepean Bay Conservation Park incorporates the shoreline and outer beach ridges of the Morrison Beach barrier system.

Nepean Bay beach (**KI-15**) is a 4.9 km long, curving, north facing beach consisting of a narrow, moderately steep, shelly-sand high tide beach, fronted by 200 to 400 m of vegetated sand flats. It extends from a tidal creek in the west to the foreland, on the other side of which is the 6.4 km long **Morrison Beach (KI-16)**. Morrison Beach receives slightly higher waves, averaging about 0.5 m, which have built a 300 to 600 m wide sand flat containing six to twelve low amplitude ridges (Fig. 5.3). The ridges are apparently maintained by periodic higher waves. The beach is narrow and moderately steep and in most places backed by dense vegetation. The beach terminates at the prominent, 20 m high scarped bluffs, called Red Banks.

Figure 5.3 *Multiple low sand ridges lie along the eastern side of low energy Morrison Beach.*

Swimming: These are two safe beaches, which can only be used at high tide, owing to the shallow sand flats.

Surfing: None.

Fishing: Most of the locals use the beach to launch and moor their small tin boats and fish the deeper waters of the bay.

Summary: A low energy, quiet beach, with an equally quiet settlement at the western end.

KI-17 – KI-23 **RED BANKS**

Unpatrolled			
No.	Name	Rating Type	Length
KI-17	Red Banks (1)	2 R+sand ridges	950 m
KI-18	Red Banks (2)	2 R+sand ridges	600 m
KI-19	Red Banks (3)	2 R+sand ridges	300 m
KI-20	Red Banks (4)	2 R+sand ridges	350 m
KI-21	Red Banks (5)	2 R+sand ridges	50 m
KI-22	Red Banks (6)	3 LTT	1.1 km
KI-23	Beach KI-23	3 LTT	1.5 km

Red Banks is a prominent 20 to 30 m high, 2.5 km long escarpment lying immediately east of Morrison Beach (Fig. 5.4). The bluffs are composed of unconsolidated

Quaternary outwash material, backed by cleared farmland. There is only public vehicle access to Red Banks (3), with the other beaches requiring a hike along the shore. The bluffs are scarped and undercut by occasional high waves, and a near continuous high tide beach lies at their base, with protruding parts of the bluffs breaking the beach into six sections (**KI-17 to KI-22**).

Figure 5.4 *Red Banks is a prominent section of scarped red bluffs, fronted by low energy beaches and sand flats.*

Each of the beaches has a narrow sand and shelly high tide beach fronted by shallow sand flats, averaging 100 m off all the beaches except Red Banks (2) and (4) where they narrow to less than 50 m. Red Banks (4) is a 50 m pocket of sand contained between protruding 20 m high bluffs. The coast swings slightly to the east past the bluffs and slightly higher waves arrive at both Red Banks (6) and beach **KI-23**, resulting in low surf and a narrow, continuous attached bar.

Swimming: Generally too shallow, with the last two beaches offering slightly deeper water.

Surfing: None.

Fishing: Also too shallow for most shore-based fishing.

Summary: These are seven relatively safe, but isolated and difficult to find and access beaches. They are worth exploring for the views more than the water. However be careful as the bluffs are steep and unstable.

outcropping off the beach.

Surfing: Only occasional beach breaks at Norma Cove and Newland Bay.

Fishing: Each beach is bordered by rocky shoreline which is only used by the locals for fishing.

Summary: Five relatively isolated pocket beaches, accessible only through farm land.

KI-24 – KI-28	**POINT MORRISON, NORMA COVE, NEWLAND BAY**

Unpatrolled				
No.	Name	Rating	Type	Length
KI-24	Pt Morrison (W)	3	R+rocks	50 m
KI-25	Pt Morrison (S 1)	3	R+rock flats	120 m
KI-26	Pt Morrison (S 2)	3	R+rock flats	150 m
KI-27	Norma Cove	5	TBR+rocks	500 m
KI-28	Newland Bay	5	TBR+rocks	250 m

Point Morrison forms the northern tip of a 5 km wide, 100 m high bedrock plateau that separates Western Cove from Eastern Cove. The 50 m high point is composed of Cambrian sedimentary and metasedimentary rocks, which also dominate the shore south into American River. Five beaches are found around the point; four backed by cleared grazing land and farms with no public access, while Newland Bay is backed by dense vegetation with farmland on the backing plateau.

Beach **KI-24** lies 2 km west of the point, and consists of a 50 m long pocket of high tide sand bordered and partly fronted by bedrock, with a small creek draining across the beach.

Two small beaches lie within 1 km of Point Morrison. The first beach (**KI-25**) is immediately east of the point and extends to the south for 120 m. It is bordered by bedrock and boulders, and consists of a sandy high tide beach grading into a more rocky low tide beach. A small creek fed by a farm dam drains across the beach. The second beach (**KI-26**) is a 150 m long pocket of high tide sand fronted by continuous low tide rocks, with a stream crossing the northern end of the beach. It is bordered by grassy, 40 m high sloping bedrock.

Norma Cove (KI-27) is a 500 m long, northeast facing beach with a sand high tide beach and bar, however they are separated by a band of low tide rocks. During periodic higher waves up to five rips form along the bar, in addition to rips running out against the rocks. A small stream and elongate lagoon back the northern end of the beach, with cleared slopes to either side.

Newland Bay (KI-28) is a 250 m long, north facing strip of high tide sand, with low tide rocks then a sandy bar, which is cut by two beach rips when waves are breaking, with headland-controlled rips to either side. The beach is backed by steep, densely vegetated slopes rising to 100 m, with a 4WD track leading down to the northern end of the beach.

Swimming: None of these beaches are suitable for safe swimming, owing to their isolated location and particularly the prevalence of rocks both bordering and

KI-29 – KI-33	**AMERICAN RIVER, BUICK POINT**

Unpatrolled				
No.	Name	Rating	Type	Length
KI-29	American R (N)	2	R+tidal flats	1.95 km
KI-30	American R	2	R+tidal flats	1 km
KI-31	Buick Pt (N)	2	R+tidal flats	750 m
KI-32	Buick Pt (S 1)	2	R+tidal flats	120 m
KI-33	Buick Pt (S 2)	2	R+tidal flats	160 m

American River settlement spreads along the southwestern shore of Eastern Cove, facing into the American River inlet, which connects Pelican Lagoon with the cove. This is a low energy, southeast facing shoreline, well protected from waves and westerly winds by its orientation and location deep inside the cove.

Ballast Head forms the western boundary of the cove, and extending south of the head is the first American River beach (**KI-29**). It is a 2 km long, irregular, discontinuous strip of high tide sand that follows the base of the 80 m high, densely vegetated bluffs. The southeast facing beach is fronted by 50 m wide intertidal rock flats, then partly vegetated sand flats extending for another 150 m. The beach terminates at a small point which was the site of a long jetty, only the ruins of which now remain. The second beach (**KI-30**) continues on in a similar manner for another 1 km, only its slopes are occupied by a tree-covered foreshore reserve, the main road and the American River settlement.

Buick Point is a cuspate-shaped sandy point that forms the western side of the American River inlet. Between the low point and the rocky bluffs to the west is a 750 m long sand beach (**KI-31**) fronted by 250 m wide tidal flats in the north, which narrow toward the deep, 100 m wide tidal channel at the point. Most of the beach is backed by a grassy foreshore reserve, which in the north abuts a small creek and salt marsh, while in the south is the main jetty for the settlement, with some boats moored in the channel.

On the south side of the jetty are two small beaches (**KI-32** & **KI-33**), 120 m and 160 m long respectively, which form the sandy southern shore of Buick Point. They are separated in the centre by a boat ramp that extends 60 m across the fronting tidal flats to reach the deeper water of the tidal channel. The main road runs right round the point, providing good access to all three beaches.

Swimming: While conditions are usually calm at these beaches, they are generally unsuitable for swimming owing to the shallow and often exposed tidal flats.

Surfing: None.

Fishing: The jetty, boat ramp and tidal channel are the most popular spots.

Summary: American River is a relatively quiet settlement, with usually placid waters fronting the beaches, apart from the tidal currents which flow through the inlet.

KI-34 – KI-38 STRAWBRIDGE PT, ISLAND BEACH, ROCKY PT, BROWN BEACH

Unpatrolled				
No.	Name	Rating	Type	Length
KI-34	Strawbridge Pt	1	R+tidal flats	1.7 km
KI-35	Island Beach	3	LTT(TBR)	5.5 km
KI-36	Rocky Point	3	LTT(TBR)	2.2 km
KI-37	Brown Beach	3	LTT+rock flats	300 m
KI-38	Brown Beach (N)	3	LTT+rock flats	200 m

The southern and eastern shore of Eastern Cove is occupied by more exposed beaches which, during strong westerly winds, receive low to moderate wind waves, capable of producing low surf and small rips on some of the beaches.

Strawbridge Point lies directly opposite the American River settlement and forms the sandy eastern side of the inlet. Sweeping south of the point for 1.7 km is a low energy sandy beach (**KI-34**) fronted by tidal flats, which widen from 100 m at the point to 350 m at the southern end. The deeper tidal channel fronts the flats. The Island Beach Road terminates at the point, with 4WD access possible along the beach. A small tidal creek also drains across the tip of the point.

Island Beach (**KI-35**) is a 5.5 km long, northeast to north facing beach that extends from Strawbridge Point to low Rocky Point in the east. A gravel road backs the beach, servicing a 2 km long stretch of houses and shacks along the centre of the beach, in addition to numerous pedestrian and vehicle access tracks. The low gradient sandy beach is fronted by a 100 m wide surf zone consisting of rips spaced every 50 to 100 m along the shore, with two more rhythmic bars seaward. These bars and rips are inactive

much of the time, with waves only breaking and rips flowing when waves exceed 1 m.

Rocky Point beach (**KI-36**) is a 2.2 km long, north to northwest facing sand beach, fronted by a continuous low tide bar (Fig. 5.5), which is cut by several rips during periodic higher waves. Toward the northern end the bar is replaced by intertidal rock flats. The main road runs 200 m behind the beach and, while there are private houses on the point, the best public access is at the northern Brown Beach parking and camping area.

Figure 5.5   *The usually low energy beach at Rocky Point.*

Brown Beach consists of two calcarenite bluff-controlled beaches; the first (**KI-37**) runs for 300 m north of the car park which overlooks the beach to a 10 m high rocky bluff. On the north side of the bluff, the second beach (**KI-38**) continues on for another 200 m to a second low bluff. Both beaches are backed by a foredune and the Brown Beach camping area, which is nestled above the bluffs in amongst the trees. The beaches have a sandy high tide beach and bar, interspersed with low tide rock flats.

Swimming: Island and Rocky Point beaches are some of the better and safer beaches on the island, particularly under normal low wave conditions, and while the Brown beaches have lower waves, be careful of the rock flats.

Surfing: Only during strong westerly winds.

Fishing: Best off Strawbridge Point into the tidal channel, and in the small rip holes following bigger waves.

Summary: Five accessible beaches offering the long, north facing Island Beach and more bluff-controlled Rocky Point and Brown beaches, all with usually low wave conditions.

KI-39 – KI-43   **AMERICAN BEACH**

Unpatrolled				
No.	Name	Rating	Type	Length
KI-39	American Beach (1)	3	R+sand flats	1.1 km
KI-40	American Beach (2)	3	R+sand flats	120 m
KI-41	American Beach (3)	3	LTT	800 m
KI-42	American Beach (4)	4	R+rocks	800 m
KI-43	American Beach (5)	3	R/LTT	1.3 km

American Beach refers to both a beach and a small holiday settlement nestled on the bluffs 2 km south of the beach (Fig. 5.6). Surrounding the settlement are five low wave beaches. The beaches and settlement are serviced by the main Hog Bay Road, which runs 100 to 300 m inland.

Figure 5.6   *American Beach is a small subdivision with low energy beaches to either side.*

Beach **KI-39** borders the western side of the settlement. It runs north for 1.1 km as a low energy, undulating sand beach, fronted by a bar which widens to the north, to 100 m wide sand flats in lee of a small vegetated islet off High Dune Point. A foredune backs the beach, with a few houses behind the northern half of the beach.

Beach **KI-40** is directly in lee of the islet and is a 120 m pocket of sand lying between the sandy foreland formed in lee of the islet and the rocky bluffs fronting the northern side of the settlement. The beach is fronted by a 50 m wide mixture of sand and rock flats, and backed by a 10 to 15 m high foredune, then the settlement.

East of **High Dune Point**, the coast faces northwest and is dominated by bluffs for 2 km, along the base of which are two 800 m long high tide sand beaches. The first (**KI-41**) forms the northern boundary of the settlement and is the site of a boat ramp, located at the end of the main settlement road. Three metre high, eroding calcarenite bluffs back the narrow high tide beach, which is fronted by a narrow sand bar. Deep Creek drains across the eastern end of the beach. East of the creek the bluffs continue

for another kilometre, along the base of which is an 800 m long, more discontinuous narrow high tide beach (**KI-42**).

American Beach (**KI-43**) lies 2 km northeast of the settlement and consists of a straight, 1.3 km long sand beach, fronted by a narrow bar, then seagrass meadows. It is bordered by low calcarenite bluffs in the west, by 50 m high sandstone bluffs in the east, and is backed by a 10 m high foredune, a grassy car park and the main Hog Bay Road.

Swimming: These are five relatively low wave beaches which are better at high tide, owing to the shallow bars and sand flats exposed at low tide. Watch for rocks and rock flats on some of the beaches.

Surfing: None.

Fishing: Better off the bluffs at high tide.

Summary: American Beach is a small settlement with no facilities other than the boat ramp, and the surrounding beaches.

KI-44, KI-45 **CONGONY, KANGAROO HEAD**

Unpatrolled				
No.	Name	Rating	Type	Length
KI-44	Congony	3	R+rocks	40 m
KI-45	Kangaroo Head	5	R+rock flats	120 m

Kangaroo Head is a 50 m high, sloping sandstone headland that forms the eastern entrance to Eastern Cove. Rocky coast extends south of the head for 4 km to American Beach and east for 2 km to Penneshaw. In amongst the rocks are two small beaches. To the south is 40 m long **Congony Beach (KI-44)**, a pocket of high tide sand fronted by a low tide boulder beach, with a patch of sand offshore. A vehicle track leads to a solitary fishing shack at the beach, which is surrounded by sloping farm land. One kilometre east of the head is a 120 m long high tide sand beach (**KI-45**) fronted by a 50 m wide intertidal rock platform and deep water. Neither beach is accessible to the public and Kangaroo Head beach in particular is hazardous owing to the rock platform.

KI-46, KI-47 **CHRISTMAS COVE, HOG BAY**

Unpatrolled				
No.	Name	Rating	Type	Length
KI-46	Christmas Cove	2	R+sand flats	300 m
KI-47	Hog Bay	4	LTT	600 m

Hog Bay is the main beach for **Penneshaw**, the major point of entry for vehicles and most people to Kangaroo Island. The western end of the beach terminates at the small breakwater built for the ferry terminal, and consequently it is the first beach seen, and often the first stop, for most visitors to the island.

The small town of Penneshaw sits on 20 m high bluffs, the eastern side of which overlook Hog Bay. The bay was initially discovered by the French in 1802. Flinders also landed here and called it Spring Cove. However, the name Hog Bay came from the early sealers who settled here and obtained many wild pigs (hogs) in the area.

Christmas Cove (KI-46), on the western side of the town, has been converted into a small boat harbour through the construction of two small harbour walls and a dredged basin (Fig. 5.7). The 300 m long high tide beach curves around the shore of the very protected, almost circular bay, surrounded by 20 m high bluffs. The bay was originally filled with the fronting sand flats, the western side of which have been dredged for a boat ramp and moorings. The cove is an important geological site, with Palaeozoic glacial erratics lying on the beach.

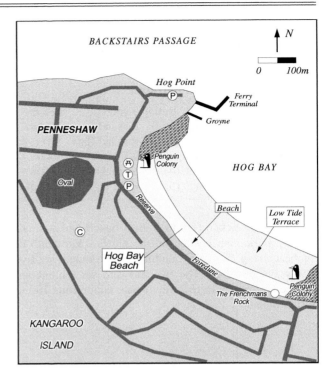

Figure 5.8   *Penneshaw is located on Hog Point, overlooking Hog Bay. It is the site of the ferry terminal and the popular Hog Bay beach.*

Figure 5.7   *Christmas Cove is a well protected, north facing cove that has been dredged and is partly protected by a breakwater, used for launching and mooring small boats.*

Hog Bay beach (**KI-47**) is 600 m long and faces the northeast. It is located between 20 m high Hog Point in the west, and the rocks and 10 m high grassy bluffs of Frenchmans Rock in the east (Figs. 5.8 & 5.9). There is a reserve and low foredune running the length of the beach, with toilets and picnic facilities at the western end, and a monument to the French explorers at the eastern end. A road backs the length of the beach, providing good access and parking. Penguin colonies occupy the small dune and rocks at each end.

Figure 5.9   *Hog Bay is the most popular swimming beach on the north coast of Kangaroo Island, and is located adjacent to the second main ferry terminal (right) at Penneshaw.*

The beach receives waves averaging 0.5 m which, together with the fine beach sand, produces a low, 50 m wide high tide beach, fronted by a 100 m wide, shallow attached bar. Rips are usually absent, only occurring when waves exceed 1 m.

Swimming: Christmas Cove is unsuitable for swimming, owing to the shallow sand flats and boat traffic. Hog Bay is usually a safe beach, so long as you stay on the bar. Strong tidal currents are encountered on the outer perimeter of the bay.

Surfing: Waves are usually too low to surf, with rideable waves only occurring during big swell and strong northerly winds.

Fishing: The jetty is the most popular fishing spot, together with the rocks at high tide.

Summary: An attractive and relatively safe beach at the main entrance to Kangaroo Island; the island's most popular and most visited beach.

KI-48 – KI-53 ALEX LOOKOUT, CUTTLEFISH BAY, SNAPPER PT, CAPE COUTTS

Unpatrolled				
No.	Name	Rating	Type	Length
KI-48	Beach KI-48	4	R+rocks	40 m
KI-49	Alex Lookout (S)	5	R+rocks	30 m
KI-50	Cuttlefish Bay	3	R	50 m
KI-51	Snapper Pt (W)	4	R/cobble	20 m
KI-52	Cape Coutts (S1)	5	LTT	30 m
KI-53	Cape Coutts (S2)	5	LTT	30 m

East of Hog Bay the coast is dominated by metasedimentary bedrock, which forms a rocky shoreline and rises as vegetated slopes to heights of 150 to 200 m. These rocks extend for 15 km east and then southeast to the Chapman River mouth. Along this entire section are only six small pockets of sand, totalling just 200 m in length. All are backed by usually cleared farmland, with no public access.

The first beach (**KI-48**) is just 1 km east of Hog Bay. It is a 40 m long pocket of high tide sand and cobbles fronted by mid to low tide rocks and boulders. This beach can be reached on foot from Hog Bay.

Alex Lookout is a 100 m high headland backed by grassy slopes. On the south side of the headland, set in a steep, V-shaped grassy valley with high ridges to either side, is a 30 m long high tide cobble beach (**KI-49**) fronted by sloping intertidal rocks.

Cuttlefish Bay (**KI-50**) is a northeast facing gap in a section of high rocky coast. It contains a 50 m long sandy beach, backed by a steep grassy valley, with a solitary shack behind the beach, and a small creek running down the western side.

Snapper Point is a prominent, 160 m high grassy headland. Five hundred metres east of the headland is a 50 m wide opening in the rocks, at the base of which is a 20 m long sand-cobble beach (**KI-51**) fronted by rocks and boulders.

At **Cape Coutts** the bedrock coast turns a right angle and trends due south into Antechamber Bay. Five hundred metres before the sandy bay beach is reached are two pockets of white sand (**KI-52, KI-53**) at the foot of densely vegetated, 100 m high bedrock slopes. These are essentially an extension of the Antechamber Bay beach, joined in the sandy nearshore, but separated by bedrock at the shoreline. The two 30 m long beaches are both bordered by steeply dipping rocks.

Summary: None of these five beaches are easily accessible. All are backed by farmland and steep slopes and, while calm conditions often prevail, their isolated location and dominance of rock make them unsuitable for safe swimming.

Lashmar Conservation Park
Established: 1996
Area: 188 ha
Beach: KI-54

Lashmar Conservation Park straddles the mouth of the Chapman River and incorporates the recreation and camping reserve on the north side of the mouth.

KI-54, KI-55 ANTECHAMBER BAY, RED HOUSE BAY

Unpatrolled				
No.	Name	Rating	Type	Length
KI-54	Antechamber Bay	4	LTT	4.2 km
KI-55	Red House Bay	3	R-LTT	2.3 km

Antechamber Bay is an open, northeast facing bay, located at the eastern extremity of the island. It is bordered by Cape Coutts and Cape St Albans, and contains two sandy beaches. The Cape Willoughby Road provides access to the western end of the bay at the Chapman River mouth picnic and camping area (Fig. 5.10), while the central and eastern section is backed by private farm land.

Figure 5.10 *The blocked Chapman River mouth (centre) lies at the north end of Antechamber Bay and is a popular camping and swimming area.*

Antechamber Bay beach (**KI-54**) is a 4.2 km long, northeast facing sand beach. It extends from the southern rocks of Cape Coutts, past the usually closed mouth of the Chapman River, down to a 50 m high bluff that separates it from Red House Bay. Most of the beach is backed by a 10 m high foredune. It usually receives waves less than 1 m high and has a moderately steep beach fronted by a narrow continuous bar then seagrass meadows, with rips absent unless waves exceed 1 m. The Chapman River crosses the northern end of the beach, with the mouth usually blocked. A recreation reserve is located along the sides of the river mouth and includes two car parks, a picnic area and a camping area.

Red House Bay occupies the eastern third of the larger Antechamber Bay. The bay shore contains a 2.3 km long, north facing, lower energy sandy beach (**KI-55**), with a steep beach, no bar and seagrass meadows just off the beach. Seagrass debris often accumulates along the beach. The beach is backed by a steep, 20 to 30 m high bluff, then grassy slopes rising to 100 m.

Swimming: The Chapman River mouth and adjoining beach is a popular summer and holiday camping and swimming area. The beach is relatively safe under normal low wave conditions.

Surfing: Usually no surf.

Fishing: The river when closed or open is always a popular spot, as well as the rocks at the northern end of the main beach.

Summary: An isolated but nonetheless relatively popular spot during the holiday periods.

KI-56 – KI-61 CAPE WILLOUGHBY to CAPE HART

Unpatrolled				
No.	Name	Rating	Type	Length
KI-56	Pink Bay	3	LTT/TBR	40 m
KI-57	Cape Willoughby	4	Boulder+TBR	150 m
KI-58	Windmill Bay	10	Boulder	100 m
Cape Hart Conservation Park				
KI-59	Cape Hart (1)	10	Boulder	200 m
KI-60	Cape Hart (2)	10	Boulder	80 m
KI-61	Cape Hart (3)	10	Boulder	40 m

Cape Willoughby is the easternmost point of the island. It is a protruding, 50 m high metasedimentary headland, capped by the Cape Willoughby lighthouse and its three lighthouse-keeper houses. The now automatic light is a major tourist attraction and viewing point for the eastern end of the island. The rugged coast to either side is predominantly composed of steep, 100 m high vegetated bluffs in Moncrieff Bay to the north, while to the south

is 10 km of generally sheer, 60 to 80 m high cliff down to Cape Hart.

There are two small beaches along the 5 km long Moncrieff Bay shoreline. **Pink Bay** (**KI-56**) is a 40 m pocket of sand wedged inside a narrow, V-shaped bay. It faces north and is somewhat protected, but does receive low swell. On the north side of the lighthouse is a second beach (**KI-57**), a 150 m long high tide boulder beach fronted by a sand bar and surf, with usually one rip draining the small beach.

On the south side of the lighthouse is the exposed southern shore. This section is devoid of sand and contains only four small boulder beaches in amongst the cliffs. The first is in **Windmill Bay**, adjacent to the cape, which houses a 100 m long, large boulder beach (**KI-58**), which usually has a heavy surf (Fig. 5.11).

Figure 5.11 *Windmill Bay (foreground) has an exposed boulder beach, which gives way to the 100 m high cliffs of the Cape Hart Conservation Park.*

Cape Hart Conservation Park
Established: 1973
Area: 1030 ha
Coast length: 2.5 km
Beaches: KI-59 & 60
Cape Hart Conservation Park incorporates both the 100 m high headland of the cape and a dramatic, 2.5 km long section of rugged coast, which contains two small boulder beaches.

Two kilometres to the south is the **Cape Hart Conservation Park**, a 1030 ha park covering 2 km of cliffed coast and the backing bluffs. There are two small beaches in the park and one just to the south. The first (**KI-59**) is located just south of the northern park boundary and is a 200 m long boulder beach at the base of 70 m high bluffs. A 4WD track leads to the top of the cliffs just north of the beach, with a steep track down the vegetated bluffs and over the rocks to the beach. The second (**KI-60**) is located just inside the southern boundary, with the boundary track leading to a parking area above the southern end of the 80 m long, south facing boulder beach. The third (**KI-61**) is located 500 m south of the track, and is a 40 m long boulder beach surrounded by 60 m high vegetated bluffs.

Swimming: The two Moncrieff Bay beaches are relatively safe when waves are low, however watch for rips if there is any surf. All the boulder beaches are extremely hazardous.

Surfing: There is only a chance of a surf at the northern Cape Willoughby beach in a larger swell.

Fishing: There are numerous rocks around the cape, however be very careful owing to the usually high swell breaking over the rocks.

Summary: Most visitors come to the cape to view the scenery and visit the lighthouse, with little use of the beaches.

KI-62 – KI-66 **FALSE CAPE, BLACK HEAD, MURRAY FLAT**

Unpatrolled					
No.	Name	Rating	Type Inner	Outer Bar	Length
KI-62	False Cape	10	R+platform		500 m
KI-63	False Cape Beach	9	RBB	RBB	2.2 km
KI-64	Black Head	9	RBB	RBB	3 km
KI-65	Murray Flat	9	RBB	RBB	300 m
KI-66	Beach KI-66	10	R+platform		200 m

At **False Cape** the rocky coast turns west and faces directly into the high southwesterly waves and wind. The entire southern coast of Kangaroo Island is a high energy, wave and wind-dominated shoreline, with high energy rip-dominated beaches, backed by massive sand dunes extending in places several kilometres inland and to elevations of well over 100 m. Between False Cape and the Wilson River, 10 km to the east, are four exposed, high energy beaches, typical of this coastline.

The first lies at the base of 50 to 80 m high, cleared bluffs 1 km east of the cape. It is a high tide sand and cobble beach (**KI-62**), backed by a small foredune and fronted by a 50 m wide rock platform, composed of steeply dipping sandstone.

False Cape Beach (KI-63) runs for 2.2 km from the 50 m high cape to Black Head, both of which protrude about 400 m seaward. The south facing sand beach is backed by massive vegetated Holocene dune transgressions that have moved several kilometres inland. Pleistocene dune calcarenite outcrops on the beach and in the surf forming reefs. The 2 m high waves maintain a 300 m wide double bar system, with up to seven strong

beach and reef-controlled rips along the beach. Toward the western end, the calcarenite becomes more dominant as the bluffs also encroach on the beach. A solitary 4WD track terminates above the western bluffs.

Black Head is a prominent, 500 m wide, 40 m high sandstone headland. A 3 km long beach (**KI-64**) runs due east of the head and terminates at a small protruding bluff fronted by calcarenite reefs. The beach receives waves averaging 2 m which maintain rip-dominated inner and outer bars, together with additional reef-controlled rips in the centre and western end of the beach. During normal wave conditions, the rip heads can flow up to 1 km seaward of the beach. Three 4WD tracks wind through the backing vegetated dunes to the rear of the beach and western bluffs.

The **Wilson River** has cut a deep, V-shaped valley which has been partially filled by river, beach and dune sands to form the 300 m long Murray Flat beach (**KI-65**). The beach is accessible via a road, then a farm track which follows the Wilson River and terminates on the western headland, providing excellent views of the beach and surf. Waves average 2 m and, together with the beach and boundary rocks and headland, maintain two strong permanent rips which flow well seaward (Fig. 5.12). The river is usually closed, but when open provides an additional hazard.

Figure 5.12   *Murray Flat beach, at the mouth of the Wilson River, is exposed to persistent high swell and usually has strong rips crossing the wide surf, such as the one seen here in the foreground.*

Two kilometres west of the river mouth is a 200 m long high tide sand beach (**KI-66**), fronted by a continuous, 50 m wide calcarenite rock platform and deeper reefs. Waves break over the reefs up to 200 m off the beach, with a large rip draining the surf. A 4WD track passes 100 m behind the beach and its backing dunes.

Swimming: These are five very hazardous beaches. The three sand beaches should not be used for swimming, unless rare calm conditions prevail, but even then deep rip holes, gutters and reef abound.

Surfing: There are energetic breaks across the inner and outer bars along three beaches, but beware as there are strong rips and reefs.

Fishing: These are all excellent locations for rock and beach fishing, with persistent and permanent deep holes and gutters along the beaches.

Summary: Four remote beaches, mainly used by local fishers and occasional surfers.

KI-67, KI-68 **SANDHURST, BEACH KI-68**

No.	Name	Rating	Type	Length
Unpatrolled				
KI-67	Sandhurst	9	TBR+reefs	1.8 km
KI-68	Beach KI-68	10	R+platform	200 m

Sandhurst beach (**KI-67**) is a 1.8 km long, south facing beach which has supplied the dune sand for the extensive Cave Sandhills, which back the beach and eastern bluffs. The beach is bordered by 20 to 30 m high calcarenite bluffs, with vegetated dunes climbing to 50 m and extending 3 km inland backing most of the beach. Two 4WD tracks wind through the dunes to the centre and eastern bluff. The beach consists of a sandy swash and high tide beach, fronted by a near continuous, 50 m wide calcarenite reef, then a 100 m wide surf zone dominated by several large rips, including permanent rips against the boundary rocks. At the eastern end of the beach is a 100 m long pocket of sand fronted by tidal pools in the reef and then surf.

Beach **KI-68** is a 200 m long high tide sand beach fronted by an irregular, 50 m wide reef, then a 200 m wide surf zone dominated by a strong rip system. This beach lies 1 km south of the Hog Bay Road, with a 4WD track leading from the road to the rear of the beach.

Swimming: These are two hazardous, reef and rip-dominated beaches and are unsuitable for swimming.

Surfing: Although waves are large, neither is suitable for safe surfing owing to the reefs bordering the beaches.

Fishing: Beach KI-68 is a readily accessible and more popular spot for locals to fish the deep holes in the reefs off the beach. Sandhurst has similar holes but is more difficult to access.

Summary: Two exposed wave, reef and rip-dominated beaches.

KI-69 – KI-72   **PENNINGTON BAY**

No.	Name	Rating	Type	Length
Unpatrolled				
KI-69	Pennington Bay (1)	9	TBR+rocks	200 m
KI-70	Pennington Bay (2)	8	TBR+rocks	250 m
KI-71	Pennington Bay (3)	7	TBR	500 m
KI-72	Pennington Bay (4)	8	TBR	350 m

Pennington Bay is the main surfing beach for Kangaroo Island. It is located midway between the two major settlements of Kingscote and Penneshaw and is readily accessible off the main road that links the towns. A gravel road runs to the calcarenite bluffs above the centre of the beach, where there is a car park with steps leading down to the main swimming and surfing area (Fig. 5.13). There are no other facilities.

Figure 5.13 *Pennington Bay is the most popular surfing beach on Kangaroo Island, however be wary if swimming as strong rips persist.*

The beach faces almost due south. It is 1.3 km long and bordered by prominent, 30 to 60 m high calcarenite headlands, with western Point Reynolds extending 3 km seaward. In addition, 20 m high calcarenite bluffs, with fronting rock platforms and reefs, divide the beach into four sections (Fig. 5.14).

The easternmost section (**KI-69**) is 200 m long and has a narrow high tide beach, fronted by a rock platform. The waves break on a bar seaward of the platform, before finally closing out on the rocks.

The second section (**KI-70**) is located immediately below the car park. It is a 250 m long sandy pocket below the bluffs, with rocks scattered over the beach and in the surf. A single strong rip usually flows out against the western rocks. This is a more sheltered and popular spot for sunbathers, but very hazardous in the surf.

The main beach section (**KI-71**) is reached by wooden steps leading down from the car park. It is 500 m long and backed by a steep foredune and sand dunes. This is the most popular surfing beach, with surfers paddling out in the strong, permanent rip that runs out against the eastern rocks.

Summary:　A picturesque, natural beach, well worth visiting just for the views; and a popular, if hazardous, surfing and swimming spot. There are no facilities at the beach, so bring anything you need.

KI-73, KI-74 **FLOUR CASK BAY, D'ESTREES BAY**

Unpatrolled			
No.	Name	Rating	Type	Length
KI-73	Flour Cask Bay	8	RBB-TBR	6.3 km
KI-74	D'Estrees Bay	7→2	TBR→R	12.3 km

D'Estrees Bay is a 15 km wide, southeast facing, open bay that has 18 km of sandy shoreline between Cape Reynolds and Point Tinline. Wave energy is high at Point Reynolds, gradually decreasing to near zero along the southern third of the bay, where seagrass meadows grow right to the shore. A gravel road reaches the southern end of the bay, providing public access to the shoreline. Apart from shoreline camping sites, there are no facilities.

The first 6 km of the bay is called **Flour Cask Bay**. This section has continuous, 20 to 30 m high, dune-capped calcarenite-bedrock bluffs fronted by a narrow beach (**KI-73**), which varies in width depending on the extent of protrusion of the bluffs and is often awash at high tide (Fig. 5.15). The beach receives high southwest swell and is fronted by a 300 m wide surf zone dominated by large rips approximately every 400 m, together with a few reefs and a permanent rip against Point Reynolds. There is 4WD access to the top of the bluffs, but otherwise the beach is largely inaccessible.

A protruding bluff separates Flour Cask Bay from the remainder of D'Estrees Bay beach (**KI-74**). The beach initially faces southeast and receives waves averaging over 1.5 m high, which maintain several rips, in addition to numerous reefs in the surf zone. As the shore swings to face due east, and a shallow, seagrass-covered bay floor fronts the beach, ocean waves gradually decrease to zero (Fig. 5.16). Occasional storms pile seagrass on the beach, which is a persistent feature of the southern half of the beach.

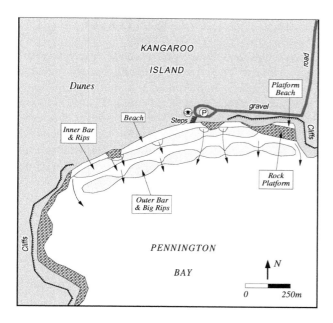

Figure 5.14　*Pennington Bay is bounded and cut by calcarenite cliffs and reefs into four beach sections. Both permanent headland rips and beach rips cut the inner bar and the continuous outer bar.*

The western section (**KI-72**) can be reached on foot along the main beach and around a low bluff. It is a 350 m long beach with bluffs at each end and a steep dune behind. Strong permanent rips run out against both bluffs.

Swimming:　This is a very hazardous beach even under the normal waves, which average 1.5-2 m. Strong permanent rips are a feature of all sections, with very strong rips against all the bluffs, together with rocks and reefs in parts of the surf zone, particularly to the east. Use extreme care if swimming here. Stay well inshore, on the shallow parts of the bars and well clear of the rips. Stick to the car park area where people can see you if you get into trouble.

Surfing:　This is the premier surfing spot on the island, owing to its good access and usually good bar and reef breaks. There are a number of breaks surfed at all four sections. However surfers should also use caution as the rips are very strong and some waves break onto shallow reefs.

Fishing:　A very popular spot for both rock and beach fishing. Deep rip gutters and holes are a permanent feature. However beware of large waves when fishing off the rocks.

The D'Estrees Bay Road reaches the bay about halfway along the beach and follows the southern 6 km of shoreline, providing access to the shore and informal camping sites (permit required), however there are no facilities, other than a store 2 km back along the access road. The road continues on past Point Tinline to the southern beaches.

Swimming:　Flour Cask Bay and northern D'Estrees Bay are wave, rip and reef-dominated and are unsafe, while the southern half of D'Estrees Bay is too shallow for swimming and dominated by seagrass meadows and debris.

Surfing: There are remote, energetic and dangerous reef breaks along Flour Cask and northern D'Estrees Bays, with the safest surf along the central D'Estrees Bay.

Fishing: The best beach and reef fishing is along Flour Cask Bay, however access is difficult.

Summary: A long and variable bay ranging from high to very low energy beaches, with little development other than the access road to the southern half of the bay.

Figure 5.15 *Flour Cask Bay is exposed to high waves, which break along an irregular beach, backed and occasionally crossed by calcarenite-capped bedrock bluffs and rocks.*

Figure 5.16 *The southern end of D'Estrees Bay is typified by calm conditions and seagrass debris littering the beach.*

KI-75 – KI-79   **OSMANLI REEF, POINT TINLINE**

Unpatrolled			
No	Name	Rating Type	Length
KI-75	Osmanli Reef	3 R+reef flats	350 m
KI-76	Pt Tinline (1)	3 R	950 m
KI-77	Pt Tinline (2)	8 R+platform	150 m
KI-78	Pt Tinline (3)	3 LTT→R	1.3 km
KI-79	Pt Tinline (4)	3 R+reef	300 m

At Point Tinline the coast turns and trends south-southwest for 5 km to Cape Linois. While exposure to the southerly swell increases south of the point, the presence of numerous bedrock reefs off the coast lowers waves at the shoreline. There are five beaches along this section, all linked by the gravel road from D'Estrees Bay. The first two lie on either side of Point Tinline.

Osmanli Reef lies 400 m north of Point Tinline and in its lee is a 350 m long, east facing beach (**KI-75**), bordered by the narrow, protruding point in the south and reef flats which lie off most of the beach. The curving beach is backed by crumbling, 10 m high bluffs, with the bay road running along the top of the grassy bluffs.

Point Tinline is a low, irregular jumble of rocks that extends 100 m to the east. On its south side is a 950 m long, curving beach (**KI-76**), that terminates at a 20 m high headland and is fronted by continuous reefs of varying depth. The reefs lower waves to about 0.5 m at the shore, where there is a continuous sandy beach fronted by deeper sand and reef. The beach is backed by a 5 to 10 m high foredune, in lee of which runs the bay road.

The first headland south of Point Tinline occupies 500 m of shore, with an embayed central section which contains a 150 m long sandy high tide beach (**KI-77**), fronted by the 100 m wide rock platform of the headland. The beach is stranded at low tide, while at high tide waves break across the jagged rocky platform.

Beach **KI-78** extends for 1.3 km south of the second headland, to a third low headland fronted by a 300 m wide rock platform, covered by numerous boulders. The beach faces east and is protected from high waves by the southern platform and reefs off the centre of the beach. Waves average less than 1 m at the shore, where they maintain a crenulate beach fronted by a continuous bar which narrows to the south.

Beach **KI-79** is the southernmost beach along this section of the bay and lies at the end of the 2WD road from D'Estrees Bay. The 300 m long, curving beach is set within a nearly circular embayment, with headlands and rock platforms along the side and reef across the 200 m wide mouth. As a result, waves average about 0.5 m at the beach, maintaining a narrow, moderately steep beach fronted by patches of sand and shallow rock reefs.

Swimming: The best swimming is along the three southern sandy beaches (KI-76, 77 and 78), where there is some sand off the beach and low waves.

Surfing: The best chance is on the reef breaks just north of Point Tinline.

Fishing: A popular spot for rock and beach fishing, with deep holes in the reefs off the beach.

Summary: Four readily accessible, lower energy, natural beaches with no facilities.

KI-80 – KI-83   CAPE LINOIS to CAPE GANTHEAUME

Unpatrolled			
No. Name	Rating	Type	Length
KI-80 Cape Linois (N)	10	Boulder+platform	500 m
KI-81 Cape Linois (S)	10	Boulder+rock reefs	400 m
KI-82 Cape Gantheaume (1)	10	R+reefs	400 m
KI-83 Cape Gantheaume (2)	5	R+rock reefs	100 m

South of **Cape Linois** is the wilderness zone of the conservation park, and the gravel road becomes a walking trail. This trail follows 15 km of spectacular, 50 to 100 m high sea cliffs out to Cape Gantheaume, the southernmost tip of the island. The coast is composed of steeply dipping

sandstone rocks for the entire distance, with only four 'beaches' lying in small embayments at the base of the cliffs.

KI-80 is a 500 m long boulder beach lying at the base of 40 m high bluffs, 1.5 km northeast of Cape Linois. It is fronted by a 50 to 100 m wide rock platform. Beach **KI-81** lies 1 km west of the cape and consists of two crescents of boulder beaches, totalling 400 m in length, backed by 70 m high cliffs and fronted by discontinuous rock platforms and reefs (Fig. 5.17). There is no foot access to this beach.

Figure 5.17   *Beach KI-81 consists of two crescentic boulder beaches backed by 40 m high bluffs. It typifies this exposed, rocky section of coast east of Cape Linois.*

Beach **KI-82** lies 3.5 km north of Cape Gantheaume. It consists of two 200 m long pockets of sand and cobbles, fronted by a mixture of sand and reefs. High waves break across the reefs and sand and maintain two strong permanent rips. The beach is inaccessible on foot owing to the backing 80 m high cliffs. Beach **KI-83** is a 100 m pocket of east facing sand, protected by a 40 m high, 300 m long headland. As a consequence it receives low waves, however it is fronted by continuous intertidal rock reefs.

Summary: These are four dangerous beaches that are difficult to impossible to access. Worth viewing from the clifftop, but otherwise very hazardous.

KI-84 – KI-88   CAPE GANTHEAUME

Unpatrolled			
No. Name	Rating	Type	Length
KI-84 Cape Gantheaume (W 1)	9	RBB+reefs	900 m
KI-85 Cape Gantheaume (W 2)	10	platform+RBB	60 m
KI-86 Cape Gantheaume (W 3)	10	platform+RBB	120 m
KI-87 Cape Gantheaume (W 4)	10	platform+RBB	200 m
KI-88 Cape Gantheaume (W 5)	10	platform+RBB	250 m

Cape Gantheaume is the southernmost tip of Kangaroo Island and one of the more isolated points on the island.

The 4WD/walking track from Cape Linois terminates at the cape, with the coast between the cape and Bales Beach 13 km to the west only accessible on foot. The 50 m high cape is composed of steeply dipping Cambrian metasedimentary rocks, capped by Pleistocene dune calcarenite, both of which extend for 500 m off the cape as reefs. The cape and rocky coast to the west are exposed to the full force of the Southern Ocean waves and winds which have battered the shore, driving most sand onshore as massive dunes extending up to 15 km inland and to heights of over 100 m. The once more continuous beaches which supplied these dunes have been reduced to a few sand beaches in more protected embayments, or to pockets of sand at the base of the near continuous cliffs.

The first 5 km west of the cape has one sand beach, immediately to the west, followed by continuous cliffs with four pockets of sand at their base. However, seaward of the cliffs, rock platforms and reefs, there is sufficient sand to form a near continuous rip and reef-dominated surf zone.

Beach **KI-84** is a 900 m long, southwest facing sand beach, backed by a foredune with active blowouts and parabolics extending kilometres inland (Fig. 5.18). Calcarenite reefs outcrop on the beach; increasingly so toward the western end. Waves average 2 m and maintain a rip-dominated double bar system, with two beach rips and two headland rips in the inner surf zone, feeding into larger outer bar rips.

Figure 5.18 *Beach KI-84, immediately west of Cape Gantheaume, is an exposed, high energy sandy beach. Note the two permanent headland rips and two central beach rips crossing the 200 m wide surf zone.*

Beach **KI-85** is a 60 m pocket of sand at the base of 70 m high cliffs, fronted by a 50 m wide intertidal platform and then a rip-dominated surf zone. Beach **KI-86** consists of two strips of high tide sand totalling 120 m in length at the base of 60 m high cliffs, with a fronting 50 m wide platform, then surf zone. Beach **KI-87** is two pockets of sand, 200 m in total length, connected by their fronting

platform at low tide. They are backed by a small foredune at the base of 40 m high cliffs, and fronted by the rip-dominated, 200 m wide surf zone. Beach **KI-88** is a 250 m long sand beach located in a low spot in the cliffs, which has allowed active blowout-parabolics to extend 400 m in from the beach. The beach has a near continuous, 100 m wide platform-reef, then a 200 m wide surf zone dominated by a large permanent rip.

Swimming: These are five very dangerous beaches dominated by high waves, rips, reefs and rocks. They are unsuitable for swimming.

Surfing: There are beach and reef breaks off all the beaches. However this is a very hazardous stretch of coast and should only be surfed by very experienced surfers.

Fishing: The entire section is dominated by deep reef and rip holes. However rock fishing is extremely hazardous, owing to the cliffs and wave-washed platforms.

Summary: An isolated, energetic, wave and rock-dominated section of coast, containing remnants of a once large beach system.

KI-89 – KI-92   **CAPE GANTHEAUME (west)**

Unpatrolled			
No. Name	Rating	Type	Length
KI-89 Cape Gantheaume (W 6)	10	R+platform+RBB	40 m
KI-90 Cape Gantheaume (W 7)	10	RBB+reef	170 m
KI-91 Cape Gantheaume (W 8)	10	RBB+reef	300 m
KI-92 Cape Gantheaume (W 9)	10	R+platform+RBB	160 m

Between the prominent headland 6 km west of Cape Gantheaume and the first of the three Bales beaches is a 4 km section of southwest facing, cliffed coast, with four small beaches wedged in amongst the base of the cliffs. The only access to this section of coast is on foot, either from Cape Gantheaume or Bales Beach.

Beach **KI-89** is a 40 m long pocket of high tide sand at the base of 50 m high cliffs, fronted by a continuous calcarenite platform and then a 200 m wide surf zone. Beach **KI-90** occupies a break in the reefs and is a 170 m long beach, with backing dunes that have climbed up onto the 60 m high cliffs. The beach is fringed by reefs, with an 80 m long sandy surf zone drained by one large rip. Beach **KI-91** consists of three pockets of near continuous sand, 300 m in length, at the base of 30 m high cliffs. It is fronted by a near continuous rock platform, with a gap in the western end providing access to the 200 m wide surf zone. Two strong reef-controlled rips front the beach. Beach **KI-92** is a high tide sand beach wedged in a cusp in the cliffs, with sufficient room for a small backing dune below the 30 m high cliffs. It has a continuous intertidal calcarenite platform, then a 200 m wide surf zone, usually drained by one large rip.

Swimming: Four isolated and very hazardous beaches, unsuitable for swimming.

Surfing: The only beach providing access to the surf is KI-91, however all are very hazardous.

Fishing: Excellent, though hazardous, beach and reef holes right along the shore.

Summary: A section of coast to be viewed rather than utilised for recreation, owing to its exposed and hazardous nature.

Seal Bay Conservation Park
Established: 1967
Area: 1 911 ha
Beaches: KI-92 to KI-103
Length of coast: 14 km

Seal Bay Conservation Park is set around the Australian sea lion colony in Seal Bay. The park extends along the coast for 14 km, inland for 1 to 2 km, and seaward of the shore for 2 km. The park is visited daily by hundreds of tourists who come to take guided tours of the sea lion colony on Seal Bay beach.

For further information:
Park Ranger
Murray's Lagoon
Phone: (0848) 28 233

KI-93 – KI-95   **BALES BEACH**

Unpatrolled				
No.	Name	Rating	Type	Length
KI-93	Bales Beach (1)	5	R+reefs	850 m
KI-94	Bales Beach (2)	7	TBR	1.9 km
KI-95	Bales Beach (3)	8	RBB+reefs	1.9 km

Bales Beach is one of the few accessible surfing beaches along the south coast of the island. It is located just 1 km off the Seal Beach Road and has a small car park providing access to the main beach (KI-95) but no other facilities. Extending east of the main beach are two additional sandy beaches, each bordered by prominent headlands.

Bales Beach (1) (**KI-93**) lies 3 km east of the car park. It is a semi-circular shaped, 850 m long, low energy beach, bordered by prominent headlands protruding 300 m

seaward, with reefs occupying much of the semi-enclosed bay, lowering waves at the shore to less than 1 m. These maintain a moderately steep reflective beach, with no bar or surf at the shore. Waves do however break on reefs in the bay and a permanent rip runs out against the western headland. Both active and stable dunes back the beach.

Bales Beach (2) (**KI-94**) is located 1.5 km east of the car park and is a 1.9 km long, semi-circular beach, bordered by prominent 200 m long headlands and fronted by 50 to 100 m wide calcarenite rock platforms. High waves move through the 800 m wide gap between the headlands and disperse along the beach, resulting in lower waves to either end, reaching a maximum in the centre of about 1.5 m. These usually maintain a 100 m wide surf zone, with three beach rips in the centre and a continuous bar toward each headland. Larger waves also break on some reefs between the headlands, and during big seas a large rip drains the entire embayment. The beach is backed by dunes, which increase in activity toward the east.

The main Bales Beach (**KI-95**) is a slightly curving, 1.9 km long beach, bordered by 20 m high headlands, with a 500 m long platform west of the car park and scattered reefs lying off the beach. This more exposed beach receives waves averaging almost 2 m, which break both on the outer reefs and outer bar, and again on the inner bar, which has two beach rips, together with three reef-controlled rips. The Bales Beach car park is located toward the western end of the beach, with moderately active dunes backing the central-eastern section.

Swimming: The main Bales Beach (3) is an accessible and frequently visited beach. However if you stop here to swim, be very careful as deep rip holes and strong rips dominate. The best location is the rock pool on the inner platform directly in front of the car park, otherwise use the inner part of the surf zone, clear of the rips.

Surfing: Both beach and reef breaks occur along the beach, however beware of the strong rips.

Fishing: A very accessible and popular spot for both beach and rock fishing.

Summary: Many visitors to Seal Bay also drive out and view Bales Beach. It is still a little used beach, which must be approached with extreme caution if swimming.

KI-96 – KI-100   BALES BEACH to SEAL BAY

Unpatrolled				
No.	Name	Rating	Type	Length
KI-96	Bales Beach-Seal Bay (1)	10	R+platform	70 m
KI-97	Bales Beach-Seal Bay (2)	8	R+reef	260 m
KI-98	Bales Beach-Seal Bay (3)	8	LTT+reef	130 m
KI-99	Bales Beach-Seal Bay (4)	8	LTT+reef	100 m
KI-100	Bales Beach-Seal Bay (5)	10	TBR+reef	250 m

Between Bales Beach and Seal Bay is 2 km of highly crenulate cliffed coast, along the base of which are five small pocket beaches, for the most part fronted by a combination of calcarenite platforms and/or reefs. All are difficult to access and none are suitable for recreation.

Beach **KI-96** is a 70 m long pocket of high tide sand, backed by 40 m high cliffs and fronted by a 50 m wide platform, then a 100 m wide surf zone dominated by a permanent rip. This beach is accessible around the rocks from Bales Beach during low waves at low tide.

Beach **KI-97** is a 260 m long high tide sand beach set in three scallops in the backing 30 m high cliffs. It is fronted by a mixture of platforms, reefs and some sand, with waves breaking on the reefs up to 300 m off the beach and a large rip draining out the western side of the beach. Beach **KI-98** is located in an adjacent breach in the cliff line. It is a 130 m long, 30 m wide beach, which is slightly protected by reefs 500 m off shore and consequently receives moderate waves, which break across a narrow sand bar fronted by deeper reefs. Its solitary rip links with the rip from KI-97 to drain the two beaches during moderate to high waves.

Beach **KI-99** is a separate pocket of sand surrounded on three sides by 20 m high cliffs and afforded moderate protection by offshore reefs and its south-southeast orientation. The beach is 100 m long, 50 m wide and fronted by a small bay and then deeper reefs. Waves average 1-1.5 m, with one rip draining the small beach.

Beach **KI-100** lies immediately east of Seal Bay beach. It is a 250 m long, southwest facing, exposed beach, bordered and backed by 50 m high cliffs and headlands, the latter extending 100 to 300 m seaward as reefs. In addition, shallow reefs lie along the centre of the beach. Waves average 1.5 to 2 m and maintain a 200 m wide surf zone with permanent rips at either end of the beach.

Swimming: All five beaches are very hard to access, isolated and hazardous. Only Beach KI-99 offers moderate protection from the usually high swell.

Surfing: The only surf is on the reefs, which lie up to a few hundred metres off the beaches.

Fishing: There are plenty of good spots to fish the deep rip holes from the beaches and rocks, but access is a problem.

Summary: Five beaches close to Seal Bay and Bales Beach, but out of sight and rarely visited.

KI-101, KI-102   SEAL BAY

Unpatrolled				
No.	Name	Rating	Type	Length
KI-101	Seal Bay	10	TBR+reefs	700 m
KI-102	Seal Bay (W)	10	R+platform+TBR	700 m

Seal Bay is one of the most popular tourist destinations on the island. Each day hundreds of people arrive to be taken on guided tours of the beach by park rangers, in order to see close-up the Australian sea lions which inhabit the beach and backing dune area. The bay is serviced by a gravel road, large car park, ranger station and elevated walkways to the beach and viewing areas. Swimming and fishing at the beach is prohibited in order to protect the sea lions. In addition, it is an extremely hazardous beach.

Seal Bay beach (**KI-101**) is 700m long, extending from a protruding eastern headland and platform; the latter cut in steeply dipping strata. The cliffs decrease in height from 40 to 20 m along the beach, and a small dune area lies between the beach and bluffs. Scattered reefs lie off the eastern half of the beach, increasing in density to the west (Fig. 5.19). The sea lions utilise the dunes and beach and use the surf for access. An elevated walkway crosses the dune to the beach for visitor viewing.

Figure 5.19 *Seal Bay, during unusually low wave conditions, is backed by dune-draped bluffs, bordered by calcarenite headlands, and fronted by extensive calcarenite reefs. The ranger station and walkways are visible to the right.*

At the western end of the beach is a small rocky headland, on the other side of which is the western beach (**KI-102**), a 700 m long, narrow, crenulate high tide sand beach fronted by a continuous, 50 to 100 m wide calcarenite platform, then a 300 m wide surf zone.

Swimming: Swimming is not only prohibited because of the seals but also because the beach is exceedingly hazardous, owing to the waves, rips and reefs.

Surfing: Also prohibited.

Fishing: Prohibited.

Summary: Kangaroo Island's most visited beach and a good example of an exposed (windy), high energy south coast beach.

KI-103 – KI-108   **SEAL BAY to VIVONNE BAY**

Unpatrolled				
No.	Name	Rating	Type	Length
KI-103	Seal-Vivonne Bays (1)	10	LTT+reef	180 m
KI-104	Seal-Vivonne Bays (2)	10	TBR	90 m
KI-105	Seal-Vivonne Bays (3)	10	TBR	120 m
KI-106	Seal-Vivonne Bays (4)	10	TBR	300 m
KI-107	Seal-Vivonne Bays (5)	10	R+platform	200 m
KI-108	Seal-Vivonne Bays (6)	10	R+platform	170 m

West of Seal Bay, the crenulate, cliffed calcarenite coast continues on for another 7 km to the eastern end of Vivonne Bay. Along this predominantly rocky coast are six small sand beaches located below the cliffs. All are extremely difficult to access and only suitable for viewing.

Beach **KI-103** lies immediately west of Seal Bay and is a curving, 180 m long beach encased in 50 m high cliffs, with access via a steep track down a rock fall toward the western end. It consists of a narrow beach awash at high tide, littered with boulders and debris from the overhanging cliffs, and fronted by a calcarenite platform, then a reef strewn surf zone.

Beach **KI-104** is 1.5 km west of Knobby Island. It is a 90 m pocket of high tide sand, lying at the base of 60 m high cliffs, fronted by a 50 m wide calcarenite platform, then a 200 m wide surf zone with permanent rips on either side of a central bar. Beach **KI-105** is 600 m to the west and is a 120 m long sliver of sand in a slight indentation in the cliffs. It is fully exposed to the high swell and has a 300 m wide sand and reef-dominated surf zone drained by one large rip.

Beach **KI-106** lies 300 m east of a prominent calcarenite headland, which can be reached along a straight 4WD track. The 300 m long beach is backed by 50 m high cliffs, which produce the rock debris that litters the beach and inner surf. The 300 m wide surf zone is dominated by two large rips and more reefs further offshore. On the western side of the headland is beach **KI-107**, a 200 m long strip of high tide sand, backed by 40 m high cliffs and fronted by a continuous, 50 to 100 m wide platform, then reefs extending 200 m offshore. A 4WD track runs along the top of the cliffs but there is no foot access to the beach.

Beach **KI-108** lies in a 300 m deep embayment and consists of a 170 m long, semi-circular high tide sand beach, wedged at the base of 40 m high cliffs and fronted by an 80 m wide platform, then a reef-littered, exposed embayment.

Summary: All six beaches are very difficult to locate and access and are not recommended for anything other than viewing.

KI-109 – KI-112   **VIVONNE BAY**

Unpatrolled				
No.	Name	Rating	Type	LengthInner/Outer Bar
KI-109	Vivonne Bay (E 2)	10	RBB RBB	600 m
KI-110	Vivonne Bay (E 1)	8	RBB RBB	450 m
KI-111	Vivonne Bay	8	RBB-LTT	5 km
KI-112	Vivonne Bay jetty	7	LTT	800 m

Vivonne Bay is the only township and port on the entire south coast of Kangaroo Island. Point Ellen, at the western end of the bay, provides sufficient shelter for the fishing fleet to moor in its lee, and for the 100 m long jetty. Most of the bay swings around to the east and extends for 6 km, terminating at high cliffs. There are four beaches in the bay area, the eastern two being accessible by road, the western two requiring a walk along the beach.

Beach **KI-109** lies 4 km east of the main Vivonne Bay beach car park. It is a 600 m long, south facing beach bordered by a prominent eastern headland and smaller western head, together with reefs off the eastern end and calcarenite beachrock fronting the swash zone. It receives waves averaging 2 m, which maintain a rip-dominated double bar system, with usually one central rip and rips adjacent to each headland. There is a 4WD track leading to the rear of the beach

Beach **KI-110** is located between two small protruding bluffs. It is 450 m long, with a continuous low tide strip of beachrock, then a rip-dominated double bar system, with two headland controlled rips.

Vivonne Bay beach (**KI-111**) is one of the most popular surfing beaches on the south coast. The small fishing and holiday settlement of Vivonne Bay backs both the beach and western headland shore behind the jetty, together with a store on the South Coast Road 1.5 km to the north. The 5 km long white sand beach begins in the east as a high energy double bar system (Fig. 5.20), before passing the usually closed mouth of the Eleanor River, beyond which wave height decreases, but calcarenite outcrops on the beach and in the surf. At the main beach car park, the surf has a single continuous bar cut by rips every 200 to 300 m. The rips generally disappear at the usually closed Harriet River mouth, as waves decrease to less than 1 m, and a continuous narrow bar continues on to a low bluff, which separates it from the jetty beach. There is access to the beach from the main car park and from the western side of the Harriet River, where there is also a camping area with basic facilities.

Figure 5.20 *The eastern higher energy section of Vivonne Bay is dominated by usually high, wide surf and strong rips.*

Vivonne Bay jetty beach (**KI-112**) extends from the low bluff to the jetty. It is an increasingly narrow, crenulate, east facing beach, backed by low calcarenite bluffs and sheltered by Ellen Point, with waves usually less than 1 m and decreasing toward the jetty. A road runs along the top of the bluffs to the jetty and on to Point Ellen. The 100 m long jetty services the fishing boats moored in lee of the point. Small tenders are winched up out of the swell onto the jetty.

Swimming: The best swimming is west of the main car park and along the western jetty beach, where waves are usually less than 1 m and rips absent. Be careful, however, as rips can extend to the Harriet River mouth when waves exceed 1 m at the beach. The beaches east of the car park are exposed and dominated by reefs and rips and are unsuitable for swimming.

Surfing: This is one of the better beach breaks on the island and a very popular spot, particularly out in front of the car park. During big swell there is a right breaking along the north end of the jetty beach, as well as a fullish right on the tip of Ellen Point.

Fishing: The jetty is the most popular spot, however there is also excellent beach fishing into the rip holes.

Summary: The largest and most accessible bay and beach on the south coast, offering basic facilities for tourists and visitors and the best beach for swimming and surfing.

Vivonne Bay Conservation Park
Established: 1971
Area: 1 481 ha
Coast length: 12 km
Beaches: KI-113 - 116
Vivonne Bay Conservation Park commences at the western end of the bay and includes 12 km of rugged, exposed rock shore and the backing Pleistocene dune-covered heathland. There are four small, exposed rock and reef-bound beaches in amongst the rocks.

KI-113 – KI-121   **POINT ELLEN, CAPE KERSAINT, SUN'SAIL BOOM**

Unpatrolled			
No.	Name	Rating Type	Length
KI-113	Point Ellen (W 1)	10 R+platform	200 m
KI-114	Point Ellen (W 2)	8 R+reef	300 m
KI-115	Point Ellen (W 3)	10 R+platform	140 m
KI-116	Point Ellen (W 4)	8 LTT+reef	220 m
KI-117	Cape Kersaint (W 1)	9 R+platform	200 m
KI-118	Cape Kersaint (W 2)	10 R+platform	250 m
KI-119	Cape Kersaint (W 3)	10 R+platform	100 m
KI-120	Sun'sail Boom	8 RBB	210 m
KI-121	Beach KI-121	8 R+reefs	220 m

West of **Point Ellen** the rocky coast trends southwest for 11 km to Cape Kersaint, then runs due west for 10 km to the mouth of the Sun'sail Boom River. Along this 21 km of rocky cliffed coast are eight small beaches, including one blocking the river mouth. The only access to these beaches is via a few rough 4WD tracks that periodically follow the top of the cliffs, with some of the beaches being very difficult to access on foot.

Beach **KI-113** lies 1 km southwest of Point Ellen. The 200 m long beach faces southeast and receives waves that refract around Cape Kersaint. The beach sits at the base of 20 m high, vegetated bluffs, and is fronted by a continuous, 50 m wide rock platform, partly covered in large boulders. Beach **KI-114** is 3.5 km from the point, faces due east and as a consequence is moderately protected, receiving waves averaging just over 1 m, which break over bedrock reefs extending 200 m offshore. Steep, vegetated, 30 m high bluffs back the 300 m long beach, which is bordered by extensive bedrock platforms.

Beach **KI-115** faces southwest into the high swell and consists of a 140 m long pocket of sand, surrounded on three sides by steep, 40 m high bluffs and bordered by bedrock platforms, which also occur as reefs off the beach. Waves averaging 1.5 to 2 m break over a bar and reef, and one permanent rip drains the small embayment. Beach **KI-116** lies 1 km to the west. It faces southeast and receives some protection from Cape Kersaint. The 220 m long beach has a western low energy section dominated by a reef-filled surf zone, and an eastern high energy section with two permanent rips draining the 200 m wide surf zone. The beach is backed by 20 m high bluffs in the west and partially active dunes extending 300 m inland in the east.

Beach **KI-117** lies 3 km west of Cape Kersaint and faces southeast toward the cape. It is moderately protected and waves refract around the southern headland and over the rock reefs, which extend 100 m off the beach. The 200 m long high tide beach is backed by steep, partly vegetated, 30 m high bluffs. Beach **KI-118** is located 500 m to the west on the exposed, southwest facing side of the headland, and receives high waves which break over a continuous calcarenite platform, with a bedrock headland and 50 m wide rock platform bordering each side. The beach is 250 m long and backed by moderately sloping, vegetated, 30 m high bluffs.

Beach **KI-119** is a 100 m long patch of sand set high on a rock platform, behind a small, reef-filled bay. It is bordered in the east by a 150 m wide platform and backed by a 5 to 15 m high bluff.

The **Sun'sail Boom River** reaches the coast in a 200 m wide, calcarenite-capped, bedrock controlled valley. The 210 m long, southwest facing beach (**KI-120**) blocks the mouth of the river, with the river flowing across the beach following heavy rain. The beach receives waves averaging 2 m, which maintain one to two rips draining the 200 to 300 m wide surf zone.

One kilometre west of the river is an isolated, southwest facing, 220 m long high tide sand beach (**KI-121**), fronted by near continuous rock platforms and rock reefs. A 4WD track leads to the centre of the beach and down the sand-covered, 20 m high bluffs that top the beach. While the reefs do lower the waves along the western half of the beach, the entire system is drained by one permanent rip that tends to run out against the western headland.

Summary: These are eight isolated and difficult to locate and access beaches. Only the Sun'sail Boom beach offers anything like a normal beach, but even this is a very hazardous, rip-dominated location.

Cape Bougher Wilderness Protection Area

Established:	1993
Area:	5 530 ha
Coast length:	14 km
Beaches:	KI-122 to124

The Cape Bougher Wilderness Protection Area covers a remote and rugged section of rocky coast, centred on Cape Bougher and the backing densely vegetated woodlands. In the west it borders Kellys Hill Conservation Park. The park extends to the coast and has 16 km of predominantly rocky shoreline, with one small beach in the east and four in the west around the South West River mouth. The park extends up to 7 km inland, including densely vegetated bedrock in the east and vegetated dunes in the west.

Kelly Hill Conservation Park

Established:	1967
Area:	2 189 ha
Length of coast:	16 km
Beaches:	none (in lee of Cape Bougher WPA)

Kelly Hill refers to a calcarenite dune located 5 km from the coast, which is underlain by a cave system. The caves are open to the public.

KI-122 – KI-128   **HANSON BAY**

Unpatrolled			
No.	Name	Rating	Type	Length
---	---	---	---	---
KI-122	Hanson Bay (E 3)	10	R+reefs	180 m
KI-123	Hanson Bay (E 2)	8	RBB	450 m
KI-124	Hanson Bay (E 1)	8	RBB	650 m
KI-125	Hanson Bay	5	LTT+reefs	250 m
KI-126	Hanson Bay (W 1)	8	R+platform	600 m
KI-127	Hanson Bay (W 2)	8	TBR+reefs	250 m
KI-128	Hanson Bay (W 3)	7	R+reefs	60 m

Hanson Bay is an open, 8 km wide, south facing embayment containing 12 km of predominantly cliffed, rocky shoreline, with the South West River entering roughly the centre of the bay. It is bordered in the east by Cape Bougher and in the west by Cape Younghusband. The only public access to the bay is along the gravel road to the river, with 4WD tracks also providing access to the three western beaches. There is a car park, water, toilet and sand boat ramp across the beach at the river mouth, together with a few fishing shacks and some cabins, but no other facilities. The eastern beaches can only be reached on foot from the river mouth.

Beach **KI-122** is located 2 km east of the river mouth. It is a 180 m long, high tide sand and rock beach, backed by a storm boulder beach then steep, partially vegetated, 40 m high bluffs, which partially encircle the beach. A continuous rock platform fronts the beach, which then drops off into deeper water. Higher waves break 100 m seaward of the platform, with one strong rip draining the embayment.

Beach **KI-123** is one of three high energy sandy beaches which share a common surf zone, but are separated by calcarenite bluffs. The first is a 450 m long, south facing beach bordered by 20 to 30 m high calcarenite bluffs. The eastern bluff is cliffed and fronted by a 30 m wide rock platform, while the western bluff barely reaches the shoreline. Two permanent rips are located at either end of the 300 m wide surf zone. On the western side of the bluff is a 650 m long beach (**KI-124**) which extends to the low bluff and more extensive reef that separates it from the river mouth beach. This beach is also dominated by the wide, high energy surf zone and two permanent boundary rips.

South West River mouth is blocked by a 250 m long beach (**KI-125**). Despite facing due south, the beach receives considerable protection from calcarenite reefs which extend up to 300 m offshore and, together with the western headland, form a 100 m wide channel at the mouth of the embayment (Fig. 5.21). Consequently waves average only 1 m at the beach, which usually has a continuous shallow bar. While conditions are relatively safe at the beach, a single large rip flows out though the channel into the high surf breaking over the reefs.

West of the river mouth, the cliffed and rocky coast continues all the way to Cape Younghusband, except for three small beaches. The first (**KI-126**) is located 1km west of the river mouth. It is a 600 m long sandy beach, backed by a low foredune and vegetated transgressive dunes, but fronted by a near continuous rock platform and reefs, with only one sand patch open to the surf. Waves break heavily over the reefs and three reef-controlled rips drain the beach and platform. A 4WD track leads to the 10 m high bluffs separating this beach from the neighbouring beach (**KI-127**). This is a 250 m long beach located below bluffs which rise from 10 m in the east to 30 m at the western end. It has a surf zone composed predominantly of reefs in the east and sand to the west.

Waves are slightly decreased by protection from western headlands, but reef-controlled rips still dominate the 100 m wide surf.

The westernmost beach in the bay is **KI-128**, a 60 m long pocket of sand backed by steep, 50 m high bluffs and fronted by a 50 m wide platform, which has been breached on its eastern side to permit waves to reach the small beach. A permanent rip drains out of the breach.

Figure 5.21 *Hanson Bay lies at the mouth of the South West River, with the beach protected by both the headlands and extensive reefs across the mouth, however a strong permanent rip drains the small bay.*

Swimming: The only beach moderately safe for swimming is at the river mouth. The river itself offers the safest swimming. All other beaches in the bay are exposed to high waves and are dominated by rips and/or reefs.

Surfing: The two beaches east of the river mouth are the most accessible and usually have breaks out over the bars. However, only surf here if you are very experienced.

Fishing: The river mouth and neighbouring beaches offer excellent access to deep rip holes, as well as rock platforms of varying degrees of exposure.

Summary: The river mouth is well worth a visit, with most day-trippers strolling along one or two beaches, while the locals come to fish, and the more experienced launch boats to fish the reefs.

KI-129 – KI-131   **SANDERSON BAY**

Unpatrolled			
No.	Name	Rating Type	Length
KI-129	Sanderson Bay (1) 8	TBR+rock	60 m
KI-130	Sanderson Bay (2) 8	TBR+rock	130 m
KI-131	Sanderson Bay (3) 6	TBR+rock	240 m

<table>
<tr><td colspan="2">Flinders Chase National Park</td></tr>
</table>

Established:	1919
Area:	32 600 ha
Length of coast:	40 km
Beaches:	129 to144 (16 beaches)

Flinders Chase National Park is the largest on the island and one of the largest coastal national parks in the state. It is also one of the oldest, having been established in 1919. Today the park covers the western third of the island; in all, 74 km², including 40 km of predominantly cliffed coast and 15 beaches along the south, west and north coasts.

Camping at: Rocky River Park, Snake Lagoon and West Bay.

Sanderson Bay is a 3 km wide bay, bounded by Cape Younghusband in the east and a protruding, unnamed headland in the west, which provides some protection to the western half of the bay. Tucked into the apex of the bay are three small beaches, all lying below cliffs and bluffs. Apart from a 4WD track that terminates above the cliff 400 m east of the first beach, there is no vehicle access to the shore.

Beach **KI-129** is a 60 m long pocket of sand located at the base of sloping, sand-covered, 20 m high bluffs, and bordered by rock platform-fringed headlands. A single bay-controlled rip drains the 100 m wide surf. One hundred metres to the west is beach **KI-130**, a similar, 130 m long beach, with two headland-controlled rips under low waves, and one large rip during higher waves. Beach **KI-131** lies 200 m further west and is in a more protected location, backed by 60 m high bluffs and bordered by massive granite headlands. It faces east and receives waves between 1 and 1.5 m high. The southern half of the beach is fronted by rock reef, with a small sand surf zone in the northern corner which has a permanent rip against the rocks.

Summary: Three small, difficult to access beaches dominated by the backing and bordering rocks and reefs. Unsuitable for swimming, with only beach KI-130 offering a beach break.

KI-132 – KI-135 **WEIRS COVE, CAPE DU COUEDIC**

<td colspan="5" align="center">Unpatrolled</td>

No.	Name	Rating	Type	Length
KI-132	Weirs Cove (1)	8	TBR+rocks	1.8 km
KI-133	Weirs Cove (2)	8	TBR	300 m
KI-134	Weirs Cove (3)	6	R/boulder	80 m
KI-135	Cape Du Couedic	10	R/boulder	90 m

Weirs Cove is a roughly semi-circular shaped, 500 m wide, east facing bay, surrounded by 70 to 130 m high calcarenite cliffs, sitting on a granite base. The partially protected cove has the ruins of a small jetty which was used as a tenuous landing to supply the Cape Du Couedic lighthouse, with a near vertical rail connecting the landing with the 80 m high clifftop.

Between the cove and Kirkpatrick Point 3 km to the east are three cliff-dominated beaches. The first (**KI-132**) occupies 1.8 km of the shore between the point and the cove. It is a moderately exposed, south facing beach lying at the base of 60 to 120 m high cliffs (Fig. 5.22). Cliff debris covers the back of the beach and provides a steep track down to the beach. Debris, rocks and rock reefs also litter the surf zone, so much so that they control the location of the five to six rips that usually form along the beach.

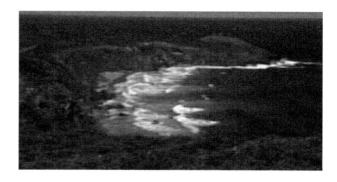

Figure 5.22 *Remarkable Rocks in the background sit atop Kirkpatrick Point, which forms the eastern boundary of a narrow beach lying at the base of the calcarenite bluffs, and which is littered by rock-falls from the bluff.*

The two beaches in Weirs Cove receive increasing protection from Cape Du Couedic, with waves usually less than 1.5 m high. Beach **KI-133** is a 300 m long strip of sand, usually awash at high tide, lying at the base of 100 m high cliffs. It can only be accessed with difficulty around the rocks from the landing beach. The landing beach (**KI-134**) is an 80 m long, east facing boulder beach located 100 m north of the landing, and is backed by steep, vegetated, 100 m high cliffs and debris.

On the very tip of Cape Du Couedic, right next to the Admirals Arch, is a 90 m long boulder beach (**KI-135**) which is used by the sea lions for coming ashore and resting. This moderately exposed and hazardous beach is part of the sea lion colony and is off limits to the public.

Summary: These four beaches are bordered by the popular Cape Du Couedic and Remarkable Rocks, with hundreds of tourists daily driving along the backing cliff tops. However all four are very difficult and dangerous to access, with equally hazardous conditions at each beach.

KI-136 – KI-139   **MAUPERTUIS BAY**

Unpatrolled				
No.	Name	Rating	Type	Length
KI-136	Maupertuis Bay (1)	10	R+platform	500 m
KI-137	Maupertuis Bay (2)	10	R+platform	300 m
KI-138	Maupertuis Bay (3)	10	R+platform	800 m
KI-139	Maupertuis Bay (4)	9	RBB+reefs	1.8 km

At Cape Du Couedic the coast turns 90 degrees and runs essentially north to Cape Borda. High calcarenite cliffs dominate the entire west coast, with four disjointed beaches along the southern Maupertuis Bay section, and six located at the mouths of rivers and creeks. In all, there are only ten beaches totalling 5 km in length along this 50 km of shoreline.

Maupertuis Bay is an open, 16 km wide, southwest facing bay, which is bordered by Cape Du Couedic in the south and Cape Bedout to the north. The bay affords no protection to the predominantly rocky shore, nor its seven beaches. The first four lie south of Rocky River, along the base of a relatively straight, 10 km long section of cliffs, which reach 100 m in height to the south.

The first three beaches (KI-136, 137 and 138) are all located at the base of steep cliffs, and consist of discontinuous high tide strips of sand partially covered with rock debris, and fronted by an irregular calcarenite platform. The first (**KI-136**) is 500 m long, with no access other than down the cliffs. The second (**KI-137**) is 300 m long with several small bluffs protruding onto the beach, while the third (**KI-138**) is 800 m long, with some sandy debris slopes providing access from the backing 40 m high cliffs. All three beaches are exposed to waves averaging 2 m. These maintain a 200 to 300 m wide surf zone that breaks over sandbars and reefs and is dominated by reef-controlled rips.

The fourth beach (**KI-139**) is a more continuous, 1.8 km long sand beach backed by a foredune and active dunes extending up to 1 km inland. The beach has prominent crenulations, induced by reefs extending 300 m off the southern half of the beach (Fig. 5.23). Five large reef-controlled rips flow across the surf zone. This beach is only accessible on foot from Rocky River beach, located 1 km to the north.

Swimming: Only possible at the northern beach, however be very careful as strong rips dominate the surf zone.

Surfing: There are a variety of breaks across the northern bars and reefs along the northern beach.

Fishing: Excellent deep rip holes occur right off the northern beach.

Summary: These four beaches are usually only viewed by hikers trekking along the coast. Only the northern beach offers sandy access to the very hazardous surf zone.

Figure 5.23 *Maupertuis Bay is an exposed, west facing, open bay with a high energy shoreline composed of a mixture of sand, rocks and reefs, and swept by a wide surf zone.*

KI-140 – KI-144   **ROCKY RIVER, SANDY, KNAPMANS CK, BREAKNECK RIVER, WEST BAY**

Unpatrolled				
No.	Name	Rating	Type	Length
KI-140	Rocky River	9	RBB+rocks	60 m
KI-141	Sandy Beach	9	RBB	500 m
KI-142	Knapmans Ck	9	RBB	80 m
KI-143	Breakneck R	9	RBB	150 m
KI-144	West Bay	8	RBB	470 m

Rocky River is the first of six large creeks that drain off the western plateau. Each of the creeks has carved deep, V-shaped valleys into the calcarenite-capped bedrock. At the coast, the now drowned valleys have been partially filled with marine sand delivered by the persistently high waves. The result is a series of small beaches periodically breached by the creeks and floods, while constantly battered by the waves, wind and rips, and each bordered by high calcarenite-capped headlands. The first five beaches are all accessible off the West Bay Road, which provides vehicle access to West Bay; the remainder requiring a 1.5 to 4 km walk.

Rocky River flows through a narrow, 2 km long gorge to reach the coast in a 100 m wide gap bordered by 80 m high valley sides. The creek meanders through the gorge to emerge on a 60 m wide sand beach (**KI-140**) which is awash with creek flow and waves. The valley sides extend out into the surf for 100 m, with the 300 m wide surf zone continuing either side of the bay (Fig. 5.24). A strong permanent rip flows seaward from the beach. The beach can be reached along a 1.5 km walking track from the Snake Lagoon car park. It is well worth the walk out, however use extreme caution if entering the surf.

Figure 5.24　*Rocky River beach is awash with both the river flow that drains across the beach and the high surf that prevails along the west coast.*

Sandy Beach (**KI-141**) lies at the mouth of Sandy River and can be reached along a 1.5 km walking track from the West Bay Road. This 500 m long beach is backed by active dunes, which both climb the backing 50 m high slopes and blow along the river bed, resulting in the 'sandy' name. The beach is bordered by steeply dipping metasedimentary bedrock headlands which extend seaward of the 200 m wide surf zone. Two permanent rips flow out either end of the beach, with a detached bar usually located in the centre (Fig. 5.25).

Figure 5.25　*Sandy Beach is backed by cliff-top dunes and the river mouth, while a wide, rip-dominated surf fills the bay.*

Knapmans Creek (**KI-142**) is a steep creek that drops down to the coast in a narrow, V-shaped valley, which is blocked by the small, 80 m wide beach that also extends 200 m into the valley as overwash, dune and creek sand.

Steeply dipping bedrock borders the beach, and a solitary rip flows out across the 150 m wide surf zone.

Breakneck River (**KI-143**) is one of the larger streams on the west coast. It descends from the plateau along a several kilometre long, V-shaped valley to reach the shore in a 150 m wide gap in the 50 m high cliffs. The beach fills the gap and a mixture of wave, dune and creek sand extends 500 m into the valley, impounding a small lagoon. The river mouth and beach can be accessed along a 3 km long walking track off the West Bay Road.

West Bay (**KI-144**) is the only beach accessible by vehicle on the west coast, with a 22 km drive from Rocky River. The beach, at 470 m, is the longest of the river mouth beaches. It blocks a 400 m wide valley mouth, with the sandy creek bed running along the southern side of the valley and onto the beach. Prominent calcarenite-capped bedrock headlands extend 2 km along the south side of the bay, providing slight protection to the southwest facing bay. Two rips usually flow out either side of the bay, with one to two central beach rips (Fig. 5.26).

Figure 5.26　*West Bay is the only readily accessible west coast beach and one of the more protected, with surf usually lower than its neighbours. However waves still average over 1 m and rips still prevail, so be careful if swimming here.*

Swimming: West Bay is the least hazardous of these five beaches, however all are exposed and dominated by strong rips, so use extreme care if waves are breaking.

Surfing:　West Bay is the most accessible and popular, with good breaks also off Rocky River.

Fishing:　Most fishers head for the beach rips and rocky shore of West Bay, with all beaches offering deep rip holes close inshore, as well as bordering rocky shoreline.

Summary:　All five beaches are interesting, with their own character, and are worth the drive or hike out to the coast for the spectacular view of the cliffs and usually wild surf.

Ravine des Casoars Wilderness Protection Area

Established: 1993
Area: 41 320 ha
Coast Length: 31 km
Beaches: KI-145 to 147

The Ravine des Casoars Wilderness Protection Area was formerly party of Flinders Chase National Park. Its protection status has been upgraded to a wilderness area, in order to ensure the long term preservation of this large area of undisturbed native flora and fauna.

Camping: Harveys Return

KI-145 RAVINE DES CASOARS

Unpatrolled				
No.	Name	Rating	Type	Length
KI-145	Ravine des Casoars	9	RBB	120 m

The **Ravine des Casoars** is a 3 km long, 100 m deep, steep sided valley that reaches the coast wedged between two 120 m high, calcarenite-capped headlands (Fig. 5.27). A small, 120 m long beach (**KI-145**) separates the headlands, while a creek (with a small lagoon) flows down the northern side of the valley. Beach and dune sand extends 600 m into the valley and dunes climb the southern side to an elevation of 100 m. The beach is accessible along a 3.5 km walking track that skirts both the sheltered valley and exposed headland. The base of the cliffs have been eroded, forming caves along the northern side. The beach is exposed to persistent high waves and one large rip usually flows out by the southern headland.

Figure 5.27 *The Ravine des Casoars is a deeply incised valley that has been partially filled with beach and dune deposits, with the strong westerlies also blowing sand up the southern side of the valley.*

Swimming: Be very careful if entering the surf, as a rip flows right along the beach and around the rocks, even under low wave conditions.

Surfing: Can have a good break over the bar, but again watch the rip which can take you along the adjoining cliffed sections.

Fishing: Always a deep rip gutter along the beach, in addition to the creek which usually flows across the beach.

Summary: A very worthwhile walk out to the ravine just to see the coastal views and caves.

KI-146 – KI-149   SCOTT COVE, HARVEYS RETURN, CAPE TORRENS

Unpatrolled				
No.	Name	Rating	Type	Length
KI-146	Scott Cove	5	R/boulder	50 m
KI-147	Harveys Return	5	Boulder/LTT	40 m
KI-148	Cape Torrens (W)	6	R/boulder	500 m
KI-149	Cape Torrens (E)	6	R/boulder	150 m

At **Cape Borda** the coast turns 90° and heads east along the north coast of the island. For 50 km high cliffs dominate the shore, with only an occasional pocket or boulder beach lying at the base of the cliffs. The first four lie within 14 km of the cape, with only one being reasonably accessible while the other three lie at the base of high cliffs.

Scott Cove (KI-146) is a 500 m wide indentation in the 80 m high cliffs, containing a 50 m long beach fronted by a mixture of boulders and sand. A car park and lookout overlook the beach, but there is no safe access down the cliffs.

Harveys Return lies 1 km east of Scott Cove, at the base of a steep valley. The beach was used as a landing site for material to construct the Cape Borda lighthouse. Today there is a car park and camping area on top of the 80 m high cliffs, with a steep track leading down to the beach. The 40 m long beach (**KI-147**) consists of a high tide boulder beach fronted by a narrow sand low tide bar. Jagged rock platforms border each side, and the remains of the jetty are located on the western headland.

Cape Torrens lies 12.5 km east of Cape Borda and is a prominent, 200 m high bedrock headland with near vertical cliffs. To either side are two pocket boulder beaches lying at the base of high cliffs. On the west is a 500 m long boulder beach (**KI-148**) fronted by rock reefs, while to the east is a 150 m long boulder beach (**KI-149**) at the base of 200 m high cliffs (Fig. 5.28). Neither beach is accessible on foot.

Figure 5.28 *Beach KI-149 is typical of the narrow boulder beaches lying at the base of the 200 m high cliffs which extend between Cape Borda and Cape Torrens.*

Swimming: Only suitable at Harveys Return and even here, although waves are usually low, beware of the numerous rocks and reefs off the beach.

Surfing: Usually flat.

Fishing: The rock platforms at Harveys Return offer the best access to the deeper water and rock reefs off the beach.

Summary: Go to Scott Cove lookout to enjoy the view, and down to Harveys Return if you are after a steep walk to picturesque little beach and rock formations.

Cape Torrens Wilderness Protection Area

Established: 1993
Area: 751 ha
Coastal length: 7 km
Beaches: KI-149 & KI-150

The 8 km of steep, 150 to 200 m high cliffs between Cape Torrens and Cape Forbin are preserved in the Cape Torrens Wilderness Protection Area. This is a remote and very difficult to access section of coast dominated by high cliffs, cut in places by steep gullies.

KI-150 – KI-154   **CAPE FORBIN, KANGAROO GULLY, SNUG COVE**

Unpatrolled				
No.	Name	Rating	Type	Length
KI-150	Cape Forbin	5	boulder+LTT	70 m
KI-151	Kangaroo Gully (W)	5	LTT	30 m
KI-152	Kangaroo Gully	5	LTT/TBR	110 m
KI-153	Snug Cove (W)	5	R+boulders	50 m
KI-154	Snug Cove	4	LTT+rocks	170 m

Cape Forbin is a prominent, 70 m high, irregular headland formed of steeply dipping metasedimentary rocks. It is typical of the cliff-dominated northern coast, where the 50 to 200 m high cliffs have been heavily dissected by small creeks forming small, V-shaped valleys, some of which have sand or boulder beaches. Waves average about 1 m along the north coast, though they can be considerably higher during strong westerly winds.

The Cape Forbin beach (**KI-150**) is located 1 km south of the cape at the mouth of the small De Mole River. The 70 m long beach is bordered by cliffs rising to 80 m. Boulders deposited by the De Mole River and washed off the cliffs form a high tide beach, which is fronted by a small, 100 m wide sand bar. The creek drains out along the western side of the beach and when waves are breaking, a rip also runs out along the western rocks. A partially cleared valley and a farm house lie east of the beach.

Kangaroo Gully creek flows through a cleared, V-shaped valley into a 200 m wide rocky cove, at the base of which is the main beach (**KI-152**) and a smaller beach tucked in amongst the western rocks (**KI-151**). The western beach is a 30 m pocket of sand encased in grassy slopes and rocks, that faces northeast and shares its surf zone with the main beach. The main beach is 110 m long and faces north. It receives waves averaging about 1 m which maintain a 60 m wide bar, with a permanent rip running out along the western rocks and past the western beach. Apart from the surrounding farm land and tracks, there is no public access.

Snug Cove is, as the name implies, a protected little cove containing a sand beach, with a smaller boulder beach 500 m to the west. The western beach (**KI-153**) is 50 m long and consists of a high tide sand beach fronted by a low tide boulder beach, and bordered and fronted by bedrock platforms and reefs. Waves are low at the beach, but rock dominates the water. The main Snug Cove beach (**KI-154**) is a 170 m long, low energy sand beach that faces northwest, but which is located deep within the 200 m long cove with just an 80 m wide entrance

(Fig. 5.29). Waves are usually less than 1 m at the beach, which has a narrow continuous bar, cut, however, by strips of dipping bedrock. The beach is serviced by a steep vehicle track that runs down to the rear of the small dunes behind the beach. A creek flows through the dunes and across the beach. The beach and backing valley are surrounded by private property and not open to the public.

Figure 5.29   *Snug Cove has a protected beach located deep inside a narrow opening between the steep bordering headlands.*

Swimming: None of these beaches are accessible to the public. Snug Cove offers the safest swimming, with Kangaroo Gully and Cape Forbin moderately safe under normal low wave conditions.

Surfing: Only low beach breaks, usually at Cape Forbin and Kangaroo Gully.

Fishing: All locations offer excellent rock fishing.

Summary: Five isolated and generally inaccessible little north coast beaches.

KI-155 – KI-161   **SEAL BEACH to CASTLE GULLY**

Unpatrolled				
No.	Name	Rating	Type	Length
KI-155	Seal Beach (W)	5	boulder	60 m
KI-156	Seal Beach (E)	6	boulder	100 m
KI-157	Waterfall Ck	6	boulder	80 m
KI-158	Goat Hill Ck (1)	6	boulder	100 m
KI-159	Goat Hill Ck (2)	6	boulder	200 m
KI-160	Castle Gully (W)	6	boulder	200 m
KI-161	Castle Gully (E)	6	boulder	60 m

Between Seal Beach and Sheoak Gully is 6 km of high cliffed coast, dissected in places by small creeks, and with small bedrock protrusions and bedrock debris littering the base of the cliffs. In amongst the rocks and debris are seven small boulder beaches, none of which are easily accessible.

Seal Beach consists of two beaches formed on either side of a protruding, 200 m long, tombolo-like series of bedrock stacks. The western beach (**KI-155**) is 60 m long and faces west, while the eastern beach (**KI-156**) is 100 m long and faces east. Both are composed of cobbles and boulders and are bordered and fronted by rocks. A steep, densely vegetated valley runs down to the eastern beach.

Waterfall Creek flows down a densely vegetated, V-shaped valley to a small, 80 m long boulder beach (**KI-157**) at its base. The beach is bordered by steep rock platforms and cliffs rising to over 100 m. Goat Hill Creek flows down the adjoining eastern valley, immediately east of which are the **Goat Hill Creek** beaches. They consist of a 100 m (**KI-158**) and 200 m (**KI-159**) strip of boulder beach at the base of 100 to 150 m high cliffs.

Castle Gully flows parallel to the coast and reaches the shore down a steep sided, vegetated valley. On either side of the valley are two boulder beaches. Four hundred metres to the west is a 200 m long boulder beach (**KI-160**) at the base of 100 m high cliffs, while 100 m to the east is a 60 m long boulder beach (**KI-161**), largely surrounded by protruding rock and rock reefs. A farm track leads to the rear of the beach.

Summary: These are seven virtually inaccessible boulder beaches. They are a hazard owing to both their rocky nature and also rock-falls from the backing high, crumbling cliffs.

Western River Wilderness Protection Area	
Established:	1993
Area:	2 373 ha
Coast length:	7 km
Beaches:	KI-157 to KI-162

Western River Conservation Park	
Established:	1971
Area:	167 ha
Beaches:	KI-163 & KI-164

The Western River area contains the publicly accessible conservation park, which straddles the river mouth, and the wilderness area that extends along the steep, rugged coast for 7 km west of the river.

KI-162 – KI-164 VALLEY CREEK, WESTERN RIVER COVE

Unpatrolled				
No.	Name	Rating	Type	Length
KI-162	Valley Creek	6	boulder	80 m
KI-163	Western R Cove	3	R/LTT	160 m
KI-164	Western R Cove (E)	6	boulder	50 m

Western River Cove is the only publicly accessible beach between Snelling Beach, 8 km to the east, and Harveys Return, 30 km to the west. One kilometre west of the cove, Valley Creek flows through a cleared, V-shaped valley to a small, 80 m long boulder beach (**KI-162**). The rear and eastern side of the beach have been excavated to form a small area for launching boats, with a 4WD track winding down the valley to the landing. Jagged rock platforms and rock reefs border either end of the beach.

Western River Cove is one of the nicer beaches on the entire island, and well worth the winding drive out to the coast. The 200 m wide, north facing cove is accessible by car, with a car park and picnic area located behind the beach. A foot bridge crosses the creek to the backing dunes and a track runs along the western rocks to the point. The cove contains two beaches. The main beach (**KI-163**) is a 160 m long, north facing sand beach, backed by dunes extending 150 m inland over the valley flats. The small river flows down the eastern side of the beach, while the beach itself is moderately steep, with a narrow continuous bar (Fig. 5.30). Boats moor in the cove at times and use the beach to land their dinghies. On the eastern side of the cove is a 50 m long boulder beach (**KI-164**), bordered and fronted by rock platforms and bedrock reefs.

Figure 5.30   *Western River Cove beach is often calm, though strong onshore winds have blown sand up to 150 m in from the beach.*

Swimming: Western River Cove is usually relatively safe under normal low waves, while the boulder beaches are unsuitable for swimming.

Surfing: Usually too small.

Fishing: Good fishing off the rocks around all three beaches.

Summary: A small beach and cove well worth visiting for a picnic, a swim or to fish.

KI-165, KI-166 PEBBLY BEACH, SNELLINGS BEACH

Unpatrolled				
No.	Name	Rating	Type	Length
KI-165	Pebbly	5	boulder+LTT+reef	130 m
KI-166	Snellings	5	TBR	660 m

Pebbly Beach (**KI-165**) is a relatively isolated pebble (boulder) and sand beach, located 3 km west of Snellings Beach, but surrounded by private farm land. The 130 m long, north facing beach lies at the mouth of a partially cleared, steep, V-shaped valley, with a farm track following the creek to the river mouth. The small beach blocks the creek and consists of a high tide cobble-boulder beach, with a narrow low tide bar and jagged rock reefs and platforms to either side.

Snellings Beach (**KI-166**) is the most visible and most popular on the north coast. The North Coast Road runs right past the western end of the beach, providing parking and beach access, while a lookout on the road provides an excellent view of the entire beach (Fig. 5.31). The 660 m long, northwest facing beach receives waves averaging over 1 m, which maintain a 100 m wide surf zone dominated by four beach rips and permanent rips against the rocks at each end. The waves tend to increase in height toward the western end, increasing the rip intensity. A 10 m high foredune backs the beach, with the Middle River meandering through the backing floodplain and exiting across the northern end of the beach. There is a barbeque, water and toilet at the beach but no other facilities.

Swimming: Snellings Beach is a popular summer beach for swimming and surfing. It is moderately safe under normal low waves, however strong rips develop when waves exceed 1 m, particularly along the central and western section of the beach.

Surfing: Usually a low beach break at Snellings.

Fishing: Both beaches have good holes off the rock platforms, with rip holes also along Snellings.

Summary: Snellings is highly visible and easily accessible and is a popular stopping spot for day-trippers and surfers.

Figure 5.31   *Snellings Beach is the most popular on the north coast. It is right on the main road and offers the best surf on the north coast.*

KI-167 – KI-176   **MIDDLE RIVER to KING GEORGE RIVER**

Unpatrolled				
No.	Name	Rating	Type	Length
KI-167	Beach KI-167	6	boulder+reef	80 m
KI-168	Beach KI-168	6	boulder	50 m
KI-169	Beach KI-169	6	boulder+platform	200 m
KI-170	Beach KI-170	6	boulder+platform	150 m
KI-171	King George (W 5)	6	boulder+platform	220 m
KI-172	King George (W 4)	6	boulder+platform	30 m
KI-173	King George (W 3)	6	boulder+reef	110 m
KI-174	King George (W 2)	6	boulder+reef	120 m
KI-175	King George (W 1)	6	boulder+platform	150 m
KI-176	King George	5	LTT/TBR	220 m

Between the mouths of Middle River and King George River are 5 km of rocky, irregular, cliffed coast containing ten small boulder beaches and the solitary sandy King George Beach. The only public access is to the boulder beaches located 1 km west of the King George River mouth. The remainder are backed by a buffer of cliffs, then an approximately 100 m wide cliff-top strip of scrub, then cleared farmland.

Beach **KI-167** is an 80 m long pocket boulder beach located 200 m past the Middle River mouth and bounded by steeply dipping rock platforms. Beach **KI-168** is a 50 m pocket boulder beach set within a 200 m deep indentation in the rocky coast. Beach **KI-169** is a discontinuous, 200 m long boulder beach cut by protruding rock bluffs and fronted by an irregular rock reef, with jagged headlands formed by the steeply dipping metasedimentary rocks. Beach **KI-170** is a 150 m long, north facing boulder

beach with rocks and reefs dominating the western half of the beach.

The **King George Beach** road reaches the coast at the rear of a steep cobble-boulder beach. Two hundred metres west of the road is beach **KI-171**, a 220 m long, curving boulder beach fronted by a continuous rock platform and exposed rocks. Beach **KI-172**, its neighbour, is a 30 m pocket boulder beach encased in 20 m high bedrock bluffs. Beach **KI-173** lies at the end of the road. There is limited parking at the rear of the beach, together with four fishing shacks, whose occupants use the beach to launch their boats at high tide (Fig. 5.32). The beach is very steep, but fronted by a partially narrow sand bar and deeper rock reefs. Immediately to the east is beach **KI-174**, a 120 m long, steep high tide boulder beach, backed by 20 m high cliffs, with a surf zone containing some sand but dominated by scattered rocks and reefs. Beach **KI-175** consists of three pockets of boulders totalling 150 m in length, backed by 20 m high bluffs and fronted by a continuous rock platform.

Figure 5.32   *The western King George beaches (KI-173 & 174) are rock-bound boulder and sand beaches, good for fishing but unsuitable for swimming.*

The main King George Beach (**KI-176**) is located at the river mouth. It is a 220 m long white sand beach, backed by a 100 m wide foredune and farm land, with the river draining across the eastern end of the beach. The beach receives waves averaging about 1 m which maintain a 50 m to occasionally 100 m wide bar, usually with a rip against the eastern rocks. Jagged rock platforms and backing cliffs border the beach.

Swimming: The only beach suitable for swimming is the main King George River mouth beach, but this is surrounded by private farmland with no public access. All the other beaches are dominated by rocks and reefs.

Surfing: Too rocky with low waves.

Fishing: Good fishing spots abound along this rocky-reefy shoreline.

Summary: The best place to see an example of a cobble-boulder beach is at the King George Beach access point. This location also provides good access to rocky fishing locations.

KI-177 – KI-182   DUTTON PARK to STOKES BAY

No.	Name	Rating	Type	Length
\multicolumn{5}{c}{Unpatrolled}
KI-177	Dutton Park	6	cobble+platform	50 m
KI-178	Cape Dutton (W)	4	boulder+reef	110 m
KI-179	Sandy Ck	5	boulder	80 m
KI-180	Springy Ck	5	boulder	120 m
KI-181	Springy Ck (E)	5	boulder+platform	120 m
KI-182	Beach KI-182	5	LTT+reefs	100 m

East of the King George River, the rocky coast continues for 10 km to Stokes Bay. This section contains 20 to 50 m high bedrock cliffs, including 60 m high Cape Dutton. There are also five boulder beaches and one sand beach located at the mouths of creeks or at the base of the cliffs. All are backed by cliffs, scrub, then farm land, with no public access.

Dutton Park beach (**KI-177**) lies 400 m east of the King George River. It is a 50 m wide cobble beach fronted by a small rock reef. The beach lies in a steep sided, cleared valley with a fishing shack and farm track at the rear of the beach.

The **Cape Dutton** beach (**KI-178**) lies 700 m southwest of the cape and consists of a 110 m long boulder beach, located at the base of 40 m high cliffs and fronted by a 50 m wide reef.

Sandy Creek flows down a small valley to an 80 m wide cobble-boulder beach (**KI-179**) at its base, bordered by rocks and fronted by a deeper reef. A steep farm track leads to the beach, which has a shack and boat ramp. **Springy Creek** 1 km to the east is backed by lower, cleared farm land, with the 120 m long cobble-boulder beach (**KI-180**) with patches of sand, while reefs lie along the shoreline. Five hundred metres to the east is a second beach (**KI-181**), consisting of a 120 m long high tide sand beach fronted by a 50 m wide rock platform and then reefs.

Beach **KI-182** lies at the base of scrub-topped, 70 m high cliffs. It is a sand beach fronted by a shallow bar, but with rock ridges running across the beach and into the water, and a storm boulder beach at the base of the cliff.

Summary: These six beaches are all backed by private farmland and in some cases by high cliffs.

KI-183 – KI-185   STOKES BAY

No.	Name	Rating	Type	Length
\multicolumn{5}{c}{Unpatrolled}
KI-183	Stokes Bay (1)	5	boulder flat	400 m
KI-184	Stokes Bay (2)	5	TBR	500 m
KI-185	Stokes Bay (3)	5	boulder	110 m

Stokes Bay is an open, north facing, 2.5 km long bay along the eastern shore of which are three beaches. The North Coast Road runs right past the centre of the bay, with a car park, camping area and café located immediately behind the eastern half of the first beach, while several shacks line the rear of the western half. The Stokes Bay Creek flows out in between.

The first beach (**KI-183**) is a 400 m long boulder flat, together with patches of high tide sand. Besides the backing facilities, it is used to launch boats at high tide. A walking track leads along the eastern half of the beach to a broken headland, through which there is a narrow-tunnelled walking track to the adjoining main beach.

The main Stokes Bay beach (**KI-184**) is a 500 m long, north facing sand beach bordered and backed by 20 to 50 m high, steep, vegetated bedrock bluffs. The beach receives waves averaging just over 1 m, which maintain a 100 m wide surf zone that usually has one central beach rip, with permanent rips to either end. In addition there is a natural rock-fringed tidal pool at the western end (Fig. 5.33).

Figure 5.33 *The main Stokes Bay beach offers both moderate surf and a natural rock pool (foreground).*

Just beyond the eastern end of the beach is the third bay beach (**KI-185**). This is a 110 m long high tide boulder beach, wedged up against 70 m high cliffs and fronted by a jagged, 50 m wide rock platform.

Swimming: The main Stokes Bay beach is a popular swimming and surfing location, and relatively safe under normal low waves. However strong rips develop whenever waves exceed 1 m, and there are signs warning of the dangerous rips.

Surfing: The main beach usually has a gentle beach break.

Fishing: The best fishing is over the rock and reef flats at high tide, or off the rock platforms.

Summary: One of the more interesting and popular north coast beaches.

KI-186 – KI-189   **BEACHES KI-186 & KI-187, HUMMOCKY GORGE**

No.	Name	Unpatrolled Rating	Type	Length
KI-186	Beach KI-186	4	R/LTT	140 m
KI-187	Beach KI-187	4	LTT	240 m
KI-188	Hummocky Gorge (W)	4	boulder	100 m
KI-189	Hummocky Gorge	4	R/LTT	100 m

Two kilometres east of Stokes Bay is an open, northwest facing bay containing four small beaches, all backed by hilly, cleared farm land. Beach **KI-186** is a 140 m long, north facing sand beach fronted by a narrow, continuous low tide bar with rocks outcropping in the surf and on the beach, and forming the eastern boundary. On the other side of the rocks is beach **KI-187**, a straight, 240 m long, north facing beach, backed by a farm house. It receives waves averaging about 1 m, which break across an 80 m wide continuous bar, with rips only forming against the boundary rocks and headlands during higher waves.

The **Hummocky Gorge** beaches lie 600 m to the east. The first beach (**KI-188**) is a 100 m long, north facing boulder beach, fringed by jagged rock platforms. Its neighbour lies across the Hummocky Gorge creek mouth. This is a 100 m long, northwest facing sand beach (**KI-189**), with a small, 50 m wide bar and fringing rock platforms and reefs. A small dune and a house back the beach, with the creek crossing the eastern end of the beach.

Swimming: All four beaches are relatively safe during normal low waves, with rips forming during higher waves.

Surfing: Only when the waves are up, and even then, only a beach break.

Fishing: Excellent spots off the bordering rock platforms.

Summary: Four small beaches surrounded by private farm land.

KI-190 – KI-201   **CAPE CASSINI, DASHWOOD BAY**

No.	Name	Unpatrolled Rating	Type	Length
KI-190	Cape Cassini (W 6)	6	boulder	260 m
KI-191	Cape Cassini (W 5)	6	boulder	220 m
KI-192	Cape Cassini (W 4)	6	boulder	100 m
KI-193	Cape Cassini (W 3)	6	boulder	80 m
KI-194	Cape Cassini (W 2)	6	boulder	240 m
KI-195	Cape Cassini (W 1)	6	boulder	310 m
KI-196	Cape Cassini	6	boulder	550 m
KI-197	Cape Cassini (E)	6	boulder	90 m
KI-198	Dashwood Bay	6	boulder	400 m
KI-199	Dashwood Bay (E 1)	6	boulder	50 m
KI-200	Dashwood Bay (E 2)	6	boulder	100 m
KI-201	Dashwood Bay (E 3)	6	boulder	140 m

The 14 km of coast centred on Cape Cassini is dominated for the most part by 50 to 100 m high cliffs cut into steep metasedimentary rocks, with jagged rock platforms and cliff debris littering much of the shore. Only at Cape Cassini and in Dashwood Bay do the slopes decrease and provide a grassy, rising backdrop. The cliffs are backed by a mixture of scrub and farmland, with no public access to this section of coast.

Beach **KI-190** is a 260 m long, northeast facing boulder beach backed by 50 m high cliffs that drop to the mouth of a small valley at the north end. Two hundred metres to the south is the isolated beach **KI-191**, a 220 m long, north facing boulder beach backed by 100 m high cliffs. Two kilometres to the east are beaches **KI-192** and **KI-193**, two neighbouring boulder beaches. The first is 100 m long and backed by 100 m high cliffs, but with a sandy seabed in front. The second is 80 m long and fronted by rock reef.

One kilometre west of **Cape Cassini** is beach **KI-194**, a 240 m long, northwest facing boulder beach, with grassy-topped, 100 m high cliffs behind, and an intertidal rock platform paralleling the beach. Its neighbouring beach **KI-195** is a curving, 310 m long boulder beach with slightly more open access to the bedrock seabed. It is backed by partially vegetated, 20 to 50 m high cliffs.

At Cape Cassini the cliffs are replaced by sloping, grassy farm land, which is bordered at the shore by an extensive intertidal rock platform grading into a shallow rock reef. At the rear of the platform is a discontinuous, crenulate boulder beach (**KI-196**) that extends for about 550 m along the shore. Three shacks are located at the centre of the beach.

Beach **KI-197** is an isolated, 90 m long boulder beach located below sloping, vegetated, 50 m high bluffs. At the base of the boulders is a patch of sandy seabed, with rocks to either side.

Dashwood Bay is an open, 2.5 km wide, north facing bay, which has been carved by the small Duncan River that enters the western side of the bay. Along the bay shore are four beaches. The first (**KI-198**) is the main bay beach. It is 400 m long and consists of a 200 m long boulder section, including raised Pleistocene boulder beaches, and a 200 m long sand section fronted by a rock-littered 50 m wide bar. A farm track reaches the rear at this part of the beach. Two hundred metres to the east the cliffs rise again, and the first of two boulder beaches is located. Beach **KI-199** is a 50 m strip of boulders, while its neighbour beach **KI-200** is 100 m long. Finally, one kilometre further on, is the last of the bay beaches (**KI-201**), an isolated, 140 m long boulder beach at the base of 100 m high cliffs.

Summary: Only the sandy part of the main Dashwood Bay beach offers anything resembling a swimming beach. All the others are dominated by boulders, rocks and reefs, with most backed by unstable high cliffs. They are not suitable or safe for swimming, and most are also dangerous for fishing owing to rock-falls.

KI-202 – KI-204   **SMITH BAY**

Unpatrolled				
No.	Name	Rating	Type	Length
KI-202	Smith Bay (W)	6	boulder	2 km
KI-203	Smith Bay (marina)	2	R	100 m
KI-204	Smith Bay (E)	6	boulder	700 m

Smith Bay is a 5 km wide, open, north facing bay, backed by cliffs rising to 100 m at either end, with the lower central 3 km section occupied by a continuous boulder beach, which is now cut in the centre by a small marina. The North Coast Road comes within 400 m of the centre of the bay, with public access via the marina road.

Smith Bay (west) (**KI-202**) is a 2 km long boulder beach, consisting of both active and higher relict boulders, the latter deposited during a higher stand of the sea. The highest inactive boulders are weathered grey, those in the central occasionally active section are covered in an orange algae, while the lower, more active, boulders are brown in colour. The western boulders run from approximately the small mouth of Freestone Creek to the marina wall.

The Smith Bay marina-boat ramp lies at the mouth of a second small creek. It consists of two training walls extending 60 m into the bay and forming a 20 m wide

entrance, with a 100 m long sand beach (**KI-203**) locked inside the small harbour.

Immediately east of the marina, the boulder beach (**KI-204**) continues for another 700 m, terminating as the cliffs begin to rise to the east.

Summary: None of these beaches are suitable for swimming, owing to the large boulders along the main beaches and the boat traffic in the marina.

KI-205 – KI-212   **EMU BAY**

Unpatrolled				
No.	Name	Rating	Type	Length
KI-205	Emu Bay (W 3)	4	R/LTT	400 m
KI-206	Emu Bay (W 2)	4	R+platform	60 m
KI-207	Emu Bay (W 1)	3	R+groyne	50 m
KI-208	Emu Bay	6	TBR/RBB	5.1 km
KI-209	Emu Bay (E 1)	5	LTT+rocks	80 m
KI-210	Emu Bay (E 2)	5	R+rocks	160 m
KI-211	Emu Bay (E 3)	5	R+rocks	60 m
KI-212	Emu Bay (E 4)	5	boulder+rocks	500 m

Emu Bay is bordered by Cape D'Estaing in the west and White Point in the east. It is 8 km wide and open to the north, though protected in the west from the prevailing westerly winds. The small settlement of Emu Bay is perched on the western bluffs and houses extend along the bluffs behind the western three beaches. There is a small jetty and boat ramp at the western end of the main beach, together with a picnic area, car park and a few shacks. There is no development to the east.

Beach **KI-205** is the northernmost of the Emu Bay beaches and is a 400 m long, northeast facing sandy beach, bordered and backed by grassy slopes which are slowly being subdivided for housing. A foreshore reserve runs the length of the beach. It usually receives low waves and has a narrow bar that is usually free of rips.

Beach **KI-206** is a high tide pocket of sand wedged in an inflection in the rocky shoreline and fronted by low tide rocks. It lies at the end of the main road through the settlement, with a 50 m walk from the road to the beach. Beach **KI-207** is a 50 m long accumulation of sand wedged between the jetty groyne and the rocks. The beach is used for launching boats, with a large car park at the rear of the beach (Fig.5.34).

The main Emu Bay beach (**KI-208**) is a sweeping, 5 km long, north facing, white sand beach backed by a near continuous, 10 to 20 m high foredune. The best access is at the western end, where there is a car park, picnic area and camping area right behind the most sheltered part of the beach. The long beach receives waves averaging just over 1 m, with occasional larger waves producing a 100 m wide surf zone, which, when waves exceed 1 m, can have rips cutting the bar every 100 m, resulting in up to 50 small rips along the beach.

Figure 5.34 *The small Emu Bay settlement is located at the more protected western end of the bay, with a jetty and boat ramp crossing the beach.*

The main beach terminates at some rocky bluffs which continue on for 3.5 km to **White Point**. Along the first 1.5 km of the bluffs are four small rock-controlled beaches. The first beach (**KI-209**) is an 80 m long sand beach, with a rock platform and rock reefs protruding 100 m off each end. It is backed by sloping, 60 m high bluffs, with a storm cobble beach at the base of the bluffs. Its neighbouring beach **KI-210** is similar, 160 m long and with more rocks and reef and a wider backing cobble beach. Beach **KI-211** is a 60 m long cobble-boulder beach surrounded by sloping, 60 m high bluffs, with two small rocky headlands, and rock reefs off the beach.

Beach **KI-212** is a curving, north facing boulder beach, backed by sloping, vegetated, 50 m high bluffs, with some low tide sand along the centre of the beach.

Swimming: Only the main beach and westernmost beach are suitable for safe swimming, with the western end of the main beach the most popular, as it usually offers low waves and a shallow surf zone. Be careful if waves exceed 1 m as rips can dominate the surf. All the other beaches are dominated by rocks and reefs.

Surfing: The main beach can provide some wide beach breaks during higher wave conditions.

Fishing: The jetty and groyne are the best locations, while the more adventurous can try the more remote, rocky parts of the bay.

Summary: Emu Bay is a popular summer holiday spot, offering a north facing, protected western end of the bay and beach, located only a few minutes from Kingscote.

KI-213 – KI-218 BOXING BAY, POINT MARSDEN

Unpatrolled				
No.	Name	Rating	Type	Length
KI-213	White Pt (W)	6	boulder+platform	200 m
KI-214	Boxing Bay (1)	4	R+rocks	200 m
KI-215	Boxing Bay (2)	4	R+boulder	1.6 km
KI-216	North Cape (1)	5	R+boulder	80 m
KI-217	North Cape (2)	3	R	100 m
KI-218	Pt Marsden (W)	4	R+boulder	260 m

Boxing Bay is an open, 2.5 km wide bay bordered by White Point in the west and North Cape, the northernmost tip of the island, to the east. Within the bay are four beaches, with two additional beaches on either side of the boundary points. The North Cape Road terminates at a gate 1 km south of the bay, with a restricted road leading up to the lighthouse on North Cape, otherwise the beaches are backed by bluffs and farm land, with no public access.

White Point beach (**KI-213**) extends 200 m west of the sand-capped, 50 m high bedrock point. It is a boulder beach fronted by an intertidal rock platform.

Within Boxing Bay are two beaches composed of both sand and boulders. The first (**KI-214**) lies at the western end of the bay, at the end of the vehicle track from the North Cape Road. It is a 200 m long, predominantly sand beach, backed by sloping, grassy bluffs and bordered by extensive boulder flats. The main beach (**KI-215**) runs for 1.6 km to the east and occupies much of the bay shore. It is a sand beach backed, and in places replaced, by a high tide boulder beach, together with cemented Pleistocene boulders along the low tide line, all the while backed by steep, sand-draped bluffs that rise to 70 m at the eastern end.

Just below 40 m high North Cape are two small, north facing pocket beaches (Fig. 5.35). The first (**KI-216**) is an 80 m pocket of high tide sand fronted and bordered by low tide rock-boulder flats. Its neighbour, **KI-217**, is 100 m long, with bordering rock platforms, but a sandy foreshore.

Point Marsden lies 1 km southeast of North Cape and lying immediately west of the point is a 260 m long sand beach (**KI-218**), bordered by 50 m wide rock platforms, together with scattered rocks lying along the shoreline.

Swimming: The only beaches suitable for safe swimming are the sandy parts of the two Boxing Bay beaches. However be careful if there are waves, as rock-controlled rips will form.

Surfing: Only low beach breaks during higher wave periods.

Fishing: The best fishing is off the rock platforms which border each beach.

Summary: An isolated bay mainly used by local fishers.

Figure 5.35 *The two northernmost beaches on the island at North Cape are both backed by dune-draped bluffs.*

GLOSSARY

bar (sand bar) - an area of relatively shallow sand upon which waves break. It may be attached to or detached from the beach, and may be parallel (longshore bar) or perpendicular (transverse bar) to the beach.

barrier - a long term (1 000s of years) shore-parallel accumulation of wave, tide and wind deposited sand, that includes the beach and backing sand dunes. It may be 100's to 1000's of metres wide and backed by a lagoon or estuary. The beach is the seaward boundary of all barriers.

beach - a wave deposited accumulation of sediment (sand, cobbles or boulders) lying between modal wave base and the upper limit of wave swash.

beach face - the seaward dipping portion of the beach over which the wave swash and backwash operate.

beach type - refers to the type of beach that occurs under wave dominated (6 types), tide-modified (3 types) and tide-dominated (3 types) beach systems. Each possesses a characteristic combination of hydrodynamic processes and morphological character, as discussed in chapter 2.

beach types
  wave dominated (abbreviations, see Figures 2.4 & 2.5)
     R - reflective
     LTT - low tide terrace
     TBR - transverse bar and rip
     RBB - rhythmic bar and beach
     LBT - longshore bar and trough
     D - dissipative
  tide-modified   (abbreviations, see Figure 2.13)
     R+LTT – reflective + low tide terrace
     R+BR + low tide bar and rip
     UD – ultradissipative
  tide-dominated (abbreviations, see Figure 2.15)
     R+SR – beach + ridged sand flats
     R+SF – beach + sand flats
     R+TSF – tidal sand flats

beach hazards - elements of the beach environment that expose the public to danger or harm. Specifically: water depth, breaking waves, variable surf zone topography, and surf zone currents, particularly rip currents. Also includes local hazards such as rocks, reefs and inlets.

beach hazard rating - the scaling of a beach according to the hazards associated with its beach type as well as any local hazards.

beachrock – see calcarenite

berm - the nearly horizontal portion of the beach, deposited by wave action, lying immediately landward of the beach face. The rear of the berm marks the limit of spring high tide wave action.

blowout - a section of dune that has been destabilised and is now moving inland. Caused by strong onshore winds breaching and deflating the dune.

calcarenite – cemented calcareous sand grains. May form in coastal sand dunes due to pedogenesis, the formation of calcrete soils, commonly called dune calcarenite or dunerock. In beaches it forms from precipitation of calcium carbonate in the intertidal zone and is known as beach calcarenite or beachrock.

cusp - a regular undulation in the high tide swash zone (upper beach face), usually occurring in series with spacing of 10 to 40 m. Produced during beach accretion by the interaction of swash and sub-harmonic edge waves.

dunerock – see calcarenite

foredune - the first sand dune behind the beach. In South Australia it is usually vegetated by spinifex and other dune grasses and succulents, especially cakile maritima.

gutter - a deeper part of the surf zone, close and parallel to shore. It may also be a rip feeder channel.

hole - a localised, deeper part of the surf zone, usually close to shore. It may also be part of a rip channel.

Holocene - the geological time period (or epoch) beginning 10 000 years ago (at the end of the last Glacial or Ice Age period) and extending to the present.

lifeguard - in Australia this refers to a professional person charged with maintaining public safety on the beaches and surf area that they patrol. Also known as *beach inspectors.*

lifesaver - an Australian term referring to an active volunteer member of Surf Life Saving Australia, who patrol the beach to maintain public safety in the surf.

megacusp - a longshore undulation in the shoreline and swash zone, with regular spacings between 100 and 500 m, which match the adjacent rips and bars. Produced by wave scouring in lee of rips (megacusp or rip embayment) and shoreline accretion in lee of bars (megacusp horn).

megacusp embayment - see megacusp

megacusp horn - see megacusp

parabolic dune - a blowout that has extended beyond the foredune and has a U shape when viewed from above.

Pleistocene - the earlier of the two geological epochs

comprising the Quaternary Period. It began 2 million years ago and extends to the beginning of the Holocene Epoch, 10 000 years ago.

reef – a submerged or partly submerged object; in South Australia usually formed of bedrock or calcarenite.

rip channel - an elongate area of relatively deep water (1 to 3 m), running seaward, either directly or at an angle, and occupied by a rip current.

rip current - a relatively narrow, concentrated seaward return flow of water. It consists of three parts: the *rip feeder current* flowing inside the breakers, usually close to shore; the *rip neck*, where the feeder currents converge and flow seaward through the breakers in a narrow 'rip'; and the *rip head*, where the current widens and slows as a series of vortices seaward of the breakers.

rip embayment - see megacusp

rip feeder current - a current flowing along and close to shore, which converges with a feeder current arriving from the other direction, to form the basis of a rip current. The two currents converge in the rip (megacusp) embayment, then pulse seaward as a rip current.

rock platform - a relatively horizontal area of rock, lying at the base of sea cliffs, usually lying above mean sea level and often awash at high tide and in storms. The platforms are commonly fronted by deep water (2 to 20 m).

rock pool - a wading or swimming pool constructed on a rock platform and containing sea water.

sea waves - ocean waves actively forming under the influence of wind. Usually relatively short, steep and variable in shape.

set-up - rise in the water level at the beach face resulting from low frequency accumulations of water in the inner surf zone. Seaward return flow results in a *set-down*. Frequency ranges from 30 to 200 seconds.

shore platform - as per rock platform.

swash - the broken part of a wave as it runs up the beach face or swash zone. The return flow is called *backwash*.

swell - ocean waves that have travelled outside the area of wave generation (sea). Compared to sea waves, they are lower, longer and more regular.

trough - an area of deeper water in the surf zone. May be parallel to shore or at an angle.

tidal pool - a naturally occurring hole, depression or channel in a shore platform, that may retain its water during low tide.

wave (ocean) - a regular undulation in the ocean surface produced by wind blowing over the surface. While being formed by the wind it is called a *sea* wave; once it leaves the area of formation or the wind stops blowing it is called a *swell* wave.

wave bore - the turbulent, broken part of a wave that advances shoreward across the surf zone. This is the part between the wave breaking and the wave swash and also that part caught by bodysurfers. Also called *whitewater*.

wave refraction - the process by which waves moving in shallow water at an angle to the seabed are changed. The part of the wave crest moving in shallower water moves more slowly than other parts moving in deeper water, causing the wave crest to bend toward the shallower seabed.

wave shoaling - the process by which waves moving into shallow water interact with the seabed causing the waves to refract, slow, shorten and increase in height.

REFERENCES AND ADDITIONAL READING:

Boreen, T, James, N, Wilson, C and Heggie, D, 1993, Surficial cool-water carbonate sediments on the Otway continental margin, southeastern Australia. *Marine Geology*, 112, 35-56.

Drexel, J F and Preiss, W V, 1995, *The Geology of South Australia. Volume 2, The Phanerozoic*. South Australian Geological Survey, Bulletin 54. Mines and Energy South Australia, 347 pp plus map.

Easton, A K, 1970, *Tides of the Continent of Australia*. Report 57, Horace Lamb Centre, Flinders University, Adelaide, 326 pp.

Hails, J R and Gostin, V A (eds) 1984, The Spencer Gulf Region. *Marine Geology*, 61, 111-424.

Laughlin, G, 1997, *The Users Guide to the Australian Coast*. Reed New Holland, Sydney, 213 pp. An excellent overview of the Australian coastal climate, winds, waves and weather.

Oppermann, A, 1999, *A Biological Survey of the South Australian Coastal Dune and Clifftop Vegetation*. Coast and Marine Section, EPS, DEHAA, Adelaide, 334 pp.

Parkin, L W, 1969, *Handbook of South Australian Geology*. Geological Survey of South Australia, Adelaide, 268 pp.

Readers Digest, 1983, Guide to the Australian Coast. Readers Digest, Sydney, 479 pp. Excellent aerial photographs and information on the more popular spots along the South Australian coast.

Ross, J (editor), 1995, *Fish Australia*. Viking, Melbourne, 498 pp. An excellent coverage of all South Australian coastal fishing spots.

Short, A D, 1993, *Beaches of the New South Wales Coast*. Australian Beach Safety and Management Program, Sydney, 358 pp. The New South Wales version of this book.

Short, A D, 1996, *Beaches of the Victorian Coast and Port Phillip Bay*. Australian Beach Safety and Management Program, Sydney, 298 pp. The Victorian version of this book.

Short, A D (editor), 1999, *Beach and Shoreface Morphodynamics*. John Wiley & Sons, Chichester, 379 pp. For those who are interested in the science of the surf.

Short, A D, 2000, *Beaches of the Queensland Coast:Cooktown to Coolangatta*. Australian Beach Safety and Management Program, Sydney, 360 pp. The Queensland version of this book.

Short, A D and Hesp, P A, 1984, *Beach and dune morphodynamics of the South East coast of South Australia*. Coastal Studies Unit Technical Report 84/1, Coastal Studies Unit F09, University of Sydney, 142 pp.

Short, A D and Fotheringham, D G, 1986, *Coastal morphodynamics and Holocene evolution of the Kangaroo Island coast, South Australia*. Coastal Studies Unit Technical Report 86/1, Coastal Studies Unit F09, University of Sydney, 112 pp.

Short, A D, Fotheringham, D G and Buckley, R C, 1986, *Coastal morphodynamics and Holocene evolution of the Eyre Peninsula coast, South Australia*. Coastal Studies Unit Technical Report 86/2, Coastal Studies Unit F09, University of Sydney, 178 pp.

Surf Life Saving Australia, 1991, *Surf Survival; The Complete Guide to Ocean Safety*. Surf Life Saving Australia, Sydney, 88 pp. An excellent guide for anyone using the surf zone for swimming or surfing.

Twidale, C R, Tyler, M J, and Davies, M, 1985, *Natural History of Eyre Peninsula*. Royal Society of South Australia, 229 pp.

Twidale, C R, Tyler, M J and Webb, B P, 1976, *Natural History of the Adelaide Region*. Royal Society of South Australia, 189 pp.

Tyler, M J, Twidale, C R and Ling, J K, 1982, *Natural History of Kangaroo Island*. Royal Society of South Australia, 184 pp.

Warren, M, 1998, Mark Warren's Atlas of Australian Surfing. Angus & Robertson, Sydney, 232 pp. Covers main South Australian surfing spots.

Williamson, J A, Fenner, P J, Burnett, J W and Rifkin, J F, (eds), 1996, *Venomous and Poisonous Marine Animals - a Medical and Biological Handbook*, University of New South Wales Press, Sydney, 504 pp. The definitive book on marine stingers.

APPENDIX 1　　SOUTH AUSTRALIAN ISLANDS

	Region Name	approx coastline km	approx area ha	Inhabited
	Southeast			
1	Penguin	0.8	6	
2	Godfrey Islands	2.3	35	
3	Granite	2	30	
4	Pullen	0.7	5	
5	Seal	0.5	2	
6	Wright	0.7	4	
7	West	1.5	20	
		8.5	102	
	Kangaroo Island			
8	The Pages N Page	0.9	5	
9	S Page	0.8	4	
10	Page 3	0.8	2	
11	'Cape Younghusband"	0.5	5	
12	Busby Islet	1.1	10	
13	Casuarina Islets 1	1	7	
14	2	1.3	12	
		6.4	45	
	Yorke Peninsula			
15	Troubridge	1.1	9	
16	Haystack	1.2	5	
17	Seal Island	0.7	5	
18	Althorpe Islands 1	0.4	2	
19	2	0.7	3	
20	3	4	90	
	Neptune Islands			
21	North Neptune	5	40	
22	S Neptune 1	7	75	
23	S Neptune 2	2	25	
	Gambier Islands			
24	North Island	3.5	170	
25	Wedge Island	14	7500	
26	Peaked rock	0.2	1	
27	South West Rock	-	-	
28	Middle	1.9	300	
29	Wardang	40	1700	
30	Goose	0.7	6	
31	Little Goose	0.15	50	
32	White Rock	0.4	2	
33	Green Is	0.2	1	
34	Island Pt	0.5	3	
35	Bird	-	-	
36	Shag	-	-	
		83.65	10347	
	Eyre Peninsula			
37	Tumby Island	1.5	30	
	Sir Joseph Banks Group			
38	Winceby	1.5	30	
39	Reevesby	20.0	600	x
40	Roxby	4.0	150	
41	Spilsby	10.0	500	x
42	Stickney	3.0	100	
43	Marum	0.5	20	
44	Partney	1.5	60	
45	Kirkby	1.5	50	
46	Dalby	0.7	10	
47	Langton	1.5	40	
48	Lusby	0.8	30	
49	Hareby	2.5	100	
50	Boucout	0.5	20	
51	Sisbey	2.0	30	
52	English	0.5	5	
53	Louth	7.0	200	x
54	Rabbit	1.2	30	
55	Boston	23.0	1000	x

56	Bickers	1.0	10	
57	Donnington	0.2	5	
58	Grantham	1.0	60	
59	Owen	0.6	20	
60	Taylor	8.0	300	
61	Grindal	3.0	100	
62	Little	0.3	5	
63	Lewis	1.0	20	
64	Hopkins	5.0	200	
65	Thistle	47.0	4000	x
66	Smith	1.5	40	
67	Williams	6.0	200	
68	Albatross	1.0	30	
69	Cutra Rocks	3.0	50	
	Neptune Islands			
70	North Neptune	6.0	200	
71	South Neptune	4.0	150	x
72	Liguanea	5.0	150	
	Whidbey Islands			
73	Price	2.0	50	
74	Perforated	4.0	100	
75	Golden	2.0	30	
76	Four Hummocks (x4)	5.0	150	
77	Avoid Bay	0.6	20	
78	Greenly	6.0	250	
79	Rocky	1.0	30	
80	Cap	1.0	20	
	Investigator Group			
81	Flinders	30.0	2500	x
82	Top Gallant (x2)	3.0	50	
83	Ward	1.5	20	
84	Pearson Isles	8.0	200	
85	Veteran Isles	1.0	20	
86	Dorothee	2.0	50	
	Waldegrove Islands			
87	West	2.0	30	
88	East	10.0	300	
89	Olives	1.0	30	
90	Eyre	15.0	1100	
91	Little Eyre	1.5	20	
92	Franklin Islands	13.0	500	
	Nuyts Archipelago			
93	St Peter	42.0	4000	
94	Goat	7.0	400	
95	Lound	1.0	20	
96	Purdie Islands (x3)	4.0	100	
97	Evans	5.0	150	
98	Lacy	4.0	150	
99	Egg	1.5	50	
100	Feeling	1.0	20	
101	Dog	2.0	50	
102	St Francis	15.0	800	x
103	Masillon	1.5	50	
104	Fenelon	2.0	80	
105	West	1.5	40	
106	Hard	0.5	10	
		374.4	**18 195**	**8**
5	South East-Fleurieu	8.5	102	
7	Kangaroo Island	6.4	45	
22	Yorke Peninsula	83.65	10347	
72	Eyre Peninsula	374.4	18 195	8
106	**SA Total**	**472.9**	**28 689**	**8**

BEACH INDEX

Note: Patrolled beaches in **BOLD**

See also:
GENERAL INDEX   p. 337
SURF INDEX   p. 346

GENERAL INDEX

SURF INDEX

Also see surfing locations by beach
name in BEACH INDEX

BEACHES OF THE AUSTRALIAN COAST

Published by the Sydney University Press for the
Australian Beach Safety and Management Program
a joint project of

Coastal Studies Unit, University of Sydney and Surf Life Saving Australia

by

Andrew D Short
Coastal Studies Unit, University of Sydney

BEACHES OF THE NEW SOUTH WALES COAST
Publication: 1993   **ISBN:**   0-646-15055-3
358 pages, 167 original figures, including 18 photographs; glossary, general index, beach index, surf index.

BEACHES OF THE VICTORIAN COAST & PORT PHILLIP BAY
Publication: 1996   **ISBN:**   0-9586504-0-3
298 pages, 132 original figures, including 41 photographs; glossary, general index, beach index, surf index.

BEACHES OF THE QUEENSLAND COAST: COOKTOWN TO COOLANGATTA
Publication: 2000   **ISBN**   0-9586504-1-1
369 pages, 174 original figures, including 137 photographs, glossary, general index, beach index, surf index.

BEACHES OF THE SOUTH AUSTRALIAN COAST & KANGAROO ISLAND
Publication: 2001   **ISBN**   0-9586504-2-X
346 pages, 286 original figures, including 238 photographs, glossary, general index, beach index, surf index.

BEACHES OF THE WESTERN AUSTRALIAN COAST: EUCLA TO ROEBUCK BAY
Publication: 2005   **ISBN**   0-9586504-3-8
433 pages, 517 original figures, including 408 photographs, glossary, general index, beach index, surf index.

Order online from **Sydney University Press** at

http://www.sup.usyd.edu.au/marine

Forthcoming titles:

BEACHES OF THE TASMANIAN COAST AND ISLANDS (publication 2006)   1-920898-12-3

BEACHES OF NORTHERN AUSTRALIA: THE KIMBERLEY, NORTHERN TERRITORY AND CAPE YORK (publication 2007)   1-920898-16-6

BEACHES OF THE NEW SOUTH WALES COAST (2nd edition, 2007)   1-920898-15-8

SYDNEY UNIVERSITY PRESS